www.wadsworth.com

wadsworth.com is the World Wide Web site for Wadsworth Publishing Company and is your direct source to dozens of online resources.

At *wadsworth.com* you can find out about supplements, demonstration software, and student resources. You can also send e-mail to many of our authors and preview new publications and exciting new technologies.

wadsworth.com
Changing the way the world learns®

SOCIOLOGICAL FOOTPRINTS

SOCIOLOGICAL FOOTPRINTS

Introductory Readings in Sociology

EIGHT EDITION

LEONARD CARGAN

Wright State University

JEANNE H. BALLANTINE

Wright State University

Wadsworth
Thomson Learning

Australia • Canada • Denmark • Japan • Mexico • New Zealand • Philippines • Puerto Rico
Singapore • South Africa • Spain • United Kingdom • United States

Publisher: Eve Howard
Assistant Editor: Ari Levenfeld
Editorial Assistant: Bridget Schulte
Marketing Assistant: Kelli Goslin
Project Editor: Jerilyn Emori
Print Buyer: Karen Hunt
Permissions Editor: Robert Kauser

Production: Rachel Youngman/Hockett
  Editorial Service
Copyeditor: Heidi Thaens
Cover Design: Margarite Reynolds
Cover and Interior Photos: Alan McEvoy
Compositor: ColorType/San Diego
Printer: Malloy Lithographing, Inc.

For permission to use material from this
text, contact us by
  web: www.thomsonrights.com
  fax: 1-800-730-2215
  phone: 1-800-730-2214

**Library of Congress
Cataloging-in-Publication Data**

Sociological footprints : introductory readings
in sociology / [compiled by] Leonard Cargan,
Jeanne H. Ballantine.—8th ed.
      p.   cm.
    ISBN 0-534-56502-6
    1. Sociology.   I. Cargan, Leonard.
II. Ballantine, Jeanne H.
   HM51.S66328   1999
   301—dc21   99-19047

Wadsworth/Thomson Learning
10 Davis Drive
Belmont, CA 94002-3098
USA
www.wadsworth.com

International Headquarters
Thomson Learning
290 Harbor Drive, 2nd Floor
Stamford, CT 06902-7477
USA

UK/Europe/Middle East
Thomson Learning
Berkshire House
168-173 High Holborn
London WC1V 7AA
United Kingdom

Asia
Thomson Learning
60 Albert Street #15-01
Albert Complex
Singapore 189969

Canada
Nelson/Thomson Learning
1120 Birchmount Road
Scarborough, Ontario M1K 5G4
Canada

CONTENTS

PREFACE

THE PRIMARY OBJECTIVE OF THIS ANTHOLOGY is to provide a link between theoretical sociology and everyday life by presenting actual samples of both classic and current sociological studies. If students are to grasp the full meaning of sociological terms and topics, they must be able to translate the jargon of sociology into real and useful concepts that are applicable to everyday life. To this end, *Sociological Footprints* presents viewpoints that demonstrate the broad range of sociological applications and the value of sociological research.

Selecting the readings for the Eighth Edition involved a number of important steps. As with the previous seven editions, we constantly received feedback from hundreds of students and were pleased that most of the readings were approved overwhelmingly by the students. Feedback was also requested from colleagues who are knowledgeable about the various topics of this anthology. An exhaustive search of the literature was conducted for additional material that was interesting and highly readable, that presented concepts clearly, that represented both recent and classic sociology, and that featured authors of diverse backgrounds. In meeting these criteria we often had to replace popular readings with more comprehensive and up-to-date ones. As a final step, we utilized reviewers' comments to make the anthology relevant and useful. In this manner, each edition of *Sociological Footprints* becomes the strongest possible effort in producing a sociologically current, interesting, and highly readable collection.

The relevance of this collection is seen in the variety of topics and in our effort to cover subjects found in Introductory Sociology courses. Among the classical sociological readings are Peter Berger's "An Invitation to Sociology" (2), C. Wright Mills's "The Promise" (3), Kingsley Davis's "Final Note on a Case of Extreme Isolation" (5), Horace Miner's "Body Ritual of the Nacirema" (10), D. L. Rosenhan's "On Being Sane in Insane Places" (50), two articles by the foremost of the early sociologists, Emile Durkheim: "What Is a Social Fact?" (1) and "The Elementary Forms of the Religious Life" (30). There are also two articles by one of the foremost current sociologists, Robert K. Merton: "Discrimination and the American Creed" (42) and "Social Structure and Anomie" (47).

Besides covering a wide spectrum of classic sociological issues, the anthology covers such contemporary sociological topics as AIDS (4), domestic violence (13), the changing family (23), educational inequality (29), anti-abortion (33), the future of work (35), credit cards (34), PACs (40), criminal justice (49), and environmental issues (52).

Not ignored are such important sociological variables as gender, culture, cross-cultural issues, intergroup behavior, class, crime, and the major institutions of marriage and family, education, religion, economics, and politics.

In sum, this anthology contains a balance of readings in each major section that, according to students, instructors, and reviewers, make the collection a valuable excursion into sociology. Although several of the readings have been condensed, the original material in them has in no way been altered. To emphasize key points, digressions, repetitions, and detailed descriptions of quantitative data were omitted. We hope this new edition of *Sociological Footprints* will be as valuable to teachers as it is to students—an intention reflected in the book's organization. First, each major part has an introduction that covers the major themes of the topic area, noting how each reading relates to those themes. Second, each reading is also introduced by a comment noting important points in the reading. Third, although anthologies do not usually define concepts used in their readings, we included a glossary of important terms where necessary to give a basic understanding of special terminology.

New to This Edition: InfoTrac College Edition, an online library, is available with this book. InfoTrac College Edition provides electronic access to over 900 periodicals, including scholarly journals and popular magazines.

Articles in this book are supported by the Wasdsworth web site at: **http://sociology. wadsworth.com.** Here students and instructors will find links to further web sites that are broken down by the chapter to which they pertain.

Instructor's Manual: The *Instructor's Manual* presents an abstract of each article, along with multiple-choice and essay questions. A password-protected version of the *Instructor's Manual* is available on the Wadsworth web site.

We wish to thank all those who made this edition of *Sociological Footprints* possible. The reviewers of this edition:

Irene Fiala, Baldwin-Wallace College
Michael Goslin, Tallahassee Community College
Susan F. Greenwood, University of Maine
William J. Miller, Ohio University
Wilbert Nelson, Phoenix College
Carol Ray, San Jose State University
George Siefert, Daemen College
Debbie A. Storrs, University of Idaho
Donna Trent, Eckerd College
Assata Zerai, Syracuse University

We also thank the reviewers of previous editions: Philip Berg, University of Wisconsin–La Crosse; Kevin J. Christiano, University of Notre Dame; Rodney B. Coates, University of North Carolina at Charlotte; Thomas F. Courtless, George Washington University; Robert W. Duff, University of Portland; Martin Monto, University of Portland; Dan J. Pence, Southern Utah University; Ralph Peters, Floyd College; David R. Rudy, Moorehead State University; Eldon E. Snyder, Bowling Green State University; Larry L. Stearley, Charles Stewart Mott Community College; and Jerry Stockdale, University of Northern Iowa; the many students who took the time to give us their opinions; the departmental secretary and aides who helped to assemble and type the material; our proofreader; and to all those good people at Wadsworth who aided in the production of this anthology. To all we give a most heartfelt "Thank You!"

TO THE STUDENT

THE PURPOSE OF THIS ANTHOLOGY IS to introduce you, the beginning student in sociology, to a wide range of sociological perspectives and to demonstrate their relevance to real-life situations. As you apply sociological perspectives to everyday events, you will begin to realize that sociology is more than jargon, more than dry statistics, more than endless terminology to be memorized. It is an exciting and useful field of study. Unfortunately, no textbook can fully describe the many applications of sociology. This anthology should help to fill the gap by supplying classical readings balanced with readings on current research.

From our experience in teaching introductory sociology, we know some of the problems that anthologies can present to the student: unexplained terms, readings seemingly unrelated to the text, and different emphases from those of the instructor's lectures. Therefore, to enjoy and benefit fully from *Sociological Footprints,* you should take the following steps:

1. Read and study the related textbook chapter and lecture materials. You must be familiar with the concepts and perspectives before you can clearly observe their daily application.
2. Read the introductions to the assigned sections in the anthology. They are designed to summarize the primary themes of the topic area and relate them to specific readings. In fact, the introductions will not only make the readings easier to understand, they will facilitate your application of the readings to other class materials and real-life situations.
3. Use the glossary before you read each reading. Knowing the terms will make the reading more interesting and understandable.
4. Read each reading through. Note the problem or issue being discussed, the evidence the author supplies in support of his or her contentions, and the conclusions drawn from this evidence.
5. Summarize the main ideas of each reading in your own terms, relating them to other material in the course and to your own everyday experiences.

Step 5 is particularly important. Many of the readings address topics of current interest: the political role of fundamentalist religion, population problems, environmental issues, the women's movement, and more. Because these are contemporary problems, you will see related materials in newspapers and magazines and on television. By applying what you have learned from the lectures and this anthology, you should develop a clearer understanding of current issues and of how sociology has aided you in this understanding.

We feel strongly that sociology is a field of study highly relevant to your world and that it can give you a fuller comprehension of day-to-day living. Our aim has been to provide you with a readable, understandable, and enlightening anthology that will convey this relevance.

The Essential Wisdom of Sociology

(Paraphrased from a paper by Earl Babbie from 1989 ASA Annual Meeting)
"I say this by way of a disclaimer. The essential wisdom of sociology may have twelve or thirteen points, but I'm going to quit at ten."

1. Society has a *sui generis* existence or reality.

"You can't fully understand society by understanding individual human beings who comprise it. For example, few people want war, but we have wars all the time."

2. It is possible to study society scientifically.

"Society can be more than learned beliefs of 'common sense.' It is actually possible to study society scientifically, just as we study aspects of the physical world."

3. Autopoesis: society creates itself.

"Autopoesis (Huberto Maturana's term) might be defined as 'self-creating.' A powerful statement that sociology has to offer is that society is autopoetic: society creates itself."

4. Cultural variations by time and place.
"Gaining awareness that differences exist is only the beginning, however. . . . Our second task in this regard is to undermine . . . implicit ethnocentrism, offering the possibility of tolerance."

5. Relation of individual and society.

"[One might] want to skirt the edge of suggesting that individuals are merely figments of society. Without going quite that far [one might suggest that] individual identity is strongly sociogenetic."

6. System imperatives.

"Society is an entity [and] as a system, it has 'needs.'"

7. The inherent conservatism of institutions.

"The first function of an institution is institutional survival."

8. Determinism.

". . . We operate with a model that assumed human behavior is determined by forces and factors and circumstances that the individual actors cannot control and/or are unaware of."

9. Paradigms.

". . . paradigms are ways of looking at life, but not life itself. They focus attention so as to reveal things, but they inevitably conceal other things, rather like microscopes or telescopes, perhaps. They allow us to see things that would otherwise be hidden from us, but they do that at a cost."

10. Sociology is an idea whose time has come.

"Finally . . . on a possible chauvinist note: All the major problems that face us as a society and as a world are to be found within the territory addressed by sociology. I say this in deliberate contrast to our implicit view that most of our problems will be solved by technology."

SOCIOLOGICAL FOOTPRINTS

INTRODUCTION: WHY STUDY SOCIOLOGY?

WHAT IS THIS SUBJECT CALLED SOCIOLOGY? What will I learn from studying sociology? Why should I take sociology? What work do sociologists do? How is sociology useful to me or to the world? If I major in sociology, what can I do when I graduate? These are some of the questions that may be in the back of your mind as you approach your study of sociology. Perhaps you are reading this because you are curious about the subject, or because sociology is a required course, or because you had sociology in high school and wanted to find out more about it, or because your instructor assigned the book and this article. Whatever the reasons, you will find an introduction to the field of sociology in the discussion that follows.

What you read in the next few pages will only begin to answer the questions just posed. As you learn more about sociology, pieces that at first seemed fragmentary will start to come together like pieces in a puzzle. These pages provide the framework into which those pieces can be placed to answer the opening question: Why study sociology?

What Is This Subject Called Sociology?

First questions first: Sociology is the study of people in groups, of people interacting with each other, even of nations interacting during peace or war. Sociologists' interests are sparked when they see two or more people with a common interest talking or working together. They are interested in how groups work and in how nations of the world relate to each other. When two or more people are interacting, sociologists have the tools to study the process. It could be a married couple in conflict or a teacher and students in a classroom situation; it could be individuals interacting in a work group, sports teams on a playing field, or negotiating teams discussing nuclear disarmament.

Sociology shares a common bond with other social sciences. All are concerned with human behavior in society; they share the perspective of the scientific method and some of the same data collection methods to study their subject matter. Sociology is the broadest of the social sciences; its main concern is with predicting human group behavior.

"That's a lot to be interested in," you may be saying. In fact, most sociologists specialize. No one sociologist is likely to be an expert in everything from studies of a few people or small group interaction (*microlevel* sociology), to large numbers of people in big groups like organizations or nations (*macrolevel* sociology). Consider the following examples of sociological specializations:

- determining the factors that lead to marital longevity
- identifying effective teachers by classroom observation
- examining public attitudes about the Clinton presidency and its policies
- locating satisfaction and problems in certain jobs

The results of these diverse interests lead sociologists into many different areas. Some sociologists specialize in *social psychology,* a field that considers such questions as how individuals behave in groups, who leaders are and what types of leaders are effective, why some groups accomplish more than other groups, why individuals usually conform to group expectations, and many other topics involving individuals as functioning members of groups. Another area of specialization is *political sociology,* which studies political power, voting behavior, bureaucracy, and political behavior of individuals and groups. *Anthropology* examines the culture of different groups; so does sociology. But the methods of study and primary focus differ. Anthropologists often study preliterate groups, whereas sociologists focus primarily on modern groups. Another area that concerns sociologists is *social history,* which emphasizes the use of history to understand social situations. These are only a few examples of the diverse interests of sociologists and how sociology shares its interests with some other social sciences.

What Will I Learn From Studying Sociology?

Consider that in some societies premarital sex is not only allowed but expected; in others, premarital sex is cause for banishment and death. Even though sociologists, like everyone else, have personal opinions, the task of the sociologist is not to judge which social attitude is right or wrong but to understand *why* such divergent practices have evolved. We all have opinions. Usually they come from our experiences, common sense, and family teaching. Some opinions are based on stereotypes or prejudices, some on partial information about an issue. Through systematic scientific study, sociologists gain insight into human behavior in groups, insight not possible through common sense alone. They attempt to understand all sides of an issue; they refrain from making judgments on issues of opinion but try instead to deal objectively with human behavior.

Consider the person who is going through the anguish of a divorce. Self-blame or hostility toward the spouse are often reactions to this personal crisis. Sociology can help us move beyond "individual" explanations to consider the social surroundings that influence the situation: economic conditions, disruptions caused by changing sex roles, and pressures on the family to meet the emotional needs of its members. Thus, sociology teaches us to look beyond individual explanations of our problems to group explanations for behavior; this practice broadens our worldview and gives us a better understanding of why events take place.

A typical college sociology program starts with a basic course introducing the general perspective of sociology; sociological terminology and areas of study; how sociologists get their information, that is, their methods; and the ideas, or theories, that lay the foundations for sociological study. Further sociology courses deal in greater depth with the major components of all societies: family, religion, education, politics, and economics. The sociology department may also offer courses on social processes such as social problems, deviance and corrections, stratification, socialization, and change or on other areas of social life such as medical, community, urban, sports, or minority sociology.

Family sociology, for instance, usually considers the family social life cycle: young people breaking away from their parents' home, forming a home of their own by selecting a spouse through the courtship process, marrying, selecting a career, making parenting decisions, raising a family, having their children leave home, retiring, and moving into old age.

Students who major in sociology generally take courses in *theory*—the basic ideas of the field—and *methods*—how sociologists approach the social world objectively and do their research. Some sociology departments offer practical experiences where students can use their sociological skills in a job setting.

These are a few examples of what you will learn from the study of sociology and how you will learn it. There is much more to the field of sociology than this, however.

Why Should I Take Sociology?

Whether you take a number of sociology courses or only one, you will profit in a number of ways. You will gain personal knowledge, new perspectives, skills needed by employers, background training useful in entering other fields, personal growth and development, new perspectives on the world, and a new way of looking at your relations with others and your place in society. You will gain tolerance for and fascination with the variety of people in the world around you and their cultural systems. You will be able to understand your interactions with your family and friends better; you will be able to watch the news or read the paper with keener perception. You will have an understanding of how to obtain information to answer questions you or your boss need answered. And the more sociology you take, the more ability you will have to express your thoughts logically, objectively, and coherently.

It is nice to know that the subjects you take in college will have some personal relevance and professional usefulness. Sociology should provide you with a number of "life skills," such as

1. Ability to view the world more objectively
2. Tools to solve problems by designing studies, collecting data, and analyzing results
3. Ability to understand group dynamics
4. Ability to understand and evaluate problems
5. Ability to understand your personal problems in a broader social context

We know from studies that employers value those applicants with the broad training of such fields as sociology because of the skills they provide. The following are skills employers look for, in order of importance:

1. Ability to work with peers
2. Ability to organize thoughts and information
3. Self-motivation

   4. Ability to plan effectively
   5. Willingness to adapt to the needs of the organization
   6. Ability to interact effectively in group situations
   7. Self-confidence about job responsibilities
   8. Ability to handle pressure
   9. Ability to conceptualize problems clearly
   10. Effective problem-solving skills
   11. Effective leadership skills
   12. Ability to listen to others

Although a college graduate in engineering, computer sciences, or business may enter the job market with a higher salary, the sociology liberal arts major is more likely to rise through the managerial and professional ranks to positions of responsibility and high pay. Businesses and organizations value the skills listed here. In today's rapidly altering society, many of us will change jobs or careers several times during a lifetime. Sociological skills can help us adapt to the expectations of new situations.

Because of the knowledge and skills learned in sociology courses, study in this area provides excellent preparation for other undergraduate and graduate fields. From nursing, business, and education to law and medicine, the knowledge of sociology can be applied to a wide variety of group situations. For instance, a current concern of sociologists who study educational settings is what characteristics make schools effective; by singling out certain characteristics, sociologists can make recommendations to improve schools. Teachers and educational administrators profit from this information.

If we are curious about understanding ourselves and our interactions with others and about why our lives take certain directions, sociology can help us understand. For instance, sociologists are interested in how our social-class standing affects how we think, how we dress, how we speak, what our interests are, whom we are likely to marry, what religion (if any) we belong to, and what our "life chances" are, including how long we will live and what we are likely to do in life. Sociologists have even examined how individuals from different social-class backgrounds raise their children, and implications of child-rearing techniques for our lifestyles. Some use physical punishment and others moral chastisement, but the end result is likely to be a perpetuation of the social class into which we are born.

What Work Do Sociologists Do?

The most obvious answer is that sociologists *teach;* this is primarily at the higher education level, but high school sociology courses are also offered as part of the social science curriculum. There would be nothing to teach if sociologists were not actively engaged in learning about the social world. Their second major activity is to conduct *research* about questions concerning the social world.

Many sociologists work in business organizations, government agencies, and social service agencies. *Practicing sociologists* are engaged in a variety of activities. Some do family counseling with the whole family group; some conduct market research for companies or opinion polls for news or other organizations; some do surveys for the government to determine what people think or need; some work with juvenile delinquents, prison programs and reforms, and police; some predict how population changes will affect schools and communities.

Applied sociologists use their sociological knowledge to help organizations. They assess organizational needs, plan programs to meet those needs, and evaluate the effectiveness of programs. For instance, a community may want to know how many of its elderly citizens need special services to remain at home rather than be moved to nursing institutions. Sociologists assess this need, help plan programs, and evaluate whether programs are meeting the needs they set out to meet.

The position a sociology major ultimately gets depends in part on the degree he or she holds in sociology. The following are some examples of jobs students have gotten with a B.A. or B.S. degree: director of county group home, research assistant, juvenile probation officer, data processing project director, public administration/district manager, public administration/health coordinator, law enforcement, labor relations/personnel, police commander/special investigations, trucking dispatcher, administrator/social worker, counselor, child caseworker, substance abuse therapist, medical social worker, data programming analyst, activities director at senior citizens center, director of student volunteer program, area sales manager, jury verdict research editor, insurance claims adjustor, employment recruiter, tester for civil service, unemployment office manager, child services houseparent, crisis worker volunteer, advertising copywriter, probation officer, travel consultant, recreation therapist, public TV show hostess, adult education coordinator, research and evaluation specialist, neighborhood youth worker.

Sociologists holding an M.A. or Ph.D. degree are more skilled in sociological theory and methods than B.A. degree holders. They are often involved in research, teaching, or clinical work with families and other clients.

How Is Sociology Useful to Me and to the World?

Technology is rapidly changing the world. New policies and programs are being implemented in government and private organizations—policies that affect every aspect of our lives. Because sociologists study *social* processes, they are able to make concrete contributions to the planning of orderly change. Sociological knowledge can also be useful to legislators and courts in making policy decisions. For example, sociologists can assist a juvenile facility to design programs to help young people convicted of crime redirect their energies; how successful such programs are in achieving their goals can be studied by evaluation research.

In summary, sociology is the broadest of the social sciences and, unlike other disciplines, can give us an understanding of the social world. The knowledge and tools of sociology make students of this field valuable in a number of settings, from business to social service to government to education. As you embark on this study, keep in mind that sociology helps us have a deeper understanding of ourselves and our place in the world as well.

Sociology is a study of all people, for all people. To enjoy your encounter with the field and to make the most use of your time in sociology, try to relate the information you read and hear to your own life and relationships with others within the broader context of your social world.

WHAT CAN YOU DO WITH A SOCIOLOGY DEGREE?*

*All positions are actual jobs held by sociology majors from The Ohio State University.

Graphics compliments of Scott, Foresman and Company

Part I

THE DISCIPLINE
OF SOCIOLOGY

Although the term sociology *may be familiar, many students are unaware of the areas of study included in the field. Sociology ranges from the study of small groups—perhaps two people—to that of such large entities as corporations and societies. It is the study of interactions within, between, and among groups; and these group interactions encompass all areas of human behavior, as noted in the following reading.*

. . . Leading thinkers in all ages have been concerned about society, human conduct, and the creation of a social order that would bring forth the best man is capable of. But, the study of these problems with the techniques and approaches of science (sociology) is only a little more than a century old.

It was only about 125 years ago that Auguste Comte published his *Cours de philosophie positive* which first included sociology as one of the scientific disciplines. No course in sociology was available in an American university until 1876 at Yale. . . . Before 1900 all the men who identified themselves as professional sociologists were trained originally in other fields such as history, politics, economics, law, and religion. Today, undergraduate students can obtain training in sociology in almost all American four-year liberal arts colleges and in many agricultural colleges and specialized schools; more than 70 schools offer a doctoral program in sociology and many additional schools offer master's degree programs.

Before World War I the opportunities for employment of men and women with professional training in sociology were largely limited to college teaching and research. Besides teaching and research, sociologists today are engaged in more than 25 different kinds of work in professional schools, in local, state, federal, and private agencies, and in business. They work in the fields of education, medicine, law, theology, corrections, agricultural extension, welfare, population study, community development, health, technological change, and the like. In short, sociologists are working on almost all the problems that concern man in relation to his fellow man and the consequences of this relationship for himself and others.* (Reprinted by permission of the Sociology Department, University of Kentucky.)

*This statement was written before it became correct to use gender-neutral terms. It should, however, be understood that it refers to humans in relation to other humans . . . for themselves.

CHAPTER 1

Science and Sociology

IS IT JUST COMMON SENSE?

SOCIOLOGISTS ARE INTERESTED IN DISCOVERING THE realities of group interactions. To distinguish their conclusions from everyday, commonsense observations or intuitions, sociologists use scientific data-gathering techniques such as surveys, participant observation, content analysis, and projective tests. Common sense is the feeling one has about a situation without systematic analysis; that is, appearances alone are accepted as the criteria for truth. In reality, what is commonsense truth for one person may not be that for another. Good examples of such "truths" are found in advice columns such as "Ann Landers" and "Dear Abby" or in such "truisms" as "Absence makes the heart grow fonder" versus "Out of sight, out of mind." Despite their common sense, the following beliefs have all been shown to be false: "Humans have a natural instinct to mate with the opposite sex," "A major proportion of people on welfare could work if they wanted to," "American Catholics are less likely than Protestants to enter interfaith marriages or be divorced and more likely to oppose birth control," and "A belief found in every society is romantic love." In contrast, the scientific approach may be defined as a procedure for analyzing data according to an *objective, logical, systematic method* to produce *reliable* knowledge. By examining the five major terms in this definition, we can clarify the difference between commonsense and scientific observations.

The first major term in the definition is *objective*. To be objective, we must be aware of the difference between social behavior and other types of behavior. In the first reading of this chapter, Emile Durkheim indicates what are social facts. As Durkheim and Berger note in the first two readings in this chapter, the rules of the scientific method are an essential part of the discipline of sociology; these rules allow us to recognize what is a social fact and help in reducing the influence of personal biases on our research experiences. Complete freedom from bias is difficult to achieve because we are products of our society and times, and what we "see" is influenced by these conditions. This difficulty is noted by C. Wright Mills in his classic piece from *The Sociological Imagination* (Reading 3). Mills states that our perceptions are limited by our family, work, and other social experiences. In short, what we observe is bounded by these particular experiences and therefore may lead to biased statements based on commonsense beliefs. However, Mills believes that such a limited perspective can be expanded by means of a "sociological imagination," which allows us to note relationships that exist between our personal experiences and general social issues.

The self-fulfilling prophecy, a hypothesis developed by Robert Merton, extends Mills' ideas by also recognizing that one's definition of a situation becomes an integral part of the situation since it affects one's subsequent behavior. As Kain shows in Reading 4, on AIDS, young people's beliefs that AIDS is only a disease of gay people and drug addicts leads to their taking little or no precautions in their own sexual behavior.

The next two major terms in our definition are *logical* and *systematic*. *Logical* refers to the arrangements of facts and their interrelations according to accepted rules of reasoning. *Systematic* means that there is consistency in the internal order of presentation of materials. As a means of being logical, sociologists use theories as a guide to possible relationships. A theory is a formulation of apparent relationships of observed phenomena.

The term *method* refers to the techniques used in the research effort as a means for testing the preceding hypotheses and theories. You will note the use of variations on the major sociological methods—the experiment, survey research, field research, analysis of existing data—throughout the articles in this anthology. Even a humorous result can occur as shown by Horace Miner in his reading on the Nacirema (Reading 10).

The final term in our definition is *reliability*, which means that the knowledge produced is dependable because it is retestable: the observations can be repeated with similar data for a predictably similar outcome. Sociological techniques can be used to make predictions about human behavior. However, unlike mechanical behavior, human behavior is subject to people's whims and fads; consequently, such behavior can be predicted only within a reliable range.

As these readings argue, adherence to the rules of scientific method produces reliable knowledge that can be used to prove or disprove commonsense observation. This approach can also expand perspectives limited by the material boundaries of time and space and help us deal efficiently with society's needs.

Finally, the publication of the results of scientific investigation may lead to the recognition of other truths, resulting in developments beneficial to society. This would be true, for example, of reforms in aiding family maintenance (Reading 25), improving schools (Reading 28), making the political system more representative (Reading 40), and making the criminal justice system more effective (Reading 49). As you read this chapter, keep these ideas in mind and apply them to your own commonsense beliefs — that is, challenge your commonsense beliefs by using your sociological imagination.

1
What Is a Social Fact?

EMILE DURKHEIM

In reading this article, be sure to note what behavior is social and why—that is, what is a social fact.

As you read, ask yourself the following questions:

1. *What facts are called social?*
2. *What social facts would you like to see changed?*

GLOSSARY

Aphorism A general truth embodied in a short saying.

BEFORE INQUIRING INTO THE method suited to the study of social facts, it is important to know which facts are commonly called "social." This information is all the more necessary since the designation "social" is used with little precision. It is currently employed for practically all phenomena generally diffused within society, however small their social interest. But on that basis, there are, as it were, no human events that may not be called social. Each individual drinks, sleeps, eats, reasons; and it is to society's interest that these functions be exercised in an orderly manner. If, then, all these facts are counted as "social" facts, sociology would have no subject matter exclusively its own, and its domain would be confused with that of biology and psychology.

But in reality there is in every society a certain group of phenomena which may be differentiated from those studied by the other natural sciences. When I fulfil my obligations as brother, husband, or citizen, when I execute my contracts, I perform duties which are defined, externally to myself and my acts, in law and in custom. Even if they conform to my own sentiments and I feel their reality subjectively, such reality is still objective, for I did not create them; I merely inherited them through my education. How many times it happens, moreover, that we are ignorant of the details of the obligations incumbent upon us, and that in order to acquaint ourselves with them we must consult the law and its authorized interpreters! Similarly, the church-member finds the beliefs and practices of his religious life ready-made at birth; their existence prior to his own implies their existence outside of himself. The system of signs I use to express my thought, the system of currency I employ to pay my debts, the instruments of credit I utilize in my commercial relations, the practices followed in my profession, etc., function independently of my own use of them. And these statements can be repeated for each member of society. Here, then, are ways of acting, thinking, and feeling that present the noteworthy property of existing outside the individual consciousness.

These types of conduct or thought are not only external to the individual but are, moreover, endowed with coercive power, by virtue of which they impose themselves upon him, independent of his individual will. Of course, when I fully consent and conform to them, this constraint is felt only slightly, if at all, and is therefore unnecessary. But it is, nonetheless, an intrinsic characteristic of these facts, the proof thereof being that it asserts itself as soon as I attempt to resist it. If I attempt to violate the law, it reacts against me so as to prevent my act before its accomplishment, or to nullify my violation by restoring the damage, if it is accomplished and reparable, or to make me expiate it if it cannot be compensated for otherwise.

In the case of purely moral maxims, the public conscience exercises a check on every act which offends it by means of the surveillance it exercises over the conduct of citizens, and the appropriate penalties at its disposal. In many cases the constraint is less violent, but nevertheless it always exists. If I do not submit to the conventions of society, if in my dress I do not conform to the customs observed in my country and in my class, the ridicule I provoke, the social isolation in which I am kept, produce, although in an attenuated form, the same effects as a punishment in the strict sense of the word. The constraint is nonetheless efficacious for being indirect. I am not obliged to speak French with my fellow-countrymen nor to use the legal currency, but I cannot possibly do otherwise. If I tried to escape this necessity, my attempt would fail miserably. As an industrialist, I am free to apply the technical methods of former centuries; but by doing so, I should invite certain ruin. Even when I free myself from these rules and violate them successfully, I am always compelled to struggle with them. When finally overcome, they make their constraining power sufficiently felt by the resistance they offer. The enterprises of all innovators, including successful ones, come up against resistance of this kind.

Here, then, is a category of facts with very distinctive characteristics: it consists of ways of acting, thinking, and feeling, external to the individual, and endowed with a power of coercion, by reason of which they control him. These ways of thinking could not be confused with biological phenomena, since they consist of representations and of actions; nor with psychological phenomena, which exist only in the individual consciousness and through it. They constitute, thus, a new variety of phenomena; and it is to them exclusively that the term "social" ought to be applied. And this term fits them quite well, for it is clear that, since their source is not in the individual, their substratum can be no other than society, either the political society as a whole or some one of the partial groups it includes, such as religious denominations, political, literary, and occupational associations, etc. On the other hand, this term "social" applies to them exclusively, for it has a distinct meaning only if it designates exclusively the phenomena which are not included in any of the categories of facts that have already been established and classified. These ways of thinking and acting therefore constitute the proper domain of sociology. It is true that, when we define them with this word "constraint," we risk shocking the zealous partisans of absolute individualism. For those who profess the complete autonomy of the individual, man's dignity is diminished whenever he is made to feel that he is not completely self-determinant. It is generally accepted today, however, that most of our ideas and our tendencies are not developed by ourselves but come to us from without. How can they become a part of us except by imposing themselves upon us? This is the whole meaning of our definition. And it is generally accepted, moreover, that social constraint is not necessarily incompatible with the individual personality.[1]

Since the examples that we have just cited (legal and moral regulations, religious faiths, financial systems, etc.) all consist of established beliefs and practices, one might be led to believe that social facts exist only where there is some social organization. But there are other facts without such crystallized form which have the same objectivity and the same ascendency over the individual. These are called "social currents." Thus the great movements of enthusiasm, indignation, and pity in a crowd do not originate in any one of the particular individual consciousness. They come to each one of us from without and can carry us away in spite of ourselves. Of course, it may happen that, in abandoning myself to them unreservedly, I do not feel the pressure they exert upon me. But it is revealed as soon as I try to resist them. Let an individual attempt to oppose one of these collective manifestations, and the emotions that he denies will turn against him. Now, if this power of external coercion asserts itself so clearly in cases of resistance, it must exist also in the first-mentioned cases, although we are unconscious of it. We are then victims of the illusion of having ourselves created that which actually forced itself from without. If the complacency with which we permit ourselves to be carried along conceals the pressure undergone, nevertheless it does not abolish it. Thus, air is no less heavy because we do not detect its weight. So, even if we ourselves have spontaneously contributed to the production of the common emotion, the impression we have received differs markedly from that which we would have experienced if we had been alone. Also, once the crowd has dispersed, that is, once these social influences have ceased to act upon us and we are alone again, the emotions which have passed through the mind appear strange to us, and we no longer recognize them as ours. We realize that these feelings have been impressed upon us to a much greater extent than they were created by us. It may even happen that they horrify us, so much were they contrary to our nature. Thus, a group of individuals, most of whom are perfectly inoffensive, may, when gathered in a crowd, be drawn into acts of atrocity. And what we say of these transitory outbursts applies similarly to those more permanent currents of opinion on religious, political, literary, or artistic matters which are constantly being formed around us, whether in society as a whole or in more limited circles.

To confirm this definition of the social fact by a characteristic illustration from common experience, one need only observe the manner in which children are brought up. Considering the facts as they are and as they have always been, it becomes immediately evident that all education is a continuous effort to impose on the child ways of seeing, feeling, and acting which he could not have arrived at spontaneously. From the very first hours of his life, we compel him to eat, drink, and sleep at regular hours; we constrain him to cleanliness, calmness, and obedience; later we exert pressure upon him in order that he may learn proper consideration for others, respect for customs and conventions, the need for work, etc.

If, in time, this constraint ceases to be felt, it is because it gradually gives rise to habits and to internal tendencies that render constraint unnecessary; but nevertheless it is not abolished, for it is still the source from which these habits were derived. . . . What makes these facts particularly instructive is that the aim of education is, precisely, the socialization of the human being; the process of education, therefore, gives us in a nutshell the historical fashion in which the social being is constituted. This unremitting pressure to which the child is subjected is the very pressure of the social milieu which tends to fashion him in its own image, and of which parents and teachers are merely the representatives and intermediaries.

It follows that sociological phenomena cannot be defined by their universality. A thought which we find in every individual consciousness, a movement repeated by all individuals, is not thereby a social fact. If sociologists have been satisfied with defining them by this characteristic, it is because they confused them with what one might call their reincarnation in the individual. It is, however, the collective aspects of the beliefs, tendencies, and practices of a group that characterize truly social phenomena. As for the forms that the collective states assume when refracted in the individual, these are things of another sort. This duality is clearly demonstrated by the fact that these two orders of phenomena are frequently found dissociated from one another. Indeed, certain of these social manners of acting and thinking acquire, by reason of their repetition, a certain rigidity which on its own account crystallizes them, so to speak, and isolates them from the particular events which reflect them. They thus acquire a body, a tangible form, and constitute a reality in their own right, quite distinct from the individual facts which produce it. Collective habits are inherent not only in the successive acts which they determine but, by a privilege of which we find no example in the biological realm, they are given permanent expression in a formula which is repeated from mouth to mouth, transmitted by education, and fixed even in writing. Such is the origin and nature of legal and moral rules, popular aphorisms and proverbs, articles of faith wherein religious or political groups condense their beliefs, standards of taste established by literary schools, etc. None of these can be found entirely reproduced in the applications made of them by individuals, since they can exist even without being actually applied.

No doubt, this dissociation does not always manifest itself with equal distinctness, but its obvious existence in the important and numerous cases just cited is sufficient to prove that the social fact is a thing distinct from its individual manifestations. Moreover, even when this dissociation is not immediately apparent, it may often be disclosed by certain devices of method. Such dissociation is indispensable if one wishes to separate social facts from their alloys in order to observe them in a state of purity. Currents of opinion, with an intensity varying according to the time and place, impel certain groups either to more marriages, for example, or to more suicides, or to a higher or lower birthrate, etc. These currents are plainly social facts. At first sight they seem inseparable from the forms they take in individual cases. But statistics furnish us with the means of isolating them. They are, in fact, represented with considerable exactness by the rates of births, marriages, and suicides, that is, by the number obtained by dividing the average annual total of marriages, births, suicides, by the number of persons whose ages lie within the range in which marriages, births, and suicides occur.[2] Since each of these figures contains all the individual cases indiscriminately, the individual circumstances which may have had a share in the production of the phenomenon are neutralized and, consequently, do not contribute to its determination. The average, then, expresses a certain state of the group mind (*l'âme collective*).

Such are social phenomena, when disentangled from all foreign matter. As for their individual manifestations, these are indeed, to a certain extent, social, since they partly reproduce a social model. Each of them also depends, and to a large extent, on the organopsychological constitution of the individual and on the particular circumstances in which he is placed. Thus they are not sociological phenomena in the strict sense of the word. They belong to two realms at once; one could call them sociopsychological. They interest the sociologist without constituting the immediate subject matter of sociology. There exist in the interior of organisms similar phenomena, compound in their nature, which form in their turn the subject matter of the "hybrid sciences," such as physiological chemistry, for example. . . .

We thus arrive at the point where we can formulate and delimit in a precise way the domain of sociology. It comprises only a limited group of phenomena. A social fact is to be recognized by the power of external coercion which it exercises or is capable of exercising over individuals, and the presence of this power may be recognized in its turn either by the existence of some specific sanction or by the resistance offered against every individual effort that tends to violate it. One can, however, define it also by its diffusion within the group, provided that, in conformity with our previous remarks, one takes care to add as a second and essential characteristic that its own existence is independent of the individual forms it assumes in its diffusion. This last criterion is perhaps, in certain cases, easier to apply than the preceding one. In fact, the constraint is easy to ascertain when it expresses itself externally by some direct reaction of society, as is the case in law, morals, beliefs, customs, and even fashions. But when it is only indirect, like the constraint which an economic organization exercises, it cannot always be so easily detected. Generality combined with externality may, then, be easier to establish. Moreover, this second definition is but another form of the first; for if a mode of behavior whose existence is external to individual consciousness becomes general, this can only be brought about by its being imposed upon them.[3]

But these several phenomena present the same characteristic by which we defined the others. These "ways of existing" are imposed on the individual precisely in the same fashion as the "ways of acting" of which we have spoken. Indeed, when we wish to know how a society is divided politically, of what these divisions themselves are composed, and how complete is the fusion existing between them, we shall not achieve our purpose by physical inspection and by geographical observations; for these phenomena are social, even when they have some basis in physical nature. It is only by a study of public law that a comprehension of this organization is possible, for it is this law that determines the organization, as it equally determines our domestic and civil relations. This political organization is, then, no less obligatory than the social facts mentioned above. If the population crowds into our cities instead of scattering into the country, this is due to a trend of public opinion, a collective drive that imposes this concentration upon the individuals. We can no more choose the style of our houses than of our clothing—at least, both are equally obligatory. The channels of communication prescribe the direction of internal migrations and commerce, etc., and even their extent. Consequently, at the very most, it should be necessary to add to the list of phenomena which we have enumerated as presenting the distinctive criterion of a social fact only one additional category, "ways of existing"; and, as this enumeration was not meant to be rigorously exhaustive, the addition would not be absolutely necessary.

Such an addition is perhaps not necessary, for these "ways of existing" are only crystallized "ways of acting." The political structure of a society is merely the way in which its component segments have become accustomed to live with one another. If their relations are traditionally intimate, the segments tend to fuse with one another, or, in the contrary case, to retain their identity. The type of habitation imposed upon us is merely the way in which our contemporaries and our ancestors have been accustomed to construct their houses. The methods of communication are merely the channels which the regular currents of commerce and migrations have dug, by flowing in the same direction. To be sure, if the phenomena of a structural character alone presented this permanence, one might believe that they constituted a distinct species. A legal regulation is an arrangement no less permanent than a type of architecture, and yet the regulation is a "physiological" fact. A simple moral maxim is assuredly somewhat more malleable,

but it is much more rigid than a simple professional custom or a fashion. There is thus a whole series of degrees without a break in continuity between the facts of the most articulated structure and those free currents of social life which are not yet definitely molded. The differences between them are, therefore, only differences in the degree of consolidation they present. Both are simply life, more or less crystallized. No doubt, it may be of some advantage to reserve the term "morphological" for those social facts which concern the social substratum, but only on condition of not overlooking the fact that they are of the same nature as the others. Our definition will then include the whole relevant range of facts if we say: *A social fact is every way of acting, fixed or not, capable of exercising on the individual an external constraint;* or again, *every way of acting which is general throughout a given society, while at the same time existing in its own right independent of its individual manifestations.*[4]

NOTES

1. We do not intend to imply, however, that all constraint is normal. We shall return to this point later.

2. Suicides do not occur at every age, and they take place with varying intensity at the different ages in which they occur.

3. It will be seen how this definition of the social fact diverges from that which forms the basis of the ingenious system of M. Tarde. First of all, we wish to state that our researches have nowhere led us to observe that preponderant influence in the genesis of collective facts which M. Tarde attributes to imitation. Moreover, from the preceding definition, which is not a theory but simply a résumé of the immediate data of observation, it seems indeed to follow, not only that imitation does not always express the essential and characteristic features of the social fact, but even that it never expresses them. No doubt, every social fact is imitated; it has, as we have just shown, a tendency to become general, but that is because it is social, i.e., obligatory. Its power of expansion is not the cause but the consequence of its sociological character. If, further, only social facts produced this consequence, imitation could perhaps serve, if not to explain them, at least to define them. But an individual condition which produces a whole series of effects remains individual nevertheless. Moreover, one may ask whether the word "imitation" is indeed fitted to designate an

effect due to a coercive influence. Thus, by this single expression, very different phenomena, which ought to be distinguished, are confused.

4. This close connection between life and structure, organ and function, may be easily proved in sociology because between these two extreme terms there exists a whole series of immediately observable intermediate stages which show the bond between them. Biology is not in the same favorable position. But we may well believe that the inductions on this subject made by sociology are applicable to biology and that, in organisms as well as in societies, only differences in degree exist between these two orders of facts.

2

An Invitation to Sociology*

PETER L. BERGER

What are the interests and activities of a sociologist? The answer is best summed up by Berger's comments that the sociologist "is interested in the action of human conduct" and that "the sociological perspective makes us see in a new light the very world in which we live." This new light comes about, according to Mills (Reading 3), through the use of sociological imagination.

As you read, ask yourself the following questions:

1. What are the items that sociologiest are interested in?

2. What new images of society surprised you?

GLOSSARY

Commensalism Close association that benefits one of two organisms without benefiting or harming the other.

Connubium Marriage.

Culture shock The impact of adjusting to a totally new culture.

Polygamy Having two or more mates.

Puberty State of physical development when sexual reproduction becomes possible.

. . . THE SOCIOLOGIST . . . IS SOMEONE concerned with understanding society in a disciplined way. The nature of this discipline is scientific. This means that what the sociologist finds and says about the social phenomena he studies occurs within a certain rather strictly defined frame of reference. One of the main characteristics of this scientific frame of reference is that operations are bound by certain rules of evidence. As a scientist, the sociologist tries to be objective, to control his personal preferences and prejudices, to perceive clearly rather than to judge normatively. This restraint, of course, does not embrace the totality of the sociologist's existence as a human being, but is limited to his operations *qua* sociologist. Nor does the sociologist claim that his frame of reference is the only one within which society can be looked at. For that matter,

*This reading was written before it became correct to use gender-neutral terms. Male nouns and pronouns should be understood to refer to males and females alike.

very few scientists in any field would claim today that one should look at the world only scientifically. The botanist looking at a daffodil has no reason to dispute the right of the poet to look at the same object in a very different manner. There are many ways of playing. The point is not that one denies other people's games but that one is clear about the rules of one's own. The game of the sociologist, then, uses scientific rules. As a result, the sociologist must be clear in his own mind as to the meaning of these rules. That is, he must concern himself with methodological questions. Methodology does not constitute his goal. The latter, let us recall once more, is the attempt to understand society. Methodology helps in reaching this goal. In order to understand society, or that segment of it that he is studying at the moment, the sociologist will use a variety of means. . . .

We daresay that this conception of the sociologist would meet with very wide consensus within the discipline today. But we would like to go a little bit further here and ask a somewhat more personal (and therefore, no doubt, more controversial) question. We would like to ask not only what it is that the sociologist is doing but also what it is that drives him to it. Or, to use the phrase Max Weber used in a similar connection, we want to inquire a little into the nature of the sociologist's demon. In doing so, we shall evoke an image that is not so much ideal-typical in the above sense but more confessional in the sense of personal commitment. Again, we are not interested in excommunicating anyone. The game of sociology goes on in a spacious playground. We are just describing a little more closely those we would like to tempt to join our game.

We would say then that the sociologist (that is, the one we would really like to invite to our game) is a person intensively, endlessly, shamelessly interested in the doings of men. His natural habitat is all the human gathering places of the world, wherever men come together. The sociologist may be interested in many other things. But his consuming interest remains in the world of men, their institutions, their history, their passions. And since he is interested in men, nothing that men do can be altogether tedious for him. He will naturally be interested in the events that engage men's ultimate beliefs, their moments of tragedy and grandeur and ecstasy. But he will also be fascinated by the commonplace, the everyday. He will know reverence, but this reverence will not prevent him from wanting to see and to understand. He may sometimes feel revulsion or contempt. But this also will not deter him from wanting to have his questions answered. The sociologist, in his quest for understanding, moves through the world of men without respect for the usual lines of demarcation. Nobility and degradation, power and obscurity, intelligence and folly—these are equally *interesting* to him, however unequal they may be in his personal values or tastes. Thus his questions may lead him to all possible levels of society, the best and the least known places, the most respected and the most despised. And, if he is a good sociologist, he will find himself in all these places because his own questions have so taken possession of him that he has little choice but to seek for answers.

It would be possible to say the same things in a lower key. We could say that the sociologist, but for the grace of his academic title, is the man who must listen to gossip despite himself, who is tempted to look through keyholes, to read other people's mail, to open closed cabinets. . . . What interests us is the curiosity that grips any sociologist in front of a closed door behind which there are human voices. If he is a good sociologist, he will want to open that door, to understand these voices. Behind each closed door he will anticipate some new facet of human life not yet perceived and understood.

The sociologist will occupy himself with matters that others regard as too sacred or as too distasteful for dispassionate investigation. He will find rewarding the company of priests or prostitutes, depending not on his personal preferences but on the questions he happens to be asking at the moment. He will also concern himself with matters that others may find much too boring. He will be interested in the human interaction that goes with warfare or with great intellectual discoveries, but also in the relations between people employed in a restaurant or between a group of little girls playing with their dolls. His main focus of attention is not the ultimate significance of what men do, but the action in itself, as another example of the infinite richness of human conduct. So much for the image of our playmate.

In these journeys through the world of men the sociologist will inevitably encounter other professional Peeping Toms. Sometimes these will resent his presence, feeling that he is poaching on their preserves. In some places the sociologist will meet up with the economist, in others with the political scientist, in yet others with the psychologist or the ethnologist. Yet chances are that the questions that have brought him to these same places are different from the ones that propelled his fellow-trespassers. The sociologist's questions always remain essentially the same: "What are people doing with each other here?" "What are their relationships to each other?" "How are these relationships organized in institutions?" "What are the collective ideas that move men and institutions?" In trying to answer these questions in specific instances, the sociologist will, of course, have to deal with economic or political matters, but he will do so in a way rather different from that of the economist or the political scientist. The scene that he contemplates is the same human scene that these other scientists concern themselves with. But the sociologist's angle of vision is different. When this is understood, it becomes clear that it makes little sense to try to stake out a special enclave within which the sociologist will carry on business in his own right. . . .

Any intellectual activity derives excitement from the moment it becomes a trail of discovery. . . . The excitement of sociology is usually of a different sort. Sometimes, it is true, the sociologist penetrates into worlds that had previously been quite unknown to him — for instance, the world of crime, or the world of some bizarre religious sect, or the world fashioned by the exclusive concerns of some group such as medical specialists or military leaders or advertising executives. However, much of the time the sociologist moves in sectors of experience that are familiar to him and to most people in his society. He investigates communities, institutions and activities that one can read about every day in the newspapers. Yet there is another excitement of discovery beckoning in his investigations. It is not the excitement of coming upon the totally unfamiliar, but rather the excitement of finding the familiar becoming transformed in its meaning. The fascination of sociology lies in the fact that its perspective makes us see in a new light the very world in which we have lived all our lives. This also constitutes a transformation of consciousness. Moreover, this transformation is more relevant existentially than that of many other intellectual disciplines, because it is more difficult to segregate in some special compartment of the mind. The astronomer does not live in the remote galaxies, and the nuclear physicist can, outside his laboratory, eat and laugh and marry and vote without thinking about the insides of the atom. The geologist looks at rocks only at appropriate times, and the linguist speaks English with his wife. The sociologist lives in society, on the job and off it. His own life, inevitably, is part of his subject matter. Men being what they are, sociologists too manage to segregate their professional insights from their everyday affairs. But it is a rather difficult feat to perform in good faith.

The sociologist moves in the common world of men, close to what most of them would call real. The categories he employs in his analyses are only refinements of the categories by which other men live — power, class, status, race, ethnicity. As a result, there is a deceptive simplicity and obviousness about some sociological investigations. One reads them, nods at the familiar scene, remarks that one has heard all this before and don't people have better things to do than to waste their time on truisms — until one is suddenly brought up against an insight that radically questions everything one had previously assumed about this familiar scene. This is the point at which one begins to sense the excitement of sociology.

Let us take a specific example. Imagine a sociology class in a Southern college where almost all the students are white Southerners. Imagine a lecture on the subject of the racial system of the South. The lecturer is talking here of matters that have been familiar to his students from the time of their infancy. Indeed, it may be that they are much more familiar with the minutiae of this system than he is. They are quite bored as a result. It seems to them that he is only using more pretentious words to describe what they already know. Thus he may use the term "caste," one commonly used now by American sociologists to describe the Southern racial system. But in explaining the term he shifts to traditional Hindu society, to make it clearer. He then goes on to analyze the magical beliefs inherent in caste tabus, the social dynamics of commensalism and connubium, the economic interests concealed within the system, the ways in which religious beliefs relate to the tabus, the effects of the caste system upon the industrial development of the society and vice versa—all in India. But suddenly India is not very far away at all. The lecture then goes back to its Southern theme. The familiar now seems not quite so familiar any more. Questions are raised that are new, perhaps raised angrily, but raised all the same. And at least some of the students have begun to understand that there are functions involved in this business of race that they have not read about in the newspapers (at least not those in their hometowns) and that their parents have not told them—partly, at least, because neither the newspapers nor the parents knew about them.

It can be said that the first wisdom of sociology is this—things are not what they seem. This too is a deceptively simple statement. It ceases to be simple after a while. Social reality turns out to have many layers of meaning. The discovery of each new layer changes the perception of the whole.

Anthropologists use the term "culture shock" to describe the impact of a totally new culture upon a newcomer. In an extreme instance such shock will be experienced by the Western explorer who is told, halfway through dinner, that he is eating the nice old lady he had been chatting with the previous day—a shock with predictable physiological if not moral consequences.

Most explorers no longer encounter cannibalism in their travels today. However, the first encounters with polygamy or with puberty rites or even with the way some nations drive their automobiles can be quite a shock to an American visitor. With the shock may go not only disapproval or disgust but a sense of excitement that things can *really* be that different from what they are at home. To some extent, at least, this is the excitement of any first travel abroad. The experience of sociological discovery could be described as "culture shock" minus geographical displacement. In other words, the sociologist travels at home—with shocking results. He is unlikely to find that he is eating a nice old lady for dinner. But the discovery, for instance, that his own church has considerable money invested in the missile industry or that a few blocks from his home there are people who engage in cultic orgies may not be drastically different in emotional impact. Yet we would not want to imply that sociological discoveries are always or even usually outrageous to moral sentiment. Not at all. What they have in common with exploration in distant lands, however, is the sudden illumination of new and unsuspected facets of human existence in society. This is the excitement and, as we shall try to show later, the humanistic justification of sociology.

People who like to avoid shocking discoveries, who prefer to believe that society is just what they were taught in Sunday School, who like the safety of the rules and the maxims of what Alfred Schuetz has called the "world-taken-for-granted," should stay away from sociology. People who feel no temptation before closed doors, who have no curiosity about human beings, who are content to admire scenery without wondering about the people who live in those houses on the other side of that river, should probably also stay away from sociology. They will find it unpleasant or, at any rate, unrewarding. People who are interested in human beings only if they can change, convert or reform them should also be warned, for they will find sociology much less useful than they hoped. And people whose interest is mainly in their own conceptual constructions will do just as well to turn to the study of little white mice. Sociology will be satisfying, in the long run, only to those who can think of nothing more entrancing than to watch men and to understand things human.

It may now be clear that we have, albeit deliberately, understated the case in the title of this chapter.

3

The Promise

C. WRIGHT MILLS

Here Mills completes the task of defining what the sociologist does, how he or she does it, and how all subjects of human behavior are in the sociological purview. Mills does this by indicating what the sociological mind helps individuals accomplish — what is going on in the world and happening to themselves.

As you read, ask yourself the following questions:

1. What does the sociological imagination allow people to accomplish?

2. What changes in your personal milieu might be better understood by looking at changes in the society?

GLOSSARY

Sociological imagination The capacity to understand the most impersonal and remote changes in terms of their effect on the human self and to see the relationship between the two.

Personal trouble A private matter that occurs within the character of an individual and within the range of that individual's immediate relations with others.

Public issue A matter that transcends the local environment of an individual and the range of that individual's inner life.

NOWADAYS MEN* OFTEN FEEL their private lives are a series of traps. They sense that within their everyday worlds, they cannot overcome their troubles, and in this feeling, they are often quite correct: What ordinary men are directly aware of and what they try to do are bounded by the private orbits in which they live; their visions and their powers are limited to the close-up scenes of job, family, neighborhood; in other milieux, they move vicariously and remain spectators. And the more aware they become, however vaguely, of ambitions and of threats which transcend their immediate locales, the more trapped they seem to feel.

Underlying this sense of being trapped are seemingly impersonal changes in the very structure of continent-wide societies. The facts of contemporary history are also facts about the success and the failure of individual men and women. When a society is industrialized, a peasant becomes a worker; a feudal lord is liquidated or becomes a businessman. When classes rise or fall, a man is employed or unemployed; when the rate of investment goes up or down, a man takes

Abridged from "The Promise," The Sociological Imagination *by C. Wright Mills. Copyright © 1959 by Oxford University Press, Inc. Renewed 1987 by Yaraslava Mills. Reprinted by permission of the publisher.*

*This reading was written before it became correct to use gender-neutral terms. Male nouns and pronouns should be understood to refer to males and females alike.

new heart or goes broke. When wars happen, an insurance salesman becomes a rocket launcher; a store clerk, a radar man; a wife lives alone; a child grows up without a father. Neither the life of an individual nor the history of a society can be understood without understanding both.

Yet men do not usually define the troubles they endure in terms of historical change and institutional contradiction. The well-being they enjoy, they do not usually impute to the big ups and downs of the societies in which they live. Seldom aware of the intricate connection between the patterns of their own lives and the course of world history, ordinary men do not usually know what this connection means for the kinds of men they are becoming and for the kinds of history-making in which they might take part. They do not possess the quality of mind essential to grasp the interplay of man and society, of biography and history, of self and world. They cannot cope with their personal troubles in such ways as to control the structural transformations that usually lie behind them.

Surely it is no wonder. In what period have so many men been so totally exposed at so fast a pace to such earthquakes of change? That Americans have not known such catastrophic changes as have the men and women of other societies is due to historical facts that are now quickly becoming "merely history." The history that now affects every man is world history. Within this scene and this period, in the course of a single generation, one-sixth of mankind is transformed from all that is feudal and backward into all that is modern, advanced, and fearful. Political colonies are freed; new and less visible forms of imperialism installed. Revolutions occur; men feel the intimate grip of new kinds of authority. Totalitarian societies rise, and are smashed to bits—or succeed fabulously. After two centuries of ascendancy, capitalism is shown up as only one way to make society into an industrial apparatus. After two centuries of hope, even formal democracy is restricted to a quite small portion of mankind. Everywhere in the underdeveloped world, ancient ways of life are broken up, and vague expectations become urgent demands. Everywhere

in the overdeveloped world, the means of authority and of violence become total in scope and bureaucratic in form. Humanity itself now lies before us, the super-nation at either pole concentrating its most coordinated and massive efforts upon the preparation of World War III.

The very shaping of history now outpaces the ability of men to orient themselves in accordance with cherished values. And which values? Even when they do not panic, men often sense that older ways of feeling and thinking have collapsed and that newer beginnings are ambiguous to the point of moral stasis. Is it any wonder that ordinary men feel they cannot cope with the larger worlds with which they are so suddenly confronted? That they cannot understand the meaning of their epoch for their own lives? That—in defense of selfhood—they become morally insensible, trying to remain altogether private men? Is it any wonder that they come to be possessed by a sense of the trap?

It is not only information that they need—in this Age of Fact, information often dominates their attention and overwhelms their capacities to assimilate it. It is not only the skills of reason that they need—although their struggles to acquire these often exhaust their limited moral energy.

What they need, and what they feel they need, is a quality of mind that will help them to use information and to develop reason in order to achieve lucid summations of what is going on in the world and of what may be happening within themselves. It is this quality, I am going to contend, that journalists and scholars, artists and publics, scientists and editors are coming to expect of what may be called the sociological imagination.

I

The sociological imagination enables its possessor to understand the larger historical scene in terms of its meaning for the inner life and the external career of a variety of individuals. It enables him to take into account how individuals, in the welter of their daily experience, often become falsely conscious of their social positions. Within that welter, the framework of modern society is sought, and within that framework the psychologies of a variety of men and women are formulated. By such means the personal uneasiness of individuals is focused upon explicit troubles and the indifference of publics is transformed into involvement with public issues.

The first fruit of this imagination—and the first lesson of the social science that embodies it—is the idea that the individual can understand his own experience and gauge his own fate by locating himself within his period; that he can know his own chances in life only by becoming aware of those of all individuals in his circumstances. In many ways it is a terrible lesson; in many ways a magnificent one. We do not know the limits of man's capacities for supreme effort or willing degradation, for agony or glee, for pleasurable brutality or the sweetness of reason. But in our time we have come to know the limits of "human nature" are frighteningly broad. We have come to know that every individual lives, from one generation to the next, in some society; that he lives out a biography, and that he lives it out within some historical sequence. By the fact of his living he contributes, however minutely, to the shaping of this society and to the course of its history, even as he is made by society and by its historical push and shove.

The sociological imagination enables us to grasp history and biography and the relations between the two within society. That is its task and its promise. To recognize this task and this promise is the mark of the classic social analyst. It is characteristic of Herbert Spencer—turgid, polysyllabic, comprehensive; of E. A. Ross—graceful, muckraking, upright; of Auguste Comte and Emile Durkheim; of the intricate and subtle Karl Mannheim. It is the quality of all that is intellectually excellent in Karl Marx; it is the clue to Thorstein Veblen's brilliant and ironic insight, to Joseph Schumpeter's many-sided constructions of reality; it is the basis of the psychological sweep of W. E. H. Lecky no less than of the profundity and clarity of Max Weber. And it is the signal of what is best in contemporary studies of man and society.

No social study that does not come back to the problems of biography, of history, and of their intersections within a society has completed its intellectual journey. Whatever the specific problems of the classic social analysts, however limited or however broad the features of social reality they have examined, those who have been imaginatively aware of the promise of their work have consistently asked three sorts of questions:

1. What is the structure of this particular society as a whole? What are its essential components, and how are they related to one another? How does it differ from other varieties of social order? Within it, what is the meaning of any particular feature for its continuance and for its change?
2. Where does this society stand in human history? What are the mechanics by which it is changing? What is its place within and its meaning for the development of humanity as a whole? How does any particular feature we are examining affect, and how is it affected by, the historical period in which it moves? And this period—what are its essential features? How does it differ from other periods? What are its characteristic ways of history-making?
3. What varieties of men and women now prevail in this society and in this period? And what varieties are coming to prevail? In what ways are they selected and formed, liberated and repressed, made sensitive and blunted? What kinds of "human nature" are revealed in the conduct and character we observe in this society in this period? And what is the meaning of "human nature" of each and every feature of the society we are examining?

Whether the point of interest is a great power state or a minor literary mood, a family, a prison, a creed—these are the kinds of questions the best social analysts have asked. They are the intellectual pivots of classic studies of man in society—and they are the questions inevitably raised by any mind possessing the sociological imagination. For that imagination is the capacity to shift from one perspective to another—from the political to the psychological; from examination of a single family to comparative assessment of the national budgets of the world; from the theological school to the military establishment; from considerations of an oil industry to studies of contemporary poetry. It is the capacity to range from the most impersonal and remote transformations to the most intimate features of the human self—and to see the relations between the two. Back of its use there is always the urge to know the social and historical meaning of the individual in the society and in the period in which he has his quality and his being.

That, in brief, is why it is by means of the sociological imagination that men now hope to grasp what is going on in the world, and to understand what is happening in themselves as minute points of the intersections of biography and history within society. In large part, contemporary man's self-conscious view of himself as at least an outsider, if not a permanent stranger, rests upon an absorbed realization of social relativity and of the transformative power of history. The sociological imagination is the most fruitful form of this self-consciousness. By its use men whose mentalities have swept only a series of limited orbits often come to feel as if suddenly awakened in a house with which they had only supposed themselves to be familiar. Correctly or incorrectly, they often come to feel that they can now provide themselves with adequate summations, cohesive assessments, comprehensive orientations. Older decisions that once appeared sound now seem to them products of a mind unaccountably dense. Their capacity for astonishment is made lively again. They acquire a new way of thinking, they experience a transvaluation of values; in a word, by their reflection and by their sensibility, they realize the cultural meanings of the social sciences.

II

Perhaps the most fruitful distinction with which the sociological imagination works is between "the personal troubles of milieu" and "the pub-

lic issues of social structure." This distinction is an essential tool of the sociological imagination and a feature of all classic work in social science.

Troubles occur within the character of the individual and within the range of his immediate relations with others; they have to do with his self and with those limited areas of social life of which he is directly and personally aware. Accordingly, the statement and the resolution of troubles properly lie within the individual as a biographical entity and within the scope of his immediate milieu—the social setting that is directly open to his personal experience and to some extent his willful activity. A trouble is a private matter: values cherished by an individual are felt by him to be threatened.

Issues have to do with matters that transcend these local environments of the individual and the range of his inner life. They have to do with the organization of many such milieux into the institutions of an historical society as a whole, with the ways in which various milieux overlap and interpenetrate to form the larger structure of social and historical life. An issue is a public matter: some value cherished by publics is felt to be threatened. Often there is a debate about what that value really is and about what it is that really threatens it. This debate is often without focus if only because it is the very nature of an issue, unlike even widespread trouble, that it cannot very well be defined in terms of the immediate and everyday environment of ordinary men. An issue, in fact, often involves a crisis in institutional arrangements, and often too it involves what Marxists call "contradictions" or "antagonisms."

In these terms, consider unemployment. When, in a city of 100,000, only one man is unemployed, that is his personal trouble, and for its relief we properly look to the character of the man, his skills, and his immediate opportunities. But when in a nation of 50 million employees, 15 million men are unemployed, that is an issue, and we may not hope to find its solution within the range of opportunities open to any one individual. The very structure of opportunities has collapsed. Both the correct statement of the problem and the range of possible solutions require us to consider the economic and political institutions of the society, and not merely the personal situation and character of a scatter of individuals.

Consider war. The personal problem of war, when it occurs, may be how to survive it or how to die in it with honor; how to make money out of it; how to climb into the higher safety of the military apparatus; or how to contribute to the war's termination. In short, according to one's values, to find a set of milieux and within it to survive the war or make one's death in it meaningful. But the structural issues of war have to do with its causes; with what types of men it throws up into command; with its effects upon economic and political, family and religious institutions, with the unorganized irresponsibility of a world of nation-states.

Consider marriage. Inside a marriage a man and a woman may experience personal troubles, but when the divorce rate during the first four years of marriage is 250 out of every 1000 attempts, this is an indication of a structural issue having to do with the institutions of marriage and the family and other institutions that bear upon them.

Or consider the metropolis—the horrible, beautiful, ugly, magnificent sprawl of the great city. For many upper-class people, the personal solution to "the problem of the city" is to have an apartment with a private garage under it in the heart of the city, and forty miles out, a house by Henry Hill, garden by Garrett Eckbo, on a hundred acres of private land. In these two controlled environments—with a small staff at each end and a private helicopter connection—most people could solve many of the problems of personal milieux caused by the facts of the city. But all this, however splendid, does not solve the public issues that the structural fact of the city poses. What should be done with this wonderful monstrosity? Break it all up into scattered units, combining residence and work? Refurbish it as it stands? Or, after evacuation, dynamite it and build new cities according to new plans in new places? What should those plans be? And who is to decide and to accomplish whatever choice is made? These are structural issues; to confront them and to solve them require us to consider

political and economic issues that affect innumerable milieux.

Insofar as an economy is so arranged that slumps occur, the problem of unemployment becomes incapable of personal solution. Insofar as war is inherent in the nation-state system and in the uneven industrialization of the world, the ordinary individual in his restricted milieu will be powerless—with or without psychiatric aid—to solve the troubles this system or lack of system imposes upon him. Insofar as the family as an institution turns women into darling little slaves and men into their chief providers and unweaned dependents, the problem of a satisfactory marriage remains incapable of purely private solution. Insofar as the overdeveloped megalopolis and the overdeveloped automobile are built-in features of the overdeveloped society, the issue of urban living will not be solved by personal ingenuity and private wealth.

What we experience in various and specific milieux, I have noted, is often caused by structural changes. Accordingly, to understand the changes of many personal milieux we are required to look beyond them. And the number and variety of such structural changes increase as the institutions within which we live become more embracing and more intricately connected with one another. To be aware of the idea of social structure and to use it with sensibility is to be capable of tracing such linkages among a great variety of milieux. To be able to do that is to possess the sociological imagination.

III

What are the major issues for publics and the key troubles of private individuals in our time? To formulate issues and troubles, we must ask what values are cherished yet threatened, and what values are cherished and supported, by the characterizing trends of our period. In the case both of threat and of support we must ask what salient contradictions of structure may be involved.

When people cherish some set of values and do not feel any threat to them, they experience *well-being*. When they cherish values but *do* feel them to be threatened, they experience a crisis—either as a personal trouble or as a public issue. And if all their values seem involved, they feel the total threat of panic.

But suppose people are neither aware of any cherished values nor experience any threat? That is the experience of *indifference,* which, if it seems to involve all their values, becomes apathy. Suppose, finally, they are unaware of any cherished values, but still are very much aware of a threat? That is the experience of *uneasiness,* of anxiety, which, if it is total enough, becomes a deadly unspecified malaise.

Ours is a time of uneasiness and indifference—not yet formulated in such ways as to permit the work of reason and the play of sensibility. Instead of troubles—defined in terms of values and threats—there is often the misery of vague uneasiness; instead of explicit issues there is often merely the beat feeling that all is somehow not right. Neither the values threatened nor whatever threatens them has been stated; in short, they have not been carried to the point of decision. Much less have they been formulated as problems of social science.

In the thirties there was little doubt—except among certain deluded business circles—that there was an economic issue which was also a pack of personal troubles. In these arguments about "the crisis of capitalism," the formulations of Marx and the many unacknowledged reformulations of his work probably set the leading terms of the issue, and some men came to understand their personal troubles in these terms. The values threatened were plain to see and cherished by all; the structural contradictions that threatened them also seemed plain. Both were widely and deeply experienced. It was a political age.

But the values threatened in the era after World War II are often neither widely acknowledged as values nor widely felt to be threatened. Much private uneasiness goes unformulated; much public malaise and many decisions of enormous structural relevance never become public issues. For those who accept such inherited values as reason and freedom, it is the uneasiness itself that is the trouble; it is the indif-

ference that is the issue. And it is this condition, of uneasiness and indifference, that is the signal feature of our period.

All this is so striking that it is often interpreted by observers as a shift in the very kinds of problems that need now to be formulated. We are frequently told that the problems of our decade, or even the crisis of our period, have shifted from the external realm of economics and now have to do with the quality of individual life—in fact with the question of whether there is soon going to be anything that can properly be called individual life. Not child labor but comic books, not poverty but mass leisure, are at the center of concern. Many great public issues as well as many private troubles are described in terms of "the psychiatric"—often, it seems, in a pathetic attempt to avoid the large issues and problems of modern society. Often this statement seems to rest upon a provincial narrowing of interest to the Western societies, or even to the United States—thus ignoring two-thirds of mankind; often, too, it arbitrarily divorces the individual life from the larger institutions within which that life is enacted, and which on occasion bear upon it more grievously than do the intimate environments of childhood.

Problems of leisure, for example, cannot even be stated without considering problems of work. Family troubles over comic books cannot be formulated as problems without considering the plight of the contemporary family in its new relations with the newer institutions of the social structure. Neither leisure nor its debilitating uses can be understood as problems without recognition of the extent to which malaise and indifference now form the social and personal climate of contemporary American society. In this climate, no problems of "the private life" can be stated and solved without recognition of the crisis of ambition that is part of the very career of men at work in the incorporated economy.

It is true, as psychoanalysts continually point out, that people do often have "the increasing sense of being moved by obscure forces within themselves which they are unable to define." But it is *not* true, as Ernest Jones asserted, that "man's chief enemy and danger is his own unruly nature and the dark forces pent up within him." On the contrary: "Man's chief danger" today lies in the unruly forces of contemporary society itself, with its alienating methods of production, its enveloping techniques of political domination, its international anarchy—in a word, its pervasive transformations of the very "nature" of man and the conditions and aims of his life.

It is now the social scientist's foremost political and intellectual task—for here the two coincide—to make clear the elements of contemporary uneasiness and indifference. It is the central demand made upon him by other cultural workmen—by physical scientists and artists, by the intellectual community in general. It is because of this task and these demands, I believe, that the social sciences are becoming the common denominator of our cultural period, and the sociological imagination our most needed quality of mind.

4

Some Sociological Aspects of HIV Disease

EDWARD L. KAIN

At first glance, the problem of AIDS might not appear to be an issue related to sociology. But if, as Berger noted in the first reading, the sociologist is "interested in the action of human conduct," then AIDS—a disease spread through human behavior—is an object of sociological interest. Kain explains this relationship.

As you read, ask yourself the following questions:

1. The social construction of AIDS can be seen from what two perspectives?

2. How would you deal with the problem of AIDS in the United States? In Africa?

GLOSSARY

Role of the other Label and expectations attached by one individual to another individual or group.

Social construction Process individuals use to create their reality through social interaction.

Stigma Negative label that changes an individual's self-concept and social identity.

ONE OF THE MOST crucial social problems emerging in the last two decades of the twentieth century is the epidemic of HIV disease. From an illness that at first was perceived as affecting only a small number of gay and bisexual men in a few U.S. cities, it has grown to a pandemic (a worldwide epidemic). Both the number of AIDS cases and the estimated number of people who are infected with HIV have grown at a staggering rate.[1] In April 1991, scarcely a decade into the epidemic, the World Health Organization (WHO) estimated that approximately 9 million people were infected worldwide (World Health Organization, 1991). By early 1993 that estimate had reached more than 12 million (WHO,

This reading was reprepared for the 8th edition of Sociological Footprints, *edited by Leonard Cargan and Jeanne H. Ballantine, editors. Research for the paper was funded, in part, by a fellowship from the Brown Foundation of Houston, Texas. The author would like to thank Dan Hilliard for his helpful comments on an earlier draft of the paper.*

as reported in Cowley, 1993) and by the beginning of 1998 it had grown to over 30 million (UNAIDS and WHO, 1998). This last source estimates that, worldwide, some 11.7 million people had already died from HIV disease by the beginning of 1998.

The epidemic has implications far beyond these numbers. Worldwide there are approximately 16,000 additional infections per day—resulting in nearly 5.8 million new infections in 1997 alone. When adults die from HIV disease, they often leave behind children who must be cared for by others. This pandemic has left over 8 million orphans worldwide, and that number will increase rapidly over the next decade (UNAIDS and WHO, 1998).

Although disease is a biological concept, illness is a social construct. As such, it has an impact on how well a society is able to combat the disease as well as on the personal experience of individuals who contract it. Discussion of HIV disease ultimately points to important sociological concepts from a number of areas within the discipline. The next few pages examine some of the ways in which a sociological perspective can help us understand the social impact of AIDS and HIV disease.

A full sociological analysis of the epidemic is beyond the scope of this reading. This discussion focuses on three topics: (1) the social construction of illness, (2) HIV disease and stigma, and (3) shifting definitions of HIV disease.

The Social Construction of Illness

The current social construction of HIV disease and AIDS must be understood in the context of historical changes in the broader social definition of disease. Rosenberg (1988) suggests that in late eighteenth- and early nineteenth-century America illness was conceived primarily in terms of the individual. "Even epidemic disease was understood to result from an unbalanced state in a particular individual . . . thus the conventional and persistent emphases on regimen and diet in the cause and cure of sickness" (Rosenberg,

1988, p. 17). Using the example of cholera, Rosenberg points out that certain groups of people were seen as "predisposed" to falling ill—those who had poor nutrition, who were dirty, or who were gluttonous. Such an approach had the function of reducing the seeming randomness with which the epidemic struck.

This approach to disease in the West slowly shifted throughout the nineteenth century. The Paris clinical school argued that disease was something "lesion-based," which played itself out in each person who was afflicted. By the end of the century, the germ theory of disease gave such a model an explanatory mechanism. At the same time, a broader range of behaviors, previously linked to concepts of sin and deviance, came under the purview of medicine. The medicalization of alcoholism, mental illness, and a variety of sexual behaviors expanded the authority of medicine to deal with behavior previously not thought of as illness.

One result of medicalization is to reduce the amount of individual blame. Defining alcoholism, for example, as a disease transforms its very character. The alcoholic is now someone who has an affliction rather than someone who is of weak moral character. The blame begins to shift from the individual to the disease.

BLAME

Blame never shifts entirely away from the individual, however. In the West, although we may be able to trace an historical shift along a continuum from blaming the individual to blaming an outside causal agent of some type (whether it be a germ, a gene, or a virus), this shift has not occurred equally for various types of diseases. Illness or injury resulting from behaviors defined by the society as morally wrong receive much greater attribution of individual responsibility than those resulting from behaviors that do not have the same degree of moral censure.

Ultimately, many types of illness can be linked to individual behavior. Smoking and overeating both are behaviors that are "chosen" by individuals. People who develop lung cancer, strokes, or heart attacks related to these behaviors are not, by and large, blamed for their illness. Sports, like football (or, more obviously, hot-dog skiing), lead to a large number of serious injuries and deaths each year. Yet these are viewed as "accidents." Typically, those who suffer the results are not defined as deserving their fate.

Sexually transmitted diseases are much more likely to elicit reactions of blame and deservedness because of their linkage to behavior that is defined as morally wrong by the community. Furthermore, there is a long cultural history of dividing the innocent victims[2] (spouses and children who are unknowingly infected by a partner, usually a husband, who strayed) from guilty sufferers who deserve their illness as the wages of sin.

The stigmatization of those who are ill and a search for someone to blame are not contemporary phenomena. During the 1656 outbreak of bubonic plague in Rome, for example, foreigners, the poor, and the Jews were blamed for the epidemic. Similarly, the poor and immigrants were blamed for the 1832 cholera outbreak in New York City. In this case, the moral failings that led to poverty were seen as the root cause of the illness. Indeed, in his examination of past responses to epidemics (1988, p. 57), Guenter Risse concludes that "in the face of epidemic disease, mankind has never reacted kindly . . . the response to disease is a powerful tool to buttress social divisions and prejudices."

THE ROLE OF THE OTHER

The cause of a disease is often understood so as to shift blame for an epidemic on the other. Whether this other is an ethnic or racial group, a religious or social category, or a group stigmatized for behavior that is labeled as deviant, this conception of the other is powerful in shaping the social response to disease.

The most extreme social response is total isolation of those who are sick, in an attempt to stop the spread of the illness from one segment of society to another. Historically, quarantine has been tried as a method for coping with a variety of illnesses. With tuberculosis, yellow fever, cholera, and leprosy, all efforts to quarantine large numbers of people have been failures. Rather than effective public health measures, these mass quarantines have been expressions of fear of the other — attacks on the civil liberties of groups not accepted by the general public (Musto, 1988).

In the case of HIV disease, the complex interactions between blame, fear, discrimination, and stigma (a concept discussed in the next section) have led to a social construction of the disease in which "us" and "them" play a central role. The worldwide social impact of the epidemic, however, makes it clear that *all* of us are living with HIV disease (see Gilmore and Somerville, 1994, for an excellent discussion of this issue).

THE ROLE OF "DESERVING TO HAVE THE DISEASE"

The search for a group to blame for illness, combined with the tendency to blame the "other," often creates the conception that those who are sick deserve to have the disease. This is reflected in early cultural constructions of AIDS in the United States. Those who contracted the disease were often divided into "innocent victims" of AIDS and those who somehow deserved their illness. This ascription of personal responsibility for their illness affects not only people with AIDS but also their caregivers (see Sosnowitz and Kovacs, 1992).

HIV Disease and Stigma

A key to understanding the social construction of HIV disease is the concept of stigma. Erving Goffman defines stigma as "an attribute that is deeply discrediting." He goes on to say that there are three types of stigma — what he calls "abominations of the body" (physical deformities); "blemishes of individual character" (some examples include a weak will, dishonesty, alcoholism, and other addictions, homosexuality, radicalism); and "the tribal stigma of race, nation, and religion" (Goffman, 1963).

HIV disease has the potential of developing all three types of stigma described by Goffman. Some of the opportunistic infections associated with end-stage HIV disease (AIDS) can be disfiguring—the skin lesions of Kaposi's sarcoma being a prime example. The extreme weight loss associated with end-stage HIV disease also creates what Goffman called an "abomination of the body." The social and cultural construction of HIV disease involving blame, as noted earlier, further links it to a number of "blemishes in individual character"—drug use, homosexuality, and inability to control one's own behavior in a safe manner. Finally, what Goffman called the "tribal stigma" of race, nation, and religion, also apply to the cultural construction of AIDS and HIV disease. In the early years of the epidemic, one of the major "risk groups" was Haitians. A number of authors have suggested that the cultural constructions of AIDS are tinged with racism. In *AIDS and Accusation,* for example, Paul Farmer argues that ethnocentrism and racism in the United States were key factors in the theories about a Haitian origin for AIDS (Farmer, 1992).

WHY IS HIV DISEASE PARTICULARLY STIGMATIZED?

Most of the literature on HIV disease and AIDS talks about the importance of stigma. From this literature, we find a number of characteristics that predict whether or not a disease will be particularly stigmatized. One of the first analyses of the stigma associated with AIDS was by Peter Conrad (1986). After a discussion of the public hysteria surrounding this illness, Conrad identifies four social aspects of AIDS that lead to its peculiar status.

First, and foremost, throughout the early years of the epidemic, AIDS was associated with "risk groups" that were both marginal and stigmatized. Because early cases of AIDS were found in homosexual men and intravenous drug users, a powerful cultural construction emerged that defined the illness as a gay disease.

Second, because a major mode of transmission involves sexual activity, the disease is thus further stigmatized. In his insightful book on the history of venereal disease in the United States, Allan Brandt suggests that human societies have never been very effective in dealing with sexually transmitted diseases. He begins his book by noting that "the most remarkable change in patterns of health during the last century has been the largely successful conquest of infectious disease." The striking exception to this pattern has been an explosion in sexually transmitted diseases. He asks, "Why, if we have been successful in fighting infectious disease in this century, have we been unable to deal effectively with venereal disease?" His answer lies in an examination of the social and cultural responses to sexuality and sexually transmitted disease. He argues that venereal disease was a social symbol for "a society characterized by a corrupt sexuality" and was used as "a symbol of pollution and contamination." The power of this social construction has rendered efforts to control sexually transmitted diseases ineffective (Brandt, 1987).

Third, Conrad points out that contagion plays a major role in whether or not a disease stigmatizes the individual who is infected. If a disease is contagious, or if it is *perceived* as contagious, then the stigma of the illness increases.

Fourth, Conrad argues that the fact that AIDS is a deadly disease also adds to its stigma. Because of these social characteristics, Conrad concludes that AIDS is a disease with a triple stigma: "it is connected to stigmatized groups, it is sexually transmitted, and it is a terminal disease."

One of the best sociological discussions on the stigma associated with HIV disease is found in Rose Weitz's *Life with AIDS* (1991). She says that stigma is greatest when an illness evokes the strongest blame and dread. She delineates six conditions that increase blame and dread. Like Conrad, Weitz includes (1) linkage to stigmatized groups, (2) an association with sexuality, (3) if the illness is perceived to be contagious and if there is no vaccine available, and (4) if the illness is "'consequential,' producing death or extensive disability and appearing to threaten not just scattered individuals but society as a whole"

(Weitz, 1991, pp. 45–48). To Conrad's list she adds two more—if the illness creates dehumanizing or disfiguring changes that "seem to transform the person into something beastly or alien" and "if mysteries remain regarding their natural history."

To this list I would add two more factors. First, industrial societies have not had to deal with fatal infectious diseases in several decades. We have come to expect that young people will not be struck down by disease in the prime of their life. Polio was the last great infectious disease to affect modern industrial societies, and much of the population has grown up in a world with no experience of consequential infectious diseases. This increases the fear and dread associated with HIV disease.

Finally, if a disease has an impact on mental functioning, it increases stigma. HIV can directly infect brain cells. As prophylactic measures such as the use of inhalant pentamidine and antiviral drugs such as AZT increase the time between exposure to HIV and first opportunistic infections, the number of people with HIV disease who have impaired mental functioning because of infection with the virus will increase . . . thus increasing the potential for stigma. Indeed dementia is now the most common neuropsychiatric problem found among HIV patients (Buckingham, 1994).

THE FUTURE OF STIGMA AS IT RELATES TO HIV DISEASE

Just as the relative stigma of various illnesses can be predicted by examining the eight characteristics just delineated, the stigma of a particular illness varies by culture and will change over time within any particular culture as these variables change. In cultures where sexual behavior is more openly discussed and where attitudes are more tolerant of homosexuality, the stigma associated with HIV disease will be less. Similarly in cultures where drug use is conceptualized differently (the Netherlands is an instructive example here), the stigma of HIV will be lessened.

The point here, however, is that there is no other disease that has such a high potential for

stigma on all eight characteristics, making HIV disease the prototypical example of stigmatized illness in modern times.

Shifting Definitions of HIV Disease

There is no single social definition of HIV disease and AIDS. Public perception of the disease varies between cultures and has changed over time. It also varies considerably from one segment of the population to another and from region to region in a population. Much of this variation is linked to epidemiology.

The social epidemiology of HIV disease in Africa, for example, is radically different from our country. Rather than being associated with certain high-risk groups, it is more equally distributed among males and females, and appears to be as common in the heterosexual population as it is in the homosexual and bisexual population (World Health Organization, 1991). These differences clearly have a major impact on the social definition of the disease in different parts of the world, as well as on the treatment and ultimate social consequences of the disease.

It is interesting to note that coverage of the disease by the popular press shifted both in magnitude and in tone when it became clear that AIDS could be contracted by so-called innocent people—babies, hemophiliacs, and recipients of blood transfusions (Altman, 1986; Kain and Hart, 1987). Indeed the social definition of AIDS shifted considerably over the first decade of its existence. Some of the early literature referred to the illness as GRID (Gay-Related Immune Deficiency), and there is evidence that when it was defined as a "gay plague" the scientific community joined the general public in its reluctance to take the disease as seriously as was warranted (Shilts, 1987).

Panic, fear, and rumor were common elements of popular press reports in the early 1980s. Indeed, many scientists and AIDS educators were frustrated by continued misconceptions about modes of HIV transmission. These early responses to AIDS reflect basic principles in research on collective behavior. Rumors are most likely to develop when there is ambiguity about something. They help to clarify the situation when data are unavailable (Shibutani, 1966; Macionis, 1991). The early years of the HIV epidemic fit this description perfectly. In addition, because the disease is fatal people have a high degree of interest in the topic. Fear of contagion is a very predictable response.

FROM ACUTE TO CHRONIC

Before the causal agent of AIDS was understood, and before effective treatments had been developed for some of the opportunistic infections, AIDS was defined as a short-term disease that led to death in a relatively short time. As the etiology of the disease has become more clear, it is becoming defined as a long-term chronic illness (see Fee and Fox, 1992). This shift has implications for the cost of treatment, the stigma associated with the disease, and calculations of the social impact of the epidemic.

One change resulting from the shift to defining HIV disease as a chronic illness is the emergence of new issues for the delivery of health care. Chronic illness has typically been associated with the elderly. Long-term care facilities must rethink their methods of patient care when working with younger persons with HIV (Zablotsky and Ory, 1995). Further, policy makers and clinicians may view this shift to HIV as a chronic disease differently. Thus a simple redefinition of the disease as chronic may be inadequate for planning patient care in the case of HIV (Clarke, 1994).

CHANGING EPIDEMIOLOGY

As the epidemiology of the disease changes, so will social definitions. In the United States, there has been a shift over time in which a larger proportion of new AIDS cases are linked to drug use rather than homosexual and bisexual activity. Further, African-Americans and Hispanics are disproportionately affected by the disease. Recent years have also seen an increase in cases among women and heterosexuals (National Research Council, 1990). Because HIV has not been defined as a women's disease in the West, and because of women's lower social status, women

with HIV have a shorter survival time than men. As the epidemiology changes, approaches to women's health issues must also change (Lea, 1994). The combination of race and gender are also important to examine. Women of color make up well over two-thirds of all cases of HIV-infected women in the United States (Land, 1994). By 1997, women had four in ten of the new adult HIV infections in the world and constituted a similar proportion of all adults living with AIDS. Nearly 45% of the people who died of AIDS in 1997 were women (UNAIDS and WHO, 1998).

Higher rates of infection among African-Americans and Hispanics have led researchers to explore ways in which HIV education programs may need to be targeted to specific populations. Different intervention techniques may be more effective with one population than another, and ethnographic analysis can help identify the best programs to use for a particular ethnic or racial group (Goicoechea-Balbona, 1994; Bletzer, 1995).

The history of the epidemic has also seen a shift in the worldwide distribution of reported cases of AIDS as well as projections concerning HIV infection. When the early reported cases were concentrated in North America, AIDS was defined as an American disease. In a number of communist countries, it was defined as a disease of Western capitalism. As data on the epidemic improved, it became clear that large numbers of people were infected in the Third World. Current estimates suggest that the continent of Africa has more than half of all worldwide HIV infections. It appears that the locus of the epidemic in the future may be Asia, where the number of new adult HIV infections each year will surpass Africa before the turn of the century (World Health Organization, 1991).

As the epidemiology of HIV disease shifts, so will cultural constructions of the disease. As women and heterosexuals become larger proportions of the HIV+ population in Western industrialized countries, the stigma will decrease. In addition, infection rates are actually beginning to decline in some of the industrialized West, and treatments using mixtures of "cocktail" drugs have been very promising in slowing (or even reversing) the replication of HIV within infected individuals. As the disease becomes more treatable, it will become less stigmatized. Worldwide, the epidemiology will continue to evolve. While Asia currently has low infection rates, the rate of spread is rapid. In Latin America and the Caribbean the overwhelming proportion of infections are found in marginalized groups, which predicts high stigmatization in those regions. In Eastern Europe most cases are associated with drug use—again predicting stigmatization of the disease.

Data on patterns and trends in HIV infection change relatively rapidly. Luckily there are several excellent resources on the Internet which provide up-to-date information both for the United States and other countries. The Centers for Disease Control maintain a number of web sites related to HIV/AIDS. For general information, statistics, etc., go to:

http://www.cdcnac.org/nachome.html

For daily updates on popular press coverage, go to: **http://www.smartlink.net/martinjh/cdcnews.txt**

The United States Census Bureau web site maintains an international data base which includes infection rates for various countries around the world at **http://www.census.gov/ipc/aidsdb**

Finally, the HIV infoweb is an online library and search engine on the topics of HIV/AIDS. It provides access to an amazing array of data on every state in the U.S., and every country in the world. It is found at **http://www.infoweb.org/**

Concluding Comments

A full sociological analysis of the HIV epidemic would examine a wide variety of issues. The economic and demographic impact of the disease has already been devastating in a number of central African nations. Gains in infant mortality that have taken four decades to achieve have been wiped out in less than a decade. Worldwide, women are particularly vulnerable to HIV disease both because of their lack of power in sexual relationships and the relative ease of viral

transmission between sexual partners (Panos Institute, 1990). Indeed, issues of social stratification are central to understanding the epidemic. In most societies, race, class, and gender are critical variables in predicting who is more likely to become infected, and once infected, who will receive adequate treatment. Worldwide, the poorest countries are among the hardest hit in the epidemic. Unfortunately, much of the progress made in the developed countries relies upon very expensive drug treatments, which will be unavailable to the majority of those infected with HIV throughout the world.

Although this reading has not covered nearly all the issues involved in such an analysis of the HIV pandemic, it has illustrated how a sociological perspective informs our understanding of one of the greatest social problems facing the world today.

NOTES

1. This reading makes a distinction between AIDS (acquired immune deficiency syndrome) and HIV disease. HIV disease begins when a person is infected with the human immunodeficiency virus. For most adults there is an extended period of approximately seven years during which there are very few symptoms of infection. AIDS is the name associated with the end stages of HIV disease.

2. The use of the word *victim* is politically charged because of its implication of powerlessness. Much of the literature on HIV disease and AIDS suggests that "HIV-positive," "seropositive," and "persons with AIDS (PWA)" be used, rather than *victim*. In this context, I use the word *victim* because the popular consciousness clearly separates "innocent victim" from "guilty" sufferers.

REFERENCES

Altman, Dennis, 1986. *AIDS in the Mind of America*. Garden City, NY: Anchor Press/Doubleday.

Bletzer, Keith V., 1995. Use of ethnography in the evaluation and targeting of HIV/AIDS education among Latino farm workers. *AIDS Education and Prevention* 7(2): 178–91.

Brandt, Allan M., 1987. *No Magic Bullet: A Social History of Venereal Disease in the United States Since 1880*. Expanded ed. New York: Oxford University Press.

Buckingham, Stephan L., 1994. HIV-associated dementia: A clinician's guide to early detection, diagnosis, and intervention. *Families in Society* 75(6): 333–45.

Clarke, Aileen, 1994. What is a chronic disease? The effects of a re-definition in HIV and AIDS. *Social Science and Medicine* 39(4): 591–97.

Conrad, Peter, 1986. The social meaning of AIDS. *Social Policy,* Summer 1986, pp. 51–56.

Cowley, Geoffrey, 1993. The future of AIDS. *Newsweek,* March 22, 1993.

Farmer, Paul, 1992. *AIDS and Accusation: Haiti and the Geography of Blame.* Berkeley: University of California Press.

Fee, Elizabeth, and Daniel M. Fox, 1992. *AIDS: The Making of a Chronic Disease.* Berkeley: University of California Press.

Gilmore, Norbert, and Margaret A. Somerville, 1994. Stigmatization, scapegoating and discrimination in sexually transmitted diseases: Overcoming "them" and "us." *Social Science and Medicine* 39(9): 1339–58.

Goffman, Erving, 1963. *Stigma: Notes on the Management of Spoiled Identity.* Englewood Cliffs, NJ: Prentice-Hall.

Goicoechea-Balbona, Anamaria, 1994. Why we are losing the AIDS battle in rural migrant communities. *AIDS and Public Policy Journal* 9(1): 36–48.

Kain, Edward L., and Shannon Hart, 1987. AIDS and the family: A content analysis of media coverage. Presented to the National Council on Family Relations, Atlanta.

Land, Helen, 1994. AIDS and women of color. *Families and Society* 75(6): 355–61.

Lea, Amandah, 1994. Women with HIV and their burden of caring. *Health Care for Women International* 15(6): 489–501.

Macionis, John J., 1991. *Sociology.* Englewood Cliffs, NJ: Prentice-Hall.

Musto, David F., 1988. Quarantine and the Problem of AIDS. In Elizabeth Fee and Daniel M. Fox (eds.), *AIDS: The Burdens of History.* Berkeley: University of California Press.

National Research Council, 1990. *AIDS: The Second Decade.* Washington, DC: National Academy Press.

Panos Institute, 1990. *Triple Jeopardy: Women & AIDS.* London: PANOS Institute.

Risse, Guenter B., 1988. Epidemics and history: Ecological perspectives and social responses. In Elizabeth Fee and Daniel M. Fox (eds.), *AIDS: The Burdens of History.* Berkeley: University of California Press.

Rosenberg, Charles E., 1988. Disease and social order in America: Perceptions and expectations. In Elizabeth Fee and Daniel M. Fox (eds.), *AIDS: The Burdens of History.* Berkeley: University of California Press.

Shibutani, Tomotsu, 1966. *Improvised News: A Sociological Study of Rumor.* Indianapolis: Bobbs-Merrill.

Shilts, Randy, 1987. *And the Band Played On: Politics, People, and the AIDS Epidemic.* New York: St. Martin's Press.

Sosnowitz, Barbara G., and David R. Kovacs, 1992. From burying to caring: Family AIDS support groups. In Joan Huber and Beth E. Schneider (eds.), *The Social Context of AIDS.* Newbury Park, CA: Sage.

UNAIDS and WHO, 1998. *Report on the Global HIV/AIDS Epidemic.*

Weitz, Rose, 1991. *Life with AIDS.* New Brunswick, NJ: Rutgers University Press.

World Health Organization, 1991. *Current and Future Dimensions of the HIV/AIDS Pandemic: A Capsule Summary.*

Zablotsky, Diane L., and Marcia G. Ory, 1995. Fulfilling the potential: Modifying the current long-term care system to meet the needs of persons with AIDS. *Research in the Sociology of Health Care* 12: 313–28.

QUESTIONS FOR DISCUSSION

For further discussion of this topic, see the Wadsworth Sociology Resource Center, "Virtual Society," *http://sociology.wadsworth.com,* under *Sociological Footprints,* by Cargan and Ballantine. You can respond to the discussion questions there or enter your own comments in the online chat forum.

SUGGESTED READINGS AND
SOCIOLOGY INTERNET RESOURCES

See the Wadsworth Sociology Resource Center, "Virtual Society," *http://sociology.wadsworth. com,* for additional links, suggestions for further reading, and learning tools related to this chapter.

Either from the "Virtual Society" website or directly from your web browser, you may access InfoTrac College Edition, an online university library that includes over 700 popular and scholarly journals in which you can find articles related to the topics in this chapter.

Part II

BECOMING A MEMBER
OF SOCIETY

A N INFANT LEARNS QUICKLY THAT CERTAIN actions bring responses from other peo-ple. In this way, the transformation of the newborn child into a social being begins. We call this transformation the *socialization process.*

The family is the first agent in this process. It protects and cares for the helpless infant, providing the training the child needs in order to survive and become a partic-ipating member of society. As the years pass, the child will come into contact with other agents of the socialization process, including peer groups, schools, and religious organizations.

The socialization process is lifelong. We must continuously learn to deal with the changes that occur throughout our lifetime, in our family circumstances, occupational roles, and in other important aspects of life including aging. What we learn is part of the complex whole known as culture. Culture provides us with guidelines for the val-ues, norms, knowledge, and materials necessary for survival in our society. Learning our culture may also have complications, as we try to master such subtleties as the non-verbal "silent language."

Only through our interaction with groups such as the family, peers, schools, and re-ligious organizations do we become social beings. However, this learning process is not always smooth, nor does society always function in perfect order. For example, families provide nurturance, but violence may also be part of our family experience. Peers may provide security and sense of belonging, but they also demand conformity and adherence to strict expectations. Schools train us in the essentials we need to fit into societal positions, but they also function to track students by social class and other factors. Religious organizations may present us with contradictory messages about love and tolerance of those different from ourselves.

Group contact provides needed social interaction and a sense of belonging, but it may also be frustrating. Chance places us in a rich or poor family, which in turn greatly influences our life opportunities. Group membership is our vehicle for carrying out the process of socialization and maintaining the social system, whatever that system might be. Although specifics of each of these three elements—socialization, culture, and groups—vary from society to society, each element is found in every society.

One result of the process of becoming a member of society must be mentioned because it influences how and what we learn. Every society ranks its members according to its own values: this ranking creates differing life opportunities, styles of living, and distribution of power within the society. We call this creation of varying status levels *stratification*. Through socialization, children internalize the class values and beliefs held by their families.

As you read this section, consider the processes that influence the infant and turn the child into a productive member of groups within a particular culture. Also consider the different experiences the infant might undergo if reared in another social class or culture.

CHAPTER 2

Socialization

A LIFELONG LEARNING PROCESS

HOW DO WE BECOME WHO WE are, with our particular values, attitudes and behaviors? Through the process of socialization that takes place within a particular society such as the United States, France, Kenya, or Japan. Each society has a culture or way of life, which has developed over time and which dictates appropriate, acceptable behavior. It is within our families that initial socialization takes place. Here we develop a self, learn to be social beings, and develop into members of our culture. From the day we are born, socialization shapes us into social beings, teaching us the behaviors and beliefs that make social existence possible. Through interaction with others, we develop our self-concepts. This process begins at an early age, when young children interact with others in a process called "symbolic interaction." Very simply, the individual (whether a baby or adult) initiates contact—a cry or words—and receives a response. By interpreting and reacting to that response, the individual learns what brings positive reactions; those actions that receive positive responses and rewards are likely to be continued. Actions receiving negative responses are likely to be dropped.

The process of socialization takes place through interaction with others—interaction that is vital to our social development. The initial "agent," or transmitter of socialization, is the family. Here we begin to learn our roles for participation in the wider world, an important aspect of which is sex-role socialization. The process continues in educational and religious institutions. When children enter school, they face new challenges and expectations. No longer do they receive unconditional love as they do in most families; now they are judged in a competitive environment, their first introduction to the world outside the protection of home. Socialization typically takes place in a series of developmental stages from birth through old age. Some sociologists focus on childhood stages, others on male or female socialization, and others on middle to old age.

Informal agents of socialization—those whose primary purpose is other than socialization—can have a major impact on the process; for instance, the mass media, books, and advertising all send out powerful messages about desirable and appropriate behaviors by presenting role models and lifestyles. Peer groups also affirm or disapprove behaviors of children. The boy who does not engage in "masculine" activities may be ridiculed, for instance.

It might be asked, though, if socialization is really so powerful? Isn't social development a natural outgrowth of physical maturity? To answer that question, some social scientists have focused their studies on cases of social isolation. In rare instances, children do not experience early socialization in the family. For example, some orphanages provide only minimal care and human contact; children growing up in such environments have been found to show a higher percentage of physical and mental retardation. Only a handful of cases of almost total isolation have been available for study. Kingsley Davis describes the case histories of Anna and Isabelle and considers the impact of social isolation on their mental development. He writes that severe retardation is likely to occur when consistent contact with other human beings is absent, and he concludes that even though socialization can take place after prolonged periods of isolation, some effects of isolation may be permanent.

The socialization process can be very rigid and difficult to change when it involves ascribed roles. Socialization into gender roles provides examples of both the socialization process and the acquisition and tenacious nature of stereotypes. In the first reading in this chapter, Clyde Franklin addresses the socialization experience of men and boys, and how they learn masculine gender roles. Each agent of socialization adds to the process of learning attitudes and behaviors: The family provides role models and reinforcement for "proper" behaviors; teachers reward boys and girls for gender-typed behaviors; religious institutions support traditional role behaviors; peer groups pressure boys into acceptable male behaviors. Informal or "nonpurposive" agents of male socialization, including mass media, barbershops, taverns, and business meetings, also contribute to the process.

Television is one of the most powerful informal agents of socialization, especially for children, who may watch several hours a day. Concern over the impact of TV violence has prompted several studies, including the National Television Violence Study, reported on this chapter. The mass media often stereotypes gender roles, especially in ads. Jean Kilbourne points out the use of sex to sell products, which in the process paints unreal pictures of men and women—women as sex objects and men as macho, unfeeling creatures. These images do a disservice to young men who see the tough, unfeeling macho image as the way to show strength, and to young women who may see these models as the ideal and strive to be like them at the expense of their health.

Results of rigid and stereotypic socialization can be destructive. Consider a problem that afflicts primarily young women—eating disorders, specifically anorexia and bulimia. Diane Taub and Penelope McLorg discuss the social context in which these disorders occur: the societal pressures to achieve and maintain a certain body image, the source of this ideal body image, how women are socialized, and the role models presented to them by the mass media. The result of these socialization influences can be dieting and purging, which cause severe health problems.

As you read the selections in this chapter, consider aspects of socialization we have mentioned. What are the effects of isolation from normal human contact? What are the stages of socialization? How does socialization take place through agents of socialization? What are the effects of the socialization process that we take for granted, such as gender stereotyping?

5

Final Note on a Case of Extreme Isolation

KINGSLEY DAVIS

Socialization takes place in stages throughout our lives. We learn through agents of socialization such as family and education, and we are socialized into proper sex roles and other specialized roles as in athletics. Socialization requires contact with other human beings. What would we be like if we were raised in isolation with limited or no contact with other humans? Could socialization take place? Rare cases exist in which humans have grown up in partial or total isolation. This reading discusses two such cases and the results of isolation for these two girls.

As you read, ask yourself the following:

1. *Why do humans need other humans in order to develop "normally"?*
2. *What happens if a child experiences neglect or abuse during socialization?*

GLOSSARY

Socialization The process of learning cooperative group living.
Learning stage The knowledge and ability individuals are expected to have attained at a particular age.

EARLY IN 1940 THERE appeared . . . an account of a girl called Anna.[1] She had been deprived of normal contact and had received a minimum of human care for almost the whole of her first six years of life. At this time observations were not complete and the report had a tentative character. Now, however, the girl is dead, and with more information available,[2] it is possible to give a fuller and more definitive description of the case from a sociological point of view.

Anna's death, caused by hemorrhagic jaundice, occurred on August 6, 1942. Having been born on March 1 or 6,[3] 1932, she was approximately ten and a half years of age when she died. The previous report covered her development up to the age of almost eight years; the present one recapitulates the earlier period on the basis of new evidence and then covers the last two and a half years of her life.

Early History

The first few days and weeks of Anna's life were complicated by frequent changes of domicile. It will be recalled that she was an illegitimate child, the second such child born to her mother, and that her grandfather, a widowed farmer in whose house the mother lived, strongly disapproved of this new evidence of the mother's indiscretion. This fact led to the baby's being shifted about.

Two weeks after being born in a nurse's private home, Anna was brought to the family farm, but the grandfather's antagonism was so great that she was shortly taken to the house of one of her mother's friends. At this time a local minister became interested in her and took her to his house with an idea of possible adoption. He decided against adoption, however, when he discovered that she had vaginitis. The infant was then taken to a children's home in the nearest large city. This agency found that at the age of only three weeks she was already in a miserable condition, being "terribly galled and otherwise in very bad shape." It did not regard her as a likely subject for adoption but took her in for

Reprinted from American Journal of Sociology, *Vol. III, No. 5, March 1947, pp. 432–437 by permission of the author.* © *1947 by the University of Chicago Press.*

a while anyway, hoping to benefit her. After Anna had spent nearly eight weeks in this place, the agency notified her mother to come to get her. The mother responded by sending a man and his wife to the children's home with a view to their adopting Anna, but they made such a poor impression on the agency that permission was refused. Later the mother came herself and took the child out of the home and then gave her to this couple. It was in the home of this pair that a social worker found the girl a short time thereafter. The social worker went to the mother's home and pleaded with Anna's grandfather to allow the mother to bring the child home. In spite of threats, he refused. The child, by then more than four months old, was next taken to another children's home in a nearby town. A medical examination at this time revealed that she had impetigo, vaginitis, umbilical hernia, and a skin rash.

Anna remained in this second children's home for nearly three weeks, at the end of which time she was transferred to a private foster-home. Since, however, the grandfather would not, and the mother could not, pay for the child's care, she was finally taken back as a last resort to the grandfather's house (at the age of five and a half months). There she remained, kept on the second floor in an attic-like room because her mother hesitated to incur the grandfather's wrath by bringing her downstairs.

The mother, a sturdy woman weighing about 180 pounds, did a man's work on the farm. She engaged in heavy work such as milking cows and tending hogs and had little time for her children. Sometimes she went out at night, in which case Anna was left entirely without attention. Ordinarily, it seems, Anna received only enough care to keep her barely alive. She appears to have been seldom moved from one position to another. Her clothing and bedding were filthy. She apparently had no instruction, no friendly attention.

It is little wonder that, when finally found and removed from the room in the grandfather's house at the age of nearly six years, the child could not talk, walk, or do anything that showed intelligence. She was in an extremely emaciated and undernourished condition, with skeletonlike legs and a bloated abdomen. She had been fed on virtually nothing except cow's milk during the years under her mother's care.

Anna's condition when found, and her subsequent improvement, have been described in the previous report. It now remains to say what happened to her after that.

Later History

In 1939, nearly two years after being discovered, Anna had progressed, as previously reported, to the point where she could walk, understand simple commands, feed herself, achieve some neatness, remember people, etc. But she still did not speak, and, though she was much more like a normal infant of something over one year of age in mentality, she was far from normal for her age.

On August 30, 1939, she was taken to a private home for retarded children, leaving the country home where she had been for more than a year and a half. In her new setting she made some further progress, but not a great deal. In a report of an examination made November 6 of the same year, the head of the institution pictured the child as follows:

> Anna walks about aimlessly, makes periodic rhythmic motions of her hands, and, at intervals, makes guttural and sucking noises. She regards her hands as if she had seen them for the first time. It was impossible to hold her attention for more than a few seconds at a time—not because of distraction due to external stimuli but because of her inability to concentrate. She ignored the task in hand to gaze vacantly about the room. Speech is entirely lacking. Numerous unsuccessful attempts have been made with her in the hope of developing initial sounds. I do not believe that this failure is due to negativism or deafness but that she is not sufficiently developed to accept speech at this time. . . . The prognosis is not favorable.

More than five months later, on April 25, 1940, a clinical psychologist, the late Professor Francis N. Maxfield, examined Anna and reported the following: large for her age; hearing "entirely normal"; vision apparently normal; able to climb stairs; speech in the "babbling stage" and "promise for developing intelligible speech later seems to be good." He said further that "on the Merrill-Palmer scale she made a mental score of 19 months. On the Vineland social maturity scale she made a score of 23 months."[4]

Professor Maxfield very sensibly pointed out that prognosis is difficult in such cases of isolation. "It is very difficult to take scores on tests standardized under average conditions of environment and experience," he wrote, "and interpret them in a case where environment and experience have been so unusual." With this warning he gave it as his opinion at that time that Anna would eventually "attain an adult mental level of six or seven years."[5]

The school for retarded children, on July 1, 1941, reported that Anna had reached 46 inches in height and weighed 60 pounds. She could bounce and catch a ball and was said to conform to group socialization, though as a follower rather than a leader. Toilet habits were firmly established. Food habits were normal, except that she still used a spoon as her sole implement. She could dress herself except for fastening her clothes. Most remarkable of all, she had finally begun to develop speech. She was characterized as being at about the two-year level in this regard. She could call attendants by name and bring in one when she was asked to. She had a few complete sentences to express her wants. The report concluded that there was nothing peculiar about her, except that she was "feeble-minded—probably congenital in type."[6]

A final report from the school made on June 22, 1942, and evidently the last report before the girl's death, pictured only a slight advance over that given above. It said that Anna could follow directions, string beads, identify a few colors, build with blocks, and differentiate between attractive and unattractive pictures. She had a good sense of rhythm and loved a doll. She talked mainly in phrases but would repeat words and try to carry on a conversation. She was clean about clothing. She habitually washed her hands and brushed her teeth. She would try to help other children. She walked well and could run

fairly well, though clumsily. Although easily excited, she had a pleasant disposition.

Interpretation

Such was Anna's condition just before her death. It may seem as if she had not made much progress, but one must remember the condition in which she had been found. One must recall that she had no glimmering of speech, absolutely no ability to walk, no sense of gesture, not the least capacity to feed herself even when the food was put in front of her, and no comprehension of cleanliness. She was so apathetic that it was hard to tell whether or not she could hear. And all this at the age of nearly six years. Compared with this condition, her capacities at the time of her death seem striking indeed, though they do not amount to much more than a two-and-a-half year mental level. One conclusion therefore seems safe, namely, that her isolation prevented a considerable amount of mental development that was undoubtedly part of her capacity. Just what her original capacity was, of course, is hard to say; but her development after her period of confinement (including the ability to walk and run, to play, to dress, fit into a social situation, and, above all, to speak) shows that she had at least this capacity—capacity that never could have been realized in her original condition of isolation.

A further question is this: What would she have been like if she had received a normal upbringing from the moment of birth? A definitive answer would have been impossible in any case, but even an approximate answer is made difficult by her early death. If one assumes, as was tentatively surmised in the previous report, that it is "almost impossible for any child to learn to speak, think, and act like a normal person after a long period of early isolation," it seems likely that Anna might have had a normal or near-normal capacity, genetically speaking. On the other hand, it was pointed out that Anna represented "a marginal case [because] she was discovered before she had reached six years of age," an age "young enough to allow for some plasticity."[7] While admitting, then, that Anna's isolation *may* have been the major cause (and was certainly a minor cause) of her lack of rapid mental progress during the four-and-a-half years following her rescue from neglect, it is necessary to entertain the hypothesis that she was congenitally deficient.

In connection with this hypothesis, one suggestive though by no means conclusive circumstance needs consideration, namely, the mentality of Anna's forebears. Information on this subject is easier to obtain, as one might guess, on the mother's than on the father's side. Anna's maternal grandmother, for example, is said to have been college educated and wished to have her children receive a good education, but her husband, Anna's stern grandfather, apparently a shrewd, hard-driving, calculating farmowner, was so penurious that her ambitions in this direction were thwarted. Under the circumstances her daughter (Anna's mother) managed, despite having to do hard work on the farm, to complete the eighth grade in a country school. Even so, however, the daughter was evidently not very smart. "A schoolmate of [Anna's mother] stated that she was retarded in school work; was very gullible at this age; and that her morals even at this time were discussed by other students." Two tests administered to her on March 4, 1938, when she was thirty-two years of age, showed that she was mentally deficient. On the Stanford Revision of the Binet-Simon Scale her performance was equivalent to that of a child of eight years, giving her an I.Q. of 50 and indicating mental deficiency of "middle-grade moron type."[8]

As to the identity of Anna's father, the most persistent theory holds that he was an old man about seventy-four years of age at the time of the girl's birth. If he was the one, there is no indication of mental or other biological deficiency, whatever one may think of his morals. However, someone else may actually have been the father.

To sum up: Anna's heredity is the kind that *might* have given rise to innate mental deficiency, though not necessarily.

Comparison with Another Case

Perhaps more to the point than speculation about Anna's ancestry would be a case for comparison. If a child could be discovered who had been isolated about the same length of time as Anna but had achieved a much quicker recovery and a greater mental development, it would be a stronger indication that Anna was deficient to start with.

Such a case does exist. It is the case of a girl found at about the same time as Anna and under strikingly similar circumstances. A full description of the details of this case has not been published, but in addition to newspaper reports, an excellent preliminary account by a speech specialist, Dr. Marie K. Mason, who played an important role in the handling of the child, has appeared.[9] Also the late Dr. Francis N. Maxfield, clinical psychologist at Ohio State University, as was Dr. Mason, has written an as yet unpublished but penetrating analysis of the case.[10] Some of his observations have been included in Professor Zingg's book on feral man.[11] The following discussion is drawn mainly from these enlightening materials. The writer, through the kindness of Professors Mason and Maxfield, did have a chance to observe the girl in April, 1940, and to discuss the features of her case with them.

Born apparently one month later than Anna, the girl in question, who has been given the pseudonym Isabelle, was discovered in November, 1938, nine months after the discovery of Anna. At the time she was found she was approximately six-and-a-half years of age. Like Anna, she was an illegitimate child and had been kept in seclusion for that reason. Her mother was a deaf-mute, having become so at the age of two, and it appears that she and Isabelle had spent most of their time together in a dark room shut off from the rest of the mother's family. As a result Isabelle had no chance to develop speech; when she communicated with her mother, it was by means of gestures. Lack of sunshine and inadequacy of diet had caused Isabelle to become rachitic. Her legs in particular were affected; they "were so bowed that as she stood erect the soles of her shoes came nearly flat together, and she got about with a skittering gait."[12] Her behavior toward strangers, especially men, was almost that of a wild animal, manifesting much fear and hostility. In lieu of speech she made only a strange croaking sound. In many ways she acted like an infant. "She was apparently utterly unaware of relationships of any kind. When presented with a ball for the first time, she held it in the palm of her hand, then reached out and stroked my face with it. Such behavior is comparable to that of a child of six months." [13] At first it was even hard to tell whether or not she could hear, so unused were her senses. Many of her actions resembled those of deaf children.

It is small wonder that, once it was established that she could hear, specialists working with her believed her to be feeble-minded. Even on nonverbal tests her performance was so low as to promise little for the future. Her first score on the Stanford-Binet was 19 months, practically at the zero point of the scale. On the Vineland social maturity scale her first score was 39, representing an age level of two-and-a-half years. [14] "The general impression was that she was wholly uneducable and that any attempt to teach her to speak, after so long a period of silence, would meet with failure." [15]

In spite of this interpretation, the individuals in charge of Isabelle launched a systematic and skillful program of training. It seemed hopeless at first. The approach had to be through pantomime and dramatization, suitable to an infant. It required one week of intensive effort before she even made her first attempt at vocalization. Gradually she began to respond, however, and, after the first hurdles had at least been overcome, a curious thing happened. She went through the usual stages of learning characteristic of the years from one to six not only in proper succession but far more rapidly than normal. In a little over two months after her first vocalization she was putting sentences together. Nine months after that she could identify words and sentences on the printed page, could write well, could add to ten, and could retell a story after hearing it. Seven months beyond this point she had a vocabulary of 1500–2000 words and was asking complicated questions. Starting from an educational level of between one and three years (depending on what aspect one considers), she had reached a normal level by the time she was eight-and-a-half years old. In short, she covered in two years the stages of learning that ordinarily require six. [16] Or, to put it another way, her I.Q.

trebled in a year and a half. [17] The speed with which she reached the normal level of mental development seems analogous to the recovery of body weight in a growing child after an illness, the recovery being achieved by an extra fast rate of growth for a period after the illness until normal weight for the given age is again attained.

When the writer saw Isabelle a year-and-a-half after her discovery, she gave him the impression of being a very bright, cheerful, energetic little girl. She spoke well, walked and ran without trouble, and sang with gusto and accuracy. Today she is over fourteen-years-old and has passed the sixth grade in a public school. Her teachers say that she participates in all school activities as normally as other children. Though older than her classmates, she has fortunately not physically matured too far beyond their level. [18]

Clearly the history of Isabelle's development is different from that of Anna's. In both cases there was an exceedingly low, or rather blank, intellectual level to begin with. In both cases it seemed that the girl might be congenitally feeble minded. In both a considerably higher level was reached later on. But the Ohio girl achieved a normal mentality within two years, whereas Anna was still marked inadequate at the end of four-and-a-half years. This difference in achievement may suggest that Anna had less initial capacity. But an alternative hypothesis is possible.

One should remember that Anna never received the prolonged and expert attention that Isabelle received. The result of such attention, in the case of the Ohio girl, was to give her speech at an early stage, and her subsequent rapid development seems to have been a consequence of that. "Until Isabelle's speech and language development, she had all the characteristics of a feeble-minded child." Had Anna, who, from the standpoint of psychometric tests and early history, closely resembled this girl at the start, been given a mastery of speech at an earlier point by intensive training, her subsequent development might have been much more rapid. [19]

The hypothesis that Anna began with a sharply inferior mental capacity is therefore not established. Even if she were deficient to start with, we have no way of knowing how much so.

Under ordinary conditions she might have been a dull normal or, like her mother, a moron. Even after the blight of her isolation, if she had lived to maturity, she might have finally reached virtually the full level of her capacity, whatever it may have been. That her isolation did have a profound effect upon her mentality, there can be no doubt. This is proved by the substantial degree of change during the four-and-a-half years following her rescue.

Consideration of Isabelle's case serves to show, as Anna's case does not clearly show, that isolation up to the age of six, with failure to acquire any form of speech and hence failure to grasp nearly the whole world of cultural meaning, does not preclude the subsequent acquisition of these. Indeed, there seems to be a process of accelerated recovery in which the child goes through the mental stages at a more rapid rate than would be the case in normal development. Just what would be the maximum age at which a person could remain isolated and still retain the capacity for full cultural acquisition is hard to say. Almost certainly it would not be as high as age fifteen; it might possibly be as low as age ten. Undoubtedly various individuals would differ considerably as to the exact age.

Anna's is not an ideal case for showing the effects of extreme isolation, partly because she was possibly deficient to begin with, partly because she did not receive the best training available, and partly because she did not live long enough. Nevertheless, her case is instructive when placed in the record with numerous other cases of extreme isolation. This and the previous article about her are meant to place her in the record. It is to be hoped that other cases will be described in the scientific literature as they are discovered (as unfortunately they will be), for only in these rare cases of extreme isolation is it possible "to observe *concretely separated* two factors in the development of human personality which are always otherwise only analytically separated, the biogenic and the sociogenic factors."[20]

NOTES

1. Kingsley Davis, "Extreme Social Isolation of a Child," *American Journal of Sociology,* XLV (January, 1940), 554–65.

2. Sincere appreciation is due to the officials in the Department of Welfare, commonwealth of Pennsylvania, for their kind cooperation in making available the records concerning Anna and discussing the case frankly with the writer. Helen C. Hubbell, Florentine Hackbusch, and Eleanor Meckelnburg were particularly helpful, as was Fanny L. Matchette. Without their aid neither of the reports on Anna could have been written.

3. The records are not clear as to which day.

4. Letter to one of the state officials in charge of the case.

5. *Ibid.*

6. Progress report of the school.

7. Davis, *op. cit.,* p. 564.

8. The facts set forth here as to Anna's ancestry are taken chiefly from a report of mental tests administered to Anna's mother by psychologists at a state hospital where she was taken for this purpose after the discovery of Anna's seclusion. This excellent report was not available to the writer when the previous paper on Anna was published.

9. Marie K. Mason, "Learning to Speak after Six and One-Half Years of Silence," *Journal of Speech Disorders,* VII (1942), 295–304.

10. Francis N. Maxfield, "What Happens When the Social Environment of a Child Approaches Zero." The writer is greatly indebted to Mrs. Maxfield and to Professor Horace B. English, a colleague of Professor Maxfield, for the privilege of seeing this manuscript and other materials collected on isolated and feral individuals.

11. J. A. L. Singh and Robert M. Zingg, *Wolf-Children and Feral Man* (New York: Harper & Bros., 1941), pp. 248–51.

12. Maxfield, unpublished manuscript cited above.

13. Mason, *op. cit.,* p. 299.

14. Maxfield, unpublished manuscript.

15. Mason, *op. cit.,* p. 299.

16. *Ibid.,* pp. 300–304.

17. Maxfield, unpublished manuscript.

18. Based on a personal letter from Dr. Mason to the writer, May 13, 1946.

19. This point is suggested in a personal letter from Dr. Mason to the writer, October 22, 1946.

20. Singh and Zingg, *op. cit.,* pp. xxi–xxii, in a foreword by the writer.

6

Becoming "Boys," "Men," "Guys," and "Dudes"

CLYDE W. FRANKLIN II

Being female or male is not a simple matter of being born with the distinct anatomy. One must learn the behavior that a particular culture assigns to that sex. Learning one's gender roles, then, is a key component of the socialization process. This process takes place throughout the stages of socialization. Franklin discusses this process for males, and indirectly for females, by describing several agents of socialization—family, education, religion, and peer groups.

As you read, consider the following questions:

1. How do babies become boys become men?

2. Who and what determines the socialization process boys go through?

GLOSSARY

Socialization The process of learning to become a social being.

Agents Groups and organizations that help in the socialization process.

Nonpurposive agents Those activities and organizations that do not have the explicit purpose of socializing but that influence socialization.

Agents of Male Socialization

THE HUMAN MALE UNDERGOES a long socialization process whereby he becomes aware of himself as a male and develops sex role skills necessary for full functioning as a social male in society. [As important as self-development] is the sex role which the biological male must learn if he is to fulfill self and others' expectations of himself as a boy, a man, a guy, or dude. What is the nature of this sex role which must be per-

formed if the biological male is to function fully as a social male? . . .

A glance at some of the agents responsible for the development and maintenance of male socialization should contribute to an understanding of this process. The agents of socialization to be discussed are divided into those formally charged with the responsibility for male sex role socialization and those that have informal (and often latent) responsibility.

PURPOSEFUL AGENTS OF MALE SOCIALIZATION

Each newborn male in society is expected to undergo a lengthy learning process to acquire appropriate male behaviors. Responsibility for this process historically lay with such societal agents as the family, religious institutions, educational institutions, the mass media, and adolescent and young adult male peer groups. A discussion of the roles of these agents in male socialization is presented in this section. Let us begin with the family.

The Family The family is a vital agent involved in teaching males appropriate attitudes and behaviors. From the moment the newborn is identified as male, a set of cultural expectations unfolds dictating what behaviors may and may not be displayed. The agent charged with initial responsibility for insuring sex role conformity with societal expectations is the family. Studies have suggested (e.g., Schau et al. 1980 and Fu and Leach 1980) that if the newborn is male, rather rigid cultural expectations exist for him to learn to give "male performances" in social interactions. This means that the socialization process for

From Men and Society, *Chicago: Nelson-Hall, 1988.*

males is likely to be especially constraining, allowing little deviation.

Even more critical for many young males learning male sex role requirements is the presence of older males within the family who serve as role models. Often such males are the fathers of these young males, although all that seems necessary for partial male socialization is that older males are seen by the younger males performing certain roles within the family setting. Seeing older males' role performances within the family provides younger males with the opportunity to learn vicariously cultural expectations for their own behaviors.

Some studies have found that parents treat children differently depending on the sex of the child (Schau et al. 1980). Differential treatment of children according to gender has been observed in fathers who are much more likely to "rough it up" with boys than with girls (Parke and Suomi 1980; Power and Parke 1983). These differences in fathers' behaviors toward their children depending on gender often follows stereotypical directions. Interestingly, fathers' stereotypical behaviors in interactions with their children follow parents' stereotypical descriptions of their newborns. Despite the lack of significant differences in birth length, weight, and APGAR scores, parents of daughters are more likely than parents of sons to give descriptions of their babies as "dainty," "pretty," "beautiful," and "cute" (Rubin et al. 1974). Certainly such differences in descriptions of children by gender may foretell parental behavioral differences by gender in parent-child interaction. We already know, for example, that parents of boys are much *less* directing of their offsprings' play than parents of girls. Such interferences by parents in the behaviors of girls may well affect girls' creativity and interests in ways inimical to their later independence and assertiveness. In the same vein, when parents of boys are less directing of their play, this begins to prepare boys for the independent and active male sex role many expect them to assume when they become adult males.

Other studies of parents' behaviors during the socialization of their children have produced mixed findings with respect to differential treatment of children by gender. Snow et al. (1983) found that parents responded differentially to some types of sex-typed behaviors in toddlers, but not others. In another study, fathers punished boys' cross-sex play behavior while mothers were found to punish and reward boys' cross-sex behaviors (Langlois and Downs 1980). These findings regarding fathers' lack of tolerance for boys' cross-sex behaviors are consistent with our contention that male sex role socialization tends to be more restrictive than female sex socialization, especially early in life. A final study which is instructive on this point is one by Eisenberg et al. (1985). In this study of mothers' and fathers' socialization of one- and two-year-olds' sex-typed play behaviors, several findings are notable. On the variables "parental choice of toys" and "parental reinforcement," parents of boys tended to choose neutral and masculine toys more than feminine toys, while parents of girls chose neutral toys more than the other two types. However, once parents had chosen toys for their children, they did not differentially reinforce them or neutrally respond to them for sex-typed or other-sex play. Eisenberg et al. concluded that "apparently, in the home, parents exert influence over their young children's play primarily via their selection of available toys" (p. 1512). Thus, parental opportunity to select and influence behavior may be a preferred method of socializing children's sex-typed behavior. Another finding from Eisenberg et al.'s study of interest is that parents reduced positive feedback for children's toy play with age. "Parents provided less positive feedback (and thus more neutral feedback) at age 26 to 33 months than at 19 to 26 months" (p. 1512). The reduction in parental reinforcement of play with age of the child occurred only for other-sex play activities, not neutral or sex-typed behaviors. This means that boys in all likelihood are aided in the development of gender constancy by continued parental reinforcement of sex-type play throughout childhood.

Findings regarding differential parental treatment by sex of child seem to be mixed at this point. Definite conclusions about differential parental behaviors by sex of child await further research. However, differentiated parental reinforcement may not be necessary for the development of sex-differentiated behavior in children. Simply attending to behavior differentially may be enough. Consider a study by Fagot and Hagan et al. (1985). This study of thirty-four children in infant play groups revealed no sex differences in assertive acts and attempts to communicate verbally with adults at ages 13 to 14 months. However, the authors observed learning center teachers attending more to boys' assertive behaviors and more to girls' *less intense* communication attempts. The result was that eleven months later twenty-nine of the same children exhibited sex-differentiated behavior: boys were more assertive and girls talked more with adults. Thus, caregivers seemed to be responsible, in part, for the development of boys' and girls' sex-differentiated behavior by guiding infant behaviors in stereotypical directions.

Educational Institutions It is well documented that there is a significant difference in what adults observe depending on whether the persons being observed are described as males or females (Condry and Ross 1985). Purported reasons for adult differential perceptions of children's behavior by sex vary. Some feel that adults may be differentially responsive to certain types of behaviors by girls and boys. For example, because girls are expected to be more verbal than boys, are teachers more attentive to girls' verbal behaviors than boys'? By the same token, because boys are expected to have more assertive interchanges in peer activities than girls, do adults attend more to boys' assertive behaviors than girls' assertive behaviors? If the answers to both questions are yes, then differential attention to certain behaviors of boys and girls result in adults' differential perception of boys' and girls' behaviors.

Another common assumption stemming from social learning theory is that adults directly socialize children to behave in sex-typed ways through differential reinforcement and punishment. This assumption is supported by Beverly

Fagot's (1981) findings that teachers differentially reinforce boys and girls for high activity levels. Even the large school context seems to be more supportive of males than females. Males continue to hold the more prestigious positions in the school system, schoolyards remain sex segregated, and in general gender differences remain in confidence, self-concept, and problem solving behaviors. Certainly such differences are related to the educational system's reinforcement of gender differences and traditional sex role behaviors. For example, findings from Phillips' playground study (1982) were quite consistent with those of Janet Lever (1976), who had found in her analysis of boys' and girls' spontaneous games on playgrounds that the games were sex differentiated. Lever concluded that boys' games were less structured than girls' games, with less emphasis on "turn-taking" and invariable procedural rules. Moreover, girls played with fewer participants while boys' games emphasized more initiative, improvisation, and extemporaneity, encouraging within-group cooperation and between-group competition. Phillips also found in her study of school playground activities that school spaces provided for boys and girls encouraged sex-differentiated play activities. Boys had large play spaces supportive of large competitive groups for competitive games. Girls' play spaces were small and generally supportive of cooperative, dyadic, and/or triadic activities. The major play space for girls in Phillips' study was on the playground apparatus, which could be easily invaded by boys and on occasion *was* invaded by boys, with the girls submissively leaving the equipment until the boys no longer used it. Phillips concluded: "Boys' play was preparing them for future work roles that would consist of the networks of competitively based groups necessary for success and achievement in the work place" (Franklin 1984, p. 43).

Jeanne Block's (1981) summary of the effects of sex-differentiated socialization in educational institutions suggests that male socialization in the education institution (which encourages curiosity, independence, initiative, etc.) extends male experiences, while female socialization in the educational institution (which discourages exploration, emphasizes class supervision, stresses proprieties, etc.) restricts the experiences of females. While some changes in the educational institution have occurred in recent years, males and females still have sex-differentiated experiences throughout their tenures in educational institutions.

Religious Institutions Almost as influential in teaching males to assume the male sex role is another agent, the religious institution. The only reason the religious institution does not assume a more critical role in male socialization is that the typical child does not spend an inordinate amount of time in religious settings. The time that is spent, however, generally is time when gender distinctions are emphasized. Such distinctions, within Christianity for example, are seen as divinely inspired in that they support the ideal relationship between husband and wife. On this point, Patricia M. Lengermann and Ruth A. Wallace (1985, p. 239) state that calling for sex role equality, questioning patriarchy, and critiquing traditional male dominance and female submissiveness in marriage and family life are antithetical to the divine plan as visualized by many Christians. Such a posture on the part of religious agents supports traditional sex roles against gender equality. In the 1980s with the rise of evangelical Christian movements and retrenchment in Roman Catholicism, we can only conclude that the religious institutions in the United States remain staunchly supportive of traditional female and male sex roles.

Support for the above position is seen in "God Goes Back to College," an article appearing in *Newsweek*'s "On Campus" edition, November 1986, noting the fervor with which college students on campuses across the nation (those mentioned included Brown University, Arizona State University, University of Illinois—Champaign-Urbana, University of Texas, Duke University, Washington University, and Northwestern University) are embracing fundamentalist religious beliefs. Two striking implications for sex role changes are discussed. These implica-

tions center around a great deal of sentiment among religious groups in these settings to deny gays equal rights and to thwart women's attempts to pursue careers. Increased religious proselytizing on college campuses in the 1980s frequently has resulted in support for homophobia and traditional sex roles for females and males.

Peer Groups A consistent finding in the literature on children is that American children show a preference for same-sex peers by the beginning of their sixth year. This tendency toward peer-group sex segregation increases during middle childhood and reaches its peak right before adolescence (Hartup 1983). In addition, as Thompson (1985) found in his study, males in preadolescence are more peer oriented than females. Part of the reason for greater peer orientation among males undoubtedly is linked to greater encouragement of independence in males at an earlier age. What this means for male socialization is that boys at a relatively early age are more subject to peer-group influence than girls. Just as important is that such influence may be perceived positively and supported by parents as indicative of boys' independence.

If boys are susceptible to early peer-group influence, this also means that males' early-age peer groups may be responsible for a great deal of those sex role performances by boys. This is to be expected if Fagot's (1981) findings that boys who exhibit feminine behaviors receive negative feedback from peers are generalizable. Some support for peer-group influence on boys' sex role performances derive also from Eisenberg et al.'s (1985) study of a stronger match for males than females between same-sex peer interactions and neutral or sex-typed toy play for fifty-one four-year-olds. When boys play with boys they prefer sex-typed toy play. Eisenberg et al. (1985) felt that this is consistent with the notion that there is more pressure for males than females to avoid sex-inappropriate activities:

> Although initiation of and/or continuation of interaction per se may not be used consciously as a positive reinforcer by children, it could

function as one. Thus, unintentionally as well as intentionally (Lamb et al., 1980), children, especially boys, may socialize peers into sex-stereotypic play behaviors. They may do so not only by initiating play with others in possession of sex appropriate toys, but also by inducing other children to engage in same sex play. (p. 1049)

There seems to be a logical relationship between children's play behavior and their everyday role performances. Indeed, when boys' play behaviors are channeled in a decidedly stereotypical masculine direction, certainly they learn that these same behaviors are expected of them by significant others in everyday situations. After all, parental brokering, approval, support, and reinforcement by early age peer groups function to inform the boy of the importance of this early socialization agent.

An early study by Fling and Manosevitz (1972) on male socialization found that young males are encouraged to participate in activities that teach and reinforce male stereotyped roles. There is little reason to think that such participation has declined in the 1980s. Interestingly, peer-group influence over males tends to decline as the young male approaches late adolescence. While in late childhood and early adolescence, male peer groups are quite influential in boys learning competitiveness, aggression, violence, and antifemininity, young males also learn that they must become independent, self-reliant, and detached from the peer group. This latter socialization, in a sense, prepares young males for the role which must be assumed in adulthood, a role which minimizes male-male relationships. Yet, adolescent peer groups, for most males, are kinds of reference groups providing information which the sixteen or seventeen-year-old male actively filters, alters, and modifies to fit his own perspective. Typically, peer group information, standards, and values are some variant of those from other socialization agents, including nonpurposive ones discussed in the next section. Most young males experience a kind of socialization which teaches them societally approved sex roles, dysfunctional ones as well as functional ones.

The Mass Media Mass media influences on sex role socialization are thought by many to be critical in the development and support of sex role stereotypes specifically and sex role inequality generally. The link between sex role stereotypes and sex role inequality is a direct one. Sex role stereotypes (expectations about and attitudes toward the sexes) lead to sex role inequality (inequitable actions toward a person based on the sex of that person). This linkage is consistent with findings in social psychological literature suggesting that stereotypes are better predictors of behavior than of attitudes.

Yet, how do sex role stereotypes relate to male socialization? Recalling that male socialization is a dual process, involving male self development and the learning of societal "shoulds" and "should nots" for males, one can see that much male socialization actually involves learning conceptions of males' "makeup" and "places" and females' "makeup" and "places." Undoubtedly, the mass media play critical roles in this process. When females in television commercials usually perform household duties and pamper men while males typically perform active roles outside the home and do not perform domestic duties, a message is given to viewers that housework is women's work and work outside the home is men's work (Mamay and Simpson 1981).

That television may play a powerful role in gender socialization is suggested by Drabman et al. (1979) since their findings indicate that young children (first-graders), when shown videotaped presentations of males and females in counterstereotyped occupations (such as male nurse and female doctor), tended to reverse sex role information in the stereotyped direction. For Drabman et al. this finding meant that television should be used in a specific way to modify sex role socialization in a more equitable manner for boys and girls. They state: "Television programming which directly informs the child nearly all life roles are available to both sexes might prove more fruitful in attempts to alter traditional gender stereotypes" (p. 388).

Not only is television a potentially powerful agent in the sex role socialization of males and females, but newspapers, comics, movies, and popular songs may also influence conceptions of gender roles. Lengermann and Wallace (1985) feel that such mass media are "a forum where critical views on gender equality are heard and aired, can affect the thinking and beliefs of men as well as women, and can be for both a resource for new meanings" (p. 222). To the extent that men and women are affected by such mass media changes and also participate in the teaching of males, the effects on male socialization are obvious.

To be sure, there have been some changes in the mass media toward presenting male and female images in a manner more consistent with sex role equality. Lengermann and Wallace point to the inclusion of women columnists like Ellen Goodman and Mary McGrory in daily newspapers as evidence of changes in newspapers which can modify a man's thinking about a woman's place. They note also the emergence of Alan Alda, a popular television entertainer and self-described feminist, as a role model in the media for nontraditional men in contrast to more conventional male images like Bob Hope and John Wayne. Men's magazines, too, are thought to be sources of sex role changes in the mass media. Magazines such as *Esquire* and *Sports Illustrated* are thought by Lengermann and Wallace to reflect "new meanings." They cite the November 1982 issue of *Esquire* with a feature article entitled "Father Love" by Anthony Brandt and *Sports Illustrated*'s (Feb. 28, 1983) coverage of Louisiana Tech women's basketball team (a sign that women's sports are making strides toward parity) as evidence of further change in sex role meanings. . . .

In summary, some mass media changes in the last decade or so have been in the direction of sex role equality which would eventually lessen male dominance, male violence, male destruction, competition, and so on. Simultaneously, however, forces have arisen in the mass media which either support traditional sex role distinctions or at least suggest that change in men's behavior cannot oc-

cur or is trivial and of dubious value for society when it does occur. As we move into the 1990s, it is hoped that the mass media will come to portray and reflect gender, especially the valued male sex role, in a realistic manner, that is, as sets of cultural expectations that are socially constructed.

NONPURPOSIVE AGENTS OF MALE SOCIALIZATION

Families, boys' groups, educational institutions, churches, newspapers, magazines, television, and radio are not the only socializing agents teaching males to be dominant, aggressive, violent, competitive, nonintimate, and non-nurturant. There are other agents in American society which are not charged with a learning function but which, nevertheless, teach males conceptions of themselves, other males, and what males should and should not do. Some of these agents are male-centered barbershops, sports events, taverns, and business meetings, where primarily males engage in social interaction. These are the same agents forming the core of men's culture. . . . Two latent socialization consequences of the above agents are emphasized: (1) indirect socialization of young males and (2) reinforcement and support of traditional conceptions of the male sex role.

With some exceptions (e.g., unisex hair salons), male barbershops, topless taverns, male-dominated business meetings, sports events, and the like function as social settings/negotiation contexts where men negotiate masculinity. While the negotiation of masculinity is a complex process involving numerous contextual and social psychological variables, the emphasis here is on the process used by men in certain settings to arrive at conceptions of who men are and how men should and should not behave. At the same time, they also form conceptions about persons who are not men and masculine—women and others perceived as feminine.

The "particulars" of masculinity negotiations in various social settings will not be discussed here; however, a broad description of such processes includes verbal and nonverbal behavior by male participants in social settings which define appropriate male attitudes and behaviors. In such settings as barbershops and male-centered sports events, frequently young, impressionable males are present during the negotiation process. The young males learn not only what behaviors are expected from the primary male participants, but also the attitudes that they should hold about the negotiation process and what outcomes from the negotiation process are most desirable. For example, a young male attending a professional football game learns not only that the more "manly" team wins—the team that is more competitive, more aggressive subdues—but also how he is to respond to such characteristics. The young boy leaves the stadium *knowing* that dominance is a desirable trait for men to have. After all, an entire group of men have just been rewarded by a host of other men for displaying the dominance trait. Just as important, too, for the young boy is the low esteem many others hold for the losing team—the one that has been subdued.

Young boys generally do not go to topless taverns where women are seen in various stages of undress. Nevertheless, masculinity negotiations and male socialization are features of such settings. Men receive support and reinforcement from other men for certain behaviors they display. The swaggers, the yells, the obscenities, the sexual references all become permanently etched in their little minds as appropriate behavior for men. Those men who do not engage in the behavior nevertheless learn that if they want others to think of them as manly, all they have to do is display similar behavior. The "new" male in the topless tavern setting will know that he is engaging in appropriate male behavior when other males slap him on the back or shake his hand as he, too, screams and yells, "Take it off, baby."

REFERENCES

Block, J. H., 1983. Differential premises arising from differential socialization of the sexes: Some conjectures. *Child Development* 54:1335–54.

Drabman, R. S., S. J. Robertson, J. N. Patterson, G. J. Javie, D. Hammer, and G. Gordua, 1979. Children's perception of media portrayal of sex roles. *Sex Roles* 7:379–89.

Eisenberg, N., S. A. Wolchik, R. Hernandez, and J. F. Pasternack, 1985. Parental socialization of young children's play: A short-term longitudinal study. *Child Development* 56:1506–13.

Fagot, B. I., 1981. Male and female teachers: Do they treat boys and girls differently? *Sex Roles* 7:263–71.

Fagot, B. I., R. Hagan, M. D. Leinbach, and S. Kronsberg, 1985. Differential reaction to assertive and communicative acts of toddler boys and girls. *Child Development* 56:1499–1505.

Fling, S., and M. Manosevitz, 1972. Sex typing in nursery school children's play interests. *Developmental Psychology* 7:146–52.

Franklin, C. W., 1984. *The changing definition of masculinity*. New York: Plenum Press.

Fu, V. R., and D. J. Leach, 1980. Sex role preferences among elementary school children in rural America. *Psychological Reports* 46:555–60.

Hartup, W. W., 1983. The peer system. In *Handbook of child psychology*. Vol. 4: *Socialization, personality, and social development,* edited by E. M. Hetherington and P. H. Mussen (series ed.). New York: Wiley.

Langlois, J. H., and A. C. Downs, 1980. Mothers, fathers, and peers as socialization agents of sex typed play behavior in young children. *Child Development* 51:1217–47.

Lengermann, P. M., and R. A. Wallace, 1980. *Gender in America: Social control and social change*. Englewood Cliffs, N.J.: Prentice-Hall.

Lever, J., 1978. Sex differences in the complexity of children's play and games. *American Sociological Review* 43:471–83.

Mamay, P. D., and R. I. Simpson, 1981. Three female roles in television commercials. *Sex Roles* 7:1223–32.

Parke, R. D., and S. J. Suomi, 1980. Adult male-infant relationships: Human and non-primate evidence. In *Behavioral development: The Bielefeld interdisciplinary project,* edited by K. Immelmann, G. Barlow, M. Main, and L. Petrinovitch. New York: Cambridge University Press.

Phillips, B. D., 1982. Sex role socialization and play behavior on a rural playground. Unpublished master's thesis, Department of Sociology, Ohio State University, Columbus, Ohio.

Power, T. G., and R. D. Parke, 1982. Play as a context for early learning: Lab and home analysis. In *The family as a learning environment,* edited by I. Sigel and M. Laosa. New York: Plenum Press.

Rubin, J. Z., E. J. Provenzano, and Z. Luria, 1976. The eye of the beholder: Parents' views on sex of newborns. *American Journal of Orthopsychiatry* 44:512–19.

Schau, C. G., I. Kahn, J. H. Diepold, and F. Cherry, 1980. The relationship of parental expectations and pre-school children's verbal sex typing to their sex-typed toy play behavior. *Child Development* 51:266–70.

Snow, M. E., C. N. Jacklin, and E. E. Maccoby, 1981. Sex-of-child differences in father-child interaction at one year of age. *Child Development* 54:227–32.

Thompson, D. N., 1985. Parent-peer compliance in a group of preadolescent youths. *Adolescence* 20(79): 501–7.

7

The National Television Violence Study: Key Findings and Recommendations

Socialization takes place through formal agents (family, school, religion) and informal agents whose primary purpose is not socialization but that nonetheless affect the socialization processes. One powerful informal agent of socialization, one that does not have the primary purpose of socializing but still has a major impact on who we are, is television. The following reading presents findings and recommendations from a study of television violence.

Consider the following questions as you read:

1. How does television violence influence the socialization of boys and girls?

2. What do you think should be done about television violence?

GLOSSARY

TV violence Any overt depiction of use of physical force or credible threat of such force intended to harm people physically.
PSA Public Service Announcement.

PREVENTING VIOLENCE INVOLVES IDENTIFYING the combination of factors that contribute to

it, from biological and psychological causes to broader social and cultural ones. Among these, television violence has been recognized as a significant factor contributing to violent and aggressive antisocial behavior by an overwhelming majority of the scientific community.

However, it is also recognized that televised violence does not have a uniform effect on viewers. The outcome of media violence depends both on the nature of the depiction and the sociological and psychological makeup of the audience. In some cases, the same portrayal of violence may have different effects on different audiences. For example, graphically portrayed violence may elicit fear in some viewers and aggression in others. Family role models, social and economic status, educational level, peer influences, and the availability of weapons can each significantly alter the likelihood of a particular reaction to viewing televised violence.

The context in which violence is portrayed may modify the contributions to viewer behaviors and attitudes. Violence may be performed by heroic characters or villains. It may be rewarded

From Young Children, *March 1996, pp. 54–55. © 1996 by the National Association for the Education of Young Children. Reprinted by permission.*

or it may be punished. Violence may occur without much victim pain and suffering or it may cause tremendous physical anguish. It may be shown close-up on the screen or at a distance.

This study is the most comprehensive scientific assessment yet conducted of the context in which violence is depicted on television, based on some 2,500 hours of programming randomly selected from 23 television channels between 6 A.M. to 11 P.M. over a 20-week period. Television content was analyzed at three distinct levels: (1) how characters interact with one another when violence occurs (violent interaction); (2) how violent interactions are grouped together (violent scene); and (3) how violence is presented in the context of the overall program.

Violence is defined as any overt depiction of the use of physical force—or the credible threat of such force—intended to physically harm an animate being or group of beings. Violence also includes certain depictions of physically harmful consequences against an animate being or group that occur as a result of unseen violent means.

Key Findings

- *The context in which most violence is presented on television poses risks for viewers.* The majority of programs analyzed in this study contain some violence. But more important than the prevalence of violence is the contextual pattern in which most of it is shown. The risks of viewing the most common depictions of televised violence include learning to behave violently, becoming more desensitized to the harmful consequences of violence, and becoming more fearful of being attacked. The contextual patterns noted below are found consistently across most channels, program types, and times of day. Thus, there are substantial risks of harmful effects of viewing violence throughout the television environment.

- *Perpetrators go unpunished in 73% of all violent scenes.* This pattern is highly consistent across different types of programs and channels. The portrayal of rewards and punishments is probably the most important of all contextual factors for viewers as they interpret the meaning of what they see on television. When violence is presented without punishment, viewers are more likely to learn the lesson that violence is successful.

- *The negative consequences of violence are not often portrayed in violent programming.* Most violent portrayals do not show the victim experiencing any serious physical harm or pain at the time the violence occurs. For example, 47% of all violent interactions show no harm to victims and 58% show no pain. Even less frequent is the depiction of any long-term consequences of violence. In fact, only 16% of all programs portray the long-term negative repercussions of violence, such as psychological, financial, or emotional harm.

- *One out of four violent interactions on television (25%) involves the use of a handgun.* Depictions of violence with guns and other conventional weapons can instigate or trigger aggressive thoughts and behaviors.

- *Only 4% of violent programs emphasize an antiviolence theme.* Very few violent programs place emphasis on condemning the use of violence or on presenting alternatives to using violence to solve problems. This pattern is consistent across different types of programs and channels.

- *On the positive side, television violence is usually not explicit or graphic.* Most violence is presented without any close-up focus on aggressive behaviors and without showing any blood and gore. In particular, less than 3% of violent scenes feature close-ups on the violence and only 15% of scenes contain blood and gore. Explicit or graphic violence contributes to desensitization and can enhance fear.

- *There are some notable differences in the presentation of violence across television channels.* Public broadcasting presents violent programs least often (18%) and those violent depictions that appear pose the least risk of harmful effects. Premium cable channels present the highest percentage of violent programs (85%) and those depictions often pose a greater risk of harm than do most violent portrayals. Broadcast networks present violent programs less frequently (44%) than the industry norm (57%), but when violence is included its contextual features are just as problematic as those on most other channels.

- *There are also some important differences in the presentation of violence across types of television programs.* Movies are more likely to present violence in realistic settings (85%) and to include blood and gore in violent scenes (28%) than other program types. The contextual pattern of violence in children's programming also poses concern. Children's programs are the least likely of all genres to show the long-term negative consequences of violence (5%), and they frequently portray violence in a humorous context (67%).

Recommendations

These recommendations are based both on the findings of this study and extensive research upon which this study is based.

FOR THE TELEVISION COMMUNITY

- Produce more programs that avoid violence. When violence does occur, keep the number of incidents low, show more negative consequences, provide nonviolent alternatives to solving problems, and consider emphasizing antiviolence themes.
- Increase portrayals of powerful nonviolent heroes and attractive characters.
- Programs with high levels of violence, including reality programs, should be scheduled in late-evening hours when children are less likely to be watching.
- Increase the number of program advisories and content codes. In doing so, however, use caution in language so that such messages do not serve as magnets to children.

- Provide information about advisories and the nature of violent content to viewers in programming guides.
- Limit the time devoted to sponsor, station, or network identification during public service announcements (PSAs) so that it does not compete with the message.

FOR POLICY AND PUBLIC INTEREST LEADERS

- Recognize that context is an essential aspect of television violence and that the basis of any policy proposal should consider the types of violent depictions that pose the greatest concern.
- Consider the feasibility of technology that would allow parents to restrict access to inappropriate content.
- Test antiviolence PSAs, including the credibility of spokespersons, with target audiences prior to production. Provide target audiences with specific and realistic actions for resolving conflicts peacefully.
- When possible, link antiviolence PSAs to school-based or community efforts and target young audiences, 8 to 13 years old, who may be more responsive to such messages.

FOR PARENTS

- Watch television with your child. In this study, children whose parents were routinely involved with their child's viewing were more likely to avoid inappropriate programming.
- Encourage critical evaluation of television content.
- Consider a child's developmental level when making viewing decisions.
- Be aware of the potential risks associated with viewing television violence: the learning of aggressive attitudes and behaviors, fear, desensitization or loss of sympathy toward victims of violence.
- Recognize that different kinds of violent programs pose different risks.

8
Beauty and the Beast of Advertising

JEAN KILBOURNE

Another example of a powerful informal agent of socialization is advertising. Kilbourne provides a graphic description of informal socialization that takes place through advertising and discusses the sometimes negative effects that ads have on the images and treatment of women and men. Adolescents are especially vulnerable to the stereotypes of male and female images presented in ads.

Keep in mind the following questions as you read:

1. What impact does advertising have on our development as individuals?

2. How has advertising influenced your socialization and that of your friends?

GLOSSARY

Mass media Sources that reach a wide audience such as TV, magazines, newspapers.
Objectify To view one's face and body as an object separate from the real self.
Internalize Make an image part of oneself.
Self-fulfilling prophecy Actualize or make real these images by accepting them.

"YOU'RE A HALSTON WOMAN from the very beginning," the advertisement proclaims. The model stares provocatively at the viewer, her long blonde hair waving around her face, her bare chest partially covered by two curved bottles that give the illusion of breasts and a cleavage.

The average American is accustomed to blue-eyed blondes seductively touting a variety of products. In this case, however, the blonde is about five years old.

Advertising is an over $130 billion a year industry and affects all of us throughout our lives. We are each exposed to over 1500 ads a day, constituting perhaps the most powerful educational force in society. The average adult will spend one and one-half years of his/her life watching television commercials. But the ads sell a great deal more than products. They sell values, images, and concepts of success and worth, love and sexuality, popularity and normalcy. They tell us who we are and who we should be. Sometimes they sell addictions.

Advertising is the foundation and economic lifeblood of the mass media. The primary purpose of the mass media is to deliver an audience to advertisers.

Adolescents are particularly vulnerable, however, because they are new and inexperienced consumers and are the prime targets of many advertisements. They are in the process of learning their values and roles and developing their self-concepts. Most teenagers are sensitive to peer pressure and find it difficult to resist or even question the dominant cultural messages perpetuated and reinforced by the media. Mass communication has made possible a kind of nationally distributed peer pressure that erodes private and individual values and standards.

But what does society, and especially teenagers, learn from the advertising messages that proliferate in the mass media? On the most obvious level they learn the stereotypes. Advertising creates a mythical, WASP-oriented world in which no one is ever ugly, overweight, poor,

Reprinted by permission from Media & Values *(Winter 1989), published by the Center for Media Literacy, Los Angeles.*

struggling, or disabled either physically or mentally (unless you count the housewives who talk to little men in toilet bowls). And it is a world in which people talk only about products.

Housewives or Sex Objects

The aspect of advertising most in need of analysis and change is the portrayal of women. Scientific studies and the most casual viewing yield the same conclusion: Women are shown almost exclusively as housewives or sex objects.

The housewife, pathologically obsessed by cleanliness and lemon-fresh scents, debates cleaning products and worries about her husband's "ring around the collar."

The sex object is a mannequin, a shell. Conventional beauty is her only attribute. She has no lines or wrinkles (which would indicate she had the bad taste and poor judgment to grow older), no scars or blemishes — indeed, she has no pores. She is thin, generally tall and long-legged, and, above all, she is young. All "beautiful" women in advertisements (including minority women), regardless of product or audience, conform to this norm. Women are constantly exhorted to emulate this ideal, to feel ashamed and guilty if they fail, and to feel that their desirability and lovability are contingent upon physical perfection.

Creating Artificiality

The image is artificial and can only be achieved artificially (even the "natural look" requires much preparation and expense). Beauty is something that comes from without: more than one million dollars is spent every hour on cosmetics. Desperate to conform to an ideal and impossible standard, many women go to great lengths to manipulate and change their faces and bodies. A woman is conditioned to view her face as a mask and her body as an object, as *things* separate from and more important than her real self, constantly in need of alteration, improvement, and disguise. She is made to feel dissatisfied with and ashamed of herself, whether she tries to achieve "the look" or not. Objectified constantly by others, she learns to objectify herself. (It is interesting to note that one in five college-age women has an eating disorder.)

"When *Glamour* magazine surveyed its readers in 1984, 75 percent felt too heavy and only 15 percent felt just right. Nearly half of those who were actually underweight reported feeling too fat and wanting to diet. Among a sample of college women, 40 percent felt overweight when only 12 percent actually were too heavy," according to Rita Freedman in her book *Beauty Bound*.

There is evidence that this preoccupation with weight begins at ever-earlier ages for women. According to a recent article in *New Age Journal*, "even grade-school girls are succumbing to stick-like standards of beauty enforced by a relentless parade of wasp-waisted fashion models, movie stars, and pop idols." A study by a University of California professor showed that nearly 80 percent of fourth-grade girls in the Bay Area are watching their weight.

A recent *Wall Street Journal* survey of students in four Chicago-area schools found that more than half the fourth-grade girls were dieting and three-quarters felt they were overweight. One student said, "We don't expect boys to be that handsome. We take them as they are." Another added, "But boys expect girls to be perfect and beautiful. And skinny."

Dr. Steven Levenkrom, author of *The Best Little Girl in the World,* the story of an anorexic, says his blood pressure soars every time he opens a magazine and finds an ad for women's fashions. "If I had my way," he said, "every one of them would have to carry a line saying, 'Caution: This model may be hazardous to your health.'"

Women are also dismembered in commercials, their bodies separated into parts in need of change or improvement. If a woman has "acceptable" breasts, then she must also be sure that her legs are worth watching, her hips slim, her feet sexy, and that her buttocks look nuder under her clothes ("like I'm not wearin' nothin'"). This image is difficult and costly to achieve and impossible to maintain—no one is flawless and everyone ages. Growing older is the great taboo. Women are encouraged to remain little girls ("because innocence is sexier than you think"), to be passive and dependent, never to mature. The contradictory message—"sensual, but not too far from innocence"—places women in a double bind: somehow we are supposed to be both sexy and virginal, experienced and naive, seductive and chaste. The disparagement of maturity is, of course, insulting and frustrating to adult women, and the implication that little girls are seductive is dangerous to real children.

Influencing Sexual Attitudes

Young people also learn a great deal about sexual attitudes from the media and from advertising in particular. Advertising's approach to sex is pornographic: it reduces people to objects and de-emphasizes human contact and individuality. This reduction of sexuality to a dirty joke and of people to objects is the real obscenity of the culture. Although the sexual sell, overt and subliminal, is at a fevered pitch in most commercials, there is at the same time a notable absence of sex as an important and profound human activity.

There have been some changes in the images of women. Indeed, a "new woman" has emerged in commercials in recent years. She is generally presented as superwoman, who manages to do all the work at home and on the job (with the help of a product, of course, not of her husband or children or friends); or as the liberated woman, who owes her independence and self-esteem to the products she uses. These new images do not represent any real progress but rather create a myth of progress, an illusion that reduces complex sociopolitical problems to mundane personal ones.

Advertising images do not cause these problems, but they contribute to them by creating a climate in which the marketing of women's bodies—the sexual sell and dismemberment, distorted body image ideal, and children as sex objects—is seen as acceptable.

This is the real tragedy, that many women internalize these stereotypes and learn their "limitations," thus establishing a self-fulfilling prophecy. If one accepts these mythical and degrading images, to some extent one actualizes them. By remaining unaware of the profound seriousness of the ubiquitous influence, the redundant message, and the subliminal impact of advertisements, we ignore one of the most powerful "educational" forces in the culture—one that greatly affects our self-images, our ability to relate to each other, and effectively destroys awareness and action that might help to change that climate.

9

The Influence of Gender Socialization in Eating Disorders

DIANE E. TAUB AND PENELOPE A. MCLORG

Not everything we learn through the process of socialization has positive results. Consider eating disorders that result from the preoccupation many people, especially young women, have with body shape and weight. As Taub and McLorg note, in order to reduce this prevalent problem, we must consider the source of the negative body images and how to change the emphasis in the socialization process on ideal images.

As you read, think about the following:

1. What messages do women and men receive about ideal body types, and from where?

2. What could be done to change the negative results of socialization, such as abuse, neglect, and unrealistic body images?

GLOSSARY

Anorexia and bulimia nervosa Eating disorders that involve (1) self-starving and (2) binging and purging behaviors.

Internalization The process of making ideas and behavior patterns an integral part of one's repertoire of behaviors.

Gender socialization Learning the gender roles expected in society.

Agents of gender socialization Ways gender expectations are passed on.

THE OCCURRENCE OF ANOREXIA nervosa (self-starvation) and bulimia nervosa (binge-purge syndrome) has recently been called epidemic (Gordon, 1988; Wiseman and others, 1992). As a risk group, females are much more likely than males to be affected, comprising approxi-mately 90–95 percent of reported cases (Halmi, Falk, and Schwartz, 1981; Leichner and Gertler, 1988). This gender difference can be clarified by examining factors of gender socialization that relate to physical appearance.

Traditionally, more emphasis has been placed on females' appearance than on males' (Orbach, 1985; Schur, 1984). Women show awareness of this focus by being more concerned with their appearance than are men (Hayes and Ross, 1987; Pliner, Chaiken, and Flett, 1990). Physical appearance is also more crucial to self-concept among females than among males (Rodin, Silber-stein, and Striegel-Moore, 1985). Whereas in men self-image is associated with skill and achieve-ment, among women it is linked to physical characteristics (Hesse-Biber, Clayton-Matthews, and Downey, 1987). In terms of bodily appeal, males' physical attractiveness is related to physi-cal abilities, with their bodies valued for being active and functional. In contrast, females' bod-ies are judged on the basis of beauty (Mishkind and others, 1986; Rodin and others, 1985).

One important consideration in appearance is body shape. Concerns about body shape ex-pressed by women range from mild weight con-sciousness at one extreme to eating disorders at the other (Gordon, 1988; Rodin and others, 1985). As part of the socialization in the "cult of thinness" (Hesse-Biber and others, 1987, p. 512), females accept an ideal body shape and a corresponding need for weight control. Dissat-isfaction from a failure to meet slim appearance standards as well as subsequent dieting behavior

Presented at the annual meeting of the American Sociological Association, Pittsburgh, PA, 1992. Revised. Reprinted by permission of Diane E. Taub.

have been identified as risk factors for the development of eating disorders (Drewnowski and Yee, 1987; Polivy and Herman, 1986; Striegel-Moore and others, 1986).

This reading examines the relationship between eating disorders and gender socialization. Our purpose is to demonstrate the important contribution female and male socialization makes to the social context of anorexia nervosa and bulimia nervosa, including the gender difference in occurrence. To illustrate the connections between gender socialization and eating disorders, we use the following framework: (1) ideal body shape, as a representation of gender norms, (2) role models and mass media messages, as agents of gender socialization, and (3) dieting, as an expression of gender socialization. Ideal body shape affects agents of socialization, which in turn reinforce ideal body shape. Both acceptance of ideal body shape norms and exposure to agents of gender socialization are expressed through dieting behavior; eating disorders are an extreme response.

Ideal Body Shape

Current appearance expectations specify thinness for women (Chernin, 1981; Garner and Garfinkel, 1980; Garner and others, 1980; Silverstein, Peterson, and Perdue, 1986). Slim bodies are regarded as the most beautiful and worthy ones; overweight is seen as not only unhealthy but also offensive and disgusting (DeJong, 1980; Ritenbaugh, 1982; Schwartz, Thompson, and Johnson, 1982). Although both males and females are socialized to devalue fatness, women are more exposed to the need to be thin (Rodin and others, 1985).

In contrast, males are socialized to be muscular and not skinny or weak (Dwyer and others, 1969; Leon and Finn, 1984; Mishkind and others, 1986; Polivy, Garner, and Garfinkel, 1986). In ratings by preadolescent, adolescent, and college-aged males, the mesomorphic or muscular male body type is associated with socially favorable behaviors and personality traits (Guy, Rankin, Norvell, 1980; Lerner, 1969). Compared with

endomorphic (plump) and ectomorphic (slender) individuals, mesomorphs are judged more likely to assume leadership, be assertive, and be most wanted as a friend. The devaluing of a thin body for males is also reflected in the frequent desire of preadolescent and teenage boys to gain weight and/or size (Collins, 1991; Dwyer and others, 1969; Huenemann and others, 1966; Striegel-Moore, Silberstein, and Rodin, 1986).

Among females, the orientation toward slimness is so established that even when they are not overweight, they frequently perceive themselves as such (Connor-Greene, 1988; Halmi, Falk, Schwartz, 1981). For example, although over 60 percent of college females believe that they are overweight, only 2 percent actually are (Connor-Greene, 1988). Over four out of five college women report that they want to lose weight (Hesse-Biber and others, 1987). In addition, college females underestimate the occurrence of being underweight. While 31 percent of college women are measured as underweight, only 13 percent think they weigh below weight norms (Connor-Greene, 1988).

Other results indicate that the ideal body shape of college women is significantly thinner than both their actual body type and what they perceive as most attractive to males (Fallon and Rozin, 1985). In a study of families, Rozin and Fallon (1988) show that mothers and daughters both want slimmer bodies than they currently have. Furthermore, the shape these females believe most attractive to males is thinner than what males actually prefer.

Collins (1991) has recently demonstrated that gender-based ideas of attractive bodies develop in children as young as 6 or 7 years. First-through third-grade girls select illustrations of their ideal figures that are significantly thinner than their current figures; this pattern is found across all levels of actual weight. Moreover, girls choose significantly slimmer figures than boys do for the ideal girl, ideal female adult, and ideal male adult (Collins, 1991).

Among females, learning to desire thinness begins at an early age. In general, females want to be slim and are critical of their weight, regardless of their actual body size and weight. A similar concern for thinness in male bodies is not common among males (Collins, 1991; Connor-Greene, 1988; Fallon and Rozin, 1985; Hesse-Biber and others, 1987; Rodin and others, 1985; Rozin and Fallon, 1988). The inaccuracy with which females of all ages perceive their body shapes (Rodin and others, 1985) parallels the distorted body images held by individuals with eating disorders.

Agents of Gender Socialization

Images of ideal body shape affect agents of gender socialization, such as role models and mass media. In turn, these influences support expectations of body size. Reflecting gender socialization, traditional female role models and mass media messages express thinness norms for females.

ROLE MODELS

Examining patterns of ideal body shape, Garner and colleagues (1980) study the measurements of Miss America contestants over the twenty-year span from 1959 to 1978. Mazur (1986) and Wiseman and colleagues (1992) conduct similar analyses of contestants' dimensions covering the period of 1979 to the mid-1980s. Both Garner and co-workers (1980) and Mazur (1986) report decreases in bust and hip measurements of Miss America contestants over the study periods. However, waist dimensions demonstrate periods of increase, suggesting a less "hourglass-shaped" standard. Further, weight for height of Miss America contestants progressively declines (Garner and others, 1980; Mazur, 1986), with a trend from 1970 to 1978 for pageant winners to be thinner than the average contestant (Garner and others, 1980). Analyzing contestants' weight in relation to expected weight for their height and age, Wiseman and colleagues (1992) additionally find a significant decrease in the women's percentage of expected weight from 1979 to 1985.

Garner and colleagues (1980), Mazur (1986), and Wiseman and co-workers (1992) also examine the beauty ideal represented by *Playboy* centerfolds. As with Miss America contestants, bust

and hip dimensions decline and waist measurements rise from 1959 to 1978 (Garner and others, 1980). During the early 1980s, bust, waist, and hip dimensions of centerfolds decrease (Mazur, 1986). Centerfolds also show declines in weight for height between 1959 and the early 1980s (Garner and others, 1980; Mazur, 1986).

The trends of slenderization exhibited by both *Playboy* centerfolds and Miss America contestants illustrate the slimness norm. In fact, from 1979 through the mid-1980s, approximately two-thirds of these ideals of female beauty weighed 15 percent or more below their expected weight; maintaining such a weight level is one of the criteria for anorexia nervosa (Wiseman and others, 1992). Thus, the declining size of female figures considered admirable represents a body size reflective of an eating disorder.

Other female role models, such as movie stars and magazine models, have become less curvaceous over the past twenty to forty years (Morris, Cooper, and Cooper, 1989; Silverstein and others, 1986; Silverstein, Peterson, and Perdue, 1986). Hence, portrayals of females in media geared toward women as well as men demonstrate the thinness norm. The "anorectic body type" of models in major women's fashion magazines illustrates "an idealized standard of beauty and high fashion" (Gordon, 1988, p. 157). As preferred bodies, fashion models set an example of slimness that is unrealistic for most women. The majority of women are thus continually reminded of their inadequacy (Pliner and others, 1990) and kept "permanently insecure" about their appearance (Stannard, 1971, p. 125).

Women's socialization to be slim is also demonstrated by role models on television. In prime-time, top-ten Nielsen-rated television shows and their commercials, females are more likely to be thin than heavy and to be thinner than males (Kurman, 1978). A related study (Silverstein and others, 1986) of most-watched television programs demonstrates that 69 percent of the actresses and only 17.5 percent of the actors are slim. In addition, 5 percent of the women are evaluated as heavy, while over a quarter of the men are rated as such. These contrasts remain over a range of ages of the performers (Silverstein and others, 1986).

Research on newspaper and magazine advertisements similarly indicates that male figures are generally portrayed as bigger than female figures (Goffman, 1979). Such representation of the size of men symbolizes their "social weight," in power, authority, or rank (Goffman, 1979, p. 28), as well as the positive valuing of a larger body in males (Schur, 1984). In contrast, females in print advertisements, along with other role models from beauty contestants to television and film actresses to fashion models, provide continual exposure to the thin female standard.

MASS MEDIA MESSAGES

Beyond displaying slim female role models for imitation, agents of gender socialization promote consciousness of weight and diet. Mass media messages encourage virtually uniform standards of beauty (Mazur, 1986), with messages directed toward females emphasizing the thin ideal. Surveying major women's magazines, Garner and colleagues (1980) find a significant increase in diet articles from 1959–1978. From 1979 to 1988, the same women's magazines show a leveling off in number of diet articles but an increase in articles on exercise as a strategy for weight loss (Wiseman and others, 1992). In addition, dieting and weight control listings in the *Reader's Guide to Periodical Literature* almost double from 1977 to 1986 (Hesse-Biber and others, 1987).

Media emphasis on a slim body standard for females is also illustrated in comparisons of female- and male-directed magazines. An analysis (Silverstein and others, 1986) of the most popular women's as well as men's magazines indicates significant differences in content of articles and advertisements. In women's magazines, ads for diet foods, and articles and advertisements dealing with body shape or size, appear 63 times and 12 times more often, respectively, than in men's magazines (Silverstein and others, 1986).

Similarly, Andersen and DiDomenico (1992) find that the most popular magazines among women aged 18 to 24 contain 10 times more ar-

ticles and advertisements on dieting or losing weight than do the most popular magazines among men aged 18 to 24. The focus on weight control in young women's magazines may have particular importance, as the late teens are a period of onset for both anorexia nervosa and bulimia nervosa (Leichner and Gertler, 1988; Mitchell and Pyle, 1988).

In addition to advertisements and articles on weight control, women's magazines surpass men's magazines in material concerning food. Articles on food and ads for food (excluding those for diet foods) in women's magazines exceed those in men's magazines by seventy-one to one (Silverstein and others, 1986). Thus, through this printed medium, females are being presented with conflicting messages. While food advertisements and articles encourage the consumption and enjoyment of food, diet aids and body shape ads and articles reinforce control of eating and weight. Popular magazines effectively maintain women's weight control preoccupation through their dual messages of "eat" and "stay slim." Moreover, although exposure to magazine advertising may be similar for individuals with and without eating disorders, anorexics and bulimics are especially likely to believe that advertisements promote the desirability of slimness (Peterson, 1987).

As agents of socialization, mass media and role models present a consistent portrayal of thin females in material directed toward both female and male audiences. Media preference for slimness in women is additionally shown in messages encouraging females' weight control efforts. Through their selective representation of thin women, these sources not only reflect the slim ideal body shape for females, but also strengthen this gender norm for appearance. Role models and mass media effectively reinforce females' acceptance of weight consciousness. The impact of these influences is to encourage and perpetuate women's repeated attempts to conform to the thin standard. A similar promotion or expression of a slimness ideal is not an aspect of the socialization experience of males (Andersen and DiDomenico, 1992; Bachman,

Johnson, and O'Malley, 1984; Drewnowski and Yee, 1987; Silverstein and others, 1986; Taub, 1986).

Dieting

Concerns about thin body size reflect gender socialization of females. With the role obligation of being visually attractive, women alter their bodies to conform to an appearance ideal. Dieting can be viewed as a response to the gender norm of slimness in females. Weight concerns and weight-loss efforts are so common among females that they have become norms (Rodin and others, 1985; Striegel-Moore, Silberstein, and Rodin, 1986). Nasser (1988, p. 574) terms dieting a "cultural preoccupation" among females, with considerations about weight and weight control persisting even into women's elderly years (Pliner and others, 1990). Females' continual efforts toward the thinness ideal are usually unsuccessful (Bennett and Gurin, 1982; Chernin, 1981; Silverstein and others, 1986).

Frequency of dieting is related to actual body size, as well as to ideal body size and the emphasis a woman places on the importance of attractiveness (Silverstein and Perdue, 1988). A history of dieting, beginning in the teen years, is common among anorexics and bulimics (Crisp, 1977b; Lacey, Coker, and Birtchnell, 1986). In fact, researchers consider dieting a "precondition" (Polivy and Herman, 1986, p. 328) or a "chief risk factor" (Drewnowski and Yee, 1987, p. 633) of eating disorders.

As shown in a Nielsen survey, 56 percent of all women aged 24 to 54 diet during the course of the year; 76 percent do so for cosmetic, rather than health, reasons (Schwartz and others, 1982). Even twenty-five years ago, high school senior females indicated that they dieted for reasons of "beauty or good looks," rather than for "physical fitness" (Dwyer and others, 1967, p. 1051). More recently among high school females, over half had dieted by the time they entered high school; and nearly 40 percent were currently dieting (Johnson and others, 1983). Another study demonstrates an even greater occurrence

of dieting in preadolescent girls. Half of 9-year-olds and nearly 80 percent of 10- and 11-year-olds indicate that they have dieted (Stein, 1986). With dieting so common among girls and women, it is not surprising that "serious dieting" in females is seen as "normal" (Leon and Finn, 1984, p. 328).

Compared with young females, young males are much less likely to diet. In one sample, 10 percent of boys versus 80 percent of girls had been on a diet before the age of 13 (Hawkins and others, cited in Striegel-Moore and others, 1986). Similarly, 64 percent of first-year college females, but only 29 percent of first-year males, followed a reduced-calorie diet in the previous month (Drewnowski and Yee, 1987). The gender difference regarding dieting behavior is additionally shown by the more frequent use of over-the-counter diet pills by high school senior women. In a nationwide sample, 1 in 3 high school senior women had taken diet pills in the previous year, compared to 1 in 10 men. Moreover, 2 in 100 females use diet pills daily, while only 1 in 1000 males report daily use (Bachman and others, 1984). High school senior women are also much more likely than their male classmates to use prescription amphetamines without a physician's orders for the purpose of losing weight (Taub, 1986).

Overall, studies demonstrate that females of differing ages frequently diet or engage in other weight loss behavior. In addition, females are much more likely than males to pursue weight loss through various methods. The extent and persistence of females' dieting efforts indicate the acceptance of the thin ideal. As an expression of gender socialization, dieting also reveals exposure to both role model and mass media promotion of slimness. Anorexia nervosa and bulimia nervosa represent extreme responses to female socialization toward thinness, with dieting usually preceding an eating disorder.

Conclusion

The influence of gender socialization in anorexia nervosa and bulimia nervosa is suggested by the gender distribution of the syndromes, with occurrence at least ten times higher among females than males. In their connections with appearance expectations, eating disorders illustrate normative elements of female socialization. For example, females' weight loss efforts can be attributed to the greater importance of appearance in evaluations of women than of men; women's figures are more emphasized and more critically assessed (Mishkind and others, 1986; Rodin and others, 1985; Schur, 1984).

As agents of gender socialization, role models and mass media are affected by, and support, notions of ideal body shape. Beauty queens, *Playboy* centerfolds, fashion models, and female television and movie characters are predominantly slender (Garner and others, 1980; Kurman, 1978; Mazur, 1986; Morris, Cooper, and Cooper, 1989; Silverstein and others, 1986; Silverstein, Peterson, and Perdue, 1986; Wiseman and others, 1992). Such role models serve as ideals for the female body shape. In addition to promoting the slimness standard, nude layouts and beauty contests epitomize the viewing of women as objects, with women's bodies judged according to narrow beauty standards (Schur, 1984).

Also supporting the expectation of thinness in females are the numerous articles and advertisements on diet aids and body size in women's magazines (Andersen and DiDomenico, 1992; Garner and others, 1980; Silverstein and others, 1986; Wiseman and others, 1992). These media messages especially promote dieting behavior when accompanied by ample food articles and advertisements that encourage individuals to eat. Although females may enjoy "forbidden" food, they are continually reminded of the need to be thin. This double-edged message of "enjoy eating but control your weight" reinforces dieting.

A low rate of success for dieting (Bennett and Gurin, 1982; Chernin, 1981), combined with consistent pressure to be slim, results in repeated weight-loss efforts by females. Such manipulations of eating and body shape illustrate the tendency of females to view their bodies as objects, subject to modification for an attractiveness standard (Rodin and others, 1985; Schur, 1984).

While dieting reflects acceptance of gender norms for body shape, anorexia nervosa and bulimia nervosa represent extreme examples of gender socialization toward slimness.

Women's concerns with body shape range from mild weight consciousness to fully developed eating disorders (Gordon, 1988; Rodin and others, 1985). Individuals with eating disorders exemplify "weight phobia" (Crisp, 1977a), and can be viewed as extensions of the slim body ideal for females (Nasser, 1988; Rodin and others, 1985). Understanding of the gender distribution of anorexia nervosa and bulimia nervosa can be expanded by examining linkages with elements of gender socialization. Analysis of these factors is crucial for explaining the social context in which eating disorders occur.

REFERENCES

Andersen, Arnold E., and Lisa DiDomenico, 1992. Diet vs. shape content of popular male and female magazines: A dose-response relationship to the incidence of eating disorders? *International Journal of Eating Disorders* 11: 283–87.

Bachman, Jerald G., Lloyd D. Johnson, and Patrick M. O'Malley, 1984. *Monitoring the Future: Questionnaire Responses from the Nation's High School Seniors.* Ann Arbor: University of Michigan.

Bennett, William, and Joel Gurin, 1982. *The Dieter's Dilemma: Eating Less and Weighing More.* New York: Basic.

Chernin, Kim, 1981. *The Obsession: Reflections on the Tyranny of Slenderness.* New York: Harper & Row.

Collins, M. Elizabeth, 1991. Body figure perceptions and preferences among preadolescent children. *International Journal of Eating Disorders* 10: 199–208.

Connor-Greene, Patricia Anne, 1988. Gender differences in body weight perception and weight-loss strategies of college students. *Women and Health* 14(2): 27–42.

Crisp, A. H., 1977a. Anorexia nervosa. *Proceedings of the Royal Society of Medicine* 70: 464–470.

Crisp, A. H., 1977b. The prevalence of anorexia nervosa and some of its associations in the general population. *Advances in Psychosomatic Medicine* 9: 38–47.

DeJong, William, 1980. The stigma of obesity: The consequences of naive assumptions concerning the causes of physical deviance. *Journal of Health and Social Behavior* 21: 75–87.

Drewnowski, Adam, and Doris K. Yee, 1987. Men and body image: Are males satisfied with their body weight? *Psychosomatic Medicine* 49: 626–634.

Dwyer, Johanna T., Jacob J. Feldman, and Jean Mayer, 1967. Adolescent dieters: Who are they? Physical characteristics, attitudes and dieting practices of adolescent girls. *American Journal of Clinical Nutrition* 20: 1045–1056.

Dwyer, Johanna T., Jacob J. Feldman, Carl C. Seltzer, and Jean Mayer, 1969. Adolescent attitudes toward weight and appearance. *Journal of Nutrition Education* 1(Fall): 14–19.

Fallon, April E., and Paul Rozin, 1985. Sex differences in perceptions of desirable body shape. *Journal of Abnormal Psychology* 94: 102–105.

Garner, David M., and Paul E. Garfinkel, 1980. Sociocultural factors in the development of anorexia nervosa. *Psychological Medicine* 10: 647–656.

Garner, David M., Paul E. Garfinkel, Donald Schwartz, and Michael Thompson, 1980. Cultural expectations of thinness in women. *Psychological Reports* 47: 483–491.

Goffman, Erving, 1979. *Gender Advertisements*. New York: Harper & Row.

Gordon, Richard A., 1988. A sociocultural interpretation of the current epidemic of eating disorders. In B. J. Blinder, B. F. Chaitin, and R. S. Goldstein (eds.), *The Eating Disorders: Medical and Psychological Bases of Diagnosis and Treatment*. New York: PMA.

Guy, Rebecca F., Beverly A. Rankin, and Melissa J. Norvell, 1980. The relation of sex role stereotyping to body image. *Journal of Psychology* 105: 167–173.

Halmi, Katherine A., James R. Falk, and Estelle Schwartz, 1981. Binge-eating and vomiting: A survey of a college population. *Psychological Medicine* 11: 697–706.

Hayes, Diane, and Catherine E. Ross, 1987. Concern with appearance, health beliefs, and eating habits. *Journal of Health and Social Behavior* 28: 120–130.

Hesse-Biber, Sharlene, Alan Clayton-Matthews, and John A. Downey, 1987. The differential importance of weight and body image among college men and women. *Genetic, Social, and General Psychology Monographs* 113: 511–528.

Huenemann, Ruth L., Leona R. Shapiro, Mary C. Hampton, and Barbara W. Mitchell, 1966. A longitudinal study of gross body composition and body conformation and their association with food and activity in a teenage population. *American Journal of Clinical Nutrition* 18: 325–338.

Johnson, Craig L., Chris Lewis, Susan Love, Marilyn Stuckey, and Linda Lewis, 1983. A descriptive survey of dieting and bulimic behavior in a female high school population. In *Understanding Anorexia Nervosa and Bulimia: Report of the Fourth Ross Conference on Medical Research*. Columbus, OH: Ross Laboratories.

Kurman, Lois, 1978. An analysis of messages concerning food, eating behaviors, and ideal body image on prime-time American network television. *Dissertation Abstracts International* 39: 1907A–1908A.

Lacey, Hubert J., Sian Coker, and S. A. Birtchnell, 1986. Bulimia: Factors associated with its etiology and maintenance. *International Journal of Eating Disorders* 5: 475–487.

Leichner, Pierre, and A. Gertler, 1988. Prevalence and incidence studies of anorexia nervosa. In Barton J. Blinder, Barry F. Chaitin, and Renee S. Goldstein (eds.), *The Eating Disorders: Medical and Psychological Bases of Diagnosis and Treatment*. New York: PMA.

Leon, Gloria R., and Stephen Finn, 1984. Sex-role stereotypes and the development of eating disorders. In Cathy Spatz Widom (ed.), *Sex Roles and Psychopathology*. New York: Plenum.

Lerner, Richard M., 1969. The development of stereotyped expectancies of body build-behavior relations. *Child Development* 40: 137–141.

Mazur, Allan, 1986. U.S. trends in feminine beauty and overadaptation. *Journal of Sex Research* 22: 281–303.

Mishkind, Marc E., Judith Rodin, Lisa R. Silberstein, and Ruth H. Striegel-Moore, 1986. The embodiment of masculinity: Cultural, psychological, and behavioral dimensions. *American Behavioral Scientist* 29: 545–562.

Mitchell, James E., and Richard L. Pyle, 1988. The diagnosis and clinical characteristics of bulimia. In Barton J. Blinder, Barry F. Chaitin, and Renee S. Goldstein (eds.), *The Eating Disorders: Medical and Psychological Bases of Diagnosis and Treatment*. New York: PMA.

Morris, Abigail, Troy Cooper, and Peter J. Cooper, 1989. The changing shape of female fashion models. *International Journal of Eating Disorders* 8: 593–596.

Nasser, Mervat, 1988. Culture and weight consciousness. *Journal of Psychosomatic Research* 32: 573–577.

Orbach, Susie, 1985. Visibility/invisibility: Social considerations in anorexia nervosa—a feminist perspective. In Steven W. Emmett (ed.), *Theory and Treatment of Anorexia Nervosa and Bulimia: Biomedical, Sociocultural, and Psychological Perspectives*. New York: Brunner/Mazel.

Peterson, Robin T., 1987. Bulimia and anorexia in an advertising context. *Journal of Business Ethics* 6: 495–504.

Pliner, Patricia, Shelly Chaiken, and Gordon L. Flett, 1990. Gender differences in concern with body weight and physical appearance over the life span. *Personality and Social Psychology Bulletin* 16: 263–273.

Polivy, Janet, David M. Garner, and Paul E. Garfinkel, 1986. Causes and consequences of the current preference for thin female physiques. In C. Peter Herman, Mark P. Zanna, and E. Tory Higgins (eds.), *Physical Appearance, Stigma, and Social Behavior: The Ontario Symposium*. Vol. 3. Hillsdale, NJ: Lawrence Erlbaum.

Polivy, Janet, and C. Peter Herman, 1986. Dieting and binging reexamined: A response to Lowe. *American Psychologist* 41: 327–328.

Ritenbaugh, Cheryl, 1982. Obesity as a culture-bound syndrome. *Culture, Medicine and Psychiatry* 6: 347–361.

Rodin, Judith, Lisa Silberstein, and Ruth Striegel-Moore, 1985. Women and weight: A normative discontent. In Theo B. Sonderegger (ed.), *Nebraska Symposium on Motivation*. Vol. 32: *Psychology and Gender*. Lincoln: University of Nebraska.

Rozin, Paul, and April Fallon, 1988. Body image, attitudes to weight, and misperceptions of figure preferences of the opposite sex: A comparison of men and women in two generations. *Journal of Abnormal Psychology* 97: 342–345.

Schur, Edwin M., 1984. *Labeling Women Deviant: Gender, Stigma, and Social Control*. New York: Random House.

Schwartz, Donald M., Michael G. Thompson, and Craig L. Johnson, 1982. Anorexia nervosa and bulimia: The sociocultural context. *International Journal of Eating Disorders* 1(3): 20–36.

Silverstein, Brett, and Lauren Perdue, 1988. The relationship between role concerns, preferences for slimness, and symptoms of eating problems among college women. *Sex Roles* 18: 101–106.

Silverstein, Brett, Lauren Perdue, Barbara Peterson, and Eileen Kelly, 1986. The role of the mass media in promoting a thin standard of bodily attractiveness for women. *Sex Roles* 14: 519–532.

Silverstein, Brett, Barbara Peterson, and Lauren Perdue, 1986. Some correlates of the thin standard of bodily attractiveness for women. *International Journal of Eating Disorders* 5: 895–905.

Stannard, Una, 1971. The mask of beauty. In Vivian Gornick and Barbara K. Moran (eds.), *Woman in Sexist Society: Studies in Power and Powerlessness*. New York: Basic.

Stein, Jeannine, 1986. Why girls as young as 9 fear fat and go on diets to lose weight. *Los Angeles Times*, October 29, Part V: 1, 10.

Striegel-Moore, Ruth, Lisa R. Silberstein, and Judith Rodin, 1986. Toward an understanding of risk factors for bulimia. *American Psychologist* 41: 246–263.

Taub, Diane E., 1986. Amphetamine usage among high school senior women, 1976–1982: An evaluation of social bonding theory. Unpublished doctoral dissertation, University of Kentucky, Lexington.

Wiseman, Claire V., James J. Gray, James E. Mosimann, and Anthony H. Ahrens, 1992. Cultural expectations of thinness in women: An update. *International Journal of Eating Disorders* 11: 85–89.

QUESTIONS FOR DISCUSSION

For further discussion of this topic, see the Wadsworth Sociology Resource Center, "Virtual Society," *http://sociology.wadsworth.com*, under *Sociological Footprints*, by Cargan and Ballantine. You can respond to the discussion questions there or enter your own comments in the online chat forum.

SUGGESTED READINGS AND SOCIOLOGY INTERNET RESOURCES

See the Wadsworth Sociology Resource Center, "Virtual Society," *http://sociology.wadsworth. com*, for additional links, suggestions for further reading, and learning tools related to this chapter.

Either from the "Virtual Society" website or directly from your web browser, you may access InfoTrac College Edition, an online university library that includes over 700 popular and scholarly journals in which you can find articles related to the topics in this chapter.

CHAPTER 3

Culture

WHY WE DO THINGS THE WAY WE DO

HUMAN BEHAVIOR IS BOTH PATTERNED AND orderly because within our society we are taught to follow similar rules of behavior (*norms*) and to cherish similar objects and behaviors (*values*). These similarities create the culture of a society: *its total way of life*. In Chapter 2 Kingsley Davis noted that the importance of culture is indicated by the fact that most human behavior is learned within a cultural context (Reading 5).

It is through *cultural relativism*—looking at other cultures in an objective manner—that we attempt to understand learned cultural patterns and behaviors by considering the functions they serve for society. In the first reading in this chapter, Horace Miner gives us an anthropological look at what appears to be a "primitive" group by examining the societal needs served by the unusual attitude of the Nacirema toward the human body. If, while reading this selection, you feel glad to be an American while wondering about the "silly" actions of the Nacirema, then your ethnocentrism is showing. (Read the article carefully, especially the italicized words. You should find it amusing.) The reading by Miner is also a good example of the points made by Mills in Chapter 1. That is, our beliefs are represented by the social system of which we are a part, and they affect subsequent behavior.

Because we are taught the norms, values, language, and beliefs (folklore, legends, proverbs, religion) of our own culture, we frequently find it difficult to see our culture objectively. What we do routinely we accept as right without question, and possibly even without understanding. This *ethnocentrism* —the belief that one's own culture is superior to others—can make it difficult to accept the different ways of others and to change our own ways. It is those two factors of learned behavior and ethnocentrism that lead to cultural constraints on our thoughts and behavior. Ethnocentrism is common despite the fact that many, if not most, of the material items used in any given culture were neither invented nor discovered in that culture but were adopted from other societies through *cultural diffusion* —the spread of cultural behavior and materials from one society to another.

Unfortunately, ethnocentrism can carry over into hostilities toward different groups of people. This hostility exists because groups feel strongly about the rightness of their own cultural beliefs, values, attitudes, and behaviors. A good example of cultural misunderstandings resulting from ethnocentrism is seen in the reading by Marvin Harris, "India's Sacred Cow." It makes the point that cultural practices have origins and can be explained, however strange they seem to outsiders.

Even within societies there can be subcultures that differ from the dominant culture in behaviors, dress, and language. An example is the subculture of youth. Deena Weinstein reveals how this subculture developed and how it is changing in her article, "Expendable Youth."

Cultural practices are not always positive for minority members of society; they may have come into existence to protect the position of the dominant group. Consider the examples in the reading by Ruth Fischbach and Elizabeth Donnelly of domestic violence against women in many societies.

As you read this chapter, ask yourself these questions: Why do the Nacirema have an apparently "pathological" concern with the body, and what are the implications of this concern? Why does our liking of the familiar sometimes lead to violent dislike of groups with dissimilar cultures? How do subcultures differ from the dominant culture? How can stereotypes shape our images, sometimes falsely, of cultural patterns? And why do cultural patterns vary so widely, sometimes to the disadvantage of members of society?

10
Body Ritual Among the Nacirema

HORACE MINER

When sociologists and anthropologists study other cultures, they attempt to be objective in their observations, trying to understand the culture from that culture's point of view. Understanding other cultures can help us gain perspective on our own culture. Yet because individuals are socialized into their own culture's beliefs, values, and practices, other cultures may seem to have strange, even bizarre or immoral activities and beliefs.

As you read Miner's essay, keep in mind these questions:

1. What practices in the Nacirema culture appear strange to an outsider?
2. How might someone from a different culture view or interpret practices in your culture?

GLOSSARY

Body ritual Ceremonies focusing on the body or body parts.

THE ANTHROPOLOGIST HAS BECOME so familiar with the diversity of ways in which different peoples behave in similar situations that he is not apt to be surprised by even the most exotic customs. In fact, if all of the logically possible combinations of behavior have not been found somewhere in the world, he is apt to suspect that they must be present in some yet undescribed tribe. This point has, in fact, been expressed with respect to clan organization by Murdock (1949:71). In this light, the magical beliefs and practices of the Nacirema present such unusual aspects that it seems desirable to describe them as an example of the extremes to which human behavior can go.

Professor Linton first brought the ritual of the Nacirema to the attention of anthropologists twenty years ago (1936:326), but the culture of this people is still very poorly understood. They are a North American group living in the territory between the Canadian Cree, the Yaqui and Tarahumare of Mexico, and the Carib and Arawak of the Antilles. Little is known of their origin, although tradition states that they came from the east. According to Nacirema mythology, their nation was originated by a culture hero, Notgnihsaw, who is otherwise known for two great feats of strength—the throwing of a piece of wampum across the river Pa-To-Mac and the chopping down of a cherry tree in which the Spirit of Truth resided.

Nacirema culture is characterized by a highly developed market economy which has evolved in a rich natural habitat. While much of the people's time is devoted to economic pursuits, a large part of the fruits of these labors and a considerable portion of the day are spent in ritual activity. The focus of this activity is the human body, the appearance and health of which loom as a dominant concern in the ethos of the people. While such a concern is certainly not unusual, its ceremonial aspects and associated philosophy are unique.

The fundamental belief underlying the whole system appears to be that the human body is ugly and that its natural tendency is to debility and disease. Incarcerated in such a body, man's only hope is to avert these characteristics through the use of the powerful influences of ritual and ceremony. Every household has one or more shrines

Reproduced by permission of the American Anthropological Association from American Anthropologist
58 (3), 503–507, 1956. Not for further reproduction.

devoted to this purpose. The more powerful individuals in this society have several shrines in their houses and, in fact, the opulence of a house is often referred to in terms of the number of such ritual centers it possesses. Most houses are of wattle and daub construction, but the shrine rooms of the more wealthy are walled with stone. Poorer families imitate the rich by applying pottery plaques to their shrine walls.

While each family has at least one such shrine, the rituals associated with it are not family ceremonies but are private and secret. The rites are normally only discussed with children, and then only during the period when they are being initiated into these mysteries. I was able, however, to establish sufficient rapport with the natives to examine these shrines and to have the rituals described to me.

The focal point of the shrine is a box or chest which is built into the wall. In this chest are kept the many charms and magical potions without which no native believes he could live. These preparations are secured from a variety of specialized practitioners. The most powerful of these are the medicine men, whose assistance must be rewarded with substantial gifts. However, the medicine men do not provide the curative potions for their clients, but decide what the ingredients should be and then write them down in an ancient and secret language. This writing is understood only by the medicine men and by the herbalists who, for another gift, provide the required charm.

The charm is not disposed of after it has served its purpose, but is placed in the charm-box of the household shrine. As these magical materials are specific for certain ills, and the real or imagined maladies of the people are many, the charm-box is usually full to overflowing. The magical packets are so numerous that people forget what their purposes were and fear to use them again. While the natives are very vague on this point, we can only assume that the idea in retaining all the old magical materials is that their presence in the charm-box, before which the body rituals are conducted, will in some way protect the worshipper.

Beneath the charm-box is a small font. Each day every member of the family, in succession, enters the shrine room, bows his head before the charm-box, mingles different sorts of holy water in the font, and proceeds with a brief rite of ablution. The holy waters are secured from the Water Temple of the community, where the priests conduct elaborate ceremonies to make the liquid ritually pure.

In the hierarchy of magical practitioners, and below the medicine men in prestige, are specialists whose designation is best translated "holy-mouth-men." The Nacirema have an almost pathological horror of and fascination with the mouth, the condition of which is believed to have a supernatural influence on all social relationships. Were it not for the rituals of the mouth, they believe that their teeth would fall out, their gums bleed, their jaws shrink, their friends desert them, and their lovers reject them. They also believe that a strong relationship exists between oral and moral characteristics. For example, there is a ritual ablution of the mouth for children which is supposed to improve their moral fiber.

The daily body ritual performed by everyone includes a mouth-rite. Despite the fact that these people are so punctilious about care of the mouth, this rite involves a practice which strikes the uninitiated stranger as revolting. It was reported to me that the ritual consists of inserting a small bundle of hog hairs into the mouth, along with certain magical powders, and then moving the bundle in a highly formalized series of gestures.

In addition to the private mouth-rite, the people seek out a holy-mouth-man once or twice a year. These practitioners have an impressive set of paraphernalia, consisting of a variety of augers, awls, probes, and prods. The use of these objects in the exorcism of the evils of the mouth involves almost unbelievable ritual torture of the client. The holy-mouth-man opens the client's mouth and, using the above-mentioned tools, enlarges any holes which decay may have created in the teeth. Magical materials are put into these holes. If there are no naturally occurring holes in the teeth, large sections of one or more teeth are gouged out so that the supernatural substance can be applied. In the client's view, the purpose of these ministrations is to arrest decay and to draw friends. The extremely sacred and traditional character of the rite is evident in the fact that the natives return to the holy-mouth-man year after year, despite the fact that their teeth continue to decay.

It is to be hoped that, when a thorough study of the Nacirema is made, there will be careful inquiry into the personality structure of these people. One has but to watch the gleam in the eye of a holy-mouth-man, as he jabs an awl into an exposed nerve, to suspect that a certain amount of sadism is involved. If this can be established, a very interesting pattern emerges, for most of the population shows definite masochistic tendencies. It was to these that Professor Linton referred in discussing a distinctive part of the daily body ritual which is performed only by men. This part of the rite involves scraping and lacerating the surface of the face with a sharp instrument. Special women's rites are performed only four times during each lunar month, but what they lack in frequency is made up in barbarity. As part of this ceremony, women bake their heads in small ovens for about an hour. The theoretically interesting point is that what seems to be a preponderantly masochistic people have developed sadistic specialists.

The medicine men have an imposing temple, or *latipso*, in every community of any size. The more elaborate ceremonies required to treat very sick patients can only be performed at this temple. These ceremonies involve not only the thaumaturge but a permanent group of vestal maidens who move sedately about the temple chambers in distinctive costume and headdress.

The *latipso* ceremonies are so harsh that it is phenomenal that a fair proportion of the really sick natives who enter the temple ever recover. Small children whose indoctrination is still incomplete have been known to resist attempts to take them to the temple because "that is where you go to die." Despite this fact, sick adults are not only willing but eager to undergo the pro-

tracted ritual purification, if they can afford to do so. No matter how ill the supplicant or how grave the emergency, the guardians of many temples will not admit a client if he cannot give a rich gift to the custodian. Even after one has gained admission and survived the ceremonies, the guardians will not permit the neophyte to leave until he makes still another gift.

The supplicant entering the temple is first stripped of all his or her clothes. In every-day life the Nacirema avoids exposure of his body and its natural functions. Bathing and excretory acts are performed only in the secrecy of the household shrine, where they are ritualized as part of the body-rites. Psychological shock results from the fact that body secrecy is suddenly lost upon entry into the *latipso*. A man, whose own wife has never seen him in an excretory act, suddenly finds himself naked and assisted by a vestal maiden while he performs his natural functions into a sacred vessel. This sort of ceremonial treatment is necessitated by the fact that the excreta are used by a diviner to ascertain the course and nature of the client's sickness. Female clients, on the other hand, find their naked bodies are subjected to the scrutiny, manipulation, and prodding of the medicine men.

Few supplicants in the temple are well enough to do anything but lie on their hard beds. The daily ceremonies, like the rites of the holy-mouthmen, involve discomfort and torture. With ritual precision, the vestals awaken their miserable charges each dawn and roll them about on their beds of pain while performing ablutions, in the formal movements of which the maidens are highly trained. At other times they insert magic wands in the supplicant's mouth or force him to eat substances which are supposed to be healing. From time to time the medicine men come to their clients and jab magically treated needles into their flesh. The fact that these temple ceremonies may not cure, and may even kill the neophyte, in no way decreases the people's faith in the medicine men.

There remains one other kind of practitioner, known as a "listener." This witch-doctor has the power to exorcise the devils that lodge in the heads of people who have been bewitched. The Nacirema believe that parents bewitch their own children. Mothers are particularly suspected of putting a curse on children while teaching them the secret body rituals. The counter-magic of the witch-doctor is unusual in its lack of ritual. The patient simply tells the "listener" all his troubles and fears, beginning with the earliest difficulties he can remember. The memory displayed by the Nacirema in these exorcism sessions is truly remarkable. It is not uncommon for the patient to bemoan the rejection he felt upon being weaned as a babe, and a few individuals even see their troubles going back to the traumatic effects of their own birth.

In conclusion, mention must be made of certain practices which have their base in native esthetics but which depend upon the pervasive aversion to the natural body and its functions. There are ritual fasts to make fat people thin and ceremonial feasts to make thin people fat. Still other rites are used to make women's breasts larger if they are small, and smaller if they are large. General dissatisfaction with breast shape is symbolized in the fact that the ideal form is virtually outside the range of human variation. A few women afflicted with almost inhuman hypermammary development are so idolized that they make a handsome living by simply going from village to village and permitting the natives to stare at them for a fee.

Reference has already been made to the fact that excretory functions are ritualized, routinized, and relegated to secrecy. Natural reproductive functions are similarly distorted. Inter-

course is taboo as a topic and scheduled as an act. Efforts are made to avoid pregnancy by the use of magical materials or by limiting intercourse to certain phases of the moon. Conception is actually very infrequent. When pregnant, women dress so as to hide their condition. Parturition takes place in secret, without friends or relatives to assist, and the majority of women do not nurse their infants.

Our review of the ritual life of the Nacirema has certainly shown them to be a magic-ridden people. It is hard to understand how they have managed to exist so long under the burdens which they have imposed upon themselves. But even such exotic customs as these take on real meaning when they are viewed with the insight provided by Malinowski when he wrote (1948: 70):

> Looking from far and above, from our high places of safety in the developed civilization, it is easy to see all the crudity and irrelevance of magic. But without its power and guidance early man could not have mastered his practical difficulties as he has done, nor could man have advanced to the higher stages of civilization.

REFERENCES

Linton, Ralph, 1936. *The Study of Man*. New York: Appleton-Century.

Malinowski, Bronislaw, 1948. *Magic, Science, and Religion*. Glencoe, Ill.: Free Press.

Murdock, George P., 1949. *Social Structure*. New York: Macmillan.

11
India's Sacred Cow

MARVIN HARRIS

Cultures vary dramatically in their beliefs and practices, yet each cultural practice has evolved with some reason behind it. One of these practices is cow worship. In India, the cultural practice among Hindus is to treat cows with great respect, even in the face of human hunger. Harris discusses this practice, which many find curious.

Consider the following as you read:

1. Why are cows sacred? What is sacred in your country that might seem strange to others?

2. What other practices in different cultures do you find strange or unusual, and what purpose might those practices serve for the culture?

GLOSSARY

Untouchables Lowest group in the stratification (caste) system of India.

Hinduism Dominant religious belief system in India.

NEWS PHOTOGRAPHS THAT CAME out of India during the famine of the late 1960s showed starving people stretching out bony hands to beg for food while sacred cattle strolled behind undisturbed. The Hindu, it seems, would rather starve to death than eat his cow or even deprive it of food. The cattle appear to browse unhindered through urban markets eating an orange here, a mango there, competing with people for meager supplies of food.

By Western standards, spiritual values seem more important to Indians than life itself. Specialists in food habits around the world like Fred Simoons at the University of California at Davis consider Hinduism an irrational idealogy that compels people to overlook abundant, nutritious foods for scarcer, less healthful foods.

What seems to be an absurd devotion to the mother cow pervades Indian life. Indian wall calendars portray beautiful young women with bodies of fat white cows, often with milk jetting from their teats into sacred shrines.

Cow worship even carries over into politics. In 1966 a crowd of 120,000 people, led by holy men, demonstrated in front of the Indian House of Parliament in support of the All-Party Cow Protection Campaign Committee. In Nepal, the only contemporary Hindu kingdom, cow slaughter is severely punished. As one story goes, the car driven by an official of a United States agency struck and killed a cow. In order to avoid the international incident that would have occurred when the official was arrested for murder, the Nepalese magistrate concluded that the cow had committed suicide.

Many Indians agree with Western assessments of the Hindu reverence for their cattle, the zebu, or *Bos indicus,* a large-humped species prevalent in Asia and Africa. M. N. Srinivas, an Indian anthropologist, states: "Orthodox Hindu opinion regards the killing of cattle with abhorrence, even though the refusal to kill vast number of useless cattle which exist in India today is detrimental to the nation." Even the Indian Ministry of Information formerly maintained that "the large animal population is more a liability than an asset in view of our land resources." Accounts from many different sources point to the same conclusion: India, one of the world's great civilizations, is being strangled by its love for the cow.

From Human Nature Magazine *1(2), pp. 28, 30–36, February 1978. Copyright ©1978 by Human Nature, Inc.; reprinted by permission of the publisher.*

The easy explanation for India's devotion to the cow, the one most Westerners and Indians would offer, is that cow worship is an integral part of Hinduism. Religion is somehow good for the soul, even if it sometimes fails the body. Religion orders the cosmos and explains our place in the universe. Religious beliefs, many would claim, have existed for thousands of years and have a life of their own. They are not understandable in scientific terms.

But all this ignores history. There is more to be said for cow worship than is immediately apparent. The earliest Vedas, the Hindu sacred texts from the second millennium B.C., do not prohibit the slaughter of cattle. Instead, they ordain it as part of sacrificial rites. The early Hindus did not avoid the flesh of cows and bulls; they ate it at ceremonial feasts presided over by Brahman priests. Cow worship is a relatively recent development in India; it evolved as the Hindu religion developed and changed.

This evolution is recorded in royal edicts and religious texts written during the last 3,000 years of Indian history. The Vedas from the first millennium B.C. contain contradictory passages, some referring to ritual slaughter and others to a strict taboo on beef consumption. A. N. Bose, in *Social and Rural Economy of Northern India,* 600 B.C.–200 A.D., concludes that many of the sacred-cow passages were incorporated into the texts by priests of a later period.

By 200 A.D. the status of Indian cattle had undergone a spiritual transformation. The Brahman priesthood exhorted the population to venerate the cow and forbade them to abuse it or to feed on it. Religious feasts involving the ritual slaughter and the consumption of livestock were eliminated and meat eating was restricted to the nobility.

By 1000 A.D., all Hindus were forbidden to eat beef. Ahimsa, the Hindu belief in the unity of all life, was the spiritual justification for this restriction. But it is difficult to ascertain exactly when this change occurred. An important event that helped to shape the modern complex was the Islamic invasion, which took place in the eighth century A.D. Hindus may have found it

politically expedient to set themselves off from the invaders, who were beefeaters, by emphasizing the need to prevent the slaughter of their sacred animals. Thereafter, the cow taboo assumed its modern form and began to function much as it does today.

The place of the cow in modern India is every place—on posters, in the movies, in brass figures, in stone and wood carvings, on the streets, in the fields. The cow is a symbol of health and abundance. It provides the milk that Indians consume in the form of yogurt and ghee (clarified butter), which contribute subtle flavors to much spicy Indian food.

This, perhaps, is the practical role of the cow, but cows provide less than half the milk produced in India. Most cows in India are not dairy breeds. In most regions, when an Indian farmer wants a steady, high-quality source of milk he usually invests in a female water buffalo. In India the water buffalo is the specialized dairy breed because its milk has a higher butterfat content than zebu milk. Although the farmer milks his zebu cows, the milk is merely a by-product.

More vital than zebu milk to South Asian farmers are zebu calves. Male calves are especially valued because from bulls come oxen, which are the mainstay of the Indian agricultural system.

Small, fast oxen drag wooden plows through late-spring fields when monsoons have dampened the dry, cracked earth. After harvest, the oxen break the grain from the stalk by stomping through mounds of cut wheat and rice. For rice cultivation in irrigated fields, the male water buffalo is preferred (it pulls better in deep mud), but for most other crops, including rainfall rice, wheat, sorghum, and millet, and for transporting goods and people to and from town, a team of oxen is preferred. The ox is the Indian peasant's tractor, thresher, and family car combined; the cow is the factory that produces the ox.

If draft animals instead of cows are counted, India appears to have too few domesticated ruminants, not too many. Since each of the 70 million farms in India requires a draft team, it follows that Indian peasants should use 140 million animals in the fields. But there are only 83 million oxen and male water buffalo on the subcontinent, a shortage of 30 million draft teams.

In other regions of the world, joint ownership of draft animals might overcome a shortage, but Indian agriculture is closely tied to the monsoon rains of late spring and summer. Field preparation and planting must coincide with the rain, and a farmer must have his animals ready to plow when the weather is right. When the farmer without a draft team needs bullocks most, his neighbors are all using theirs. Any delay in turning the soil drastically lowers production.

Because of this dependence on draft animals, loss of the family oxen is devastating. If a beast dies, the farmer must borrow money to buy or rent an ox at interest rates so high that he ultimately loses his land. Every year foreclosures force thousands of poverty-stricken peasants to abandon the countryside for the overcrowded cities.

If a family is fortunate enough to own a fertile cow, it will be able to rear replacements for a lost team and thus survive until life returns to normal. If, as sometimes happens, famine leads a family to sell its cow and ox team, all ties to agriculture are cut. Even if the family survives, it has no way to farm the land, no oxen to work the land, and no cows to produce oxen.

The prohibition against eating meat applies to the flesh of cows, bulls, and oxen, but the cow is the most sacred because it can produce the other two. The peasant whose cow dies is not only crying over a spiritual loss but over the loss of his farm as well.

Religious laws that forbid the slaughter of cattle promote the recovery of the agricultural system from the dry Indian winter and from periods of drought. The monsoon, on which all agriculture depends, is erratic. Sometimes, it arrives early, sometimes late, sometimes not at all. Drought has struck large portions of India time and again in this century, and Indian farmers and the zebus are accustomed to these natural disasters. Zebus can pass weeks on end with little or no food and water. Like camels, they store both in their humps and recuperate quickly with only a little nourishment.

During droughts the cows often stop lactating and become barren. In some cases the condition is permanent but often it is only temporary. If barren animals were summarily eliminated, as Western experts in animal husbandry have suggested, cows capable of recovery would be lost along with those entirely debilitated. By keeping alive the cows that can later produce oxen, religious laws against cow slaughter assure the recovery of the agricultural system from the greatest challenge it faces—the failure of the monsoon.

The local Indian governments aid the process of recovery by maintaining homes for barren cows. Farmers reclaim any animal that calves or begins to lactate. One police station in Madras collects strays and pastures them in a field adjacent to the station. After a small fine is paid, a cow is returned to its rightful owner when the owner thinks the cow shows signs of being able to reproduce.

During the hot, dry spring months most of India is like a desert. Indian farmers often complain they cannot feed their livestock during this period. They maintain the cattle by letting them scavenge on the sparse grass along the roads. In the cities the cattle are encouraged to scavenge near food stalls to supplement their scant diet. These are the wandering cattle tourists report seeing throughout India.

Westerners expect shopkeepers to respond to these intrusions with the deference due a sacred animal; instead, their response is a string of curses and the crack of a long bamboo pole across the beast's back or a poke at its genitals. Mahatma Gandhi was well aware of the treatment sacred cows (and bulls and oxen) received in India. "How we bleed her to take the last drop of milk from her. How we starve her to emaciation, how we ill-treat the calves, how we deprive them of their portion of milk, how cruelly we treat the oxen, how we castrate them, how we beat them, how we overload them" [Gandhi, 1954].

Oxen generally receive better treatment than cows. When food is in short supply, thrifty Indian peasants feed their working bullocks and ignore their cows, but rarely do they abandon the cows to die. When the cows are sick, farmers worry over them as they would over members of the family and nurse them as if they were children. When the rains return and when the fields are harvested, the farmers again feed their cows regularly and reclaim their abandoned animals. The prohibition against beef consumption is a form of disaster insurance for all India.

Western agronomists and economists are quick to protest that all the functions of the zebu cattle can be improved with organized breeding programs, cultivated pastures, and silage. Because stronger oxen would pull the plow faster, they could work multiple plots of land, allowing farmers to share their animals. Fewer healthy, well-fed cows could provide Indians with more milk. But pastures and silage require arable land, land needed to produce wheat and rice.

A look at Western cattle farming makes plain the cost of adopting advanced technology in Indian agriculture. In a study of livestock production in the United States, David Pimentel of the College of Agriculture and Life Sciences at Cornell University, found that 91 percent of the cereal, legume, and vegetable protein suitable for human consumption is consumed by livestock. Approximately three-quarters of the arable land in the United States is devoted to growing food for livestock. In the production of meat and milk, American ranchers use enough fossil fuel to equal more than 82 million barrels of oil annually.

Indian cattle do not drain the system in the same way. In a 1971 study of livestock in West Bengal, Stewart Odend'hal [1972] of the University of Missouri found that Bengalese cattle ate only the inedible remains of subsistence crops — rice straw, rice hulls, the tops of sugar cane, and mustard-oil cake. Cattle graze in the fields after harvest and eat the remains of crops left on the ground; they forage for grass and weeds on the roadsides. The food for zebu cattle costs the human population virtually nothing. "Basically," Odend'hal says, "the cattle convert items of little direct human value into products of immediate utility."

In addition to plowing the fields and producing milk, the zebus produce dung, which fires the hearths and fertilizes the fields of India. Much of the estimated 800 million tons of manure produced annually is collected by the farmers' children as they follow the cows and bullocks from place to place. And when the children see the droppings of another farmer's cattle along the road, they pick those up also. Odend'hal reports that the system operates with such high efficiency that the children of West Bengal recover nearly 100 percent of the dung produced by their livestock.

From 40 to 70 percent of all manure produced by Indian cattle is used as fuel for cooking; the rest is returned to the fields as fertilizer. Dried dung burns slowly, cleanly, and with low heat — characteristics that satisfy the household needs of Indian women. Staples like curry and rice can simmer for hours. While the meal slowly cooks over an unattended fire, the women of the household can do other chores. Cow chips, unlike firewood, do not scorch as they burn.

It is estimated that the dung used for cooking fuel provides the energy-equivalent of 43 million tons of coal. At current prices, it would cost India an extra 1.5 billion dollars in foreign exchange to replace the dung with coal. And if the 350 million tons of manure that are being used as fertilizer were replaced with commercial fertilizers, the expense would be even greater. Roger Revelle of the University of California at San Diego has calculated that 89 percent of the energy used in Indian agriculture (the equivalent of about 140 million tons of coal) is provided by local sources. Even if foreign loans were to provide the money, the capital outlay necessary to replace the Indian cow with tractors and fertilizers for the fields, coal for the fires, and transportation for the family would probably warp international financial institutions for years.

Instead of asking the Indians to learn from the American model of industrial agriculture, American farmers might learn energy conservation from the Indians. Every step in an energy cycle results in a loss of energy to the system. Like a pendulum that slows a bit with each swing, each transfer of energy from sun to plants, plants to animals, and animals to human beings involves energy losses. Some systems are more efficient than others; they provide a higher percentage of the energy inputs in a final, useful form. Seventeen percent of all energy zebus consume is returned in the form of milk, traction, and dung. American cattle raised on Western rangeland return only 4 percent of the energy they consume.

But the American system is improving. Based on techniques pioneered by Indian scientists, as least one commercial firm in the United States is reported to be building plants that will turn manure from cattle feedlots into combustible gas. When organic matter is broken down by anaerobic bacteria, methane gas and carbon dioxide are produced. After the methane is cleansed of the carbon dioxide, it is available for the same purposes as natural gas — cooking, heating, electric generation. The company constructing the biogasification plant plans to sell its product to a gas-supply company, to be piped through the existing distribution system. Schemes similar to this one could make cattle ranches almost independent of utility and gasoline companies, for methane can be used to run trucks, tractors, and cars as well as to supply heat and electricity. The relative energy self-sufficiency that the Indian peasant has achieved is a goal American farmers and industry are now striving for.

Studies like Odend'hal's understate the efficiency of the Indian cow, because dead cows are used for purposes that Hindus prefer not to acknowledge. When a cow dies, an Untouchable, a member of one of the lowest ranking castes in India, is summoned to haul away the carcass. Higher castes consider the body of the dead cow polluting; if they handle it, they must go through a rite of purification.

Untouchables first skin the dead animal and either tan the skin themselves or sell it to a leather factory. In the privacy of their homes, contrary to the teachings of Hinduism, untouchable castes cook the meat and eat it. Indians of all castes rarely acknowledge the existence of these practices to non-Hindus, but most are aware that beefeating takes place. The prohibition against beefeating restricts consumption by the higher castes and helps distribute animal protein to the poorest sectors of the population that otherwise would have no source of these vital nutrients.

Untouchables are not the only Indians who consume beef. Indian Muslims and Christians are under no restriction that forbids them beef, and its consumption is legal in many places. The Indian ban on cow slaughter is state, not national, law and not all states restrict it. In many cities, such as New Delhi, Calcutta, and Bombay, legal slaughterhouses sell beef to retail customers and to restaurants that serve steak.

If the caloric value of beef and the energy costs involved in the manufacture of synthetic leather were included in the estimate of energy, the calculated efficiency of Indian livestock would rise considerably. As well as the system works, experts often claim that its efficiency can be further improved. Alan Heston [et al., 1971], an economist at the University of Pennsylvania, believes that Indians suffer from an overabundance of cows simply because they refuse to slaughter the excess cattle. India could produce at least the same number of oxen and the same quantities of milk and manure with 30 million fewer cows. Heston calculates that only 40 cows are necessary to maintain a population of 100 bulls and oxen. Since India averages 70 cows for every 100 bullocks, the difference, 30 million cows, is expendable.

What Heston fails to note is that sex ratios among cattle in different regions of India vary tremendously, indicating that adjustments in the cow population do take place. Along the Ganges River, one of the holiest shrines of Hinduism, the ratio drops to 47 cows for every 100 male animals. This ratio reflects the preference for dairy buffalo in the irrigated sectors of the Gangetic Plains. In nearby Pakistan, in contrast, where cow slaughter is permitted, the sex ratio is 60 cows to 100 oxen.

Since the sex ratios among cattle differ greatly from region to region and do not even approximate the balance that would be expected if no females were killed, we can assume that some culling of herds does take place; Indians do adjust their religious restrictions to accommodate ecological realities.

They cannot kill a cow but they can tether an old or unhealthy animal until it has starved to death. They cannot slaughter a calf but they can yoke it with a large wooden triangle so that when it nurses it irritates the mother's udder and gets kicked to death. They cannot ship their animals to the slaughterhouse but they can sell them to Muslims, closing their eyes to the fact that the Muslims will take the cattle to the slaughterhouse.

These violations of the prohibition against cattle slaughter strengthen the premise that cow worship is a vital part of Indian culture. The practice arose to prevent the population from consuming the animal on which Indian agriculture depends. During the first millennium B.C., the Ganges Valley became one of the most densely populated regions of the world.

Where previously there had been only scattered villages, many towns and cities arose and peasants farmed every available acre of land. Kingsley Davis, a population expert at the University of California at Berkeley, estimates that by 300 B.C. between 50 million and 100 million people were living in India. The forested Ganges Valley became a windswept semidesert and signs of ecological collapse appeared; droughts and floods became commonplace, erosion took away the rich topsoil, farms shrank as population increased, and domesticated animals became harder and harder to maintain.

It is probable that the elimination of meat eating came about in a slow, practical manner. The farmers who decided not to eat their cows, who saved them for procreation to produce oxen, were the ones who survived the natural disasters. Those who ate beef lost the tools with which to farm. Over a period of centuries, more and more farmers probably avoided beef until an unwritten taboo came into existence.

Only later was the practice codified by the priesthood. While Indian peasants were probably aware of the role of cattle in their society, strong sanctions were necessary to protect zebus from a population faced with starvation. To remove temptation, the flesh of cattle became taboo and the cow became sacred.

The sacredness of the cow is not just an ignorant belief that stands in the way of progress. Like all concepts of the sacred and the profane, this one affects the physical world; it defines the relationships that are important for the maintenance of Indian society.

Indians have the sacred cow, we have the "sacred" car and the "sacred" dog. It would not occur to us to propose the elimination of automobiles and dogs from our society without carefully considering the consequences, and we should not propose the elimination of zebu cattle without first understanding their place in the social order of India.

Human society is neither random nor capricious. The regularities of thought and behavior called culture are the principal mechanisms by which we human beings adapt to the world around us. Practices and beliefs can be rational or irrational, but a society that fails to adapt to its environment is doomed to extinction. Only those societies that draw the necessities of life from their surroundings inherit the earth. The West has much to learn from the great antiquity of Indian civilization, and the sacred cow is an important part of that lesson.

REFERENCES

Gandhi, Mohandas K. 1954. *How to Serve the Cow.* Bombay: Navajivan Publishing House.

Heston, Alan, et al. 1971. "An Approach to the Sacred Cow of India." *Current Anthropology* 12, 191–209.

Odend'hal, Stewart. 1972. "Gross Energetic Efficiency of Indian Cattle in Their Environment." *Journal of Human Ecology* 1, 1–27.

12

Expendable Youth:
The Rise and Fall of Youth Culture

DEENA WEINSTEIN

Cultures are constantly changing and adapting to new ideas and technology. Weinstein follows youth culture through several generations, characterizing each period and the changes that led to the next period.

As you read, consider the following:

1. Is youth culture a reaction to adult culture?
2. What else would you add to this changing picture of youth culture, and what exemplifies youth culture today?

GLOSSARY

youth A socially defined period of transition between childhood and adulthood.

IT WOULD BE A mistake to think that there is some fixed or natural definition of "youth," however convenient it might be to have one. As people continuously interpret their world, even their most crucial terms keep changing their meanings. *Youth* is no exception. As a term in our discourses, it has a history.

If you stop and think about it for a moment, you will probably find out that you are not clear about what you mean by the word *youth*. Ordinarily it has at least three distinct and sometimes divergent meanings: *youth* can mean a biological category defined by age, a distinctive social group, and a cultural construct.

In the biological sense, *youth* refers to an age group of human organisms who are going through a specific process of physical maturation. Cultures may find some ways of marking off this physiological group from others by

codes and disciplines, particularly those pertaining to sexuality, aggression, and work or preparation for work. But the biological stratum doesn't get us very far. We know there is more to youth than hormonal changes.

The social definition of youth is at the center of the modern idea of youth. There are many technologically unsophisticated cultures in which there is no separate youth age grade: one is either a child or an adult. The Jewish boys' rite of passage of the Bar Mitzvah—part of which involves the statement to the 13-year-old boy who is the center of the ceremony, "Today you are a man"—reflected the ancient Hebrews' lack of a youth age grade. (The corresponding ritual for girls is the Bat Mitzvah.)

The vast changes that gave rise to a modern life characterized by specialization, continuous innovation, an increasingly knowledge-based economy, high rates of mobility, economic surplus, and leisure created a socially defined period in the life cycle between the dependency of childhood and the responsibility of adulthood. The social definition of youth and its consequent marginalization as a group betwixt and between (not fully integrated into society) coincided with the industrial era. A socially defined period of transition from being cared for to becoming a provider originated in the upper middle class after the French Revolution and increased in length and spread throughout the population as surplus wealth, specialization requiring lengthy schooling, and the power of labor groups to restrict employment opportunities increased. Youth

Presented at the annual meetings of the American Sociological Association, Pittsburgh, Pennsylvania, August 1992.

as a social group became universal in America after World War II. Since that time, the number of years defining this age grade has increased.

Youth Culture: "Talkin' 'bout My Generation" (The Who)

When society isolates a group of individuals into a category, that group begins to be defined for itself, both by its members and by others. The group develops its own distinctive values, ideals, sentiments, and activities. As the modern period proceeded, "youth" became, by the mid-1950s, a distinctive subculture, with symbols, practices, and folkways peculiarly its own; that is, "youth" became a cultural construct as well as a biological and social category.

The flowering of this subculture in the middle of the twentieth century was related to suburbanization, an extended and universal secondary school system (Coleman, 1961), and a nationwide electronic mass media. "Youth" gained a sharp cultural configuration through its music (rock-and-roll), certain forms of attire, and a set of rituals and activities centered on leisure and entertainment.

The youth subculture was partly created by adolescents themselves and partly contrived by the consumer goods industry. During "the fifties, youth became an isolatable consumer market, with its own capital, its own desires and its own commodities" (Grossberg, 1984, p. 107).

Talcott Parsons coined the term "youth subculture," which for him developed "inverse values to [those of] the adult world of productive work and conformity to routine and responsibility" (Brake, 1985, p. 40). Other investigators indicated that the youth culture as "dress, adornment, music, dancing, and slang has the function of asserting independence from adult authority" (Roe, 1987, pp. 222–223).

Before youth had its own culture, its transitional status was obvious in its imitation of adult style. Holly Brubach describes the ingenue as the almost-grown woman before the full onset of the youth subculture. Played in movies by Audrey Hepburn, the ingenue "sought to give the impression of being experienced beyond [her] years. . . . Fashion was something girls learned by emulating their mothers" (Brubach, 1990, p. 124). By the end of the 1950s, however, young women were wearing clothing distinctively different from, not imitative of, adults.

The central feature of the youth culture, and its metonym, is its music. Rock-and-roll has always symbolized youth: "The whole adolescent milieu is penetrated at many levels by an active interest in music; . . . adolescent discourse centers around the language and terminology of rock and that music provides the core values" (Roe, 1987, p. 215).

Another scholar argues that the core of adolescents' personal identities are their musical tastes. Such preferences, as well as the salience of music as such, serve "to differentiate late-adolescents from adults" (Moffatt, 1989, p. 151). Prinsky and Rosenbaum give a similar assessment of the function of music. They contend that rock music helps teenagers identify with their peers rather than with their families: "It is created by and for young people and may function for teenagers in delineating a rebellious subculture that stands apart from the adult world" (Prinsky and Rosenbaum, 1987, p. 394).

The youth subculture, from its beginnings, was in opposition to, and not merely different from, the general (adult) culture. Markson elaborates on this point, claiming that attempts by family and school to direct and restrain adolescents were contested by them: "Rock emerged as part of the resistance to such disciplinization as a music of opposition to the enforcement of mainstream values" (Markson, 1989, p. 4). Grossberg provides a similar interpretation. He contends that rock's denunciation of the family is not due to an antifemale bias, as some have contended: "A more accurate reading would see it as an attack on the institution itself, as a resistance to the very disciplinization which constructs its youth" (Grossberg, 1984, p. 107).

The rebelliousness of 1950s rock-and-roll is shown by many of its features, from Elvis Presley's sneer to the lyrical complaints about the teenage role found in a host of songs such as Chuck Berry's "School Days" and the Coasters' "Yakety-Yak." Beyond symbolic content, the appeal of black and working-class rock-and-roll performers for middle-class white youth challenged mainstream norms. Rock-and-roll reacted against sexual and racial disciplinizations. Its sexual side was held in check during the decade by the communications industry, which coded sex as "dance" and banned now-innocuous numbers such as Mickey and Sylvia's "Love Is Strange."

The disciplinary forces had less success with the incursion of the black sensibility. Initially songs were whitened from their R&B originals in cover versions by such performers as conventionality's pin-up boy, Pat Boone. But personalities such as Little Richard and Chuck Berry were irresistible. The "percentage of best-selling records by black artists increased from 3 in 1954 to 29 in 1957" (Lipsitz, 1990, p. 126). George Lipsitz (1990, p. 120) concludes, "At the very moment that residential suburbs increased class and racial segregation, young people found 'prestige from below' by celebrating the ethnic and class interactions of the urban street."

The first decade of a distinctive youth culture is bounded by the years 1955–1964, from the rise of Chuck Berry to the ascent of his most successful pupils, the Beatles. In its second decade, youth culture remained music-centered, but transformed opposition to adult culture into self-conscious confrontation. During the first era, youth was a group *in* itself, but in this second era it also became a group *for* itself. "Somewhere in the 1960s, millions of people began to regard themselves as a class separate from mainstream society *by virtue of their youth and the sensibility that youth produced*" (Greenfield, 1987, p. 48).

This second era was marked by an alternative source of rebellious authority to the prestige from below provided by black rock-and-rollers. Blacks had become acceptable to the middle class: the popularity of Martin Luther King and the civil rights movement itself reflected and caused that change. Other evidence of the

change in the cultural image of blacks was in the mass popularity of Motown and Stax musics, created by blacks, and whitened a bit by them. The new affront to bourgeois values was drugs, especially marijuana, which had been used by the 1950s rebels, the beats. The beats were marginals who disdained the mainstream culture.

Beyond their symbolic value, the mood- and mind-altering properties of marijuana and LSD had a rebellious effect. Each privileged the interior monologue over the social conversation, and each privileged the id over the superego directives of conventional society. In the experience of both drugs, clock time was absent; the Bergsonian *durée* was the only time that could matter. Accentuating "the 'now,' and the feeling of freedom to 'walk around and feel the moment,' led to a total breakdown of conventional notions of time. Industrial and job-oriented time is crucially concerned with order" (Willis, 1975, p. 108).

By the late 1960s, the youth subculture was transformed into a counterculture. Still attached to biological and social youth, its way of life was explicitly, self-consciously understood as standing in opposition to mainstream or adult culture. Moreover, it was promoted by the young as a universally good culture, one that should be adopted by adults. "Turn on, tune in, drop out" was the official invitation, the raising of the gap between childhood dependency and adult responsibility into the human ideal. Youth culture as hegemon, in the late 1960s, expressed a loathing for aging. "I hope I die before I get old" was its rallying cry (now ashes in the mouth for more people than merely the originator of the phrase, the rather ancient and partly deaf Pete Townshend of the Who).

The Detachment of Youth Culture from Youth: "Rock'n'Roll Nursing Home" (Iron Prostrate)

The fabled "sixties" ended sometime in the first half of the 1970s, decimated by a multiplicity of events including the killings at Kent State, the end of the military draft, and the OPEC-induced economic recession. The end of the 1960s was also the end of the youth culture as centered on the demographic grouping of young people. Youth culture persisted and youth, in its biological and social dimensions, certainly did not disappear, but the cultural formation of "youth" floated free from the social group of young people. No longer restricted to adolescents, "youth" became available to all. The youth culture got co-opted into the general leisure culture, and lost its moorings in a particular group. It became what postmodernists call a "floating signifier," a designation or identification that could be taken up by anyone as the emblem of a lifestyle. It was "chic" for adults to take up aspects of the youth culture in the late 1960s, but afterward a youthful image, as defined by the leisure culture, became a normalized component of anything else that might be "chic," "trendy," "hip," or "in."

This process is not unique to "youth." Other floating signifiers are constantly being created as symbols are detached from socially relevant groups. One example is "family." The term (or claim to) *family* can be used by baseball fans ("We are family," shouted the Pittsburgh fans the year their team triumphed), by gays who demanded to be listed as dependents in couples' health insurance plans, and unrelated members of communes.

The free-float of youth culture, the detachment of a social group from its set of significant symbols, creates two newly isolated entities, one cultural and the other social. Youth, in the sense of young people in a special biological and social predicament, has become marginal to "youth" as a cultural code of beliefs, values, sentiments, and practices. Youth does not have its own "youth," but instead has the "youth" that the media gives to it and to everyone else. Young people do not have a culture that is theirs; they have become marginal to the idea of youth itself.

Youth culture became a free-floating signifier in the 1970s. Practices associated with the 1960s youth culture such as smoking pot, listening to rock music, and males wearing long hair, began

to characterize adults. Many of these people had adopted these practices when they were young, and did not abandon them when they became adults. "When I became a man, I put away childish things" (1 Corinthians 13:11) was no longer the applicable rule. Youth culture was appropriated by the middle-aged as well as by those biologically classified as children. No longer did one hear about the much ballyhooed generation gap. The music too was disconnected from adolescents; anyone was free to appreciate it.

The reasons for this severance of the biological and social grouping from the cultural are complex. Partly the process is related to the meaning of youth in the contemporary world and partly it is due to the particular content of that cultural formation.

While the old are generally venerated in so-called traditional societies, contemporary society gives the place of honor to youth. In part this is a reflection of the distinction between cultures that are oriented to the past and those that are oriented to the future. The elderly are the embodiment of the past, youth are representatives of the future. The modern era, in whose ashes we now play, was future oriented, had a sense of history rather than myth, and placed stress on the hope of progress. "In the 1950's, 'youth' came to symbolise the most advanced point of social change: youth was employed as a *metaphor* for social change" (Clarke and others, 1975, p. 71).

Youth also came to be valued because of secularization. Christianity's promise to individuals of eternal life was erased as the Nietzschean rumor of the death of God gained wider circulation. Science's view of mortality usurped religion's hope. The body became the alpha and omega of existence when transcendence was no longer imaginable. Youth began to be equated with the immortality of the body.

"Youth" thus became more than a transitional stage; it was an ideal to which all aspired. Athletic and nutritional regimens became popular as means to "staying young" or regaining a "youthful" shape. Medical technology, ever sensitive to consumer demand, made face-lifts, tummy-tucks, and silicone implants popular practices.

By the late 1960s, the negative valuation of the old, the horror of aging, was keenly felt by youth itself. "Don't trust anyone over 30," they repeated — until they reached their own thirtieth birthdays. "I hope I die before I get old" was sung lustily by those who could not imagine in their worst nightmares or acid trips their being no longer young.

Further, as the postmodern era shifted the focus away from a future, as the death, not of God but of history, was proclaimed, the most valued temporal dimension became the present. Past and future are devalued, inconsequential. Youth "had become the symbol of (post-)modernity, of the present, of the NOW, presided over by the 'brats' of the fashion, music, video-clip, and cinema world" (Chambers, 1987, p. 241).

The free-float of youth culture is also due to the content of that culture. In an abstract sense, the two central features of youth culture are its leisure and consumption focus and its opposition to mainstream culture. Both these factors helped make the culture of youth appealing to people of all ages.

The spread of the consumer ethic (as a response to late capitalism's need to replace the more producer-oriented "Protestant Ethic") coincides with the free-float of youth culture. The centrality of work, of one's occupation, to a sense of self has been replaced by one's use of leisure:

> The seeds of an American youth culture can be found in the 1920's "youthful approach to consumption." Gradually this middle-class definition of youthful leisure consumption spread to other class and age groups as well. Style, fashion, and consumption based fads became commonplace concerns. Indeed, an entire segment of the mass media was devoted to supplying the physical artifacts, cultural values, and artistic expressions of the youth culture to wider society. (Dotter, 1987, p. 37)

The oppositional feature of youth culture also coincided with a long-term social trend, interpreted from different perspectives as the "crisis of authority," the rise of mass democracy, and the power of crowds. All refer to the decline of traditional elites. In all sectors of society, political, economic, familial, religious, and academic, those on the bottom resisted the rule and definition of the situation of dominant groups. "Question authority" was a 1960s-era motto. Karl Mannheim termed this trend "leveling" and expounded on it in his "Democratization of Culture" (1971). If youth culture thumbed its collective nose at parental, school, and, in the 1960s, political authorities, its gesture resonated with all those who were demanding their rights against traditional elites. The civil rights and student movements in the 1960s, and subsequently the women's, gay rights, and the disabled persons movements all work in the direction of leveling. They are resistances to what Foucault calls "disciplinization."

Listen to the music of youth cultures. From the 1950s era of Jerry Lee Lewis's "Great Balls of Fire," through such predisco Rolling Stones' songs as "Satisfaction" and "Sympathy for the Devil," the symbolic rebellion is evident. Rock is a resistance to disciplinization, a resistance to enforcement of mainstream values. Compare this attitude to the adult form of popular music, pop, with its emphasis on romance.

Reaction to the Free-Float by Youth: "Desperate Cry" (Sepultura)

Young people have responded to the extortion of "youth" from them in a variety of ways. As the youth culture of the 1960s dissipated into the youthfulness or youthful mystique of the leisure culture in the 1970s, most young people simply followed along, losing any special distinctiveness and merging into the youthful leisure culture as its distinctive representatives. Others, however, could not or would not adjust to the lack of an identity that would set them apart from other groups in the population. They confronted a special problem. The "youth culture" of the 1960s had dissolved and could not be reclaimed, and — even more to the point — neither could a relatively confrontational youth subculture, such as that of the 1950s, be created.

The responses of the young people who became marginal are familiar and define many of the so-called problems of youth today. Some became "burnouts" and "dropouts," retreating into depression, drugs, and sometimes suicide, none of which were of great significance in discourse about "youth" before our postmodern age. The term *deracination*, used by anthropologists to categorize the elimination of the native, autochthonous culture of a people, particularly tribal peoples, can be applied to youth. The deracinated, such as American Indians on reservations or in cities who are bereft of ability to lead meaningful lives and are not able to assimilate into the dominant culture, exhibit similar symptoms worldwide: depression; the widespread use of depressant drugs, especially alcohol; and very high rates of suicide. It is in this context that the suicide statistics of youth, rising sharply since 1970, can be usefully interpreted. During the 1980s, "the suicide rate among 15- to 19-year-olds has increased about 21 percent" (Leland, 1991, p. 53).

Some young people have entered cults and authoritarian sects, which set themselves sharply off from the general culture. Such responses have complex psychological and social determinants, but it is wise to think of them first as culturally specific, as responses of the young to the loss of "youth."

Another response by young people to the usurpation of their culture has been the formation of subcultures, such as "punk" or "metal," which raised the symbolic stakes too high for the general leisure culture to co-opt them. These attempts to re-attach a youth culture to a biological and social group began shortly after youth culture floated free from its social anchor and has continued since then, centered in the realm of rock, the one sphere in which the myth of the youth culture was embedded. Rock also was one of the few cultural spheres in which youth could be a producer, not merely a consumer, of cultural forms.

Punk, which originated as a form of music in New York around 1975, spread as a youth subculture to Britain shortly thereafter. Generally interpreted within an economic framework as working-class dissent, the British punk movement should also be seen as an attempt to rein-state a genuine youth culture. Its enemy was not the rich, not even the bourgeoisie; it was adults: "Punks decried anyone or anything connected with the established social order as boring old farts (BOFs). They regurgitated the impulse behind the mod slogan of the '60s, 'I hope I die before I get old'" (Cashmore, 1987, p. 247). Much of the punk style (mohawks, neon hair colors, safety pins through cheeks) cannot be appropriated by adults, at least by those who prefer to keep their jobs. The U.S. punk scene, centered in Los Angeles around 1980, cannot be understood in economic terms either. There were no dole queue kids there. If anything, the U.S. punk fan's parents tended to be middle class.

Heavy metal, which appeared as a form of music around 1970, achieved subcultural status in Britain by 1974. This subculture too spread to the United States, and to much of the industrial world by the end of the decade. One of the most frequent terms in the lyrics of heavy metal songs, whatever their themes, is night. Night is a time of danger, obscurity, and mystery, when the forces of chaos are strongest. But it is also the time for bacchanalian revelry. Heavy metal's rhetoric and imagery puts forward Dionysian themes and themes of chaos that are related in that both conjure with powers the adult world wishes to keep at bay and to exclude even from symbolic representation. Under the cover of night, everything repressed by the respectable world can come forth. What is that respectable world? For heavy metal's youthful audience, that world is the adult world (Weinstein, 1991).

Heavy metal, like punk, raises the stakes higher than adults can reach; for example, the style of very long hair for males. The focus on power in the subculture is exemplified by an emphasis on extreme sonic volume. Both adults and young children seem to have difficulty withstanding (appreciating?) loudness. This feature is a self-acknowledged gatekeeper, reserving the music for youth only. "If it's too loud, then you're too old" is the often repeated rallying cry.

During the 1980s, there were repeated attempts by demographic youth to recapture a youth culture for themselves. Many of these efforts built on the heavy metal and punk subcultures of the prior decade. Examples include hard core, thrash metal, skate punk, and death metal. All these are sonically, physically, and lyrically too rough for little kids or adults. They are not commercial and receive no radio or MTV exposure. They are only known to cognoscenti and are not easily co-opted by the forces of commerce.

During the latter half of the 1980s, still another recapturing project was underway, which looked to the mythic past of the 1960s. Centering around the Grateful Dead, the music, clothing style (especially tie-dyed T-shirts), and drugs of choice (particularly LSD) of the counterculture found a new audience among young people in the upper middle class. Here the reattachment efforts were not aimed at excluding other demographic groups, but rather at embracing the final supernova of an attached youth culture.

By the early 1990s, a new musical style emerged that has tried to speak to and for youth. Exemplified by groups such as Jane's Addiction and Nirvana, the style has been labeled Alternative. Coming from music that was once exclusive to college radio stations, the last bastion of the 1960s' progressive, free-format FM, these ready-for-prime-time, MTV-friendly bands play music with neopsychedelic elements over a guitar grunge and strong bass guitar and drum rhythm section. During the summer of 1991, an explicit, self-conscious attempt was made to re-invoke the 1960s counterculture. A touring festival, with a variety of Alternative music bands and political and artistic sideshow, called the Lollapalooza Festival, drew an avid audience of middle-class, college-aged youth. The tour was dubbed by the *New York Times*, rather appropriately, "Woodstock for a Lost Generation" (Reynolds, 1991, p. 28).

Alternative music has had much commercial success. The accessible hit song by Nirvana "Smells Like Teen Spirit" has won heavy rotation on a wide variety of radio formats, in addition to MTV. The rapidity of the process of usurpation by commercial culture (witness rap music as used on advertisements and by M. C.

Hammer) may make Nirvana's title more accurate than they might have intended—only smelling like, rather than being, teen spirit.

Adult Reaction to Reattachment Efforts: "Welcome Home (Sanitarium)" (Metallica)

The wide range of attempts by young people to recapture a culture of their own has been met by a variety of adult reactions. The mainstream, commercial rock industry ignored or actively derided the new subcultural musical styles. The worldwide Live Aid concert in 1986 decisively excluded youth-specific musics, most notably, given its popularity at that time, heavy metal. Not until the 1990s did *Rolling Stone* magazine take any interest in metal or rap.

The sense of impending doom—ecological, economic, political, educational, and social— has replaced a sense of progress and of hope for a future world that is better than the current state of affairs. Much of the distinctively youth-based music, especially thrash metal, cogently and emotionally articulates this view. Against the commercial music's message of "Don't Worry, Be Happy," young people belonging to youth subcultures worry a lot.

The reaction against insurgent subcultures has gone as far as efforts to suppress them. Although adults have disapproved of rock music since its introduction in the mid-1950s (Martin and Segrave, 1988), during the 1980s there was a strong, concerted effort to resist it politically. This movement has been aimed not against rock music as such but at the youth-specific musics, particularly metal and rap (Weinstein, 1991, chapter 7). It is no accident that those who testified against heavy metal at U.S. Senate hearings in 1985 were representatives of parental interest groups (such as the PTA), fundamentalist ministers, and physician-owners of psychiatric hospitals specializing in the treatment of adolescents.

Even more ominous than lawsuits against bands, banning concerts, and efforts at censorship was the exceedingly large increase in the number of adolescents sent to mental hospitals. Among others, *Newsweek* concluded that part of this was a response to youth behaving rebelliously:

> The "illness" for which many teenagers are committed is usually not the kind of delusional psychosis or thought disorder commonly associated with institutionalization. Instead, the diagnoses are commonly behavioral problems: "conduct disorder," "oppositional defiant disorder," and the popular "adolescent adjustment reaction." (Darnton, 1989, p. 68)

Rosenbaum and Prinsky (1992, p. 528) examined California hospitals with adolescent care programs:

> When these hospitals were given a hypothetical situation in which the parents' main problem with their child was the music he or she listened to, the clothes he or she wore, and the posters on his or her bedroom wall, 83 percent of the facilities believed the youth needed hospitalization.

Therapy often includes requiring the elimination of the music and its associated sartorial styles. In contrast, actions such as school officials banning heavy metal T-shirts seem innocuous.

The moves to suppress youth subcultures that symbolically revolt against adult authority have had only episodic success. Indeed, these subcultures thrive on efforts to oppose them, because such endeavors vindicate the subcultures' claims to be genuinely countercultural. Not to minimize the threat of censorship, it is still true that First Amendment protections of expression are widely backed by the media and the courts, and help to shelter oppositional subcultures. The cultural war will go on, probably without either side winning the field.

Where does that leave youth? The present situation results from two failures. Firstly, the youth culture of the 1960s, as an ideal and ideology of a specific age grade, was unable to become culturally hegemonic. That is, youth did not become the leading social group. Instead, the signifier "youth" was detached from the age grade and made available to everyone. But, secondly, the free-floating signifier "youth" was unable to displace the cultural expression of young people altogether. Genuine youth subcultures have arisen that exist by marginalizing themselves from the leisure culture's free-floating definition of "youth." Thus, young people are now free to choose between an array of radically confrontational youth subcultures and the commercialized "youth" image.

That is a choice with which many are uncomfortable. They would like to be part of a generation with its own voice, as they suppose 1960s youth were. In the present situation, they can become marginal to the majority of young people by entering an exclusive subculture or they can blend in with pop fashion. Those who choose to reclaim "youth" for youth in oppositional subcultures carry the torch of rebellion, ready to light a conflagration if the circumstances ever permit a younger generation to coalesce again. At the same time, they alienate themselves from the bulk of their age mates. These young people, doubly marginalized, must criticize their own generation as well as the adult world. Youth culture(s) have not disappeared; the price of participation has simply become socially and emotionally higher. Youth is now the province and achievement of select groups of marginalized young people, who must actively establish their claim to it through cultural struggle.

REFERENCES

Brake, Michael, 1985. *Comparative Youth Culture: The Sociology of Youth Cultures and Youth Subcultures in America, Britain and Canada.* London: Routledge & Kegan Paul.

Brubach, Holly, 1990. In fashion: A certain age. *New Yorker* (November 5), pp. 122–128.

Cashmore, E. Ellis, 1987. Shades of black, shades of white. In James Lull (ed.), *Popular Music and Communication.* Newbury Park, CA: Sage.

Chambers, Iain, 1987. British pop: Some tracks from the other side of the record. In James Lull (ed.), *Popular Music and Communication.* Newbury Park, CA: Sage.

Clarke, John, Stuart Hall, Tony Jefferson, and Brian Roberts, 1975. Subcultures, cultures and class: A theoretical overview. In Stuart Hall and Tony Jefferson (eds.), *Resistance through Rituals: Youth Subcultures in Post-war Britain*. London: Hutchinson.

Coleman, James S., 1961. *The Adolescent Society*. New York: Free Press.

Darnton, Nina, 1989. Committed youth: Why are so many teens being locked up in private mental hospitals? *Newsweek* 14, no. 5 (July 31): 66–72.

Dotter, Daniel, 1987. Growing up is hard to do: Rock and roll performers as cultural heroes. *Sociological Spectrum*, pp. 25–44.

Greenfield, Jeff, 1987. They changed rock, which changed the culture, which changed us. In Janet Podell (ed.), *Rock Music in America*. New York: Wilson.

Grossberg, Lawrence, 1984. I'd rather feel bad than not feel anything at all: Rock and roll, pleasure and power. *Enclitic* 3, No. 1–2 (Spring–Fall): 94–111.

Lipsitz, George, 1990. *Time Passages: Collective Memory and American Popular Culture*. Minneapolis: University of Minnesota Press.

Mannheim, Karl, 1971. The democratization of culture. In Kurt H. Wolff (ed.), *From Karl Mannheim*. New York: Oxford University Press.

Markson, Stephen L., 1989. Claims-making, quasi-theories and the social construction of the rock and roll. Paper presented at the American Sociological Association.

Martin, Linda, and Kerry Segrave, 1988. *Anti-rock: The Opposition to Rock 'n' Roll*. Hamden, CT: Archon Book.

Moffatt, Michael, 1989. *Coming of Age in New Jersey: College and American Culture*. New Brunswick, NJ: Rutgers University Press.

Prinsky, Lorraine E., and Jill Leslie Rosenbaum, 1987. "Leer-ics" or lyrics: Teenage impressions of rock 'n' roll. *Youth and Society* 18, no. 4 (June): 384.

Reynolds, Simon, 1991. Woodstock for the lost generation. *New York Times* (August 4), section H, p. 22.

Roe, Keith, 1987. The school and music in adolescent socialization. In J. Lull (ed.), *Popular Music and Communication*. Newbury Park, CA: Sage.

Rosenbaum, Jill L., and Lorraine Prinsky, 1991. The presumption of influence: Responses of popular music subcultures. *Crime & Delinquency* 37 (4): 528–535.

Weinstein, Deena, 1991. *Heavy Metal: A Cultural Sociology*. New York: Macmillan/Lexington.

Willis, Paul E., 1975. The cultural meaning of drug use. In Stuart Hall and Tony Jefferson (eds.), *Resistance Through Rituals: Youth Subcultures in Post-war Britain*. London: Hutchinson.

13

Domestic Violence Against Women: A Contemporary Issue in International Health

RUTH L. FISHBACH AND ELIZABETH DONNELLY

This reading is disturbing! It is hard to imagine the hardships—including discrimination, war, torture, slavery, mutilation, and genocide—suffered by many people in the world. These practices illustrate the negative side of our culture—what can happen as a result of cultural beliefs and practices.

As you read this article, consider the following:

1. Why do human beings engage in practices that hurt members of their own cultural group or other groups?

2. Why are women often at the receiving end of violence?

GLOSSARY

Domestic violence Violence occurring in the home against family members, usually the wife.

Violations of human rights Actions that threaten the security of individuals and their fundamental rights to life, liberty, and freedom from fear and want.

Soncha (Mexico), pregnant with her third child, resists her husband's sexual advances, but he forces her to have sex, violently and against her will. He reminds her that she is his property so he has the right to enjoy her as he desires. Equally as devastating as his slaps and kicks are Jose's abusive words, which leave Soncha feeling demoralized and hopeless, with a pain in her heart that throbs long after the bruises fade.

Ajita (India, Pakistan) comes from a relatively poor family that has sacrificed greatly to provide a dowry payment for their daughter's husband. Soon after her marriage, Ajita dies an unnatural death in what the police file as "a kitchen accident"; her husband reports that a kerosene stove exploded while she was cooking. Ajita's is the third such "accidental" death of a new bride in 5 months in this village.

Fatou (Somalia) is 12 and her family is preparing her for marriage, the only status position available to women in this northeast African society. The central component of her rite of passage involves circumcision; a procedure in which a traditional midwife and elder female kin remove her clitoris and other parts of her external genitalia. The surgery helps ensure her premarital purity, an absolute prerequisite for marriage and social identity.

As Jing Wang (China) nears marriage age, she dreads her inevitable, arranged marriage to Rui Xiao. After an unsuccessful attempt to escape with the man she wishes to marry, the reluctant bride is held captive and beaten by her prospective husband and father-in-law, who feel humiliation at her rejection. After several days of torment, Jing Wang returns to her unsympathetic family, who resent her renunciation of traditional values and customs. Later that week, Jing Wang is found dead, an apparent suicide.

VIOLENCE AGAINST WOMEN IS manifest and legitimized in many ways within and across cultures. Gender-based violence has only recently emerged as a crucial global issue that cuts across regional, social, cultural, and economic boundaries. As the data accumulate, the pervasiveness of gender-based violence and its impact on morbidity and mortality are becoming alarmingly clear. Indeed, gender-based violence is a nearly universal phenomenon threatening the health and freedom

In Subedi and Gallagher. Society, Health, and Disease. *Prentice-Hall, Inc. 1996, pp. 316–345 (excerpts).*

of women in the streets, in the workplace, and, most troublesome, in the home.

Domestic violence must be viewed as a public health problem that directly damages women's health and well-being. The *World Development Report 1993* indicates that in 1990, domestic violence and rape caused an estimated 5% of the global health burden for women aged 15 to 44 (World Bank 1993). Hence, violence and rape account for 5% of the total years of healthy life lost. Because of underreporting, the magnitude of the global health burden from domestic violence and rape is almost certainly greater than current estimates (World Bank 1993).

Domestic violence must also be viewed as a human rights issue, one that threatens the security of women and their fundamental right to life and liberty as well as freedom from fear and want, as described in the *Universal Declaration of Human Rights* (UN 1948). Although extreme poverty, deprivation, and social and economic oppression increase the potential for the violation of human rights (Jilani 1992), these abuses can occur in any society where disadvantaged groups are oppressed by political and social forces.

This chapter focuses on violence in developing or low-income countries. These countries are dynamic, with rising populations and evolving economies. Much information needs to be exchanged. Only within the last few years have researchers and clinicians begun to compile data on the prevalence, nature, and consequences of domestic violence. Despite the disturbing findings, in countries where infectious disease, malnutrition, or war pose critical challenges to survival, protection from violence in the home has been given a low priority on the health and human rights agenda. Yet, a deeper appreciation of the nature and consequences of domestic violence would reveal its association with myriad serious health and social problems, including depression, chronic pain, substance abuse and dependence, suicide, child abuse, and homelessness. There are compelling reasons why the issue of domestic violence in the developing world is in urgent need of attention and why this issue should be a high-priority item on the health and human rights agenda:

1. Even with the paucity of systematic documentation, the magnitude of reported incidents of domestic violence throughout the developing world constitutes a significant cause of physical morbidity and mortality. And these health risks affect not only women but future generations; battered women run twice the risk of miscarriage and four times the risk of having a below-average-weight baby (World Bank 1993). One of the most effective means to reduce congenital birth defects and infant mortality is to ensure the safety of pregnant women and mothers.

2. In countries where more extensive data do exist, there are indications of a direct association between domestic violence and psychiatric morbidity. Research from the United States reveals, for example, that "battered women are four to five times more likely to require psychiatric treatment and five times more likely to attempt suicide" (Stark and Flitcraft 1991). Ethnographic data from Oceania, South America, and China provide further evidence that wife-beating is widespread and is associated with depression and suicide (Counts 1987, 1990a, 1990b; Gilmartin 1990).

3. Most violations against women are perpetrated by someone the woman knows and most often in the "privacy" of the home. Historically, in most societies, the criminal justice system has relegated authority in domestic affairs to men as heads of households. Similarly, the human rights community has failed to acknowledge the gender-specific ways in which the rights and dignity of women are violated. Too often, systematic violence and abuse are tolerated internationally as cultural patterns. These are just two examples of the systemic devaluation of women's worth.

4. Domestic violence against women is a global issue requiring global awareness and action. Despite cultural variations in the manifestations, the underlying factors that promote and perpetuate violence are remarkably similar. A cross-cultural perspective allows us to

learn not only *about* violence against women in other regions but to learn *from* the experiences of people working to define and combat this problem in their communities. These examples provide the direction and guidance needed to respond to this global problem. . . .

Lexicon for Domestic Violence

Abuse and violence take many forms, and to date there is little universal agreement on the terminology to describe the experiences of women. Often the terms used become ideological and highly politicized. Recognizing that "violence" may be defined according to particular cultural understandings (Counts 1990a), we consider violence to be "an act carried out with the intention, or perceived intention, of physically hurting another person" (Straus et al. 1980, p. 20). Other definitions focus on power and control, such as that offered by Heise (1993c): Violence against women includes "any act of force or coercion that gravely jeopardizes the life, body, psychological integrity or freedom of women." Abuse includes behaviors that cause harm without the inclusion of physical force, encompassing neglect, psychological and emotional harm, and nonviolent sexual harm. We use the term "domestic violence" in this chapter to encompass harmful behaviors by intimates directed toward females assuming the role of wife in household relationships, with or without the legal sanction of marriage. We broaden the definition to include behaviors that occur while preparing for marriage. We also focus on women, as they are the victims/survivors in more than 85% of cases of domestic violence (Koss 1990). . . .

Health care professionals and researchers are developing measures to assess the impact of abuse. According to Heise (1993a), once a woman is hit, even one time, the "effectiveness" and fear associated with threats and psychological abuse increase. Thus, women who are assaulted only rarely can suffer long-term effects from the memory of the original incident and ongoing verbal and emotional abuse. The physical and mental health consequences rise dramatically when women are subjected to repeated violence.

Epidemiology of Domestic Violence in Low-Income Countries

. . . In reviewing cross-cultural data from studies documenting domestic violence, several cautions are advised. First, prevalences of domestic violence may not be directly comparable because investigators are not consistent in the probes used to identify abuse or the nominal and operational definitions of their dependent variables (e.g., violence "over a lifetime" or "in the last year"). Furthermore, a higher rate in a particular culture does not necessarily mean that something intrinsic to that culture is conducive to domestic violence. Rather, it may be that rates in other countries appear low owing to poor ascertainment. In addition, in contrast to the perception that relying on women's self-reports inflates statistics, clinical and research experience suggest that the reverse occurs: Women *underestimate* the level of physical and psychological violence they experience in relationships. Likely explanations for this are (1) women in many cultures are socialized to accept physical and emotional chastisement as part of a husband's marital prerogative, thereby limiting the range of behavior they identify as abusive (Counts 1990a) and (2) women are averse to report abuse out of shame or reluctance to incriminate family members for fear of retaliation (Heise 1993a). Either of these factors, as well as others, may contribute to estimates lacking complete accuracy.

Data on domestic violence have been accruing for several years in the United States, and they can be used as a measure for comparison when examining the prevalence of violence against women on an international scale. These data indicate that for women in the United States, the family is a violent institution (Koss 1990). In the United States, a woman is more likely to be assaulted, raped, or killed by a male partner than by any other assailant (Browne and Williams 1989). Population-based surveys suggest that between 21% and 30% of American women are beaten by a partner at least once in their lives (Kilpatrick et al. 1985; Koss 1990). Based on national probability samples from the United States, 47% of the husbands who beat their wives do so more than three times per year (Straus et al. 1980). Assaults by husbands, ex-husbands, and lovers cause more injuries to women than motor vehicle accidents, rape, and muggings combined (Rosenberg et al. 1986; Stark and Flitcraft 1991). Studies estimate that 22% to 35% of women seeking emergency treatment do so for symptoms related to abuse (Randall 1990) and, because abuse is an ongoing cycle producing increasingly severe injuries over time, battered women are likely to see physicians frequently. One study reported that nearly one in five battered women had seen a physician at least 11 times for trauma, and another 23% had seen a physician 6 to 10 times for abuse-related injuries (Stark and Flitcraft 1979). Health officials estimate that each year more than 4 million women are battered and more than 4,000 are killed by such "intimate assaults" (Royner 1991). . . . .

Experiences Faced by Many Women

Among the many forms of domestic or gender-based violence, we examine three closely in order to illustrate the complex cultural patterns and values that promote and perpetuate the threat to women's lives and welfare. Female circumcision, marital rape, and dowry-related deaths are currently affecting women in epidemic proportions.

SOCIAL SURGERY: FEMALE
CIRCUMCISION AND INFIBULATION

It is only a decade since women in Africa and the Middle East first dared to speak publicly about a procedure known as female circumcision, the cultural practice of surgically altering female genitalia. There has been little reduction in its occurrence during that time. Three procedures exist: The Sunna type (excision) consists of removal of the tip of the prepuce of the clitoris.

The simplest, it is considered the type recommended by Islam, yet is the least practiced form of circumcision (El-Dareer 1982). Clitorectomy consists of removal of all of the clitoris. Infibulation, the most extreme form of circumcision (includes the excising of the clitoris, the labia minora, and the inner wall of the labia majora, and the suturing together of the two sides of the vulva except for a small opening for the passage of urine and menstrual blood), is practiced on nine out of ten girls in Sudan and Somalia (Seager and Olson 1986). The surgery may be performed on girls as young as a few days old or in late adolescence. Rarely are surgical instruments or anesthesia used. Gruenbaum (1988) described the operation itself as performed by a woman behind closed doors with only women taking part; men are completely excluded. Moreover, it is the women who strongly defend the propagation of this practice, not surprisingly, because they derive much of their social status and economic security from their roles as wives and mothers.

The derivation of this practice is difficult to trace (WHO 1993). Although it is generally believed that female circumcision was originally an attempt to ensure chaste or monogamous behavior or as a means of suppressing female sexuality, some believe that it served to protect adolescent girls against rape. Moslems mistakenly believe that female circumcision is demanded by the Islamic faith, although it has no basis in the Koran (Heise 1993c). Significantly, the WHO closely links this practice with poverty, illiteracy, and low status of women. In such settings, an uncircumcised woman is stigmatized and not sought in marriage, which helps explain the paradox mentioned earlier that circumcised women are among its strongest proponents, along with the traditional midwives who perform the act.

Hicks (1993) offers a series of "indigenous" explanations to justify infibulation: (a) it has religious significance; (b) it preserves virginity until the time of marriage; (c) it is hygienic and purifying; (d) it curbs the (potentially) excessive sexuality of females; (e) it promotes fertility; and

(f) it maintains general body health (p. 13). Both clitoral excision and infibulation are seen to relate to Islamic religious tradition and the absolute prerequisite of virginity for a bride-to-be. Indeed, very few men would marry a girl who has not been excised and infibulated, which precludes women's choice in the matter (Forni 1980, cited in Hicks 1993).

Health Consequences of Female Genital Surgery
Female circumcision—especially infibulation—is a major health problem in Africa and the Middle East. According to the WHO (1993), it is believed that more than 84 million females in more than 30 countries have been subject to female genital mutilation. Health consequences are considerable, including both acute problems of hemorrhage, shock, and tetanus and long-term problems such as pelvic infections leading to infertility, vesiculovaginal fistula, HIV infection from contaminated instruments, chronic urinary tract infections, cysts and abscesses, keloid and severe scar formation, lack of sexual response, and often death (e.g., 10% to 30% estimated mortality of young Sudanese girls) (Heise 1993c). In women who have been infibulated, a cut must be made to enlarge the vaginal opening immediately before marriage and before childbirth. During childbirth, the risk of maternal death is doubled, and the risk of a stillbirth increases severalfold. The long-term morbidity becomes cumulative and chronic. In one country alone, the Sudan, 30% of all Sudanese women have circumcision-related complications, and that figure rises to 83% in Sierra Leone (World Bank 1993). Maternal and infant mortality rates are particularly high where the incidence of circumcision is high (Seager and Olson 1986).

There are two perspectives on the impact of circumcision and infibulation on women. One position argues that while the *medical* consequences are obvious, genital mutilation, in addition, has staggering *mental health* consequences for women. The extent and degree of sexual and mental health problems can only be guessed at, however, because circumcised women are often hesitant to discuss the subject that both means

little to them and is embarrassing—that is, their sexuality (Seager and Olson 1986). The other position, articulated by Hicks (1993), contends that after the initial operation, the issue is closed (literally and figuratively speaking). Women do not even correlate their subsequent physical discomfort, pain, and related gynecologic and obstetric problems with having been circumcised. Such physical problems are perceived as being the common lot of women. In short, infibulation is not considered a problem but a badge of merit and identification. It accords girls the right to marriage and the protection and the status this union provides. Individual social identity is based on being infibulated. Women's collective social identity is based on *all* women being infibulated. As long as the tradition holds that men will not marry uncircumscribed women, considered to be promiscuous, unclean, and sexually untrustworthy (Heise et al. 1989), the practice will be perpetuated.

Women are subjected to other gender-related, culturally determined medical or surgical procedures. Forced sterilization and forced abortion (described by Clare 1992) are examples of government policies or cultural beliefs that condone abuses or violence against women.

In addition, certain nonsurgical/medical procedures that are gender-related have been practiced widely. The custom of footbinding ended under the Nationalists in the late 1930s in some areas of China and was outlawed all over China after 1949. Neck elongation in parts of Africa, as well as breast augmentation in the United States, are examples of women subjecting themselves or being subjected to painful and potentially harmful practices to conform to what is considered by local custom to be desirable to their men.

RAPE

Rape is a complex experience for women, as its impact is contingent on the circumstances surrounding the rape, the relationship to the perpetrator, and the cultural meaning attached to rape. It may occur in the form of an assault outside the home by a stranger, as an aftermath of war where the victor gets the spoils, and even within the margins of the marital relationship. The impact of rape on the woman is often additive. The victim of rape is victimized twice (at least)—first by the rapist, physically and emotionally, and then by the consequent stigmatization and rejection by the family, community, and culture. All too frequently, it is the victim who is blamed for the attack.

Rape is probably the most underreported, fastest growing, and least convicted crime in the world. Particularly in low-income countries, few agencies provide services as well as data from which to estimate the prevalence of rape. It has long been considered as a crime only against property—the woman as man's property. "Now rape is beginning to be recognized for what it is: a crime of violence and power, and a violation of women's civil rights" (Seager and Olson 1986).

The Latin root of rape means "theft," and most cultural responses to such violence emphasize reclaiming the woman's lost value, not prosecuting the offender. This helps explain why rape is either seen as a man's prerogative or as a crime against the honor of a woman's family or husband, rather than a violation of the woman (Heise 1993c). As Heise described (1993b), in parts of Asia and the Middle East, the stain of rape is so great that victims are sometimes killed by family members to cleanse the family honor. According to Mollica and Son, in Southeast Asian cultures, a husband often rejects his wife if she has been sexually violated because she is perceived as having been "used" or "left over" by the rapist(s) (Mollica and Son 1989, cited in Heise 1993b). The consequences of rape in societies where a young woman's worth is equated with her virginity are ruinous.

Marital Rape There is growing acknowledgment of the extent of marital rape as more women are able to report and discuss its occurrence. Researchers still find, however, that most women are not able to address sexual abuse in their marriage because of the deep shame they experience. Heise's report on rape (1993b) includes the following empirical data: In the United States, whereas 14% of all wives report being raped, the

prevalence among battered wives is at least 40%; in severely abusive relationships, forcible rape may occur as often as several times a month. Heise (1993b) also noted that in Bolivia and Puerto Rico, 58% of battered wives report being sexually assaulted by their partner, and in Colombia, the reported rate is 46%. Reports from the Philippines and Guatemala indicate that forced sex with husbands is a common experience.

Medical Consequences of Rape The medical consequences of rape, including physical trauma and risk of pregnancy, sexually transmitted diseases (STDs), and AIDS, are considerable. Reports from a Rape Crisis Center in Bangkok, Thailand, indicate that 10% of its clients contract STDs as a result of rape and 15% to 18% become pregnant, a figure consistent with data from Mexico and Korea. Where abortion is illegal or restricted, death and infertility are often the sequelae of the resorted to *illegal* abortion (World Bank 1993). Additionally, as discussed below, rape may be a risk factor for the development of risk-taking behaviors (e.g., substance abuse). In cultures where the stigma of rape and subsequent victim blaming are especially powerful, the risk of suicide (and even homicide) rises dramatically.

DOWRY-RELATED DEATHS

A deadly form of domestic violence has been occurring on the Indian subcontinent. "Dowry deaths" or "bride burning" has been reported with alarming frequency amid speculation that the exact magnitude of the mortality statistics remains unknown. This form of violence, visible and sensational, is seen by activists as the extreme symbol of a society where women are considered expendable (Jilani 1992).

By long-established custom, dowry referred to the wealth given to a woman by her family, which she subsequently brought to her husband upon her marriage. This was a Hindu practice that enabled parents to pass on family assets to their daughters, who were not allowed by law or custom to inherit property. The tradition has been transformed into a crucial premarital negotiation and now refers to the money or goods that the bride's parents agree to present to the groom and his family as a condition of the marriage settlement. No longer a gesture of love and devotion, increasingly dowry is seen as a "get rich quick" scheme by prospective husbands (Heise 1993c). Typically the bride is exploited by the husband and his family, who make ongoing demands for her wealth, demands that her parents may not be able to meet. If the demands go unmet, with increasing frequency the bride may be severely abused to a point that culminates in her unnatural death. With her death, the husband (widower) is free to pursue a more profitable marriage.

The most common form of dowry death is by burning, usually by dousing the woman with kerosene and setting her on fire. The murder is disguised, claimed by the husband to be the result of a "kitchen accident" caused by the bursting of a stove used for cooking. Investigations are rarely conducted, as police are loathe to get involved.

Because dowry deaths are notoriously undercounted (Heise 1993c), exactly how many women die as a result of this form of domestic violence can be inferred only from data that are known. In all of India, 999 deaths were registered as dowry-related deaths in 1985; as of 1991, that number had increased to 5,157 registered deaths (Kelkar 1991). Yet the Ahmedabad Women's Action Group estimates that 1,000 women may be burned alive each year in Gujarat State alone (Heise et al. 1989). Given that in two of India's cities, one of every four deaths among women aged 15 to 24 is reported to be due to "accidental burns" (Karkal 1985), one can only hypothesize what the true mortality figures might be. . . .

Research Initiatives

Using a cross-cultural perspective, we have described various facets of domestic violence, a significant social problem that appears to affect the mental and physical health of a substantial number of women around the world. Because of the complexity of this issue and the paucity of available data, the need for further research is obvious and urgent. The challenges to implement appropriate unbiased studies are great. Direct and indirect causal relationships must be postulated, and attention to multiple causal pathways must be considered (Newberger et al. 1992). Statistical data are needed from every region to document the magnitude and impact of the problem, as well as ethnographic data to enhance our understanding of domestic violence as a social process in different local contexts.

WHO IS THE VICTIM?

Our principal needs are for epidemiologic data at the descriptive level which are prospective as well as retrospective, taken from both communities and treatment facilities. Basic sociodemographic characteristics including age, marital and family status, socioeconomic and education levels, and medical and psychiatric histories should be linked with culture-specific stressful life events (e.g., barrenness, failure to have the right gender or number of children, forced marriages, social surgical procedures) and histories of violence or abuse. Men as well as women should be assessed to elicit gender-associated differences. These data should be acquired through the use of standardized research designs, instruments, and indices. An immediate need is to formulate a standard definition of domestic violence that is acceptable to varied types of researchers so that it can be applied across cultures.

Studies on domestic violence have understandably tended to concentrate on victims who have sought help and on batterers who have sought help, have been incarcerated, or have been remanded to treatment programs. To avoid Berkson's (selection) bias, we need studies of random or probability samples drawn from the community rather than the usual convenience samples recruited from groups seeking help for psychiatric or other problems or those reported to the police. Data from all social strata are essential because domestic violence is socially ubiquitous. Ultimately, both qualitative (including ethnographic) and quantitative data should be gathered.

WHO IS THE PERPETRATOR?

We stress the need for balanced research that includes the perpetrator as well as the victim. The causes and consequences of violence against women are multifactorial in nature, requiring a broad research agenda that examines the issue from varied perspectives. Basic sociodemographic data are also essential here, with links made to early experiences with violence. Studies in the United States suggest that at least 60% of men who batter grew up in homes where they were beaten, sexually abused, or witnessed parental battering, resulting in an ominous cycle of violence. Neglecting to understand the dynamics of interpersonal relations and conflict resolution between intimate partners may perpetuate "band-aid solutions" to the problem and sustain the emphasis on the victim as deviant. Further at the core of the problem lie the cultural values, mores, rights, and practices that condone physical violence as a means of conflict resolution within intimate relationships. Direct assessment of perpetrators (who in a small proportion may be women) is essential. Recruiting an adequate sample undoubtedly poses an enormous challenge.

Research is needed also to better understand domestic violence in homosexual relationships. The accumulating body of research indicates, for example, that approximately 25% to 30% of lesbian relationships have experienced abuse (Jim Shattuck, personal communication).

INVESTIGATING THE ASSOCIATION BETWEEN ABUSE AND MENTAL DISORDERS

Gender- and culture-sensitive mental health research remain neglected areas of investigation. We urge intensive examination of the association between domestic violence against women and the women's mental health disorders. Protocols need to be developed that can be applied cross-culturally to document prevalences of violence and subsequent mental health disorders so that temporal relationships essential for a causal model can be generated. Researchers must be cognizant

of particular concerns of women across cultures and cultural variations in interpretation of violence. For example, is a husband's chastisement of his wife categorized as abuse? In certain settings, focus groups have been an effective means to elicit qualitative information in a culturally sensitive way. In other settings, however, a woman might be reluctant to speak openly in front of a group of women from her neighborhood or community for fear of becoming the focus of local gossip. Survey data alone may fail to capture subtle nuances that reveal both the true impact and the cultural variations of domestic violence in the same way that a medical examination without explicit probing by a sensitized physician fails to detect a history of violence and abuse underlying myriad symptoms.

Ethical Considerations

The fundamental ethical consideration when conducting research in the area of domestic violence is to avoid exacerbating violence against women while attempting to understand it or to reduce its occurrence. *Primum no nocere*— above all, do no harm. Protecting the confidentiality of sources of the data and avoiding the medicalization of social problems experienced by women present significant challenges. By treating social problems as medical or psychiatric disorders, the sufferers lose the context of the experience, the suffering is delegitimated, and society is precluded from responding appropriately (Kleinman and Kleinman 1985).

Sigler (1989) and others have offered caveats about the link between activism and research in which the research is designed to provide support for particular activist positions. Findings may be misleading or slanted, affording only partial understanding that is shaped by activist perspectives. Those who attempt to dramatize the scope and magnitude of violence, for example, can inflate the results by defining domestic violence, broadly stating, for example, that a single act over a lifetime of marriage constitutes domestic violence. Because the results of a research project are likely to have an effect on the way in which people live and the extent to which they are labeled deviant, sensitivity is required to present data objectively, without ethnocentrism, and with respect and appreciation for the culture being studied.

Policy Implications

Domestic violence against women destroys women's physical and mental health, diminishes their dignity, and violates their fundamental human rights. This is a problem deeply embedded in the social, cultural, religious, and economic values and mores of a community. In this chapter we have attempted to emphasize the variation and pluralism in the forms and consequences of domestic violence. As local context gives shape and meaning to domestic violence and its ramifications, so too will local context shape appropriate strategies of intervention. No one approach will or should work in every setting.

We must be creative in our interventions and policy initiatives. While recognizing the need for variation in solutions, ultimately the need is for an integrative approach that brings together grassroots organizations and national and international agencies to achieve sustainable social change (World Bank 1993). This integrative approach should include the legal and medical professions, governments, the media, and public education institutions. Accepting that violence against women is a learned behavior that can be changed is a progressive step (Carrillo 1991).

Domestic violence must not be seen as "just a women's issue." Although it affects women primarily, impairment in the woman's physical, mental, and emotional capacity to function adequately resulting from exposure to violence ultimately impairs the care she can give to her family. Consequently the harm spreads to include other family members, particularly her children. Furthermore, the strong correlation between domestic violence and child abuse and mortality suggests that the best protection for a child's well-being is protecting the mother from becoming a victim of domestic violence.

Common to societies in which men exhibit violence against intimate partners is the devalued status of women. All policies that seek to alleviate domestic violence must address the social position of women. Equality of women is, in fact, essential to peace, not only between intimate partners but in the workplace, the political realm, and society in general. In their *Statement on Peace,* the Baha'i Peace Council of Canada eloquently expresses the significance of "equality between the sexes":

> The emancipation of women, the achievement of full equality between the sexes, is one of the most important, though less acknowledged prerequisites of peace. The denial of such equality perpetuates an injustice against one half of the world's population and promotes in men harmful attitudes and habits that are carried from he family to the workplace, to political life, and ultimately to international relations. There are no grounds, moral, practical, or biological, upon which such denial can be justified. Only as women are welcomed into full partnership in all fields of human endeavor will the moral and psychological climate be created in which international peace can emerge (1986, p. 8).

Domestic violence must no longer be a private affair in which public agencies cannot and do not intervene. Governments must be held accountable for the violence to which their citizens are subjected. They must acknowledge the specific experiences of women in order to provide appropriate protection to this half of their citizenry.

In most countries, women have very little legal or actual protection from abusive husbands. Traditionally, laws in most societies have denied women equal protection and have often relegated that responsibility to men as heads of households. Men's violence against "their" women is accepted (or ignored) across a wide spectrum of cultures. Activists are now pushing to have domestic violence recognized as a crime distinct from general assault laws, and for police to have powers of arrest and intervention in cases of domestic assault (Seager and Olson 1986).

The rape of a man's wife is considered a serious crime only because of the longstanding legal tradition of wife as property; for this reason wife-rape is rarely prosecutable. How can you violate the rights of someone who has no independent civil rights? (Seager and Olson 1986). In the United States, as recently as the 1970s, husbands had conjugal rights to consummate the marriage which their women did not have the right to refuse (Sigler 1989).

Legal reform is an area where women in a number of societies have succeeded in effecting change, both to extend their rights and to ensure protection of their rights. In Puerto Rico, for example, the Law for the Prevention of Domestic Violence was implemented in 1990, recognizing domestic violence as a felony. After only 7 months, 936 men were arrested on charges of domestic violence (*Women's World* 1991–2). Legal reform alone is inadequate, however, in societies where women have never experienced participation in public affairs. Educating the public about nuances of legal reform (e.g., equal rights to protection) must accompany such change.

Universal education is a key factor in raising women's status and thereby providing them with greater options and with the tools for self-reliance and access to positions of influence in society. Where social resources are limited, however, education of girls is afforded a low priority. Nonetheless, investment in the education of girls may well be the highest-return investment in the developing world (Summers 1993), as women with even an elementary education raise the standard of living in a poor country. Furthermore, studies have repeatedly shown that educating mothers is the single most effective way to reduce child mortality because it mitigates the woman's fatalism, improves her self-confidence, and changes the power balance within the family (Heise et al. 1989). In nearly every society, the medical, legal, and public professions remain male-dominated. Equal access to higher education will facilitate women's entry into such fields, where they will have a stronger voice for the concerns of women in the population they serve.

Education must extend to professionals as well. Raising awareness of police, lawyers, and judges to the magnitude of the problem and the unique needs of abused women is necessary if the law is to be an arm by which women are protected.

Health care professionals are in the best position to identify, counsel, and empower women experiencing domestic violence because the health care system represents the one institution with which almost every woman interacts during her lifetime. But most health care professionals are ill-prepared to provide assistance and need training in effective intervention strategies for caring for abused and battered women. Strategies, taken from Sugg and Inui (1992) and Star and colleagues (1981), include educating professionals to know how to do the following:

> Assess the immediate crisis; take an in-depth history, especially inquiring about abuse and violence in the home.
> Refer the woman to appropriate resources, including medical assistance, financial and legal aid, emergency shelters, and other alternatives to consider.
> Help the woman increase her self-esteem; deal with the battered woman's ambivalence about whether or not to leave the abusive spouse.
> Establish a trust relationship with the woman which provides ongoing emotional support, security, practical assistance, and empowerment options.

Viewing domestic violence in the contexts of public health and human rights exposes the pervasive and systematic exclusion of women from public policy. David Christiani links public health and human rights in the following way:

> An issue becomes a public health issue when an individual disease is shown to be more prevalent in a group or among groups of people for social, cultural, or economic reasons—for systemic reasons. A public health issue, in turn, becomes a human rights issue when basic rights to existence are systematically denied to a group or groups of people. Human rights abuses involve issues of coercion, of profound inequality . . . inequalities in the ability to control basic

conditions of daily life. When groups of people are denied control of their basic living conditions, and their health and even existence are placed in jeopardy, a public health issue expands to become a human rights issue (Christiani, quoted in Klinger 1993, p. 11).

Women must have a role in determining the policies and programs that affect them (Rodriguez-Trias 1992). Before any strategy to combat certain culturally determined behaviors is initiated, attempts to understand the nature and function of the behaviors in their social context are mandatory in order to determine the appropriateness and feasibility of the intended strategy. Sidel (1993) noted the evolution in thinking from "helping the victim" to "blaming the victim" to the current "organizing the victim." Whereas helping the victim promotes dependence and blaming the victim draws attention away from the societal causes of illness, organizing the victim gives attention to those who have traditionally been socially and economically vulnerable and the disenfranchised. Organizing increases the likelihood of the survivors gaining control over their lives and wielding power to make changes that ultimately could lead to improvements in their sense of dignity, their health, and their well-being.

Lessons from the Field

Although estimates of the prevalence of domestic violence in low-income countries and its burden on the mental and physical health of women paint a grim picture, energy, insight, and hope can be gained from examples of concerned individuals around the world unifying their efforts to combat this problem in their communities. A global perspective on domestic violence allows us to recognize and ultimately to transform the social and cultural underpinnings that condone violence against women in the home.

What follows are examples of innovative strategies from around the world which challenge, on various fronts, the values and structures that have perpetuated domestic violence against women.

India In rural India, where the stigma of rape is great and the status of women is low, women who are raped are no longer considered marriageable. But an unmarried woman in traditional rural communities cannot earn enough money to make a living. Today, rape survivors are turning to the legal system to back their demand for retribution from their rapists. They are lobbying for legal measures requiring rapists to turn over half their annual income or an equivalent in land to their victims (Parikh 1991).

Mexico "Specialized agencies" providing integrated legal, medical, and psychological services to rape survivors were established in Mexico City in 1988 by the Commission on Violence. Centralizing services makes it easier for women to be aware of and take advantage of the services that are available. The commission plans to extend such agencies to assist survivors of family violence and to reach rural areas (Shrader-Cox 1992).

Malaysia In Malaysia, women's organizations representing various disciplines have joined forces to raise awareness of violence against women and to effect legislative and social change. Their focus has been first to dispel the myths surrounding the issue of domestic violence and, second, to draw attention to the inadequacy of present laws to protect women's rights. The Joint Action Group Against Violence Against Women (JAG), established in 1985, includes representatives from a battered women's shelter as well as the Association of Women Lawyers, the University Women's Association, the Malaysian Trade Union Congress Women's Section, and the media. Their campaign has been waged largely through public education programs (Fernandez 1992).

Jamaica A women's theater group, Sistren, has been writing and producing drama-in-education workshops for schools and community groups, primarily in rural areas throughout the Caribbean, in an attempt to raise awareness about the reality of violence in women's lives. Women's Media Watch, a grass roots organization, has been lobbying both politicians and advertisers to create legislation to end sexist advertising (Parikh 1991).

Bangladesh/Chile/Thailand The Bangladesh Women's Health Coalition and the Chilean Institute of Reproductive Medicine offer integrated family planning services at the same time as child health services, and Thailand is experimenting with mobile health clinics to reach women in their homes (World Bank 1993). Comprehensive health services designed to treat or prevent domestic violence can, similarly, be made more accessible.

Brazil In Brazil, women have responded to police officers' insensitivity to raped and battered women by lobbying for, and eventually establishing, all-women police stations. The first station was established in São Paulo in 1985 and responded to 2,000 cases of abuse. Now, with six stations in the capital city alone, women police, detectives, and social workers respond to more than 7,000 domestic violence and rape cases each year (Parikh 1991).

One feature common to all of the strategies is that they began as grass roots movements that, with concerted effort, succeeded in making their way onto the agenda of national and international agencies. To date, strategies emanating from women's groups have been the most effective in educating the public about the magnitude of the problem and initiating national and international debate.

At the recent Human Rights Conference in Vienna (1993), an unprecedented representation of 950 women's organizations from around the world coalesced into the strongest and most effective lobby grappling with the common concern of violence against women. A significant achievement of the conference was the demonstration that although they came from vastly different social and cultural backgrounds, they could unite around a common issue of violence against women while collaborating on what has become the Global Campaign for Women's Human Rights.

Interaction between different levels of society is crucial to the strength of the movement to bring domestic violence to the forefront of the health and human rights agenda. Community-based grassroots organizations keep the movement alive and in touch with the people most in need. National and international agencies have the clout to effect social and political change and to secure the ever-needed funds that sustain a movement. The effort, however, starts with a steadfast few in the spirit of anthropologist Margaret Mead's well-known words: "Never doubt that a small group of thoughtful, committed citizens can change the world; indeed, it's the only thing that ever does."

REFERENCES

BAHA'I PEACE COUNCIL OF CANADA. 1986. *To the Peoples of the World—A Statement on Peace.* Ontario, Canada: Universal House of Justice.

BROWNE, A. and K. WILLIAMS. 1989. "Exploring the Effect of Resource Availability and the Likelihood of Female-Perpetrated Homicides." *Law and Society Review* 23:1.

CARRILLO, ROXANNE. 1991. "Violence Against Women: An Obstacle to Development." Pp. 19–41 in *Gender Violence: A Development and Human Rights Issue,* edited by C. Bunch and Roxanne Carrillo. New Brunswick, NJ: Center for Women's Global Leadership, Douglass College, Rutgers University.

CLARE, ANTHONY. 1992. *Medicine Betrayed: The Participation of Doctors in Human Rights Abuses.* London: Zed Books/British Medical Association.

COUNTS, DOROTHY A. 1987. "Female Suicide and Wife Abuse: A Cross-cultural Perspective." *Suicide and Life-Threatening Behavior* 17:194–204.

———. 1990a. "Domestic Violence in Oceania: Introduction." *Pacific Studies* 13(3):1–5.

———. 1990b. "Beaten Wife, Suicidal Woman: Domestic Violence in Kaliai, West New Britain." *Pacific Studies* 13(3):151–69.

EL-DAREER, A. 1982. *Women, Why Do You Weep? Circumcision and Its Consequences.* London: Zed Books.

FERNANDEZ, IRENE. 1992. "Mobilizing on All Fronts: A Comprehensive Strategy to End Violence Against Women in Malaysia." Pp. 101–20 in *Freedom from Violence: Women's Strategies from Around the World,* edited by Margaret Schuler. Washington, DC: OEF International.

FORNI, E. 1980. "Women's Role in the Economic, Social, and Political Development of Somalia." *Afrika Spectrum* 15:19–28.

GILMARTIN, CHRISTINA. 1990. "Violence Against Women in Contemporary China." Pp. 203–25 in *Violence in China: Essays in Culture and Counterculture*, edited by J. Lipman and S. Harrell. Albany, NY: State University of New York Press.

GRUENBAUM, E. 1988. "Reproductive Ritual and Social Reproduction: Female Circumcision and the Subordination of Women in Sudan." Pp. 308–25 in *Economy and Class in Sudan*, edited by N. O'Neill and J. O'Brien. Aldershot, UK: Avebury.

HEISE, LORI. 1993a. "Background Data for Establishing the Health Burden from Domestic Violence for the World Health Report of the World Bank." Unpublished.

———. 1993b. "Global Estimates of the Health Burden Resulting from Rape for the World Health Report of the World Bank." Unpublished.

———. 1993c. "Violence Against Women: The Missing Agenda." Pp. 171–95 in *The Health of Women: A Global Perspective*, edited by M. Koblinski, J. Timyan, and J. Gay. Boulder, CO: Westview Press.

HEISE, LORI, R. W. KNOWLTON, and J. STOLTENBERG. 1989. "The Global War Against Women." *Utne Reader* 36:40–49.

HICKS, E. K. 1993. *Infibulation: Female Mutilation in Islamic Northeastern Africa.* New Brunswick, NJ: Transaction Publishers.

JILANI, HINA. 1992. "Whose Laws? Human Rights and Violence Against Women in Pakistan." Pp. 63–74 in *Freedom from Violence: Women's Strategies from Around the World*, edited by Margaret Schuler. Washington, DC: OEF International.

KARKAL, M. 1985. "How the Other Half Dies in Bombay." *Economic and Political Weekly* p. 1424 (Cited in Heise, 1993c).

KELKAR, GOVIND. 1991. "Stopping the Violence Against Women: Issues and Perspectives from India." Pp. 75–99 in *Freedom from Violence: Women's Strategies from Around the World*, edited by Margaret Schuler. Washington, DC: OEF International.

KILPATRICK, D., C. BEST, L. VERONEN, et al. 1985. "Mental Health Correlates of Criminal Victimization: A Random Community Survey." *Journal of Consulting and Clinical Psychology* 53:866–73.

KLEINMAN, ARTHUR and JOAN KLEINMAN. 1985. "Somatization: The Interconnections in Chinese Society Among Culture, Depressive Experiences and the Meanings of Pain." Pp. 1–42 in *Culture and Depression: Studies in Anthropology and Cross-Cultural Psychiatry of Affect and Disorder*, edited by Arthur Kleinman and Byron Good. Berkeley: University of California Press.

KLINGER, KAREN. 1993. "A Voice for Health and Human Rights." *Harvard Public Health Review*, Spring, pp. 6–11. David Christiani quoted in text.

KOSS, MARY P. 1990. "The Women's Mental Health Research Agenda." *American Psychologist* 45:374–80.

MOLLICA, RICHARD and L. SON. 1989. "Cultural Dimensions in the Evaluation and Treatment of Sexual Trauma: An Overview." *Psychiatric Clinics of North America* 12:363–79.

NEWBERGER, E., S. E. BARKEN, E. S. LIEBERMAN, et al. 1992. "Abuse of Pregnant Women and Adverse Birth Outcome." *Journal of the American Medical Association* 267:2370–72.

PARIKH, R. 1991. "Violence Meets Its Match." *The Canadian Nurse*, October, pp. 14–15.

RANDALL, T. 1990. "Domestic Violence Intervention Calls for More than Treating Injuries." *Journal of the American Medical Association* 264:939–40.

RODRIGUEZ-TRIAS, HELEN. 1992. "Women's Health, Women's Lives, Women's Rights." *American Journal of Public Health* 82:663–64.

ROSENBERG, M. L., E. STARK, and M. A. ZAHN. 1986. "Interpersonal Violence: Homicide and Spouse Abuse." Pp. 1399–1426 in *Public Health and Preventive Medicine*, 12th ed., edited by J. M. Last. Norwalk, CT: Appleton-Century-Crofts.

ROYNER, S. 1991. "Battered Wives: Centuries of Silence." *The Washington Post*, August 20, p. 7. Cited in Rodriquez-Trias 1992.

SEAGER, J. and A. OLSON. 1986. *Women in the World: An International Atlas.* London: Pluto Press Limited.

SHATTUCK, JIM. 1993. (Director of *Men Overcoming Violence*) Personal communication.

SHRADER-COX, ELIZABETH. 1992. "Developing Strategies: Efforts to End Violence Against Women in Mexico." Pp. 175–98 in *Freedom From Violence: Women's Strategies from Around the World*, edited by Margaret Schuler.

SIDEL, VICTOR. 1993. "From 'Helping the Victims' to 'Blaming the Victims' to 'Organizing the Victims': Lessons from China, Chile and the Bronx." Pp. 205–18 in *A New Dawn in Guatemala: Toward a Worldwide Health Vision*, edited by R. Luecke. Prospect Heights, IL: Waveland Press.

SIGLER, R. T. 1989. *Domestic Violence in Context: An Assessment of Community Attitudes.* Lexington, MA: D. C. Heath.

STAR, B., C. G. CLARK, K. M. GOETZ, and L. O'MALIA, 1981. "Psychosocial Aspects of Wifebeating." Pp. 426–61 in *Women and Mental Health*, edited by

E. Howell and M. Baynes. New York: Basic Books.

STARK, E. and A. FLITCRAFT. 1979. "Medicine and Patriarchal Violence: The Social Construction of a Private Event." *International Journal of Health Services* 9:461–93, as cited in Council on Scientific Affairs, American Medical Association. 1992. "Violence Against Women." *Journal of the American Medical Association* 267:3184–95.

STARK, E. and A. FLITCRAFT. 1991. "Spouse Abuse." Pp. 123–57 in *Violence in America: A Public Health Approach,* edited by M. Rosenberg and M. Fenley. New York: Oxford University Press.

STRAUS, MURRAY A., RICHARD J. GELLES, and SUZANNE K. STEINMETZ. 1980. *Behind Closed Doors: Violence in the American Family.* Garden City, New York: Anchor Press.

SUGG, NANCY K. and THOMAS INUI. 1992. "Primary Care Physicians' Response to Domestic Violence: Opening Pandora's Box." *Journal of the American Medical Association* 267:3157–60.

SUMMERS, L. 1993. *Women's Education in Developing Countries: Barriers, Benefits and Policies.* Washington, DC: The World Bank.

UNITED NATIONS GENERAL ASSEMBLY (UN). 1948. *Universal Declaration of Human Rights.* Adopted 10 December 1948. G.A. Res. 217A(III), U.N. Doc. A/810.

WOMEN'S WORLD. 1991–2. "Wife Abuse: Law and Order." 26:24–25.

WORLD BANK. 1993. *World Development Report 1993: Investing in Health.* The World Bank, Report No. 11778, Washington, DC.

WORLD HEALTH ORGANIZATION (WHO). 1993. "World Health Assembly Calls for the Elimination of Harmful Traditional Practices." Press Release: May 12.

QUESTIONS FOR DISCUSSION

For further discussion of this topic, see the Wadsworth Sociology Resource Center, "Virtual Society," *http://sociology.wadsworth.com,* under *Sociological Footprints,* by Cargan and Ballantine. You can respond to the discussion questions there or enter your own comments in the online chat forum.

SUGGESTED READINGS AND SOCIOLOGY INTERNET RESOURCES

See the Wadsworth Sociology Resource Center, "Virtual Society," *http://sociology.wadsworth. com,* for additional links, suggestions for further reading, and learning tools related to this chapter.

Either from the "Virtual Society" website or directly from your web browser, you may access InfoTrac College Edition, an online university library that includes over 700 popular and scholarly journals in which you can find articles related to the topics in this chapter.

CHAPTER 4

Interaction and Group Behavior

LIFE IS WITH PEOPLE

GROUPS ARE THE SETTING FOR SOCIALIZATION, the lifelong process of learning the ways of society. This process takes place first in the family, then in play groups, followed by more formal group associations in institutions of education and religion, and as adults in economic and political institutions. As individuals grow, so do their group contacts and roles within these groups. Familial relations expand to other primary groups such as play and peer groups, then to more formal groups such as the school. When individuals become adults, many of their interactions take place within secondary groups or organizations characterized by more formal relations.

A major part of our interaction with other individuals or within groups takes place through nonverbal communication. In the first reading Edward and Mildred Hall describe the silent part of language—the nonverbal communication of unconscious body movements. Body language involves physical cues—glances, movement, facial expressions, and so on—as well as the social distance and personal space. Because of its subtleties and the vast range of possible nonverbal rules, silent language is more difficult to master than verbal speech. To complicate matters, as Hall and Hall show, similar movements may have different meanings in different situations or cultures.

Gender language differences also affect communications. Women seek connection and contact through communication, whereas men engage in competition and hierarchy or status determination through communication. If we understand these differences, we may avoid some fatal misunderstandings. The reading by Henley, Hamilton, and Thorne explains the impact of gender differences in verbal and nonverbal language.

Most of our interactions take place in groups. A large body of sociological literature deals with group relations, from role relationships, leadership, and decision making to the tendency toward conformity in group situations because individuals strive to get along with others. Understanding group dynamics involves us in everything from patterns of communication to roles we play within groups, as discussed in the next reading. The Adlers analyze the process of learning and carrying out the college athlete role. The intense pressure for performance, adulation from crowds and the press, and intimidation of classroom expectations combine to provide an example of socialization into rigid role expectations.

One famous observer of groups and society was Max Weber. Writing at the turn of the century, he saw a modern form of groups emerging as societies industrialized. As businesses grew larger and more technical, owners no longer hired friends and relatives, but rather hired and fired people on the basis of their training, skills, and performance. Bureaucracies as Weber described them[1] were organized as hierarchies with sets of rules and regulations, and individuals were formally contracted to carry out certain tasks in exchange for compensation. Today it would seem strange if organizations did not function in this manner. Characteristics of bureaucratic organizations include the formal contractual relationships, rules and regulations, and hiring and firing based on merit and competence.

Since Weber first wrote about bureaucracy, there have been changes in the way bureaucracies function. Technology, for instance, has created new relationships between organizations and individuals. At the same time, individual participation in groups has been changing. Today there is less personal involvement in groups and more technological involvement, which tends to isolate individuals. George Ritzer discusses "McDonaldization" in the excerpt from his book about the worldwide trends toward sameness and efficiency.

As you read the selections in this chapter, consider the definitions of primary and secondary groups and reflect on examples of these groups in modern society. What dynamics do we need to understand to interact and communicate effectively in groups? Are some problems in group dynamics outcomes of our need for belonging and acceptance? How are organizations changing? How is individual participation in groups changing?

NOTE

1. H. H. Gerth and C. Wright Mills, *From Max Weber: Essays in Sociology* (New York: Oxford University Press, 1973). First published 1946.

14

The Sounds of Silence

EDWARD T. HALL AND MILDRED REED HALL

A crucial element in human interaction and group behavior is language, both verbal and nonverbal. Nonverbal communication is quite as important in communicating as verbal language. When we study foreign language, we may learn the words, but the gestures and facial expressions are difficult to master unless we have been raised in the culture. The Halls discuss nonverbal communication patterns and differences across cultures.

Think about the following questions as you read this selection:

1. What is meant by "the sounds of silence"?
2. Can you think of times you have been uncomfortable because of misunderstandings in nonverbal communication?

GLOSSARY

Silent language Communication that takes place through gestures, facial expressions, and other body movements.
Machismo Assertive masculinity.

BOB LEAVES HIS APARTMENT at 8:15 A.M. and stops at the corner drugstore for breakfast. Before he can speak, the counterman says, "The usual?" Bob nods yes. While he savors his Danish, a fat man pushes onto the adjoining stool and overflows into his space. Bob scowls and the man pulls himself in as much as he can. Bob has sent two messages without speaking a syllable.

Henry has an appointment to meet Arthur at 11 o'clock; he arrives at 11:30. Their conversation is friendly, but Arthur retains a lingering hostility. Henry has unconsciously communicated that he doesn't think the appointment is very important or that Arthur is a person who needs to be treated with respect.

George is talking to Charley's wife at a party. Their conversation is entirely trivial, yet Charley glares at them suspiciously. Their physical proximity and the movements of their eyes reveal that they are powerfully attracted to each other.

José Ybarra and Sir Edmund Jones are at the same party, and it is important for them to establish a cordial relationship for business reasons. Each is trying to be warm and friendly, yet they will part with mutual distrust and their business transaction will probably fall through. José, in Latin fashion, moved closer and closer to Sir Edmund as they spoke, and this movement was miscommunicated as pushiness to Sir Edmund, who kept backing away from this intimacy, and this was miscommunicated to José as coldness. The silent languages of Latin and English cultures are more difficult to learn than their spoken languages.

In each of these cases, we see the subtle power of nonverbal communication. The only language used throughout most of the history of humanity (in evolutionary terms, vocal communication is relatively recent), it is the first form of communication you learn. You use this preverbal language, consciously and unconsciously, every day to tell other people how you feel about yourself and them. This language includes your posture, gestures, facial expressions, costume, the way you walk, even your treatment of time and space and material things. All people communicate on several different levels at the same time but are usually aware of only the verbal dialog and don't realize that they respond to nonverbal messages. But when a person says one thing and really believes something else, the discrepancy between the two can usually be sensed. Nonverbal communication systems are much less subject to the conscious deception that often occurs in verbal systems. When we find ourselves thinking, "I don't know what it is about him, but he doesn't seem sincere," it's usually this lack of congruity between a person's words and his behavior that makes us anxious and uncomfortable.

Few of us realize how much we all depend on body movement in our conversation or are aware of the hidden rules that govern listening behavior. But we know instantly whether or not the person we're talking to is "tuned in" and we're very sensitive to any breach in listening etiquette. In white middle-class American culture, when someone wants to show he is listening to someone else, he looks either at the other person's face, or specifically, at his eyes, shifting his gaze from one eye to the other.

If you observe a person conversing, you'll notice that he indicates he's listening by nodding his head. He also makes little "Hmm" noises. If he agrees with what's being said, he may give a vigorous nod. To show pleasure or affirmation, he smiles; he has some reservations, he looks skeptical by raising an eyebrow or pulling down the corners of his mouth. If a participant wants to terminate the conversation, he may start shifting his body position, stretching his legs, crossing or uncrossing them, bobbing his foot or diverting his gaze from the speaker. The more he fidgets, the more the speaker becomes aware that he has lost his audience. As a last measure, the listener may look at his watch to indicate the imminent end of the conversation.

Talking and listening are so intricately intertwined that a person cannot do one without the other. Even when one is alone and talking to oneself, there is part of the brain that speaks while another part listens. In all conversations, the listener is positively or negatively reinforcing the speaker all the time. He may even guide the conversation without knowing it, by laughing or frowning or dismissing the argument with a wave of his hand.

The language of the eyes—another age-old way of exchanging feelings—is both subtle and complex. Not only do men and women use their eyes differently but there are class, generation, regional, ethnic, and national cultural differences. Americans often complain about the way foreigners stare at people or hold a glance too long. Most Americans look away from someone who is using his eyes in an unfamiliar way because it makes them self-conscious. If a man looks at another man's wife in a certain way, he's asking for trouble, as indicated earlier. But he might not be ill-mannered or seeking to challenge the husband. He might be a European in this country who hasn't learned our visual mores. Many American women visiting France or Italy are acutely embarrassed because, for the first time in their lives, men really look at them—their eyes, hair, nose, lips, breasts, hips, legs, thighs, knees, ankles, feet, clothes, hairdo, even their walk. . . .

Analyzing the mass of data on the eyes, it is possible to sort out at least three ways in which the eyes are used to communicate: dominance versus submission, involvement versus detachment, and positive versus negative attitude. In addition, there are three levels of consciousness and control, which can be categorized as follows: (1) conscious use of the eyes to communicate, such as the flirting blink and the intimate nose-wrinkling squint; (2) the very extensive category of unconscious but learned behavior governing where the eyes are directed and when (this unwritten set of rules dictates how and under what circumstances the sexes, as well as people of all status categories, look at each other); and (3) the response of the eye itself, which is completely outside both awareness and control—changes in the cast (the sparkle) of the eye and the pupillary reflex.

The eye is unlike any other organ of the body, for it is an extension of the brain. The unconscious pupillary reflex and the cast of the eye have been known by people of Middle Eastern origin for years—although most are unaware of their knowledge. Depending on the context, Arabs and others look either directly at the eyes or deeply *into* the eyes of their interlocutor. We became aware of this in the Middle East several years ago while looking at jewelry. The merchant suddenly started to push a particular bracelet at a customer and said, "You buy this one." What interested us was that the bracelet was not the one that had been consciously selected by the purchaser. But the merchant, watching the pupils of the eyes, knew what the purchaser really wanted to buy. Whether he specifically knew *how* he knew is debatable.

A psychologist at the University of Chicago, Eckhard Hess, was the first to conduct systematic studies of the pupillary reflex. His wife remarked one evening, while watching him reading in bed, that he must be very interested in the text because his pupils were dilated. Following up on this, Hess slipped some pictures of nudes into a stack of photographs that he gave to his male assistant. Not looking at the photographs but watching his assistant's pupils, Hess was able to tell precisely when the assistant came to the nudes. In further experiments, Hess retouched the eyes in a photograph of a woman. In one print, he made the pupils small, in another, large; nothing else was changed. Subjects who were given the photographs found the woman with the dilated pupils much more attractive. Any man who has had the experience of seeing a woman look at him as her pupils widen with reflex speed knows that she's flashing him a message.

The eye-sparkle phenomenon frequently turns up in our interviews of couples in love. It's apparently one of the first reliable clues in the other person that love is genuine. To date, there is no scientific data to explain eye sparkle; no investigation of the pupil, the cornea, or even the white sclera of the eye shows how the sparkle originates. Yet we all know it when we see it.

One common situation for most people involves the use of the eyes in the street and in public. Although eye behavior follows a definite set of rules, the rules vary according to the place, the needs and feelings of the people, and their ethnic background. For urban whites, once they're within definite recognition distance (16–32 feet for people with average eyesight), there is mutual avoidance of eye contact—unless they want something specific: a pickup, a handout, or information of some kind. In the West and in small towns generally, however, people are much more likely to look at and greet one another, even if they're strangers.

It's permissible to look at people if they're beyond recognition distance; but once inside this sacred zone, you can only steal a glance at strangers. You *must* greet friends, however; to fail to do so is insulting. Yet, to stare too fixedly even at them is considered rude and hostile. Of course, all of these rules are variable. . . .

[A] very basic difference between people of different ethnic backgrounds is their sense of territoriality and how they handle space. This is the silent communication, or miscommunication, that caused friction between Mr. Ybarra and Sir Edmund Jones in our earlier example. We know from research that everyone has around himself an invisible bubble of space that contracts and expands depending on several factors: his emotional state, the activity he's performing at the time, and his cultural background. This bubble is a kind of mobile territory that he will defend against intrusion. If he is accustomed to close personal distance between himself and others, his bubble will be smaller than that of someone who's accustomed to greater personal distance. People of North European heritage—English, Scandinavian, Swiss, and German—tend to avoid contact. Those whose heritage is Italian, French, Spanish, Russian, Latin American, or Middle Eastern like close personal contact.

People are very sensitive to any intrusion into their spatial bubble. If someone stands too close to you, your first instinct is to back up. If that's not possible, you lean away and pull yourself in, tensing your muscles. If the intruder doesn't respond to these body signals, you may then try to protect yourself, using a briefcase, umbrella, or raincoat. . . . As a last resort, you may move to another spot and position yourself behind a desk or a chair that provides screening. Everyone tries to adjust the space around himself in a way that's comfortable for him; most often, he does this unconsciously.

Emotions also have a direct effect on the size of a person's territory. When you're angry or under stress, your bubble expands and you require more space. New York psychiatrist Augustus Kinzel found a difference in what he calls Body-Buffer Zones between violent and nonviolent prison inmates. Dr. Kinzel conducted experiments in which each prisoner was placed in the center of a small room and then Dr. Kinzel slowly walked toward him. Nonviolent prisoners allowed him to come quite close, while prisoners with a history of violent behavior couldn't tolerate his proximity and reacted with some vehemence.

Apparently, people under stress experience other people as looming larger and closer than they actually are. Studies of schizophrenic patients have indicated that they sometimes have a distorted perception of space, and several psychiatrists have reported patients who experience their body boundaries as filling up an entire room. For these patients, anyone who comes into the room is actually inside their body, and such an intrusion may trigger a violent outburst.

Unfortunately, there is little detailed information about normal people who live in highly congested urban areas. We do know, of course, that the noise, pollution, dirt, crowding, and confusion of our cities induce feelings of stress in most of us, and stress leads to a need for greater space. [People who are] packed into a subway, jostled in the street, crowded into an elevator, and forced to work all day in a bull pen or in a small office without auditory or visual privacy [are] going to be very stressed at the end of [the] day. They need places that provide relief from constant overstimulation. . . . Stress from overcrowding is cumulative and people can tolerate more crowding early in the day than later; note the increased bad temper during the evening rush hour as compared with the morning melee. Certainly one factor in people's desire to commute by car is the need for privacy and relief from crowding (except, often, from other cars); it may be the only time of day when nobody can intrude.

In crowded public places, we tense our muscles and hold ourselves stiff, and thereby communicate to others our desire not to intrude on their space and, above all, not to touch them. We also avoid eye contact and the total effect is that of someone who has "tuned out." Walking along the street, our bubble expands slightly as we move in a stream of strangers, taking care not to bump into them. In the office, at meetings, in restaurants, our bubble keeps changing as it adjusts to the activity at hand.

Most white middle-class Americans use four main distances in their business and social relations: intimate, personal, social, and public. Each of these distances has a near and a far phase and is accompanied by changes in the volume of the voice. Intimate distance varies from direct physical contact with another person to a distance of six to eighteen inches and is used for our most private activities—caressing another person or making love. At this distance, you are overwhelmed by sensory inputs from the other person—heat from the body, tactile stimulation from the skin, the fragrance of perfume, even the sound of breathing—all of which literally envelop you. Even at the far phase, you're still within easy touching distance. In general, the use of intimate distance in public between adults is frowned on. It's also much too close for strangers, except under conditions of extreme crowding.

In the second zone—personal distance—the close phase is one and a half to two and a half feet; it's at this distance that wives usually stand from their husbands in public. If another woman moves into this zone, the wife will most likely be disturbed. The far phase—two and a half to four feet—is the distance used to "keep someone at arm's length" and is the most common spacing used by people in conversation.

The third zone—social distance—is employed during business transactions or exchanges with a clerk or repairman. People who work together tend to use close social distance—four to seven feet. This is also the distance for conversation at social gatherings. To stand at this distance from one who is seated has a dominating effect (for example, teacher to pupil, boss to secretary). The far phase of the third zone—seven to twelve feet—is where people stand when someone says, "Stand back so I can look at you." This distance lends a formal tone to business or social discourse. In an executive office, the desk serves to keep people at this distance.

The fourth zone—public distance—is used by teachers in classrooms or speakers at public gatherings. At its farthest phase—25 feet and beyond—it is used for important public figures. Violations of this distance can lead to serious complications. During his 1970 U.S. visit, the president of France, Georges Pompidou, was harassed by pickets in Chicago, who were permitted to get within touching distance. Since pickets in France are kept behind barricades a block or more away, the president was outraged by this insult to his person, and President Nixon was obliged to communicate his concern as well as offer his personal apologies.

It is interesting to note how American pitchmen and panhandlers exploit the unwritten, unspoken conventions of eye and distance. Both take advantage of the fact that once explicit eye contact is established, it is rude to look away, because to do so means to brusquely dismiss the other person and his needs. Once having caught the eye of his mark, the panhandler then locks on, not letting go until he moves through the public zone, the social zone, the personal zone, and finally, into the intimate sphere, where people are most vulnerable.

Touch also is an important part of the constant stream of communication that takes place between people. A light touch, a firm touch, a blow, a caress are all communications. In an effort to break down barriers among people, there's been a recent upsurge in group-encounter activities, in which strangers are encouraged to touch one another. In special situations such as these, the rules for not touching are broken with group approval and people gradually lose some of their inhibitions.

Although most people don't realize it, space is perceived and distances are set not by vision alone but with all the senses. Auditory space is perceived with the ears, thermal space with the skin, kinesthetic space with the muscles of the body, and olfactory space with the nose. And, once again, it's one's culture that determines how his senses are programmed—which sensory information ranks highest and lowest. The important thing to remember is that culture is very persistent. In this country, we've noted the existence of culture patterns that determine distance between people in the third and fourth generations of some families, despite their prolonged contact with people of very different cultural heritages.

Whenever there is great cultural distance between two people, there are bound to be problems arising from differences in behavior and expectations. An example is the American couple who consulted a psychiatrist about their marital problems. The husband was from New England and had been brought up by reserved parents who taught him to control his emotions and to respect the need for privacy. His wife was from an Italian family and had been brought up in close contact with all the members of her large family, who were extremely warm, volatile, and demonstrative.

When the husband came home after a hard day at the office, dragging his feet and longing for peace and quiet, his wife would rush to him and smother him. Clasping his hands, rubbing his brow, crooning over his weary head, she never left him alone. But when the wife was upset or anxious about her day, the husband's response was to withdraw completely and leave her alone. No comforting, no affectionate embrace, no attention—just solitude. The woman became convinced her husband didn't love her and, in desperation, she consulted a psychiatrist. Their problem wasn't basically psychological but cultural.

Why [have people] developed all these different ways of communicating messages without words? One reason is that people don't like to spell out certain kinds of messages. We prefer to find other ways of showing our feelings. This is especially true in relationships as sensitive as courtship. . . . We work out subtle ways of encouraging or discouraging each other that save face and avoid confrontations. . . .

If a man sees a woman whom he wants to attract, he tries to present himself by his posture and stance as someone who is self-assured. He moves briskly and confidently. When he catches the eye of the woman, he may hold her glance a little longer than normal. If he gets an encouraging smile, he'll move in close and engage her in small talk. As they converse, his glance shifts over her face and body. He, too, may make preening gestures — straightening his tie, smoothing his hair, or shooting his cuffs.

How do people learn body language? The same way they learn spoken language — by observing and imitating people around them as they're growing up. Little girls imitate their mothers or an older female. Little boys imitate their fathers or a respected uncle or a character on television. In this way, they learn the gender signals appropriate for their sex. Regional, class, and ethnic patterns of body behavior are also learned in childhood and persist throughout life.

Such patterns of masculine and feminine body behavior vary widely from one culture to another. In America, for example, women stand with their thighs together. Many walk with their pelvis tipped sightly forward and their upper arms close to their body. When they sit, they cross their legs at the knee or cross their ankles. American men hold their arms away from their body, often swinging them as they walk. They stand with their legs apart (an extreme example is the cowboy, with legs apart and thumbs tucked into his belt). When they sit, they put their feet on the floor with legs apart and, in some parts of the country, they cross their legs by putting one ankle on the other knee.

Leg behavior indicates sex, status, and personality. It also indicates whether or not one is at ease or is showing respect or disrespect for the other person. Young Latin American males avoid crossing their legs. In their world of *machismo,* the preferred position for young males when with one another (if there is no old dominant male present to whom they must show respect) is to sit on the base of the spine with their leg

muscles relaxed and their feet wide apart. Their respect position is like our military equivalent: spine straight, heels and ankles together — almost identical to that displayed by properly brought up young women in New England in the early part of this century.

American women who sit with their legs spread apart in the presence of males are *not* normally signaling a come-on — they are simply (and often unconsciously) sitting like men. Middle-class women in the presence of other women to whom they are very close may on occasion throw themselves down on a soft chair or sofa and let themselves go. This is a signal that nothing serious will be taken up. Males, on the other hand, lean back and prop their legs up on the nearest object.

The way we walk, similarly, indicates status, respect, mood, and ethnic or cultural affiliation. . . . To white Americans, some French middle-class males walk in a way that is both humorous and suspect. There is a bounce and looseness to the French walk, as though the parts of the body were somehow unrelated. Jacques Tati, the French movie actor, walks this way; so does the great mime, Marcel Marceau. . . .

All over the world, people walk not only in their own characteristic way but have walks that communicate the nature of their involvement with whatever it is they're doing. The purposeful walk of North Europeans is an important component of proper behavior on the job. Any male who has been in the military knows how essential it is to walk properly. . . . The quick shuffle of servants in the Far East in the old days was a show of respect. On the island of Truk, when we last visited, the inhabitants even had a name for the respectful walk that one used when in the presence of a chief or when walking past a chief's house. The term was *sufan,* which meant to be humble and respectful.

The notion that people communicate volumes by their gestures, facial expressions, posture and walk is not new; actors, dancers, writers, and psychiatrists have long been aware of it. Only in recent years, however, have scientists begun to make systematic observations of body motions. Ray L. Birdwhistell of the University of Pennsylvania is one of the pioneers in body-motion research and coined the term *kinesics* to describe this field. He developed an elaborate notation system to record both facial and body movement, using an approach similar to that of the linguist, who studies the basic elements of speech. Birdwhistell and other kinesicists such as Albert Shellen, Adam Kendon, and William Condon take movies of people interacting. They run the film over and over again, often at reduced speed for frame-by-frame analysis, so that they can observe even the slightest body movements not perceptible at normal interaction speeds. These movements are then recorded in notebooks for later analysis. . . .

Several years ago in New York City, there was a program for sending children from predominantly black and Puerto Rican low income neighborhoods to summer school in a white upper-class neighborhood on the East Side. One morning, a group of young black and Puerto Rican boys raced down the street, shouting and screaming and overturning garbage cans on their way to school. A doorman from an apartment building nearby chased them and cornered one of them inside a building. The boy drew a knife and attacked the doorman. This tragedy would not have occurred if the doorman had been familiar with the behavior of boys from low-income neighborhoods, where such antics are routine and socially acceptable and where pursuit would be expected to invite a violent response.

The language of behavior is extremely complex. Most of us are lucky to have under control one subcultural system—the one that reflects our sex, class, generation, and geographic region within the United States. Because of its complexity, efforts to isolate bits of nonverbal communication and generalize from them are in vain; you don't become an instant expert on people's behavior by watching them at cocktail parties. Body language isn't something that's independent of the person, something that can be donned and doffed like a suit of clothes.

Our research and that of our colleagues have shown that, far from being a superficial form of communication that can be consciously manipulated, nonverbal-communication systems are interwoven into the fabric of the personality and, as sociologist Erving Goffman has demonstrated, into society itself. They are the warp and woof of daily interactions with others, and they influence how one expresses oneself, how one experiences oneself as a man or a woman.

Nonverbal communications signal to members of your own group what kind of person you are, how you feel about others, how you'll fit into and work in a group, whether you're assured or anxious, the degree to which you feel comfortable with the standards of your own culture, as well as deeply significant feelings about the self, including the state of your own psyche. For most of us, it's difficult to accept the reality of another's behavioral system. And, of course, none of us will ever become fully knowledgeable of the importance of every nonverbal signal. But as long as each of us realizes the power of these signals, this society's diversity can be a source of great strength rather than a further—and subtly powerful—source of division.

15
Womanspeak and Manspeak

NANCY HENLEY, MYKOL HAMILTON, AND BARRIE THORNE

Misunderstandings in interactions can occur in verbal and nonverbal language and in nuances of meanings in interaction; this is especially evident between people of different cultures. However, an increasing body of literature focuses on the differences in interaction and language usage between women and men in the same culture. Elaborating on the meaning of interaction and differences in socialization experiences, the authors of this reading discuss differences in female and male forms and experiences of communication.

As you read this selection, consider the following:

1. What are behaviors that cause misunderstandings when women and men are communicating?

2. Can you recall examples of gender miscommunication in your experience?

GLOSSARY

Grammarians Those who study the uses of grammar.
Deprecating Putting down others.
Demeanor How one presents oneself to others.

A WOMAN STARTS TO speak but stops when a man begins to talk at the same time; two men find that a simple conversation is escalating into full-scale competition; a junior high school girl finds it hard to relate to her schoolbooks, which are phrased in the terminology of a male culture and refer to people as "men"; a woman finds that when she uses the gestures men use for attention and influence, she is responded to sexually; a female college student from an all-girl high school finds a touch or glance from males in class intimidating.

What is happening here? First, there are differences between female and male speech styles, and the sexes are often spoken about in different ways. Male nonverbal communication also has certain elements and effects that distinguish it from its female counterpart. Moreover, females and males move in a context of sexual inequality and strongly differentiated behavioral expectations. Because interaction with others always involves communication of some sort, verbal and nonverbal, it is through communication that much of our pattern of sexist interaction is learned and perpetuated. . . .

Language has been used in the past, and is still used, to dehumanize a people into submission; it both reflects and shapes the culture in which it is embedded.

The Sexist Bias of English

Sexism in the English language takes three main forms: It ignores; it defines; it deprecates.

IGNORING

Most of us are familiar with ways in which our language ignores females. The paramount example of this is the masculine "generic," which has traditionally been used to include women as well as men. We are taught to use *he* to refer to someone whose sex is unspecified, as in the sentence, "Each entrant should do his best." We are told that using *they* in such a case ("Everyone may now take their seat") is ungrammatical; yet Bodine (1975) reports that prior to the eighteenth century, *they* was widely used in this way. Grammarians who insist that we use *he* for numerical agreement with the antecedent overlook the disagreement in gender such usage may entail. Current grammars condemn "he or she" as clumsy, and the singular "they" as inaccurate, but expect pupils to achieve both elegance of expression and accuracy by referring to women as *he*. Despite the best efforts of grammarians, however, singular *they* has long been common in informal conversation and is becoming more frequent even in formal speech and writing.

Many people who claim they are referring to both females and males when they use the word *he* switch to the feminine pronoun when they speak of someone in a traditionally feminine occupation, such as homemaker or schoolteacher or nurse, raising questions about the inclusion of females in the masculine pronoun. Although compared to specific masculine reference, the masculine "generic" occurs infrequently, but it still has a high occurrence in many of our lives. MacKay (1983) estimates that highly educated Americans are exposed to it a million times in their lifetimes. . . .

DEFINING

Language both reflects and helps maintain woman's secondary status in our society, by defining her and her "place." Man's power to define through naming is illustrated in the tradition of a woman's losing her own name, and taking her husband's when she marries; the children of the marriage also have their father's name, showing that they too are his possessions. The view of females as possessions is further evidenced in the common practice of applying female names and pronouns to material possessions such as cars ("Fill 'er up!"), machines, and ships. . . .

The fact that our language generally ignores women also means that when it does take note of them, it often defines their status. Thus "lady doctor," "lady judge," "lady professor," "lady pilot" all indicate exceptions to the rule of finding males in these occupations. Expressions like "male nurse" are much less common, because many more occupations are typed as male and because fewer men choose to enter female-typed occupations than vice versa. Even in cases in which a particular field is female-typed, males

who enter it often have a term of their own, with greater prestige, such as *chef* or *couturier*. Of course, patterns of usage subtly reinforce our occupational stereotypes, and deeper undertones further reinforce stereotypes concerning propriety and competency. . . .

DEPRECATING

The deprecation of women in the English language can be seen in the connotations and meanings of words applied to male and female things. The very word *virtue* comes from an old root meaning *man;* to be *virtuous* is, literally, to be "manly." Different adjectives are applied to the actions or productions of the different sexes: Women's work may be referred to as *pretty* or *nice;* men's work will more often elicit adjectives like *masterful, brilliant.* While words such as *king, prince, lord, father* have all maintained their elevated meanings, the similar words *queen, madam,* and *dame* have acquired debased meanings.

A woman's sex is treated as if it were the most salient characteristic of her being; this is not the case for males. This discrepancy is the basis for much of the defining of women, and it underlies much of the accompanying deprecation. Sexual insult is applied overwhelmingly to women; Stanley (1977), in researching terms for sexual promiscuity, found 220 terms for a sexually promiscuous woman, but only 22 terms for a sexually promiscuous man. Furthermore, trivialization accompanies many terms applied to females. . . . The feminine endings *-ess* and *-ette,* and the female prefix *lady,* are added to many words which are not really male-specific. Thus we have the trivialized terms *poetess, authoress, aviatrix, majorette, usherette.* Male sports teams are given names of strength and ferocity: "Rams," "Bears," "Jets." Women's sports teams often have cute names like "Rayettes," "Rockettes." As Alleen Nilsen (1972) has put it,

> The chicken metaphor tells the whole story of a girl's life. In her youth she is a *chick,* then she marries and begins feeling *cooped up,* so she goes to *hen parties* where she *cackles* with her friends. Then she has her *brood* and begins to *henpeck* her husband. Finally she turns into an *old biddy.* (p. 109)

. . . Recent research on conversational interaction reflects the attempt to conceptualize language not in terms of isolated variables nor as an abstract code, but within contexts of use, looking at features of conversation within the give-and-take of actual talk. Pamela Fishman (1983) analyzed recurring patterns in many samples of the household conversations of three heterosexual couples. Although the women tried more often than the men to initiate conversations, the women succeeded less often because of minimal responses from their male companions. In contrast, the women pursued topics the men raised, asked more questions, and did more verbal support-work than the men. Fishman concluded that the conversations were under male control, but were mainly produced by female work.

Self-Disclosure

Self-disclosure is another variable that involves language but goes beyond it. Research studies have found that women disclose more personal information to others than men do. Subordinates (in work situations) are also more likely to self-disclose than superiors. People in positions of power are required to reveal little about themselves, yet typically know much about the lives of others—perhaps the ultimate exemplar of this principle is the fictional Big Brother.

According to the research of Jack Sattel (1983), men exercise and maintain power over women by withholding self-disclosure. An institutional example of this use of power is the psychiatrist (usually male), to whom much is disclosed (by a predominantly female clientele), but who classically maintains a reserved and detached attitude, revealing little or nothing of himself. Nonemotionality is the "cool" of the professional, the executive, the poker player, the street-wise operator. Smart men—those in power, those who manipulate others—maintain unruffled exteriors. . . .

Nonverbal Communication

Although we are taught to think of communication in terms of spoken and written language,

nonverbal communication has much more impact on our actions and reactions than does verbal. One psychological study concluded, on the basis of a laboratory study, that nonverbal messages carry over four times the weight of verbal messages when both are used in interaction. Yet, there is much ignorance and confusion surrounding the subtler nonverbal form, which renders it a perfect avenue for the unconscious manipulation of others. Nonverbal behavior is of particular importance for women, because their socialization to docility and passivity makes them likely targets for subtle forms of social control, and their close contact with men—for example as wives and secretaries—entails frequent verbal and nonverbal interaction with those in power. Additionally, women have been found to be more sensitive than men to nonverbal cues, perhaps because their survival depends upon it. (Blacks have also been shown to be better than whites at interpreting nonverbal signals.) . . .

DEMEANOR

Persons of higher status have certain privileges of demeanor that their subordinates do not: the boss can put his feet on the desk and loosen his tie, but workers must be more careful in their behavior. Also, the boss had better not put her feet on the desk; women are restricted in their demeanor. Goffman (1967) observed that in hospital staff meetings, the doctors (usually male, and always of high status) had the privilege of swearing, changing the topic of conversation, and sitting in undignified positions. They could lounge on the (mostly female) nurses' counter and initiate joking sessions. Attendants and nurses, of lower status, had to be more circumspect in their demeanor. Women are also denied privileges of swearing and sitting in the undignified positions allowed to men; in fact, women are explicitly required to be more cautious than men by all standards, including the well-known double one. This requirement of propriety is similar to women's use of more proper speech forms, but the requirement for nonverbal behavior is much more compelling.

Body tension is another sex-differentiated aspect of demeanor. In laboratory studies of conversation, communicators are more relaxed with lower-status addressees than with higher-status ones, and they are more relaxed with females than with males. Also, males are generally more relaxed than females; females' somewhat tenser postures are said to convey submissive attitudes (Mehrabian 1972).

USE OF SPACE

Women's general bodily demeanor must be restrained and restricted; their femininity is gauged, in fact, by how little space they take up, while masculinity is judged by males' expansiveness and the strength of their flamboyant gestures. Males control both greater territory and greater personal space, a situation associated with dominance and high status in both human beings and animals. Both field and laboratory studies have found that people tend to approach females more closely than males, to seat themselves closer to females and otherwise intrude on their territory, and to cut across their paths. In the larger aspect of space, women are also less likely to have their own room or other private space in the home.

LOOKING AND STARING — EYE CONTACT

Eye contact is greatly influenced by sex. It has been repeatedly found that in interactions, women look more at the other person than men do and maintain mutual eye contact longer. . . . Other writers have observed that rather than stare, women tend more than men to avert the gaze, especially when stared at by men. Public staring, clothing designed to reveal the contours of the body, and public advertising which lavishly flashes women across billboards and through magazines, all make females a highly visible sex. Visual information about women is readily available, just as their personal information is available through greater self-disclosure.

SMILING

The smile is women's badge of appeasement. . . . Women engage in more smiling than men do, whether they are truly happy or not. Research has confirmed this. Erving Goffman (1979) an-

alyzed the depiction of gender in U.S. print advertising and concluded that women's smiles are ritualistic mollifiers; women smile more, and more expansively, than men. The smile is a requirement of women's social position and is used as a gesture of submission. . . . The smile is generally thought to signal to an aggressor that the subordinate individual intends no harm. In many women, and in other subordinate persons, smiling has reached the status of a nervous habit.

TOUCHING

Touching is another gesture of dominance, and cuddling to the touch is its corresponding gesture of submission. Touching is reportedly used by primates to maintain a dominance order, and it is likely that it is used by human beings in the same way. Just as the boss can put a hand on the worker, the master on the servant, the teacher on the student, the business executive on the secretary, so men more frequently put their hands on women, despite a folk mythology to the contrary. . . . Much of this touching goes unnoticed because it is expected and taken for granted, as when men steer women across the street, through doorways, around corners, into elevators, and so on. The male doctor or lawyer who holds his female client's hand overlong, and the male boss who puts his hand on the female secretary's arm or shoulder when giving her instructions, are easily recognizable examples of such everyday touching of women by men. There is also the more obtrusive touching: the "pawing" by sexually aggressive males, the pinching of waitresses and female office and factory workers, and the totally unexpected and unwelcomed tactual familiarity women are subjected to from complete strangers on the street.

Many interpret this pattern of greater touching by males as a reflection of sexual interest and of a greater level of sexuality among men than women. This explanation, first of all, ignores the fact that touching is a status and dominance signal for human and animal groups. . . . It also ignores the findings of sexual research, which gives us no reason to expect any greater sexual drive in males than in females. Rather, males in our culture have more freedom and encouragement to express their sexuality, and they are also accorded more freedom to touch others. Touching carries the connotation of possession when used with objects, and the wholesale touching of women carries the message that women are community property. . . .

INTIMACY AND STATUS IN NONVERBAL GESTURES

There is another side to touching, one which is much better understood: Touching symbolizes friendship and intimacy. To speak of the power dimension of touching is not to rule out the intimacy dimension. A particular touch may have both components and more, but it is the *pattern* of touching between two individuals that tells us most about their relationship. When touching is symmetrical—that is, when both parties have equal touching privileges—it conveys information about the *intimacy* dimension of the relationship: much touching indicates closeness, and little touching indicates distance. When one party is free to touch the other but not vice versa, we gain information about the *status,* or power, dimension: the person with greater touching privileges is of higher status or has more power. Even when there is mutual touching between two people, it is most likely to be initiated by the higher status person; e.g., in a dating relationship it is usually the male who first puts an arm around the female or begins holding hands.

GESTURES OF DOMINANCE AND SUBMISSION

We have named several gestures of dominance (invasion of personal space, touching, staring) and of submission (allowing oneself to be touched, averting the eyes, and smiling). Pointing may be interpreted as another gesture of dominance, and the corresponding submissive action is to stop talking or acting. In conversation, interruption often functions as a gesture of dominance, and allowing interruption signifies submission. Often mock play between males and females also carries strong physical overtones of dominance: the man squeezing the woman too hard, "pretending" to twist her arm, playfully

lifting her and tossing her from man to man, chasing, catching and spanking her. This type of "play" is also frequently used to control children and to maintain a status hierarchy among male teenagers.

BREAKING THE MOLD—A FIRST STEP

Women can reverse these nonverbal interaction patterns with probably greater effect than can be achieved through deliberate efforts to alter speech patterns. Women can stop smiling unless they are happy, stop lowering their eyes, stop getting out of men's way on the street, and stop letting themselves be interrupted. They can stare people in the eye, be more relaxed in demeanor (when they realize it is more a reflection of status than of morality), and touch when they feel it is appropriate. . . .

REFERENCES

Bodine, A., 1975. Androcentrism in prescriptive grammar. Singular "they," sex-indefinite "he," and "he or she." *Lang. in Soc. 4:* 129–146.

Fishman, P., 1983. Interaction: The work women do. In B. Thorne, C. Kramarae, & N. Henley, Eds., *Language, Gender and Society.* Rowley, MA: Newbury House.

Goffman, E., 1967. The nature of deferences and demeanor. In *Interaction Ritual* (pp. 47–95). New York: Anchor.

Goffman, E., 1979. *Gender Advertisements.* New York: Harper & Row.

MacKay, D. G., 1983. Prescriptive grammar and the pronoun problem. In B. Thorne, C. Kramarae, & N. Henley, Eds., *Language, Gender and Society.* Rowley, MA: Newbury House.

Mehrabian, A., 1972. *Nonverbal Communication.* Chicago: Aldine Atherton.

Nilsen, A. P., 1972. Sexism in English: A feminist view. In N. Hoffman, C. Secor, and A. Tinsley, Eds., *Female Studies VI* (pp. 102–109). Old Westbury, NY: Feminist Press.

Sattel, J., 1983. Men, inexpressiveness, and power. In B. Thorne, C. Kramarae, & N. Henley, Eds., *Language, Gender and Society.* Rowley, MA: Newbury House.

Stanley, J. P., 1977. Paradigmatic Woman: The prostitute. In B. Shores and C. P. Hines, Eds., *Papers in Language Variation* (pp. 303–321). University, AL: University of Alabama Press.

16

Backboards & Blackboards: College Athletes and Role Engulfment

PATRICIA A. ADLER AND PETER ADLER

We all carry out roles in groups to which we belong. We are socialized into school, work, and other roles. Sometimes this socialization is intense and influences all other roles. Such is the case with the achieved role of college athletes. The Adlers provide an example from the world of NCAA basketball, describing the process of socialization, role engulfment, and abandonment of former roles, and formation of team cliques that reflect earlier socialization experiences. The athletes work hard to achieve their goals and in the process are shaped by pressures from adoring fans and intimidating professors. Their roles in the group are shaped by all these pressures.

Consider the following as you read:

1. What are the pressures faced by high-profile athletes?

2. What role responsibilities have you had that involve role engulfment and abandonment of former roles?

GLOSSARY

Role engulfment Demands and rewards of athletic role supersede other roles.

Role domination Process by which athletes become engulfed in athletic roles to exclusion of other roles.

Role abandonment Detachment from investment in other roles, letting go of alternative goals and priorities.

Statuses Positions in organized groups related to other positions by set of normative expectations.

Roles Activities of people of given status.

IT WAS A WORLD of dreams. They expected to find fame and glory, spotlights and television cameras. There was excitement and celebrity, but also hard work and discouragement, a daily grind characterized by aches, pains, and injuries, and an abundance of rules, regulations, and criticism. Their lives alternated between contacts with earnest reporters, adoring fans, and fawning women, and with intimidating professors, demanding boosters, and unrelenting coaches. There was secrecy and intrigue, drama and adulation, but also isolation and alienation, loss of freedom and personal autonomy, and overwhelming demands. These conflicts and dualisms are the focus of this reading. This is a study of the socialization of college athletes.

For five years we lived in and studied the world of elite NCAA (National Collegiate Athletic Association) college basketball. Participant-observers, we fit ourselves into the setting by carving out evolving roles that integrated a combination of team members' expectations with our interests and abilities. Individually and together, we occupied a range of different positions including friend, professor, adviser, confidant, and coach. We observed and interacted with all members of the team, gaining an intimate understanding of the day-to-day and year-to-year character of this social world. From behind the scenes of this secretive and celebrated arena, we document the experiences of college athletes, focusing on changes to their selves and identities over the course of their college years.

The team we researched had many features of the average liberal arts college or state university, yet athletically it lay in the upper echelon of competitors. Although not an established dynasty, it had a newly hired coach who rapidly built a strong team through his winning reputation and energetic recruiting practices. The result was a team with an outstanding record that quickly captivated the interest and enthusiasm of the community. In this book we offer a detailed description and analysis of the players, coaches, supporters, news media, and other key personnel comprising the inner circle of the team. We depict the problems, pressures, and rewards characteristic of the intense, high-visibility, swiftly moving world of college athletics.

This world is the intricate result of complex, interdependent factors and people, all of whom operate to fulfill their own needs and ends. The way these come together creates the composite that is big-time college athletics. While we will reveal facts that will surprise both casual observers and those knowledgeable about college athletics, many of our findings are consistent with the image generally portrayed by investigative media accounts. What our long-lens perspective offers is a close, detailed, and sympathetic portrayal of the virtues and vices inherent in this social world, hidden to most people. We illuminate the processes that operate within the "black box" of college athletics, showing how and why the transformations and outcomes occur.

Existing knowledge about the world of college athletics comes from two basic sources: academic research and journalistic accounts. While the former literature has addressed important concerns and offered significant findings,[1] its methodological character has limited its scope to some extent: it is composed entirely of survey research, the analysis of secondary data, and armchair theorizing. These methods can tell us something about *what* is going on in college athletics, but they can do no more than speculate about the profound and often complex motivations, causes, and other factors underlying the social conditions they document. To date, there have been no systematic academic participant-observation studies of college athletics.[2] Through our five-year in-depth investigation, we go beyond previous correlational studies to shed light on the social dynamics and process of socialization in the world of college athletics.

Journalistic accounts, the second source of knowledge about college athletics, fall into two distinct types. Some are simply the standard institutional propaganda touting the benefits of athletic participation and community cohesion. More commonly, however, journalists interested in college athletics have leaned toward investigative reporting, producing exposés that are meant to highlight the exploitative characteristics of athletics in American higher education. Some of the best investigative reports have used unstructured interviewing and depth involvement to produce insightful and complex portrayals.[3] Many of these accounts are rich in data, but limited in other ways: reporters typically gather information over a relatively short-term period; they do not apply the canons of data-gathering required for scholarly research, and they fail to provide deeper theoretical explanations linking the specific behavior they describe to a greater understanding of human nature, human behavior, social structure, and social change.[4] In this reading we use a longitudinal approach, carefully follow ethnographic verification and validation procedures, and make conceptual links by providing a theoretically informed analysis based on the social psychological theory of symbolic interactionism. We use the experiences of college athletes to offer insights into the relation between the individual, society, and social trends.

Theoretical Approach

This is not only a study of college athletes, then, but also a study in the social psychology of the self.[5] Our observations reveal a significant pattern of transformation experienced by all our subjects: *role-engulfment*. Many of the individuals we followed entered college hoping to gain wealth and fame through their involvement with sport. They did not anticipate, however, the cost of dedicating themselves to this realm. While nearly all conceived of themselves as athletes first, they possessed other self-images that were important to them as well. Yet over the course of playing college basketball, these individuals found the demands and rewards of the athletic role overwhelming and became engulfed by it. However, in yielding to it, they had to sacrifice other interests, activities, and, ultimately, dimensions of their selves. They immersed themselves completely in the athletic role and neglected or abandoned their identities lodged in these other roles. They thus became extremely narrow in their focus. In this work we examine *role domination,* the process by which athletes became engulfed in their athletic role as it ascended to a position of prominence. We also examine the concomitant process of *role abandonment,* where they progressively detached themselves from their investment in other areas and let go of alternative goals or priorities. We analyze the changes this dual process of self-engulfment had on their self-concepts and on the structure of their selves.

The self has been approached from a variety of social psychological perspectives. In examining the experiences of college athletes, we focus on the structure and dynamic processes of their selves within the context of the surrounding culture and social structure. As athletes struggle to adapt to a confusing and conflicting social world, striving to integrate their previous expectations with their individual and joint definitions of the situation, they experience profound changes in their selves. These arise due to the interplay between improvised behavior (agency) and imposed structure, as these meet in the domain of collectively forged interaction. Our approach integrates elements from related, but divergent, models of social psychology. We draw on the structural elements of role theory to enhance a basic foundation of symbolic interactionism.[6] The structural social psychological perspective examines the composition of the self and its relation to social norms and structures. We employ several key concepts from this approach in this analysis.

Role theory focuses on the systems, or institutions, into which interaction fits. According to its tenets, *statuses* are positions in organized groups or systems that are related to other positions by a set of normative expectations. Statuses are not defined by the people that occupy them, but rather they are permanent parts of those systems. Each status carries with it a set of role expectations specifying how persons occupying that status should behave. *Roles* consist of the activities people of a given status are likely to pursue when following the normative expectations for their positions. *Identities* (or what McCall and Simmons 1978, call role-identities), are the self-conceptions people develop from occupying a particular status or enacting a role. The *self* is the more global, multirole, core conception of the real person.

Because in modern society we are likely to be members of more than one group, we may have several statuses and sets of role-related behavioral expectations. Each individual's total constellation of roles forms what may be termed a *role-set,* characterized by a series of relationships with role-related others, or role-set members (Merton 1957). Certain roles or role-identities may be called to the fore, replacing others, as people interact with individuals through them. Individuals do not invest their core feelings of identity or self in all roles equally, however. While some roles are more likely to be called forth by the expectations of others, other roles are more salient to the individual's core, or "real self,"[7] than others. They are arranged along a *hierarchy of salience* from peripheral roles to those that "merge with the self" (Turner 1978), and their ranking may be determined by a variety of factors. Role theory enhances an understanding of both the internal structure of the self and the relation between self and society; it sees this relation as mediated by the concept of role and its culturally and structurally derived expectations.

The interpretive branch of symbolic interactionism focuses on examining agency, process, and change. One of the concepts most critical to symbolic interactionism is the self. Rather than merely looking at roles and their relation to society, symbolic interactionism looks at the individuals filling those roles and the way they engage not only in role-taking, but also in active, creative role-making (Turner 1962). The self is the thinking and feeling being connecting the various roles and identities individuals put forth in different situations (Cooley 1962; Mead 1934). Symbolic interactionism takes a dynamic view of individuals in society, believing that they go beyond merely reproducing existing roles and structures to collectively defining and interpreting the meaning of their surroundings. These subjective, symbolic assessments form the basis for the creation of new social meanings, that then lead to new, shared patterns of adaptation (Blumer 1969). In this way individuals negotiate their social order as they experience it (Strauss 1978). They are thus capable of changing both themselves and the social structures within which they exist. Symbolic interactionism enhances understandings of the dynamic processes characterizing human group life and the reciprocal relation between those processes and changes in the self.

In integrating these two perspectives we show how the experiences of college athletes are both bounded and creative, how athletes integrate structural, cultural, and interactional factors, and how they change and adapt through a dynamic process of action and reaction, forging collective adaptations that both affirm and modify existing structures.

The Setting

We conducted this research at a medium-size (6000 students), private university (hereafter referred to as "the University") in the southwestern portion of the United States. Originally founded on the premise of a religious affiliation, the University had severed its association with the church several decades before, and was striving to make a name for itself as one of the finer private, secular universities in the region. For many years it had served the community chiefly as a commuter school, but had embarked on an aggressive program of national recruiting over the past five to ten years that considerably broad-ened the base of its enrollment. Most of the students were white, middle class, and drawn from the suburbs of the South, Midwest, and Southwest. Academically, the University was experimenting with several innovative educational programs designed to enhance its emerging national reputation. Sponsored by reforms funded by the National Endowment for the Humanities, it was changing the curriculum, introducing a more interdisciplinary focus, instituting a funded honors program, increasing the general education requirements, and, overall, raising academic standards to a fairly rigorous level.

Within the University, the athletic program overall had considerable success during the course of our research: the University's women's golf team was ranked in the top three nationally, the football team won their conference each season, and the basketball program was ranked in the top forty of Division I NCAA schools, and in the top twenty for most of two seasons. The basketball team played in post-season tournaments every year, and in the five complete seasons we studied them, they won approximately four times as many games as they lost. In general, the basketball program was fairly representative of what Coakley (1986) and Frey (1982a) have termed "big-time" college athletics. Although it could not compare to the upper echelon of established basketball dynasties or to the really large athletic programs that wielded enormous recruiting and operating budgets,[8] its success during the period we studied it compensated for its size and lack of historical tradition. The University's basketball program could thus best be described as "up and coming."[9] Because the basketball team (along with the athletic department more generally) was ranked nationally and sent graduating members into the professional leagues, the entire athletic milieu was imbued with a sense of seriousness and purpose.

The team's professionalism was also enhanced by the attention focused on it by members of the community. Located in a city of approximately 500,000 with no professional sports teams, the University's programs served as the primary source of athletic entertainment and identification for the local population. When the basketball program meteorically rose to prominence, members of the city embraced it with fanatical support. Concomitant with the team's rise in fortunes, the region—part of the booming oil and sun belts—was experiencing increased economic prosperity. This surging local pride and financial windfall cast its glow over the basketball team, as it was the most charismatic and victorious program in the city's history, and the symbol of the community's newfound identity. Interest and support were therefore lavished on the team members.

The People

Over the course of our research, we observed 39 players and seven coaches. Much like Becker et al. (1961), we followed the players through their recruitment and entry into the University, keeping track of them as they progressed through school. We also watched the coaches move up their career ladders and deal with the institutional structures and demands.

Players, like students, were recruited primarily from the surrounding region. Unlike the greater student population, though, they generally did not hail from suburban areas. Rather, they predominantly came from the farming and rural towns of the prairies or southlands, and from the ghetto and working class areas of the mid-sized cities.

DEMOGRAPHICS

Over the course of our involvement with the team, two-thirds of the players we studied were black and one-third white. White middle class players accounted for approximately 23 percent of the team members.[10] They came from intact families where fathers worked in such occupations as wholesale merchandising, education, and sales. Although several were from suburban areas, they were more likely to come from mid- to larger-sized cities or exurbs. The remaining white players (10 percent) were from working-

class backgrounds. They came from small factory towns or cities and also from predominantly (although not exclusively) intact families. Some of their parents worked in steel mills or retail jobs.

The black players were from middle, working, and lower class backgrounds. Those from the middle class (15 percent) came from the cities and small towns of the Midwest and South. They had intact families with fathers who worked as police chiefs, ministers, high school principals, or in the telecommunications industry. A few came from families in which the mothers also worked. Several of these families placed a high premium on education; one player was the youngest of five siblings who had all graduated from college and gone on for professional degrees, while another's grandparents had received college educations and established professional careers. The largest group of players (33 percent) were blacks from working class backgrounds. These individuals came from some small Southern towns, but more often from the mid- to larger-sized cities of the South, Midwest, and Southwest. Only about half came from intact families; the rest were raised by their mothers or extended families. Many of them lived in the ghetto areas of these larger cities, but their parents held fairly steady jobs in factories, civil service, or other blue-collar or less skilled occupations. The final group (18 percent) was composed of lower class blacks. Nearly all of these players came from broken homes. While the majority lived with their mothers, one came from a foster family, another lived with his father and sister, and a third was basically reared by his older brothers. These individuals came from the larger cities in the Southwest, South, and Midwest. They grew up in ghetto areas and were street smart and tough. Many of their families subsisted on welfare; others lived off menial jobs such as domestic service. They were poor, and had the most desperate dreams of escaping.

CLIQUES

Moving beyond demographics, the players fell into four main coalitions that served as informal social groups. Not every single member of the team neatly fit into one of these categories or belonged to one of these groups, but nearly all players who stayed on the team for at least a year eventually drifted into a camp. At the very least, individuals associated with the various cliques displayed many behavioral characteristics we will describe. Players often forged friendship networks within these divisions, because of common attitudes, values, and activities. Once in a clique, no one that we observed left it or shifted into another one. In presenting these cliques, we trace a continuum from those with the most "heart" (bravery, dedication, willingness to give everything they had to the team or their teammates), a quality highly valued by team members, to those perceived as having the least.

Drawing on the vernacular shared by players and coaches,[11] the first group of players were the *"bad niggas."*[12] All of these individuals were black, from the working or lower classes, and shared a ghetto upbringing. Members of what Edwards (1985) has called the underclass (contemporary urban gladiators), they possessed many of the characteristics cited by Miller (1958) in his study of delinquent gangs' lower class culture: trouble, toughness, smartness, excitement, fate, and autonomy.[13] They were street smart and displayed an attitude that bespoke their defiance of authority. In fact, their admiration for independence made it hard for them to adjust to domination by the coach (although he targeted those with reform potential as pet "projects"). Fighters, they would not hesitate to physically defend their honor or to jump into the fray when a teammate was in trouble. They worked hard to eke the most out of their athletic potential, for which they earned the respect of their teammates; they had little desire to do anything else.[14] These were the players with the most heart. They may not have been "choir boys," but when the competition was fierce on the court, these were the kind of players the coach wanted out there, the kind he knew he could count on. Their on-court displays of physical prowess contributed to their assertions of masculinity, along with sexual conquests and drug use. They were

sexually promiscuous and often boasted about their various "babes." With drugs, they primarily used marijuana, alcohol (beer), and cocaine. Their frequency of use varied from daily to occasional, although who got high and how often was a significant behavioral difference dividing the cliques. This type of social split, and the actual amounts of drugs team members used, is no different from the general use characteristic of a typical college population (see Moffatt 1989).

Tyrone was one of the bad niggas. He came from the ghetto of a mid-sized city in the Southwest, from an environment of outdoor street life, illicit opportunity, and weak (or absent) parental guidance. He was basically self-raised: he saw little of his mother, who worked long hours as a maid, and he had never known his father. Exceptionally tall and thin, he walked with a swagger (to show his "badness"). His speech was rich with ghetto expressions and he felt more comfortable hanging around with "brothers" than with whites. He often promoted himself boastfully, especially in speaking about his playing ability, future professional chances, and success with women. His adjustment to life at the University was difficult, although after a year or so he figured out how to "get by"; he became acclimated to dorm life, classes, and the media and boosters. When it came to common sense streetsmarts, he was one of the brightest people on the team. He neither liked nor was favored by many boosters, but he did develop a solid group of friends within the ranks of the other bad niggas.

Apollo was another bad nigga. His family upbringing was more stable than Tyrone's, as his parents were together and his father was a career government worker (first in the military, then in the postal service). They had never had much money though, and scraped by as best they could. He was the youngest of six children, the only boy, and was favored by his father because of this. Tall and handsome, he sported an earring (which the coach made him remove during the season) and a gold tooth. One of the most colorful players, Apollo had a charismatic personality and a way with words. He was a magnetic force on the team, a leader who related emotionally to his teammates to help charge and arouse them for big games. He spoke in the common street vernacular of a ghetto "brother," although he could converse in excellent "White English" when it was appropriate. He was appealing to women, and enjoyed their attention, even though he had a steady girlfriend on the women's basketball team. Like Tyrone, Apollo was intelligent and articulate; he was able to express his perspective on life in a way that was insightful, entertaining, and outrageous. He was eager to explore and experience the zest of life, traveling the world, partying heavily, and seizing immediate gratification. He disdained the boring life of the team's straighter members, and generally did not form close relations with them. Yet he managed to enjoy his partying and playing and still graduate in four years. He would never have thought about college except for basketball, and had to overcome several debilitating knee injuries, but he ended up playing professionally on four continents and learning a foreign language.

A second group of players were the "*candy-asses.*" These individuals were also black, but from the middle and working classes. Where the bad niggas chafed under the authority of the coach, the candy-asses craved his attention and approval; they tended to form the closest personal ties with both him and his family. In fact, their strong ties to the coach made them the prime suspects as "snitches," those who would tell the coaches when others misbehaved. They "browned up" to the coaches and to the boosters and professors as well. The candy-asses were "good boys," the kind who projected the public image of conscientious, religious, polite, and quiet individuals. Several of them belonged to the Fellowship of Christian Athletes. They could be counted on to stay out of trouble. Yet although they projected a pristine image, they did not live like monks. They had girlfriends, enjoyed going to discos, and occasionally drank a few beers for recreation. They enjoyed parties, but, responsibly, moderated their behavior. The candy-asses

enjoyed a respected position on the team because, like the bad niggas, they were good athletes and had heart. As much as they sought to be well rounded and attend to the student role, they did not let this interfere with their commitment to the team. They cared about the team first, and would sacrifice whatever was necessary—playing in pain, coming back too soon from an injury, relinquishing personal statistics to help the team win, diving to get the ball—for its benefit. Above all, they could be counted on for their loyalty to the coach, the team, and the game.

Rob was one such player from a large, extended working class family in a sizable Southern city. He was friendly and easy-going, with a positive attitude that came out in most of his activities. Although he was black and most of his close friends on the team were as well, Rob interacted much more easily with white boosters and students than did Tyrone. Rob transferred to the University to play for the coach because he had competed against him and liked both his reputation and style of play. Once there, he devoted himself to the coach, and was adored by the coach's wife and children. He often did favors for the family; one summer he painted the house in his spare time, and ate many of his meals there during the off-season. Rob's family was very close-knit, and both his mother and brother moved to town to be near him while he was at the University. They grew close with the coach's family, and often did things together. They also became regulars at the games, and were courted by many boosters who wanted to feel as if they knew Rob. Rob kept his academic and social life on an even keel during his college years; he worked hard in class (and was often the favorite of the media when they wanted to hype the image of the good student-athlete), and had a steady girlfriend. She was also very visible, with her young child from a previous boyfriend, in the basketball stadium. Like one or two of the others on the team, Rob spent one summer traveling with a Christian group on an around-the-world basketball tour.

Another typical candy-ass was Darian. He lived next door to Rob for two of the years they overlapped at the University, and the two were very close. Darian came from a middle class family in a nearby state and his family came to town for most of the home games. He was much more serious about life than Rob, and worked hard in everything he did. He was recruited by the team at the last minute (he had health problems that many thought would keep him from playing), yet he devoted himself to improvement. By his senior year he was one of the outstanding stars and had dreams of going pro. He wanted to make the most of his college education as well, and spent long hours in his room trying to study. Most people on the team looked up to him because he did what they all intended to do—work hard, sacrifice, and make the most of their college opportunity. His closest friends, then, were others with values like his, who were fairly serious, respectful, and who deferred their gratification in hopes of achieving a future career.

The "*whiners*" constituted a third category of players. Drawn from the middle and working classes and from both races, this group was not as socially cohesive as the previous two, yet it contained friendship cliques of mixed class and race. These individuals had neither the athletic prowess of the candy asses nor the toughness of the bad niggas, yet they wanted respect. In fact the overriding trait they shared was their outspoken belief that they deserved more than they were getting: more playing time, more deference, more publicity. In many ways they envied and aspired to the characteristics of the two other groups. They admired the bad niggas' subculture, their independence, toughness, and disrespect for authority, but they were not as "bad." They wanted the attention (and perceived favoritism) the candy-asses received, but they could not keep themselves out of trouble. Like the bad niggas, they enjoyed getting high. While their athletic talent varied, they did not live up to their potential; they were not willing to devote themselves to basketball. They were generally not the kind of individuals who would get into fights,

either on the court or on the street, and they lacked the heart of the previous two groups. Therefore, despite their potential and their intermittent complaining, they never enjoyed the same kind of respect among their teammates as the bad niggas, nor did they achieve the same position of responsibility on the team as the candy-asses.

Buck fell into this category. A young black from a rural, Southern town, Buck came from a broken home. He did not remember his father, and his mother, who worked in a factory, did not have the money to either visit him at the University or attend his games. Yet he maintained a close relationship with her over the phone. Buck had gone to a primarily white school back home and felt more comfortable around white people than most of the black players; in fact, several of his best friends on the team were white and he frequently dated several white girls. He had a solid academic background and performed well in his classes, although he was not as devoted to the books as some of the candy-asses. He liked to party, and occasionally got in trouble with the coach for breaking team rules. He spent most of his time hanging around with other whiners and with some of the bad niggas, sharing the latter's critical attitude toward the coach's authoritarian behavior. He felt that the coach did not recognize his athletic potential, and that he did not get the playing time he deserved. He had been warned by the coach about associating too much with some of the bad niggas who liked to party and not study hard. Yet for all their partying, the bad niggas were fiercer on the court than Buck, and could sometimes get away with their lassitude through outstanding play. He could not, and always had the suspicion that he was on the coach's "shit list." Like the candy-asses, he wanted to defer his gratification, do well in school, and get a good job afterward, yet he was not as diligent as they were and ended up going out more. He redeemed himself in the eyes of his peers during his senior year by playing the whole season with a debilitating chronic back injury that gave him constant pain.

James was another player in this clique, who also fell somewhere in between the bad niggas and the candy-asses. Like Buck, he socialized primarily with whiners and with some of the bad niggas, although he was white. He came from a middle class family in a small town near the school, and was recruited to play along with his brother (who was a year younger). James came to the University enthusiastic about college life and college basketball. He threw himself into the social whirl and quickly got into trouble with the coach for both his grades and comportment. He readily adapted to the predominantly black ambience of the players' peer subculture, befriending blacks and picking up their jargon. He occasionally dated black women (although this was the cause of one major fight between him and some of the football players, since black women were scarce on this campus). He had heart, and was willing to commit his body to a fight. The most famous incident erupted during a game where he wrestled with an opponent over territorial advantage on the court. Yet he had neither the speed nor the size of some other players, and only occasionally displayed flashes of the potential the coaches had seen in him.

The final group was known to their teammates as the "L-7s" (a "square," an epithet derived from holding the thumbs and forefingers together to form the "square" sign). Members of this mixed group were all middle class, more often white than black. They were the most socially isolated from other team members, as they were rejected for their squareness by all three other groups, and even among themselves seldom made friends across racial lines (the white and black L-7s constituted separate social groups). They came from rural, suburban, or exurban backgrounds and stable families. They were fairly moralistic, eschewing drinking, smoking, and partying. They attempted to project a studious image, taking their books with them on road trips and speaking respectfully to their professors. Compared with other players, they had a stronger orientation to the student population and booster community. They were, at heart, up-wardly mobile, more likely to consider basketball a means to an end than an end in itself. Because of this orientation, several of the white players landed coveted jobs, often working at boosters' companies, at the end of their playing careers. They tended to be good technical players bred on the polished courts of their rural and exurban high schools rather than on the street courts of the cities; they knew how to play the game, but it did not occupy their full attention. They had varying (usually lesser) degrees of athletic ability, but they were even less likely than the whiners to live up to their potential. In contrast to the other groups, they had decidedly the least commitment, least loyalty, and least heart.

Mark was a dirty-blond-haired white boy from the West Coast with a strong upper body, built up from surfing. He was clean cut, respectful, and a favorite of the boosters. His sorority girlfriend was always on his arm, and helped create his desired image of a future businessman. He consciously worked to nurture this image, ostentatiously carrying books to places where he would never use them, and dressing in a jacket and tie whenever he went out in public. It was very important to him to make it financially, because he came from a working class family without much money. He had arrived at the University highly touted, but his talent never materialized to the degree the coaches expected. He was somewhat bitter about this assessment, because he felt he could have "made it" if given more of a chance. Like Buck, part of his problem may have lain in the difference between his slowdown style and the fast-paced style favored by the coach. He roomed and associated with other L-7s on the road and at home, but he also spent a lot of his time with regular students. After graduation he got a job from a booster working for a life insurance company.

CONSTELLATION OF ROLE-SET MEMBERS

Basketball players generally interacted within a circle that was largely determined by their athletic environment. Due to the obligations of their position, these other role-set members fell into three main categories: athletic, academic, and social. Within the athletic realm, in addition to their teammates, athletes related primarily to the coaching staff, secretaries, and athletic administrators. The coaching staff consisted of the head coach, his first assistant (recruiting, playing strategy), the second assistant (recruiting, academics, some scouting), a part-time assistant (scouting, tape exchange with other teams, statistics during games), a graduate assistant (running menial aspects of practices, monitoring study halls, tape analysis), the trainer (injuries, paramedical activities), and the team manager (laundry, equipment). Let us portray each of them briefly.

The coaching staff was led by the head coach, a black man of enormous musculature. Resembling a football player more than a basketball player, he had the ability to bellow forcefully, like a bull, when he was angry. More often, however, he was stern but compassionate. He had high standards, he gave tough advice, and he firmly commanded respect, but he was sensitive and cared deeply about his players; once he recruited them, he felt responsible for them. Gentle by nature, he could address even the smallest child with sincerity and interest. Yet within his inner circle he was given to raucous practical jokes, magic tricks, and unrestrained guffawing, and the pranks he had gotten into with his long-time friend and first assistant coach were notorious. Extremely intelligent and insightful, he had achieved his current position, in large part, through his ability to figure things out logically, and by applying his keen intuition about people and human nature. From a poor background, he had worked his way up through the coaching ranks, starting at the high school level. Like most coaches, he was passionate about his job, and identified himself completely by it (even his wife referred to him as "Coach").

During our five-year tenure with the team, the position of first assistant coach was held by two different people. When we first came to the University, we met Stubbs, Coach's closest friend and longtime compatriot. One could not help but like this jovial and outgoing black man instantly. Warm, caring, and with a huge belly that shook whenever he laughed, Stubbs would engage anybody in conversation about basketball at any time and any place. His demeanor conveyed the flavor of his down-home Southern upbringing, with his lack of pretension and colorful expressions. He worked hard on developing rapport with prospective players and their families, and was an outstanding recruiter not only because he had a flair for recognizing talent (he had spotted many diamonds-in-the-rough), but because of his natural gregariousness. He was a good foil for Coach because he could play "good cop" to some of Coach's "bad cop" routines, and because he knew him well enough to break him out of his brooding when the heat of the season got too intense. After three seasons at the University he left the team to assume a head coaching job at another university. His post was filled by Stanger, a black man distinct from Stubbs in both character and style. Quiet and unassuming, Stanger struck people as somewhat cold and distant. Tall, thin, and dapper, he was not a typical recruiter. He was less successful attracting players to the team because he had difficulty getting prospects to warm up to him. In fact, due largely to his withdrawn and formal demeanor, no one on the team really got to know him very well, even after two years.

The second assistant, a young man named Mickey, was the highest-ranking white coach on the team. Originally brought to the University as a graduate assistant, he was promoted before the start of his (and Coach's) first season. He had the primary responsibility for players' academic schedules and performance, in addition to his recruiting duties, which increased as he gained experience. Handsome, well-dressed, muscular, and macho, he conveyed the feeling of solid Midwestern beliefs and values. He was friendly but somewhat reserved, trying to fill the role of coach despite his youth. Due partly to his inexperience, he lacked confidence for the first few years and had difficulty treading between the players, to whom he was closer in age, and the coaches, whose status he shared. His job, however, required that he make decisions and command respect. He eventually grew into the role, and became fairly authoritarian and rule-oriented. Grateful to Coach for his opportunity, and wanting to learn everything he could, he displayed unquestioning loyalty to the program.

Two different individuals also held the job of part-time assistant over the course of our five years. Fielding was tall, thin, white, bespeckled, and balding, and looked like a former basketball player. Not youthful like Mickey, he had tried for several years to leave basketball but found himself continually drawn back to the game. He therefore got a high school coaching job and kept applying for slots on college teams, anything to get his foot in the door. Fielding was a gym rat and a real student of the game. He liked both the politics of the game and talking strategy for hour upon hour. He traveled throughout the region, while he worked at the University, to talk with legendary retired coaches about their experiences, the history of basketball, and his theories of offense and defense. These people eventually advised and helped him, writing letters that assisted him in landing a coaching job at a nearby junior college. When he left, Coach hired Gordon in his place. Shorter, heavier, and more swarthy, Gordon shared Fielding's intense commitment and involvement in the sport. He had formerly been a full-time assistant coach on an unsuccessful program, but left it to work part-time for the University, because it had a successful program from which he thought he could more readily springboard into a coaching career. Beyond riding Coach's coattails, he (like Fielding) would often write to other coaches for jobs and to gain experience and sponsorship. Through his position at the University he had the opportunity to network on the road, meeting members of the coaching fraternity. At home, he had pet projects, players he had singled out to take under his wing. This came at some risk because, like Fielding, Gordon was an "X's and O's" man (that is, they both approached the game through methodical, set offenses and defenses), and this did not fit into Coach's run-and-gun style of playing (bringing the ball rapidly up court, finding an opening in the defense, and taking a quick shot or making a quick pass to the "open man").

Several graduate assistants worked with the team over five years. Todd was a fairly representative example. Fresh out of college, he was there to get his start in coaching any way he could. He would probably end up coaching high school, but might make it to the college level if he "caught a break." He was therefore working on his Master's degree, a requisite for most college jobs and some of the better high school ones. Like all of the lower ranking team personnel, Todd was white, with a background more akin to the community and students than the upper ranking coaches and players. He was a go-fer, doing all the dirty work of the coaches, but still trying to maintain some degree of status over the players. At times he appeared hyperactive, as he rushed around, trying to make himself indispensable to the coaches. And like all the coaches except the head coach, he "dipped" (chewed tobacco). At times, the row of them on the bench during games appeared almost comical, as they tried to dip discreetly, spitting their brown tobacco-spit into styrofoam cups which they kept under their folding chairs.

At the bottom of the hierarchy was the team manager. This was a position that turned over every year or two, being filled by an undergraduate student who was on a "work-study" job. Most of these individuals were scrawny little acned white kids who were both efficient and officious. The players clearly regarded and treated them as punks, as they ran around collecting the towels, shirts, underwear, and jock straps, and filling the water cooler for each game. These kids were not interested in the program as a stepping stone to a professional career, like all the others, but were there for the hourly wage and the opportunity to bask in the team's reflected glory.

The final member of the coaching staff was the trainer. This individual filled a paramedical position, applying treatments, taping ankles/ knees/ wrists, and obtaining and dispensing medicines from the team's doctors. A thorough knowledge of anatomy, the effects of drugs, and the treatment of injuries was required, and younger trainers were the most professionalized. The team's trainer was an integral part of the program, making every road trip and developing a relationship with all the players. As with several other slots, we had a chance to observe two people in this job. The first was Harrington, a dark, handsome, well-dressed athletic man in his early thirties who fancied himself a ladies' man. Some of the players mistrusted him because they thought he looked out for the team doctors' interests over theirs; he did eventually leave to accept a position with the doctors when they opened a sports injury rehabilitation clinic. Flannigan took his place and was welcomed by the team members. Fair-haired and small-boned, he was not as flashy or assertive as Harrington, but did his job quietly and competently. He was a conservative Midwesterner with solid roots, a wife and a young child. Although these men differed in appearance and demeanor, they were both referred to as "Doc."

Mothering the whole group was Coach's secretary, Thelma. A plump, older woman with local roots, she often brought her granddaughter around the coaching offices. She was not the brightest or best educated of women, but she was kind and caring. She typed the players' papers, administered Coach's summer camps, and handled both his personal and professional finances. She knew more about him than any other person, and could be absolutely trusted to keep his confidences. While she was slow and old-fashioned, she brought a homey atmosphere to the office and was very personable in answering the phone and receiving visitors.

Secondary members of the athletic role-set included boosters, fans, athletic administrators (the Athletic Director, Sports Information Director, and their staffs), and members of the media. The boosters could be divided into three main groups: the elite, the inner core, and the outer ring. The elite consisted of a small circle of the big money men and their wives, numbering about two to five couples. These people contributed around $30,000 to $50,000 a year to the program, in addition to buying large blocks of prime stadium seats and hosting parties and banquets. For this kind of money they expected to have open access to the practice court at any time they wanted. The top two supporters, one a Jew and the other an old line WASP, both around 60, also had access to the locker room in the basketball stadium before and after games and during halftime. This gave them the most intense feeling of being a part of the team's inner circle. The elite boosters also liked to socialize with Coach and his wife (although Coach always kept some distance between himself and them). This meant that they could invite Coach over to dinner and expect him to make time for them in his schedule, or they could invite him to their country club for lunch or a round of golf and he would accept. Members of this top echelon of boosters maintain a steady connection to the program, despite the replacement of one coach after another. Some may feel a personal liking or comfort with one coach more than another and vary their friendship patterns accordingly, but they remain steadfast supporters of the school's athletic effort, concentrating their attention on one sport specifically. With this level of booster, the onus is on the coach to get along with them, no matter how he feels about them.

The next group of boosters, members of the University's "Double Dribblers" club, comprised the inner core of the team's supporters. About 60 to 70 people, either alone or in couples, belonged to this organization, for which they had to pay an annual fee ($1000 per year). This entitled them to purchase season tickets in the stadium, to buy elite "slam dunk" jackets, to attend official pregame or halftime fundraising functions (such as cocktail parties or dinners) either at the stadium or in a private club or restaurant. They also attended many of the weekly banquets held during the season. Some of these people were very rich, but many of them were not. They were usually businessmen or professionals from the upper or upper-middle classes. More than money, they gave significant amounts of their time to the team. Double Dribbler members were also invited to perform favors for the coach, either at home or on the road. Some of the most devoted members went to most of the away games. Boosters who drove large recreational vehicles to these games often were asked to transport players' family members or even coaches, the team manager, or trainer. Once there, they provided transportation for the team from the hotel to restaurants or the stadiums. They enjoyed the excitement and having the extra social life during basketball season provided by membership in this group. Attending team functions enabled them to visit more frequently with the coaches and players, and to single out particular players and get to know them. They could then invite specific individuals with whom they had developed a relationship over to their house for dinners or special occasions. Many of these people were older, in their fifties and sixties, and had either grown children or no children. Their contact with the players served almost as a substitute parental relationship. Some boosters took players on fishing trips in the summer, ran over to the dorms to help if one of their favorite players was sick or injured, or cared about them sincerely and treated them like close friends or family. Others saw the relationship more in terms of status, where they could boost their role among their friends and with other members of this club by belonging and giving to the team and players.

The final group, the "Cheetah" club (named after the University's mascot), represented the outer ring of boosters. These people joined a less expensive supporter club ($250 per year) and were entitled to buy season tickets in the stadium. While they did not have a regular round of social occasions to attend, they were occasionally invited to rallies and fundraising events. These people were essentially devoted fans.

Within the academic arena, athletes' role-set members consisted of professors, tutors, and students in their classes, and, to a lesser extent, academic counselors and administrators. The players also tended to regard their families as falling primarily into this realm, although family members clearly cared about their social lives and athletic performance as well.

Socially, athletes related to girlfriends, local friends, and other students (non-athletes), but most especially to other college athletes: the teammates and dormmates (football players) who were members of their peer subculture.

NOTES

1. Academically, attention has focused predominantly on several subjects; racism and sexism (American Institute for Research 1987–88; Chu 1989; Edwards 1983; Kiger and Lorentzen 1986; Lapchick 1990; Leonard 1986b); the academic performance of college athletes (American Institute for Research 1987–88; Baumann and Henschon 1986; Brede and Camp 1987; Chu 1989; Eitzen and Purdy 1986; Figler 1987; Henschen and Frey 1984; Hochfield 1987; Kiger and Lorentzen 1986; Purdy, Eitzen, and Hufnagel 1982; 1985; Kaney, Knapp, and Small 1983; Shapiro 1984; Underwood 1984); the political economy of collegiate sport (Chu 1989; Frey 1982b; Hart-Nibbrig and Cottingham 1986; Porto 1985; Sack 1985), and policy analysis/recommendations, many of which arose in response to journalistic exposes denouncing the problems, hypocrisy, and exploitation inherent in college athletics (Edwards 1984; Gerdy 1987; Hammel 1980; Hanford 1978; Lapchick and Malekoff 1987; Lapchick and Slaughter 1990; Odenkirk 1981; Porto 1984; Sack 1984; Sage 1987).

2. The dearth of ethnographic research on this topic is especially surprising for two reasons. First, given the attention focused on the problems of college athletics, in-depth research is the best way to investigate both its deviant and multi-faceted character and solutions (Ball 1975; Jonassohn, Turowetz, and Gruneau 1981). Second, as much academic research is carried out on college students, one might expect a captive population, such as athletes, to be overstudied. One possible explanation for this ethnographic void lies in the anti-sport bias held by many academicians; sport is somehow viewed as anti-intellectual (Snyder and Spreitzer 1989), the "toy department of life" (Novak 1976). Gathering participant-observation data on this subject requires a longitudinal research commitment, one that could incur stigma for the researcher(s) (Kirby and Corzine 1981). While these negative factors cannot be overlooked, the advantages outweigh the cost.

3. These range from those focusing on drawing attention to the immediate crisis (Axthelm 1980; Golenbock 1989; Kennedy 1974; Marcin 1983; Shaw 1972; Telander 1989; Underwood 1969; 1980a; 1980b), to the travelogue mode (Feinstein 1988), to the more detailed and ethnographically textured day-in-the-life genre (Feinstein 1986).

4. See Gallmeier (1989) for an insightful elaboration of the differences between sport journalism and sport ethnography.

5. While our findings have implications for such dimensions as race, community, deviance, social control, and culture, these issues require complex treatment and fall outside of our primary concern. We will address them tangentially as they are indicated by our data, while focusing predominantly on the socialization of college athletes.

6. In this way we add to other theories bridging these two perspectives, such as structural symbolic interactionism or identity theory (Handel 1979; Heiss 1981; Stokes and Hewitt 1976; Stryker 1964, 1968, 1980) and processual role theory or role-identity theory (McCall and Simmons 1978; Turner 1962; Zurcher 1977).

7. Stryker (1968, 1980) used the term "salient," while McCall and Simmons (1978) spoke of certain roles being more "prominent" for an individual's self. The phrase "real self" is Turner's (1978).

8. At the height of its success, the University's recruiting budget swelled to around $80,000.

9. Yet the program's comparability to those at other, more average schools is evidenced by the fact that as soon as the charismatic and successful coach moved on to a more prestigious position, the University's basketball team once again lapsed into mediocrity.

10. Our estimates of team members' class backgrounds are rough, based on their parents' occupation, the neighborhoods from which they came, and their cultural upbringing.

11. The identifiers we have selected for these four typologies of players are drawn from their own terms for each other. While they are slightly pejorative in nature, this negativity is characteristic of the athletes' peer subculture, which was often sardonic in tone, and is not meant to reflect our assessments of the groups. Throughout the book, however, we draw extensively on direct quotations obtained from players and coaches during interviews and casual conversations. In transcribing these we have tried to preserve the flavor and character of the members' "argot" with as much integrity as possible. For this purpose we have created phonetically based spellings to reproduce common features of their pronunciation and left their grammar intact. This is done to pay homage to the speaking patterns we learned and appreciated during the course of this research. For an excellent discussion of this technique and its rationale, see Blauner (1989).

12. These clique names were not used at all times by all players and coaches in referring to the various groups. Team members generally regarded players as clumped into informal friendship groups whose membership was easily discernible. Occasionally these groups were identified by people according to a leading member of the clique, such as "Brandon and them," or "Marcus and those boys." At other times, however, descriptor names, such as the ones we present, were used. Because team personnel did not rigidly divide up the players into cliques or use the clique divisions on organized occasions such as in practices, there were no formal names for the groups. Instead, members of these groups were informally typologized by these monikers in several ways: by the coaches in discussing issues relating to social control (how much trouble various groups of players were going to get into), academics (who was going to class and getting decent grades), and playing ability or dependability on the court (who had heart); and by players in contrasting their friends with members of other social groups with regard to specific behavior such as hanging out (who they were hanging with and what they were going to do), people's willingness to get into a fight (who had heart), academics (who was showing them up), and gossiping about interpersonal team behavior (who was snitching on whom, who was behaving disreputably). Clique monikers were thus drawn from the most common players' and coaches' composite usage, and were usually used in complaining about or making fun of other groups, a practice common to all team personnel.

13. Fine's (1987) study of Little League also showed the importance accorded to toughness, competitiveness, effort, heart, and tolerance of pain in this youthful sport culture.

14. As Rudman (1986) has noted, this attitude is characteristic of ghetto blacks due to a combination of racial and social structural influences.

REFERENCES

Becker, Howard, Blanche Geer, Everett Hughes, and Anselm Strauss, 1961. *Boys in White.* Chicago: University of Chicago Press.

Blumer, Herbert, 1969. *Symbolic Interactionism.* Englewood Cliffs, N.J.: Prentice-Hall.

Coakley, Jay J., 1986. *Sport in Society.* Third Edition. St. Louis: Mosby. [Second edition, 1982.]

Cooley, Charles H., 1962. *Social Organization.* New York: Scribners.

Edwards, Harry, 1985. Beyond symptoms: Unethical behavior in American collegiate sport and the problem of the color line: *Journal of Sport and Social Issues* 9:3–11.

Frey, James H., 1982a. Boosterism, scarce resources and institutional control: The future of American intercollegiate athletics. *International Review of Sport Sociology.* 17: 53–70.

McCall, George J. and Jerry L. Simmons, 1978. *Identities and Interaction*. New York: Free Press.

Mead, George H., 1934. *Mind, Self and Society*. Chicago: University of Chicago Press.

Merton, Robert K., 1957. The role-set: Problems in sociological theory. *British Journal of Sociology*. 8:106–20.

Miller, Walter B., 1958. Lower class culture as a generating milieu of gang delinquency. *Journal of Social Issues*. 14:5–19.

Moffatt, Michael, 1989. *Coming of Age in New Jersey*. New Brunswick, N.J.: Rutgers University Press.

Strauss, Anselm, 1978. *Negotiations*. San Francisco: Jossey-Bass.

Turner, Ralph H., 1962. Role taking: Process versus conformity. In A. M. Rose, ed., *Human Behavior and Social Processes*, pp. 20–40. Boston: Houghton-Mifflin.

———— 1978. The role and the person. *American Journal of Sociology* 84:1–23.

17

An Introduction to McDonaldization

GEORGE RITZER

Modernization of societies means change in the groups and organizations that make up societies. The trend in modernization is toward increasing efficiency and cost-effective operations. Ritzer describes this process taking place around the world in many economic sectors; he calls this trend "McDonaldization." In the following excerpt from his book, he describes Max Weber's theory, which lies behind the trend.

As you read, consider these questions:

1. How do Weber's ideas on bureaucracy help explain the idea of McDonaldization?

2. In what places do you see evidence of McDonaldization?

GLOSSARY

Bureaucracy An organizational model designed to perform tasks efficiently.

Formal rationality Optimum means to a given end, or most efficient means of accomplishing a given task.

RAY KROC, THE GENIUS behind the franchising of McDonald's restaurants, was a man with big ideas and grand ambitions. But even Kroc could not have anticipated the astounding impact of his creation. McDonald's is one of the most influential developments in twentieth-century America. Its reverberations extend far beyond the confines of the United States and the fast-food business. It has influenced a wide range of undertakings, indeed the way of life, of a significant portion of the world. And that impact is likely to expand at an accelerating rate.

However, this is *not* a book about McDonald's, or even the fast-food business, although both will be discussed frequently throughout these pages. Rather, McDonald's serves here as the major example, the "paradigm," of a wide-ranging process I call *McDonaldization*, that is,

the process by which the principles of the fast-food restaurant are coming to dominate more

Thousand Oaks: Pine Forge Press, 1993 (excerpts pp. 1–3, 18–24).

and more sectors of American society as well as of the rest of the world.

As you will see, McDonaldization affects not only the restaurant business, but also education, work, health care, travel, leisure, dieting, politics, the family, and virtually every other aspect of society. McDonaldization has shown every sign of being an inexorable process by sweeping through seemingly impervious institutions and parts of the world.

McDonald's success is apparent: in 1993 its total sales reached $23.6 billion with profits of almost $1.1 billion. The average U.S. outlet has total sales of approximately $1.6 million in a year. Many entrepreneurs envy such sales and profits and seek to emulate McDonald's success. McDonald's, which first began franchising in 1955, opened its 12,000th outlet on March 22, 1991. By the end of 1993, McDonald's had almost 14,000 restaurants worldwide.

The impact of McDonaldization, which McDonald's has played a central role in spawning, has been manifested in many ways:

- The McDonald's model has been adopted not only by other budget-minded hamburger franchises such as Burger King and Wendy's, but also by a wide array of other low-priced fast-food businesses. Subway, begun in 1965 and now with nearly 10,000 outlets, is considered the fastest-growing of these businesses, which include Pizza Hut, Sbarro's, Taco Bell, Popeye's, and Charley Chan's. Sales in so-called "quick service" restaurants in the United States rose to $81 billion by the end of 1993, almost a third of total sales for the entire food-service industry. In 1994, for the first time, sales in fast-food restaurants exceeded those in traditional full-service restaurants, and the gap between them is projected to grow.
- The McDonald's model has also been extended to "casual dining," that is, more "upscale," higher-priced restaurants with fuller menus. For example, Outback Steakhouse and Sizzler sell steaks, Fuddrucker's offers

"gourmet" burgers, Chi-Chi's and Chili's sell Mexican food, The Olive Garden proffers Italian food, and Red Lobster purveys . . . you guessed it.

- McDonald's is making increasing inroads around the world. In 1991, for the first time, McDonald's opened more restaurants abroad than in the United States. As we move toward the next century, McDonald's expects to build twice as many restaurants each year overseas than it does in the United States. By the end of 1993, over one-third of McDonald's restaurants were overseas; at the beginning of 1995, about half of McDonald's profits came from its overseas operations. McDonald's has even recently opened a restaurant in Mecca, Saudi Arabia.
- Other nations have developed their own variants of this American institution. The large number of fast-food croissanteries in Paris, a city whose love for fine cuisine might lead you to think it would prove immune to fast food, exemplifies this trend. India has a chain of fast-food restaurants, Nirula's, which sells mutton burgers (about 80% of Indians are Hindus, who eat no beef) as well as local Indian cuisine. Perhaps the most unlikely spot for an indigenous fast food restaurant, war-ravaged Beirut of 1984, witnessed the opening of Juicy Burger, with a rainbow instead of golden arches and J.B. the Clown for Ronald McDonald. Its owners hoped that it would become the "McDonald's of the Arab world."
- Other countries with their own McDonaldized institutions have begun to export them to the United States. For example, the Body Shop is an ecologically sensitive British cosmetics chain with 893 shops in early 1993, 120 of which were in the United States, with 40 more scheduled to open that year. Furthermore, American firms are now opening copies of this British chain, such as The Limited, Inc.'s, Bath and Body Works.
- As indicated by the example of the Body Shop, other types of business are increasingly

adapting the principles of the fast-food business to their needs. Said the vice chairman of Toys Я Us, "We want to be thought of as a sort of McDonald's of toys." The founder of Kidsports Fun and Fitness Club echoed this desire: "I want to be the McDonald's of the kids' fun and fitness business." Other chains with similar ambitions include Jiffy-Lube, AAMCO Transmissions, Midas Muffler & Brake Shops, Hair Plus, H & R Block, Pearle Vision Centers, Kampgrounds of America (KOA), Kinder Care (dubbed "Kentucky Fried Children"), Jenny Craig, Home Depot, Barnes & Noble, Petstuff, and Wal-Mart (the nation's largest retailer with about 2,500 stores and almost $55 billion in sales).

- Almost 10% of America's stores are franchises, which currently account for 40% of the nation's retail sales. It is estimated that by the turn of the century, about 25% of the stores in the United States will be chains, by then accounting for a whopping two-thirds of retail business. About 80% of McDonald's restaurants are franchises.

Bureaucratization: Making Life More Rational

A *bureaucracy* is a large-scale organization composed of a hierarchy of offices. In these offices, people have certain responsibilities and must act in accord with rules, written regulations, and means of compulsion exercised by those who occupy higher-level positions. The bureaucracy is largely a creation of the modern Western world. Though earlier societies had organizational structures, they were not nearly as effective as the bureaucracy. For example, in traditional societies, officials performed their tasks on the basis of a personal loyalty to their leader. These officials were subject to personal whim rather than impersonal rules. Their offices lacked clearly defined spheres of competence, there was no clear hierarchy of positions, and officials did not have to obtain technical training to gain a position.

Ultimately, the bureaucracy differs from earlier methods of organizing work because it has a formal structure that, among other things, allows for greater efficiency. Institutionalized rules and regulations lead, even force, those employed in the bureaucracy to choose the best means to arrive at their ends. A given task is broken up into a variety of components, with each office responsible for a distinct portion of the larger task. Incumbents of each office handle their part of the task (usually following preset rules and regulations), often in a predetermined sequence. When each of the incumbents has, in order, handled the required part, the task is completed. Furthermore, in handling the task in this way, the bureaucracy has used what its past history has shown to be the optimum means to the desired end.

The roots of modern thinking on bureaucracy lie in the work of the turn-of-the-century German sociologist Max Weber. His ideas on bureaucracy are embedded in his broader theory of the rationalization process. In the latter, Weber described how the Occident managed to become increasingly rational—that is, dominated by efficiency, predictability, calculability, and nonhuman technologies that control people. He also examined why the rest of the world largely failed to rationalize. As you can see, McDonaldization is an extension of Weber's theory of rationalization. For Weber, the model of rationalization was the bureaucracy; for me, the fast-food restaurant is the paradigm of McDonaldization.

Weber demonstrated in his research that the modern Western world had produced a distinctive kind of rationality. Various types of rationality had existed in all societies at one time or another, but none had produced the type that Weber called *formal rationality*. This is the sort of rationality I refer to when I discuss McDonaldization or the rationalization process in general.

What is formal rationality? According to Weber, formal rationality means that the search by people for the optimum means to a given end is shaped by rules, regulations, and larger social structures. Individuals are not left to their own devices in searching for the best means of attaining a given objective. Weber identified this as a major development in the history of the world: Previously, people had been left to discover such mechanisms on their own or with vague and general guidance from larger value systems (religion, for example). After the development of formal rationality, they could use rules to help them decide what to do. More strongly, people existed in social structures that dictated what they should do. In effect, people no longer had to discover for themselves the optimum means to an end; rather, optimum means had already been discovered and were institutionalized in rules, regulations, and structures. People simply had to follow them. An important aspect of formal rationality, then, is that it allows individuals little choice of means to ends. Since the choice of means is guided or even determined, virtually everyone can (or must) make the same, optimal choice.

Weber praised the bureaucracy, his paradigm of formal rationality, for its many advantages over other mechanisms that help people discover and implement optimum means to ends. The most important advantages are the four basic dimensions of rationalization (and McDonaldization).

First, Weber viewed the bureaucracy as the most *efficient* structure for handling large numbers of tasks requiring a great deal of paperwork. As an example, Weber might have used the Internal Revenue Service, for no other structure could handle millions of tax returns so well.

Second, bureaucracies emphasize calculability, or the quantification of as many things as possible. Reducing performance to a series of quantifiable tasks helps people gauge success. For example, an IRS agent is expected to process a certain number of tax returns each day. Handling less than the required number of cases is unsatisfactory performance; handling more is excellence.

The quantitative approach presents a problem: little or no concern for the actual quality of work. Employees are expected to finish a task with little attention paid to how well it is handled. For instance, IRS agents may manage large numbers of cases and, as a result, receive positive evaluations from their superiors. Yet they may actually handle the cases poorly, costing the government thousands, or even millions, of dollars in uncollected revenue. Or, the agents may handle cases so quickly that taxpayers may be angered by the way the agents treat them.

Third, because of their well-entrenched rules and regulations, bureaucracies also operate in a highly *predictable* manner. Incumbents of a given office know with great assurance how the incumbents of other offices will behave. They know what they will be provided with and when they will receive it. Outsiders who receive the services the bureaucracies dispense know with a high degree of confidence what they will receive and when they will receive it. Again, to use an example Weber might have used, the millions of recipients of checks from the Social Security Administration know precisely when they will receive their checks and exactly how much money they will receive.

Finally, bureaucracies emphasize *control over people through the replacement of human with nonhuman technology*. As you will recall, nonhuman technologies (machines and rules, for example) tend to control people, while human technologies (hammers and pens, for example) tend to be controlled by people. Indeed, the bureaucracy itself may be seen as one huge nonhuman technology. Its nearly automatic functioning may be seen as an effort to replace human judgment with the dictates of rules, regulations, and structures. Employees are controlled by the division of labor, which allocates to each office a limited number of well-defined tasks. Incumbents must do those tasks, and no others, in the manner prescribed by the organization. They may not, in most cases, devise idiosyncratic ways of doing those tasks. Furthermore, by making few, if any, judgments, people begin to resemble human robots or computers. Having reduced people to this status, it is then possible to think about actually replacing human beings with machines. This has already occurred to some extent: in many settings, computers have taken over bureaucratic tasks once performed by humans. One can imagine that once the technology has been developed and priced reasonably, robots will begin replacing humans in the office.

Similarly, the bureaucracy's clients are also controlled. They may receive only certain services and not others from the organization. For example, the Internal Revenue Service can offer people advice on their tax returns, but not on their marriages. People may receive those services in a certain way only. For example, people can only receive welfare payments by check, not cash.

Thus, the bureaucracy, like the fast-food restaurant, is well-defined by four basic components of formal rationality: efficiency, predictability, quantification, and control through the substitution of nonhuman for human technology.

The bureaucracy also suffers from the *irrationality of rationality*. Like the fast-food restaurant, it is a dehumanizing place in which to work and by which to be serviced. As Ronald Takaki put it, these rationalized settings are places in which "the self was placed in confinement, its emotions controlled, and its spirit subdued." In other words, they are settings in which people cannot behave as human beings, where people are dehumanized.

But the irrationalities of bureaucracies hardly stop there. Instead of remaining efficient, bureaucracies can become increasingly inefficient because of "red tape" and the other pathologies associated with them. Bureaucracies often become unpredictable as employees grow unclear about what they are supposed to do and clients do not get the services they expect. The emphasis on quantification often leads to large amounts of poor-quality work. Because of these and other inadequacies, bureaucracies begin to lose control over those who work within and are served by them. Anger at the nonhuman technologies that can replace them often lead people to undercut or sabotage the operation of these technologies. All in all, what were designed as highly rational operations often end up quite irrational.

Although Weber was concerned about the irrationalities of formally rationalized systems such as bureaucracies, he was even more animated by what he called the "iron cage of rationality." In Weber's view, bureaucracies are cages in the sense that people are trapped in them, their basic humanity denied. Weber feared most that these systems would grow more and more rational and that rational principles would come to dominate an accelerating number of sectors of society. Weber anticipated a society of people locked into a series of rational structures, who could move only from one rational system to another. Thus, people would move from rationalized educational institutions to rationalized work places, from rationalized recreational settings to rationalized homes. Society would become nothing more than a seamless web of rationalized structures; there would be no escape.

A good example of what Weber feared is found in the contemporary rationalization of recreational activities. Recreation can be thought of as a way to escape the rationalization of daily routines. However, over the years these escape routes have themselves become rationalized, embodying the same principles as bureaucracies and fast-food restaurants. Of the many examples of the rationalization of recreation, take today's vacations. For those who wish to visit Europe, a package tour rationalizes the process. People can efficiently see, in a rigidly controlled manner, many sights while traveling in conveyances, staying in hotels, and eating in fast-food restaurants just like those at home. For those who wish to escape to the Caribbean, there are resorts such as Club Med that offer many routinized activities and where one can stay in predictable settings without ever venturing out into the unpredictability of native life on a Caribbean island. For those who wish to flee back to nature within the United States, rationalized campgrounds offer little or no contact with the unpredictabilities of nature. People can even remain within their RVs and enjoy all of the comforts of home—TV, VCR, Nintendo, CD player. These and legion other examples show that the escape routes from rationality have, to a large degree, become rationalized. With little or no way out, people do live to a large extent in the iron cage of rationality.

The fast-food restaurant can also be seen as part of a bureaucratic system; in fact, huge conglomerates now own many of the fast-food chains. Further, the fast-food restaurant has employed the rational principles pioneered by the bureaucracy. McDonald's has combined bureaucratic and other principles to help create McDonaldization.

QUESTIONS FOR DISCUSSION

For further discussion of this topic, see the Wadsworth Sociology Resource Center, "Virtual Society," *http://sociology.wadsworth.com,* under *Sociological Footprints,* by Cargan and Ballantine. You can respond to the discussion questions there or enter your own comments in the online chat forum.

SUGGESTED READINGS AND SOCIOLOGY INTERNET RESOURCES

See the Wadsworth Sociology Resource Center, "Virtual Society," *http://sociology.wadsworth. com,* for additional links, suggestions for further reading, and learning tools related to this chapter.

Either from the "Virtual Society" website or directly from your web browser, you may access InfoTrac College Edition, an online university library that includes over 700 popular and scholarly journals in which you can find articles related to the topics in this chapter.

CHAPTER 5

Stratification

SOME ARE MORE EQUAL THAN OTHERS!

THE DECLARATION OF INDEPENDENCE STATES THAT "all men are created equal"—a statement that is more helpful than factual. In the United States there exists a stratified class structure; that is, like all societies, we have a system of ranking. In the United States it is based on family background, wealth, age, sex, occupation, authority, and power. Also, we are all aware that some Americans are more equal than others. The existence of a stratified class structure raises two questions: First, what is the effect of having class rankings? Second, why do we tolerate such a system?

Class rankings significantly affect almost all facets of our lives. Unequal distribution of wealth and occupational opportunity creates inequality in life expectancy and in our chances for health and education; it also affects the kind of legal protection and justice we receive and, as we learned in the last two presidential elections, the ability to avoid the draft. Class status influences our values, our beliefs, our personalities, and hundreds of other aspects of life in society. Finally, ranking can determine our control over the life chances of others.

The first reading, by Hacker, describes this class structure and some further consequences of the system.

A related question is dealt with in the second reading of this chapter. Kerbo describes how the rich maintain the system in their own behalf.

As the above paragraph notes, class affects our beliefs. However, as Kahlenberg notes in the third reading of this chapter, a belief that may override class influences is that of race. He also notes how such appeals may cause people to vote against their best interests.

The second question in the opening paragraph asks why we tolerate such a system. A reason for the existence of inequality may lie in the meaning of the term *doublethink*. In his futuristic novel *1984*, George Orwell defined doublethink as holding two contradictory ideas at the same time and believing both of them. This is often done unconsciously, since we often do not examine our thoughts closely enough to note that they are, perhaps, contradictory. For example, an important measure of social justice is equality, and yet inequality not only exists in our society but is increasing.

Another reason for the rise in inequality is found in the distinction many make between the "deserving" and "undeserving" poor. That is, the public appears willing to help those who suffer inequality through no fault of their own. The difficulty with this idea is that both groups of poor get tainted with the same stereotypes—that the lower classes are inferior and bad. The result is discriminatory treatment, which constitutes a major deterrent to reducing inequality. In another aspect, the readings also reflect the self-fulfilling prophecy noted in Chapter 1.

A second reason for the continuation of inequality is given by Herbert Gans in the final reading in this chapter. He claims that poverty is functional; that is, it performs various important tasks for society.

As you consider the readings in this chapter, ask yourself realistically what your mobility opportunities are, what the meaning of your answer is for you and for others in various classes, and what, if anything, can be done to increase your opportunities and the opportunities of others, since they affect you.

18
Money and the World We Want

ANDREW HACKER

Two issues are covered in this reading: First, is there a class structure and what is its breakdown? Second, what are the consequences of this class structure?

As you read, ask yourself the following questions:

1. What problems result from the way the United States distributes income?

2. What changes would you make to help all people to be fully Americans?

GLOSSARY

Corollary A natural result or easily drawn conclusion.
Esoteric Understood by those with special knowledge.
Skepticism Doubt, unbelief.

THE THREE DECADES SPANNING 1940 to 1970 were the nation's most prosperous years. Indeed, so far as can happen in America's kind of economy, a semblance of redistribution was taking place. During this generation, the share of national income that went to the bottom fifth of families rose by some fractions of a point to an all-time high, while that received by the richest fifth fell to its lowest level. This is the very period of shared well-being that many people would like to re-create in this country.

Yet in no way is this possible. It was an atypical era, in which America won an adventitious primacy because of a war it delayed entering and its geographic isolation which spared it the ravages that the other combatants suffered.

America's postwar upsurge lasted barely three decades. The tide began to turn in the early 1970s. Between 1970 and 1980, family income rose less than 7 percent. And that small increase resulted entirely from the fact that additional family members were joining the workforce. Indeed, the most vivid evidence of decline is found in the unremitting drop in men's earnings since 1970. Averages and medians, however, conceal important variations, such as that some American households have done quite well for themselves during the closing decades of the century. Those with incomes of $1 million or more have reached an all-time high, as are families and individuals making over $100,000. By all outward appearances, there is still plenty of money around, but it is landing in fewer hands. Yet it is by no means apparent that people are being paid these generous salaries and options and fees because their work is adding much of substance to the nation's output. Indeed, the coming century will test whether an economy can flourish by exporting most notably action movies and flavored water.

The term *upper class* is not commonly used to describe the people in America's top income tier since it connotes a hereditary echelon that passes on its holdings from generation to generation. Only a few American families have remained at the very top for more than two or three generations, and even when they do, as have the du Ponts and Rockefellers, successive descendants slice the original pie into smaller and smaller pieces. So if America does not have a "class" at its apex, what should we call the people who have the most money? The answer is to refer to them as we usually do, as being "wealthy" or "rich."

The rich, members of the 68,064 households who in 1995 filed federal tax returns that declared a 1994 income of $1 million or more, have varied sources of income. . . . For present purposes, wealth may be considered holdings that would yield you an income of $1 million a year without your having to put in a day's work. Assuming a 7 percent return, it would take income-producing assets of some $15 million to ensure that comfort level.

Unfortunately, we have no official count of how many Americans possess that kind of wealth. It is measurably less than 68,064, since those households report that the largest segment of their incomes come from salaries. Indeed, among the chairmen of the one hundred largest firms, the median stock holding is only $8.4 million. A liberal estimate of the number of wealthy Americans would be about thirty thousand households, one-thirtieth of one percent of the national total.

This still puts almost 90 percent of all Americans between the very poor and the wealthy and the rich. One way to begin to define this majority is by creating a more realistic bottom tier than merely those people who fall below the official poverty line. Since Americans deserve more than subsistence, we may set $25,000 a year for a family of three as a minimum for necessities without frills. And even this is a pretty bare floor. Indeed, only about 45 percent of the people questioned in the Roper-Starch poll felt that their households could "get by" on $25,000 a year. In 1995, almost 20 million families—28.4 percent of the total—were living below that spartan standard. This stratum is a varied group. . . . For over a third, all of their income comes from sources other than employment, most typically Social Security and public assistance. Over 43.3 percent of the households have only one earner, who at that income level is usually a woman and the family's sole source of support. The remainder consists of families where two or more members have had jobs of some kind, which suggests sporadic employment at close to the minimum wage. Whatever designation we give to these households— poor or just getting by—their incomes leave them deprived of even the more modest acquisitions and enjoyments available to the great majority of Americans.

The question of who belongs to America's middle class requires deciding where to place its upper and lower boundaries. While it is difficult to set precise boundaries for this stratum, it is accurate to say that, by one measure or another, most Americans fit into a middle class.

While it is meaningless to classify the households in the middle of America's income distribution because this stratum is so substantial and it encompasses such a wide range of incomes, meaningful divisions can be drawn in our country's overall income distribution.

This book has sought to make clear that the prominent place of the rich tells us a great deal about the kind of country we are, as does the growing group of men and women with $100,000 salaries and households with $100,000 incomes. The same stricture applies to the poor, who, while not necessarily increasing in number, are too often permanently mired at the bottom. This noted, a three-tier division can be proposed, with the caveat that how you fare on a given income can depend on local costs and social expectations.

. . .

That the rich have become richer would seem to bear out Karl Marx's well-known prediction. The nation's greatest fortunes are substantially larger than those of a generation ago. Households with incomes exceeding $1 million a year are also netting more in real purchasing power. At a more mundane level, the top 5 percent of all households in 1975 averaged $122,651 a year; by 1995, in inflation-adjusted dollars, their average annual income had ascended to $188,962.

But Marx did not foresee that the number of rich families and individuals would actually increase over time. Between 1979 and 1994, the number of households declaring incomes of $1 million or more rose from 13,505 to 68,064, again adjusting for inflation. In 1996, *Forbes's* 400 richest Americans were all worth at least $400 million. In 1982, the year of the magazine's first list, only 110 people in the 400 had holdings equivalent to the 1996 cutoff figure. In other words, almost three-quarters of 1982's wealthiest Americans would not have made

How Families Fare: Three Tiers			
Tier	**Income Range**	**Number of Families/Percentage**	
Comfortable	Over $75,000	12,961,000	18.6%
Coping	$25,000–$75,000	36,872,000	53.0%
Deprived	Under $25,000	19,764,000	28.4%

the 1996 list. And families with incomes over $100,000—once more, in constant purchasing power—increased almost threefold between 1970 and 1995, rising from 3.4 percent to 9 percent of the total. During the same period, the group of men making more than $50,000 rose from 12 percent of the total to 17 percent.

It is one thing for the rich to get richer when everyone is sharing in overall economic growth, and it is quite another for the better off to prosper while others are losing ground or standing still. But this is what has been happening. Thus 1995 found fewer men earning enough to place them in the $25,000 to $50,000 tier compared with twenty-five years earlier. And the proportion in the bottom bracket has remained essentially unchanged. But this is not necessarily a cause for cheer. In more halcyon times, it was assumed that each year would bring a measure of upward movement for people at the bottom of the income ladder and a diminution of poverty.

The wage gap of our time reflects both the declining fortunes of many Americans and the rise of individuals and households who have profited from recent trends. Economists generally agree on what brought about static wages and lowered living standards. In part, well-paying jobs are scarcer because goods that were once produced here, at American wage scales, are now made abroad and then shipped here for sale. A corollary cause has been the erosion of labor unions, which once safeguarded generous wages for their members. Between 1970 and 1996, the portion of the workforce represented by unions fell from 27 percent to 15 percent. Today, the most highly organized occupations are on public payrolls, notably teachers and postal employees. Only 11 per-

cent of workers in the private sector belong to unions. For most of the other 89 percent this means that their current paychecks are smaller than they were in the past.

Analysts tend to differ on the extent to which the paltriness of the minimum wage has lowered living standards. Despite the 1996 increase, the minimum wage produces an income that is still below the poverty line. Even more contentious is the issue of to what degree immigrants and aliens have undercut wages and taken jobs once held by people who were born here. We can all cite chores that Americans are unwilling to do, at least at the wages customarily paid for those jobs. Scouring pots, laundering clothes, herding cattle, and caring for other people's children are examples of such tasks. At the same time, employers frequently use immigrants to replace better-paid workers, albeit by an indirect route. The most common practice is to remove certain jobs from the firm's payroll and then to hire outside contractors, who bring in their own staffs, which are almost always lower paid and often recently arrived in this country.

Most economists agree that the primary cause of diverging earnings among American workers has been the introduction of new technologies. These new machines and processes are so esoteric and complex that they require sophisticated skills that call for premium pay. In this view, the expanded stratum of Americans earning over $50,000 is made up largely of men and women who are adept at current techniques for producing goods, organizing information, and administering personnel. New technologies have reduced the number of people who are needed as telephone operators, tool and die makers, and aircraft mechanics. Between 1992 and 1996, Delta Airlines was scheduling the same number of flights each year, even though it was discharging a quarter of its employees. Closer to the top, there is a strong demand for individuals who are skilled at pruning payrolls. And this cadre has been doing its work well. . . . [I]n 1973, the five hundred largest industrial firms employed some 15.5 million men and women. By 1993, these firms employed only 11.5 million people. But this reduction in the industrial workforce amounted to more than a loss of 4 million positions. Given the increase in production that took place over this twenty-year period, it meant a comparable output could be achieved in 1993 with half as many American workers as were needed in 1973. In fact, American workers account for an even smaller share of the output, since many of the top five hundred industrial firms are having more of their production performed by overseas contractors and subsidiaries.

So what special skills do more highly educated workers have that make them eligible for rising salaries? In fact, such talents as they may display have only marginal ties to technological expertise. The years at college and graduate school pay off because they burnish students' personalities. The time spent on a campus imparts cues and clues on how to conduct oneself in corporate cultures and professional settings. This demeanor makes for successful interviews and enables a person to sense what is expected of him during the initial months on a job.

Does America's way of allocating money make any sense at all? Any answer to this question requires establishing a rationale. The most common explanation posits that the amounts people get are set in an open market. Thus, in 1995, employers offered some 14.3 million jobs that paid between $20,000 and $25,000, and were able to find 14.3 million men and women who were willing to take them. The same principle applied to the 1.7 million positions pegged at $75,000 to $85,000, and to the dozen or so corporate chairmen who asked for or were given more than $10 million. By the same token, it can be argued that

market forces operate at the low end of the scale. Wal-Mart and Pepsico's Pizza Hut cannot force people to work for $6.50 an hour, but those companies and others like them seem to attract the workers they need by paying that wage.

The last half century gave many groups a chance to shield themselves from the labor auction.

But many, if not most, of these protections are no longer being renewed. The up-and-coming generations of physicians, professors, and automobile workers are already finding that they must settle for lower pay and fewer safeguards and benefits. The most graphic exception to this new rule has been in the corporate world, where boards of directors still award huge salaries to executives, without determining whether such compensation is needed to keep their top people from leaving or for any other reason. They simply act as members of an inbred club who look after one another. Only rarely do outside pressures upset these arrangements, which is why they persist.

A market rationale also presumes that those receiving higher offers will have superior talents or some other qualities that put them in demand. Some of the reasons why one person makes more than another make sense by this standard. Of course, a law firm will pay some of its members more if they bring in new business or satisfy existing clients. Two roommates have just received master's degrees with distinction, one in education and the other in business administration. The former's first job is teaching second-graders and will pay $23,000. The latter, at an investment firm, will start at $93,000. About all that can be said with certainty is that we are unlikely to arrive at a consensus on which roommate will be contributing more to the commonweal.

Would America be a better place to live, and would Americans be a happier people, if incomes were more evenly distributed? Even as the question is being posed, the answers can be anticipated.

One side will respond with a resounding "Yes!" After which will come a discourse on how poverty subverts the promise of democracy, while allowing wealth in so few hands attests to our rewarding greed and selfishness. There would be far less guilt and fear if the rich were not so rich and no Americans were poor. But the goal, we will be told, is not simply to take money from some people and give it to others. Rather, our goal should be to create a moral culture where citizens feel it is right to have no serious disparities in living standards. Other countries that also have capitalist economies have shown that this is possible.

Those who exclaim "No way!" in response to the same question will be just as vehement. To exact taxes and redistribute the proceeds is an immoral use of official power since it punishes the productive and rewards the indolent. And do we want the government telling private enterprises what wages they can offer? The dream of economic equality has always been a radical's fantasy. Apart from some primitive societies, such a system has never worked. If you want efficiency and prosperity, and almost everyone does, then variations in incomes are part of the equation.

These different responses arise in part from disparate theories of human nature. Since the earliest days of recorded history, philosophers have disagreed over whether our species is inherently competitive or cooperative.

Also at issue is whether greater economic equality can only be achieved by giving oppressive powers to the state, either to limit incomes through heavy taxes or by setting levels of earnings.

Of course, the price that America pays for economic inequality is its persisting poverty.

In one way or another all Americans will pay the high costs of poverty. California now spends more on its prison system than it does on higher education, and other states will soon be following suit. Bolstering police forces is hardly cheap: upward of $75,000 per officer, when overtime and benefits are added in. Being poor means a higher chance of being sick or being shot, or bearing low-birth-weight babies, all of which consume medical resources and have helped to make Medicaid one of the costliest public programs. In addition, poor Americans now repre-

sent the fastest-growing group of AIDS victims: mainly drug users and the women and children they infect. Generally, $100,000 worth of medical treatment is spent on each person dying of AIDS. The poor also have more of their children consigned to "special education" classes, which most never leave and where the tab can reach three times the figure for regular pupils.

Additional expenses are incurred by families who put as much distance as possible between themselves and the poor. Doing so often entails the upkeep of gated communities, security systems, and privately supplied guards. Yet these are expenses better-off Americans readily bear. They are willing to foot the bills for more prisons and police, as well as the guns they keep in their bedrooms and the alarms for their cars. Indeed, their chief objection is to money given to non-married mothers who want to be at home with their pre-school-age children. Virtually every such penny is begrudged, followed by the demand that these mothers take jobs even before their children go to kindergarten. While the imposition of work may make some taxpayers cheerier, it will not do much to close the income gap.

How disparities in income affect the nation's well-being has long been debated by economists. Much of the argument centers on what people do with their money. The poor and those just trying to cope devote virtually all of what they have to necessities plus the few extras they can afford. If their incomes were raised, they would obviously spend more, which would create more demand and generate more jobs. It should not be forgotten that the year 1929, which was noted for a severe imbalance in incomes, gave us a devastating economic depression that lasted a decade.

The traditional reply to critics of economic inequality has been that we need not only the rich, but also a comfortably off class, who are able to put some of their incomes into investments. In other words, disparities give some people more than they "need," which allows them to underwrite the new enterprises that benefit everyone. While there is obvious validity to this argument, it should be added that much of this outlay now goes to paper contrivances, which have only a remote connection with anything productive, if any at all. Nor are the rich as necessary as they may once have been, since institutions now supply most invested capital. Metropolitan Life, Merrill Lynch, Bank America, and the California Public Employees Pension Fund put substantially more into new production than do the 68,064 families with $1 million incomes.

In one sphere, the income gap comes closer to home. Most young Americans will not live as well as their parents did. Indeed, in many instances this is already occurring. A generation ago, many men had well-paid blue-collar jobs; now their sons are finding that those jobs are no longer available. In that earlier era, college graduates could enter a growing managerial stratum; today, firms view their payrolls as bloated and are ending the security of corporate careers. . . .

The patterns of decline prevail if the entire nation is viewed as the equivalent of family. Federal programs now award nine times as much to retirees as they do to the nation's children, so senior citizens as a group fare better than younger Americans. Twice as many children as Americans over the age of sixty live in households below the poverty line. (And as death approaches, the government is more generous: almost 30 percent of the total Medicare budget is spent on the terminal year of elderly patients' lives.) Many retired persons have come to view a comfortable life as their entitlement, and have concluded that they no longer have obligations to repay. Grandparents tend to support campaigns for the rights of the elderly, not for school bonds and bigger education budgets.

In the end, the issue may be simply stated: what would be required for all Americans — or at least as many as possible — to make the most of their lives? . . .

Poverty takes its greatest toll in the raising of children. With a few exceptions, being poor consigns them to schools and surroundings that do little to widen their horizons. The stark fact is that we have in our midst millions of bright and talented children whose lives are fated to be a fraction of what they might be. And by any moderate standard, deprivation extends above the official poverty line. In most of the United States, families with incomes of less than $25,000 face real limits to the opportunities open to their children. Less than one child in ten from these households now enters and graduates from college. The statistics are apparent to outside observers, and to the children themselves, who very early on become aware of the barriers they face. And from this realization results much of the behavior that the rest of the society deplores. The principal response from solvent Americans has been to lecture the poor on improving their ways.

No one defends poverty, but ideologies differ on what can or should be done to alleviate it. Conservatives generally feel it is up to the individual: those at the bottom should take any jobs they can find and work hard to pull themselves up. Hence the opposition to public assistance, which is seen as eroding character and moral fiber. Indeed, conservatives suggest that people will display character and moral fiber if they are made to manage on their own.

Not many voting Americans favor public disbursements for the poor or even for single working mothers who cannot make ends meet. Most American voters have grown weary of hearing about the problems of low-income people. Yet even those who are unsettled by the persistence of income imbalances no longer feel that government officials and experts know how to reduce the disparities.

Of course, huge redistributions occurs everyday. Funds for Social Security are supplied by Americans who are currently employed, providing their elders with pensions that now end up averaging $250,000 above what their own contributions would have warranted. Agricultural subsidies give farmers enough extra cash to ensure that they will have middle-class comforts. The same subventions furnish farms owned by corporations with generous profit margins. . . .

In contrast, there is scant evidence that public programs have done much for the bottom tiers of American society. Despite the New Deal and the Great Society, including public works and public assistance, since 1935, share of income going to the poorest fifth of America's households has remained between 3.3 percent and 4.3 percent. Thus, if many elderly Americans have been raised from poverty, it is clear that younger people are now taking their places. . . .

All parts of the population except the richest fifth have smaller shares of the nation's income than they did twenty years ago. The gulf between the best-off and the rest shows no signs of diminishing, and by some political readings this should mean increased tensions between the favored fifth and everyone else. But declines in living standards have not been so severe or precipitous as to lead many people to question the equity of the economic system. The economy has ensured that a majority of Americans remain in moderate comfort and feel able to count their lives a reasonable success. Airline reservationists making $14,000 do not consider themselves "poor," and no one tells them that they are. Thus a majority of Americans still see themselves as middle class, and feel few ties or obligations to the minority with incomes less than their own.

Given this purview, why should the way America distributes its income be considered a problem? At this moment, certainly, there is scant sentiment for imposing further taxes on the well-to-do and doing more for the poor. As has been observed, there is little resentment felt toward the rich; if anything, greater animus is directed toward families receiving public assistance. Nor is it regarded as untoward if the well-off use their money to accumulate luxuries while

public schools must cope with outdated textbooks and leaking roofs. Although this book is about money, about why some have more and others less, it should not be read as a plea for income redistribution. The reason is straightforward: if people are disinclined to share what they have, they will not be persuaded by a reproachful tone. Rather, *Money*'s aim is to enhance our understanding of ourselves, of the forces that propel us, and the shape we are giving to the nation of which we are a part.

How a nation allocates its resources tells us how it wishes to be judged in the ledgers of history and morality. America's chosen emphasis has been on offering opportunities to the ambitious, to those with the desire and the drive to surpass. America has more self-made millionaires and more men and women who have attained $100,000 than any other country.

But because of the upward flow of funds, which has accelerated in recent years, less is left for those who lack the opportunities or the temperament to succeed in the competition. The United States now has a greater percentage of its citizens in prison or on the streets, and more neglected children, than any of the nations with which it is appropriately compared. Severe disparities—excess alongside deprivation—sunder the society and subvert common aims. With the legacy we are now creating, millions of men, women, children are prevented from being fully American, while others pride themselves on how much they can amass.

19
Upper-Class Power

HAROLD R. KERBO

It is important to note by what means the upper-class dominates the corporate structure and the political structure.

As you read, ask yourself the following questions:

1. What tactics do members of the upper class use in order to dominate?

2. What changes would you make to adjust the power of the upper class?

GLOSSARY

Common economic interests Extensive ownership by upper class families in many different corporations.
Economic elites Boards of directors and top executive officers of corporations.

Upper-Class Economic Power

IF WE HAVE AN upper class in this country that, because of its power, can be described as a governing class, by what means does it govern or dominate? . . . We will examine first how the upper class is said to have extensive influence over the economy through stock ownership, then turn to the question of economic power through extensive representation in major corporate offices.

STOCK OWNERSHIP

As some argue, the most important means of upper-class economic power lies in its ownership of the primary means of production. The upper class has power over our economy because of its control of the biggest corporations through stock ownership. . . .

Legally, the ultimate control of corporations is found not with top corporate executives, but with major stockholders. In a sense, top cor-

porate executives and boards of directors are charged with managing these corporations for the real owners—the stockholders. Stockholders have the authority to elect corporate directors who are to represent stockholder interests. These directors are then responsible for general corporate policy, including the task of filling top executive positions. The day-to-day management of the corporation is in turn the responsibility of the executive officers, who must generally answer to stockholders and their representatives on the board of directors.

Assuming for now that corporate authority actually operates this way (questions about this ideal power arrangement will be considered below), the largest stockholder or stockholders in a corporation should be in control. Thus, if we find that upper-class families have extensive stock ownership and that this stock is in major corporations, we can say that upper-class families dominate the American economy.

It is clear . . . that wealth is very unequally distributed in this country . . . more so even than family or personal income. One of the most important categories of wealth (because of its usual high return on investment) is corporate stock. . . . [One] percent of the people in this country owned *56.5 percent* of the privately held corporate stock, and only 0.5 percent of the people owned *49.3 percent* of the privately held corporate stock in the United States. Thus, from 1 to 0.5 percent of the people in this country . . . hold most of the privately owned corporate stock.

This concentration of private stock ownership is even more striking when we find that most of

In Social Stratification and Inequality, *2nd ed. McGraw-Hill, 1991. Reprinted by permission of the publisher.*

the remaining stock is controlled by large financial corporations (see U.S. Senate Committee on Governmental Affairs 1978a, 1980; Kerbo and Della Fave 1983, 1984). To the degree that the upper class also has a lot of influence over these financial corporations (such as banks with large amounts of stock control in other big corporations), the actual stock control of the upper class is much greater. . . .

Large amounts of stock held by a family afford economic power in many ways, but the most extensive power flowing from stock ownership comes when this stock is concentrated in a corporation to a sufficient degree to ensure control over the corporation. The amount of stock owned brings an equal number of votes toward electing the board of directors (who can hire and fire the managers) and deciding major issues that come before stockholders. Thus, it becomes important to know how much stock individuals or families hold in each major corporation in determining the control of that corporation.

In the early stages of industrialization in this country the control of corporations was fairly easy to estimate. We knew, for example, that the Rockefeller family controlled Standard Oil, the McCormick family controlled International Harvester, the Mellon family controlled Aluminum Company of America, and the Morgan family controlled Morgan Bank by virtue of their extensive stock ownership of these companies. But this concentration of stock ownership by specific families in one or a few corporations has changed greatly in recent decades. Few clearly family-controlled corporations . . . are found today.

Because of the wide distribution of stockholders in most corporations, government agencies and researchers agree that 5 to 10 percent ownership in a particular company by a family will often result in control of that company by the family.

A government study, however, found only thirteen of the top 122 corporations in this country to be clearly controlled by one family (see U.S. Senate Committee on Governmental Affairs 1978a:252). But we must emphasize *clearly* controlled. One of the problems in determining

control is that the ownership of stock in specific corporations is often hidden. For example, the owner of stock may be listed under meaningless names (such as street names) or under trusts and foundations (Zeitlin, 1974). To make the situation even more complex, corporations (especially banks) control stock in other corporations.

Consider the following situation: A family owns about 2 percent of the stock in corporation A with other families also owning about 2 percent each. In turn, this original family owns, say, 5 percent of the stock in corporation B (a bank) and 6 percent in corporation C (an insurance company). We find upon further investigation that company B (the bank) controls 4 percent of the stock in corporation A, and corporation C (the insurance company) controls 7 percent of the stock in corporation A. Who controls corporation A?

It *may* be that our original family does, through its stock in corporation A, as well as B and C. But other families own stock in A who in addition have much stock in corporations D and E. And (you are probably ahead of me), corporations D and E also control stock in corporation A! This example is not an exaggeration, as anyone will see in examining the data on stock ownership published in a Senate study (U.S. Senate Committee on Governmental Affairs 1978a, 1980). In the face of this complexity of wide stockholdings many researchers simply conclude that top managers of a corporation control by default. . . . But, as we will see below, this generalization also has many drawbacks.

In terms of upper-class dominance in the economy through stock ownership, all we can say is that its stock ownership is very extensive. But we cannot as yet arrive at firm conclusions about upper-class control of specific corporations. The data are hard to obtain, complex, and sometimes of questionable quality. Therefore, the arguments by ruling-class or governing-class theorists that an upper class dominates the economy through control of stock in major corporations cannot be completely confirmed.

One argument by ruling-class theorists is given more support; the upper class today has more extensive *common economic interests* than it once had. Rather than each family owning and controlling one corporation, today the families more often have extensive ownership in many corporations. They don't own each corporation in common, but they do have extensive common interests throughout top corporations in this country. There is a bond of common economic interests that helps cement a unity in defense of the top corporate structure as a whole.

UPPER-CLASS BACKGROUNDS OF ECONOMIC ELITES

Aside from actual stock ownership there is another possible means of upper-class leverage over the economy. After the authority of stockholders in a corporation we find the board of directors and top executive officers. We will call these people *economic elites*. The family backgrounds of these economic elites may be important in how they think, whom they trust, and what group interests they serve while making decisions in their positions of authority in the corporate world. Ruling-class theorists such as Domhoff believe that these economic elites often come from, or have backgrounds in, upper-class families. Thus, even if upper-class families may not own enough stock to control many corporations, their people are there in important positions of authority.

Domhoff (1967) has examined the directors from many top corporations. He has found . . . that of the top twenty industrial corporations, 54 percent of the board members were from the upper class; of the top fifteen banks, 62 percent were upper-class members; of the top fifteen insurance companies, 44 percent were upper-class members; of the top fifteen transportation companies, 53 percent were upper-class members; and of the top fifteen utility corporations, 30 percent were upper-class members. Clearly we find much overrepresentation by the upper class on these boards of directors when it is noted that the upper class accounts for only about 0.5 percent of the population.

In another study Soref (1976) took a random sample of board members from the top 121 corporations in the United States. Using Domhoff's definition of upper class, he found upper-class board members had more board positions in other companies (average of 3.49 for upper-class directors, 2.0 for others), and were more often members of board subcommittees that made important long-range decisions in the company.

Finally, in a massive study of institutional elites, Thomas Dye (1979, 1983) obtained background information on the boards of directors *and* top executive officers of the top 201 corporations in 1976 (those corporations controlling 50 percent of the assets in each type of corporation — industrial, financial, insurance, and utilities). This sample included 3,572 people who by our definition were economic elites. Using a list of 37 upper-class social clubs from Domhoff's work, Dye (1979:184) found that *44 percent* of these 3,572 economic elites were members of one or more of these upper-class clubs. . . . Thus, by Domhoff's definition, 44 percent of these people were members of the upper class. . . .

Upper-Class Political Power

The next questions of importance for ruling-class or governing-class theorists are the degree and means of political power exercised by the upper class. . . .

The potential impact of the federal government upon upper-class interests is clear. If the upper class is to maintain a position of dominance in the nation it is imperative that it have influence over the state as well as the economy. . . .

UPPER-CLASS PARTICIPATION IN GOVERNMENT

Research on direct participation by the upper class in government is focused heavily on the President's cabinet. Cabinet members are under the direction of the President, but because of the President's many concerns and lack of time in gathering all the needed information in making policy, the President must rely heavily upon cabinet members for advice and information. If these cabinet members represent the interests of the upper class, they can provide the President with information to guide his policy decisions in a way that will ensure that upper-class interests are maintained. . . .

Using his definition of upper-class membership outlined earlier, Domhoff (1967:97–99) . . . examined the backgrounds of secretaries of state, the treasury, and defense between 1932 and 1964. He found that 63 percent of the secretaries of state, 62 percent of the secretaries of defense, and 63 percent of the secretaries of the treasury could be classified as members of the upper class before assuming office. . . . As Domhoff admits, the above represents only a small part of the cabinet for a period of little more than thirty years. But with these positions we find the upper class represented in proportions far greater than their 0.5 percent of the population would suggest.

Since Domhoff's earlier work, an extensive study of cabinet members has been conducted by Beth Mintz (1975). Using Domhoff's indicators of upper-class membership, Mintz (1975, along with Peter Freitag, 1975) undertook the massive job of examining the backgrounds of *all* cabinet members (205 people) serving between 1897 and 1973. Her most interesting finding at this point is that 66 percent of these cabinet members could be classified as members of the upper class before obtaining their cabinet positions. . . . Also interesting is that the number of cabinet members coming from the upper class is fairly consistent between 1897 and 1973. . . . Mintz's data show that Republican presidents chose over 71 percent of their cabinet members from the upper class, while Democratic presidents chose over 60 percent from the upper class.

In her background research on these cabinet members Mintz also included information pertaining to the previous occupations of these people. Along with Freitag (1975), she reports that over 76 percent of the cabinet members were associated with big corporations before or after their cabinet position, 54 percent were from *both* the upper class and top corporate positions, and 90 percent either came from the upper class or were associated with big corporations. Focusing on corporate ties of cabinet members, Freitag (1975) shows that these ties have not changed much over the years, and vary only slightly by

particular cabinet position. In fact, even most secretaries of labor have been associated with big corporations in the capacity of top executives, board members or corporate lawyers.

Most ruling-class or governing-class theorists consider the cabinet to be the most important position for direct government participation by the upper class. The cabinet allows easy movement into government and then back to top corporate positions. As might be expected, Mintz (1975) found most cabinet members between 1897 and 1973 coming from outside of government rather than working their way up within government bureaucracies. . . . [T]his atypical method of political elite recruitment affords the upper class and corporate elite opportunities for political influence lacking in these other industrial nations. . . .

POLITICAL CAMPAIGN CONTRIBUTIONS

Today it costs money, lots of money, to obtain a major elective office. . . . In the 1978 U.S. congressional elections, special interest groups alone contributed $35 million to candidates. This figure increased to $55 million in 1980, and to $150 million in 1988! The average Senate campaign in 1988 cost $4 million.

In his famous work on the power elite just a little over thirty years ago, C. Wright Mills had relatively little to say about campaign contributions. But the subject can no longer be neglected. Especially in an age when political campaigns are won more through presenting images than issues, the image-creating mass media are extremely important and costly. Most presidents and congressional officeholders are wealthy, but they are not super-rich. With a few rare exceptions they cannot afford to finance their own political campaigns. Who, then, pays for these campaigns? Thousands of contributors send $25 or $50 to favored candidates. For the most part, however, the money comes from corporations and the wealthy.

With the nationwide reaction against Watergate and the many illegal campaign contributions to Nixon's reelection committee in 1972, some election reforms were undertaken by Congress in 1974. Among these reforms was the creation of a voluntary $1-per-person campaign contribution from individual income tax reports. A Presidential Election Campaign Fund was established to distribute this money to the major parties and candidates during an election year. In addition, a Federal Election Commission was established to watch over campaign spending, and people were limited to a $1,000 contribution in any single presidential election, with organizations limited to $5,000.

An interesting outcome of the campaign reform law of 1974 is that much of the illegal activity in Nixon's 1972 campaign was *made legal* as long as correct procedures are followed. For example, organizations are limited to $5,000 in political contributions per election. However, if there are more organizations, more money can be contributed. And this is precisely what happened by 1976. There was an explosion in the number of political action committees (PACs) established by large corporations and their executives, an increase far outnumbering those established by any other group, such as labor unions (Domhoff 1983:125). By the 1980 congressional elections, 1,585 corporate, health industry, and other business PACs contributed $36 million to candidates, while $13 million was contributed by 240 labor union PACs. . . .

Campaign contributions, therefore, continue to be an important means of political influence. The wealthy are not assured that their interests will be protected by those they help place in office, but they obviously consider the gamble worth taking. Usually, it is hoped that these campaign contributions are placing people in office who hold political views that lead to the defense of privilege when unforeseen challenges to upper-class interests occur along the way. . . .

Since the early 1970s a number of studies have been done on this subject (Mintz 1989). For example, Allen and Broyles (1989) examined data pertaining to the campaign contributions of 100 of the most wealthy families (629 individuals) in the United States. They found that about one-half of these individuals made

large contributions. And it was the more "visible" and active rich who made these large contributions. By this they mean that the rich were more likely to make contributions if they were corporate directors or executives, listed in "*Who's Who,*" and/or directors of nonprofit foundations. These people were more likely to contribute to Republicans, and this was especially so with the new rich, non-Jews, and people with extensive oil stocks. . . .

CONGRESSIONAL LOBBYING

If the interests of the wealthy are not ensured by their direct participation in government, and if those the wealthy helped put in office seem to be forgetting their debtors, a third force can be brought into action. The basic job of a lobbyist is to make friends among congressional leaders, provide them with favors such as trips, small gifts, and parties, and, most importantly, provide these leaders with information and arguments favoring their employers' interests and needs. All of this requires a large staff and lots of money.

Oil companies in the United States are among the largest corporations, holding six of the top ten positions in terms of industrial assets. It may not be surprising, then, that oil companies pay the lowest taxes on profits of all major corporations (Sampson, 1975:205). In 1972, for example, Exxon paid *6.5 percent* of its net profits in income taxes. . . . The actual corporate tax rate in the United States was *supposed* to be 48 percent in 1972. But, as a whole, the nineteen top oil companies paid an average of 7.6 percent of profits to taxes in this year (Blair, 1976:187). . . .

In one of the first empirical studies of the effects of certain characteristics of corporations on government policies toward these corporations (such as tax policies), Salamon and Siegfried (1977) found that the size of the corporation showed a strong inverse relation to the amount of taxes paid by the corporation. And this inverse relation between size of the corporation and the corporation's tax rate was especially upheld when examining the oil companies and including their state as well as federal taxes paid (Salamon and Siegfried 1977:1039). Thus, the bigger the corporation, the less it tends to pay in corporate taxes.

Later studies have confirmed this relationship between size (and power) and corporate tax rates. Jacobs (1988), however, measured the concentration of powerful corporations within each type of industry. The findings were similar: The more corporate concentration (meaning the size of the firms in the industry and their dominance in the industry), the less the taxes for the corporations in that industry. . . .

Lobby organizations, therefore, can be of major importance in ensuring that the special interests of a wealthy upper-class and corporate elite are served. If special favors are not completely ensured through direct participation in the cabinet and campaign contributions, the powerful lobby organizations may then move into action. The upper class and big business are not the only groups that maintain lobby organizations in Washington. The American Medical Association, the National Rifle Association, the Milk Producers Association, and many others have maintained successful lobby organizations. But when considering the big issues such as how to deal with inflation, tax policy, unemployment, foreign affairs, and many others that broadly affect the lives of people in this country, the corporate and upper-class lobbies are most important. . . .

SHAPING GOVERNMENT POLICY

Of the various means of upper-class and corporate political influence, the type least recognized by the general public is referred to as the *policy-forming process*. As scholars believe, in the long run this means of political influence is perhaps one of the most important. The basic argument is this: The federal government is faced with many national problems for which there are many possible alternative solutions. For example, consider the problem of inflation. The possible government means of dealing with this problem are varied, and a key is that different solutions to the problem may favor different class interests. Some possible solutions (such as wage and price controls) are believed to favor the working class. One important means of ensuring that the fed-

eral government follows a policy that is favorable to your class interests is to convince the government through various types of research data that one line of policy is the overall best policy. Generating the needed information and spelling out the exact policy required take a lot of planning, organization, personnel, and resources. And there must be avenues for getting this policy information to the attention of government leaders. It is no surprise, ruling-class theorists argue, that the upper class and its corporations are able to achieve the above and guide government policy in their interests . . .

At the heart of this process are (1) upper-class and corporate *money and personnel* (2) that fund and guide *research* on important questions through foundations and universities, (3) then process the information through *policy-planning groups* sponsored by the upper class (4) that make direct recommendations to government, and (5) influence the opinion-making centers, such as the media and government commissions, which in turn influence the population and government leaders in favoring specific policy alternatives. . . .

In this policy-forming process the next important link is through what has been called *policy-planning groups*. The corporate elites and upper class come together in these groups, discuss policy, publish and disseminate research, and, according to Dye and Domhoff, arrive at some consensus about what should be done in the nation. The most important of the policy groups are sponsored directly by the upper class for the purpose of linking the research information discussed above to specific policy alternatives and making certain these policy alternatives find their way to government circles.

Perhaps the most has been written about the Council on Foreign Relations (CFR) and the Committee on Economic Development (CED). Both groups are clearly upper-class institutions (as defined by Domhoff), the former specializing in foreign policy and the latter specializing in domestic issues. The Council on Foreign Relations was established shortly after World War I by upper-class members with the direct intent of influencing the United States government with

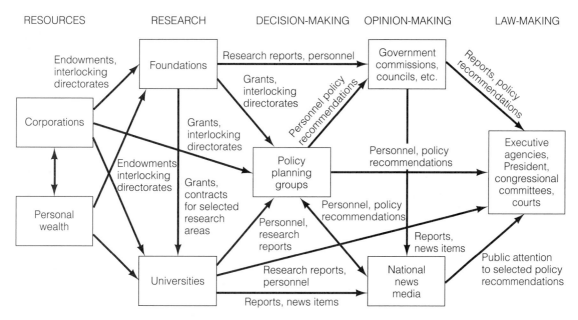

Figure 1 The Policy Formation Process.
Source: G. William Domhoff, *The Powers That Be* (New York: Vintage Press, 1979), p. 63.

respect to their business interests overseas (see Shoup, 1975). . . . Membership in the CFR is limited to 1,400 people, half of whom are members of the upper class (Domhoff, 1983). The Committee on Economic Development emerged out of the Business Advisory Council in 1942 to continue the input into government by the upper class that began with this earlier organization in the 1930s (Domhoff 1970:123–128, 1979:67–69; Collins 1977). . . .

We have finally to mention briefly the other parts of the policy-forming process described in Figure 1. Various government commissions are established from time to time to make recommendations on such things as civil disorders, the conduct of the CIA, and energy (to list some recent examples). These commissions make public the recommendations of these upper-class policy groups and provide their members with a semi-official position in the government.

As for the national news media, they are often said to have a liberal slant in their views and to possess much power in shaping public and government opinion. Recent investigations have

given some support to these charges (Halberstam 1979; Dye 1983). But the moderate conservative wing of the upper class and corporate leaders most influences the media. The major television networks, magazines, and newspapers are highly concentrated and tied to corporations (see Dye 1983:120). In terms of the background of top leaders in the national media, Dye (1983:210) found 33 percent had previous careers in big corporations and 44 percent were members of upper-class clubs. Their upper-class backgrounds are not as extensive as those of corporate leaders, but neither are the top media leaders from humble origins.

The backgrounds of media directors, the extensive corporate influence through ownership, and the huge funding from advertising all contribute to making mass media organizations cautious in presenting views that may be overly critical of the upper class and corporate interests. Information critical of these interests is not ignored by the mass media. The upper class might not even want this information to be ignored, for corrective action is often needed to prevent economic problems and corporate abuse from getting out of hand and requiring more drastic solutions that may harm more general corporate interests. . . .

One final point requires emphasis. Few theorists writing in this area (on the upper class or, more specifically, on upper-class influence in the mass media) suggest that the upper-class or corporate elites completely control the mass media in the country. Neither do most writers in this area believe that there is some kind of upper-class secret conspiracy to control the mass media — or anything else in the country, for that matter. Rather, they are trying to call attention to an economic structure that allows more influence (in the many ways outlined above) to fall into the hands of groups like the upper-class and corporate elites. Each class or economic interest group tends to have a world view or way of perceiving reality that has been shaped by its own economic and political interests. When one group has more influence over the major means of conveying information, its view of reality often comes to be accepted by more people.

In summarizing the total policy-forming process, we find an underground network in this country that is highly influenced by corporate and upper-class institutions. The federal government and Congress have the authority to adopt or reject the policy recommendations flowing from this process, but most often they accept them, leaving the government to decide exactly how the policy will be carried out (Dye 1979:226). . . .

The Upper Class: A Conclusion

. . . We have found some support for the existence of upper-class unity through interaction patterns in prep schools, social clubs, policy-formation organizations, and multiple corporate board positions. We have also found evidence of upper-class influence in the economy through stock ownership and membership on corporate boards. And we have found evidence of upper-class political influence through direct participation in government, campaign contributions, lobby organiza-

tions, and a policy-formation process. Upper-class interests are said to be maintained through these means of influence in the economy and the political system. . . .

REFERENCES

Blair, John. 1976. *The Control of Oil*. New York: Vintage Books.

Collins, Robert. 1977. "Positive Business Responses to the New Deal: The Roots of the Committee for Economic Development, 1933–1942." *Business History Review*, 22:103–119.

Domhoff, G. William. 1967. *Who Rules America?* Englewood Cliffs, N.J.: Prentice Hall.

Domhoff, G. William. 1970. *The Higher Circles*. New York: Random House.

Domhoff, G. William. 1974. *The Bohemian Grove and Other Retreats*. New York: Harper & Row.

Domhoff, G. William. 1979. *The Powers That Be*. New York: Vintage Press.

Domhoff, G. William. 1983. *Who Rules America Now?: A View for the '80s*. Englewood Cliffs, N.J.: Prentice-Hall.

Dye, Thomas R. 1979. *Who's Running America?* Englewood Cliffs, N.J.: Prentice-Hall.

Dye, Thomas R. 1983. *Who's Running America? The Reagan Years*. Englewood Cliffs, N.J.: Prentice-Hall.

Freitag, Peter. 1975. "The Cabinet and Big Business: A Study of Interlocks." *Social Problems*, 23:137–152.

Galbraith, John Kenneth. 1971. *The New Industrial State*. Boston: Houghton Mifflin.

Halberstam, David. 1979. *The Powers That Be*. New York: Alfred Knopf.

Jacobs, David. 1988. "Corporate Economic Power and the State: A Longitudinal Assessment of Two Explanations." *American Journal of Sociology*, 93:852–881.

Keller, Suzanne. 1963. *Beyond the Ruling Class: Strategic Elites in Modern Society*. New York: Random House.

Kerbo, Harold R., and L. Richard Della Fave. 1983. "Corporate Linkage and Control of the Corporate Economy: New Evidence and a Reinterpretation." *Sociological Quarterly*, 24:201–218.

Kerbo, Harold R., and L. Richard Della Fave. 1984. "Further Notes on the Evolution of Corporate Control and Institutional Investors: A Response to Niemonen." *Sociological Quarterly*, 25:279–283.

Laumann, Edward, David Knoke, and Yon-Hak Kim. 1985. "An Organizational Approach to State Policy Formation: A Comparative Study of Energy and Health Domains." *American Sociological Re-*

view, 50:1–19.

Mintz, Beth. 1975. "The President's Cabinet, 1897–1972: A Contribution to the Power Structure Debate." *Insurgent Sociologist,* 5:131–148.

Putnam, Robert. 1976. *The Comparative Study of Elites.* Englewood Cliffs, N.J.: Prentice-Hall.

Salamon, Lester, and John Siegfried. 1977. "Economic Power and Political Influence: The Impact of Industry Structure on Public Policy." *American Political Science Review,* 71:1026–1043.

Shoup, Laurence. 1975. "Shaping the Postwar World: The Council of Foreign Relations and U.S. War

Aims During WWII." *Insurgent Sociologist,* 5:9–52.

U.S. Senate Committee on Governmental Affairs. 1978b. *Interlocking Directorates among the Major U.S. Corporations.* Washington, D.C.: U.S. Government Printing Office.

U.S. Senate Committee on Governmental Affairs. 1980. *Structure of Corporate Concentration.* 2 vols. Washington, D.C.: U.S. Government Printing Office.

Zeitlin, Maurice. 1974. "Corporate Ownership and Control: The Large Corporation and the Capitalist Class." *American Journal of Sociology,* 79: 1073–1119.

20

Class, Not Race

RICHARD KAHLENBERG

In dealing with the problem of poverty, racial attitudes tend to intrude. Thus, affirmative action is considered racially biased. According to Kahlenberg, the means for correcting this bias is to realize that poverty is at one end of a class spectrum of rich and poor; therefore affirmative action should be based on class, not race.

As you read, ask yourself the following questions:

1. Why does the author claim that the problem is not between black and white but between rich and poor?

2. What would you include in a "bill of rights" for the disadvantaged?

GLOSSARY

Deus ex machina An artificial intervention intended to settle an involved situation.

IN AN ACT THAT reflected panic as much as cool reflection, Bill Clinton said recently that he is re-

viewing all federal affirmative action programs to see "whether there is some other way we can reach [our] objective without giving a preference by race or gender." As the country's mood swings violently against affirmative action, and as Republicans gear up to use the issue to bludgeon the Democratic coalition yet again in 1996, the whole project of legislating racial equality seems suddenly in doubt. The Democrats, terrified of the issue, are now hoping it will just go away. It won't. But at every political impasse, there is a political opportunity. Bill Clinton now has a chance, as no other Democrat has had since 1968, to turn a glaring liability for his party into an advantage—without betraying basic Democratic principles.

There is, as Clinton said, a way "we can work this out." But it isn't the *"Bakke* straddle," which says yes to affirmative action (race as a factor) but no to quotas. It isn't William Julius Wilson's call

From The New Republic, *April 3, 1995, pp. 21–27. Reprinted by permission of the publisher.*

to "emphasize" race-neutral social programs, while downplaying affirmative action. The days of downplaying are gone; we can count on the Republicans for that. The way out—an idea Clinton hinted at—is to introduce the principle of race neutrality and the goal of aiding the disadvantaged into affirmative action preference programs themselves; to base preferences in education, entry-level employment and public contracting, on class, not race.

Were Clinton to propose this move, the media would charge him with lurching to the right. Jesse Jackson's presidential campaign would surely soon follow. But despite its association with conservatives such as Clarence Thomas, Antonin Scalia and Dinesh D'Souza, the idea of class-based affirmative action should in fact appeal to the left as well. After all, its message of addressing class unfairness and its political potential for building cross-racial coalitions are traditional liberal staples.

For many years, the left argued not only that class was important, but also that it was more important than race. This argument was practical, ideological and politic. An emphasis on class inequality meant Robert Kennedy riding in a motorcade through cheering white and black sections of racially torn Gary, Indiana, in 1968, with black Mayor Richard Hatcher on one side, and white working-class boxing hero Tony Zale on the other.

Ideologically, it was clear that with the passage of the Civil Rights Act of 1964, class replaced caste as the central impediment to equal opportunity. Martin Luther King Jr. moved from the Montgomery Boycott to the Poor People's Campaign, which he described as "his last, greatest dream," and "something bigger than just a civil rights movement for Negroes." RFK told David Halberstam that "it was pointless to talk about the real problem in America being black and white, it was really rich and poor, which was a much more complex subject."

Finally, the left emphasized class because to confuse class and race was seen not only as wrong but as dangerous. This notion was at the heart of the protest over Daniel Patrick Moynihan's 1965 report, *The Negro Family: The Case for National Action,* in which Moynihan depicted the rising rates of illegitimacy among poor blacks. While Moynihan's critics were wrong to silence discussion of illegitimacy among blacks, they rightly noted that the title of the report, which implicated all blacks, was misleading, and that fairly high rates of illegitimacy also were present among poor whites—a point which Moynihan readily endorses today. (In the wake of the second set of L.A. riots in 1992, Moynihan rose on the Senate floor to reaffirm that family structure "is not an issue of race but of class. . . . It is class behavior.")

The irony is that affirmative action based on race violates these three liberal insights. It provides the ultimate wedge to destroy Robert Kennedy's coalition. It says that despite civil rights protections, the wealthiest African American is more deserving of preference than the poorest white. It relentlessly focuses all attention on race.

In contrast, Lyndon Johnson's June 1965 address to Howard University, in which the concept of affirmative action was first unveiled, did not ignore class. In a speech drafted by Moynihan, Johnson spoke of the bifurcation of the black community, and, in his celebrated metaphor, said we needed to aid those "hobbled" in life's race by past discrimination. This suggested special help for disadvantaged blacks, not all blacks; for the young Clarence Thomas, but not for Clarence Thomas's son. Johnson balked at implementing the thematic language of his speech. His Executive Order 11246, calling for "affirmative action" among federal contractors, initially meant greater outreach and required hiring without respect to race. In fact, LBJ rescinded his Labor Department's proposal to provide for racial quotas in the construction industry in Philadelphia. It fell to Richard Nixon to implement the "Philadelphia Plan," in what Nixon's aides say was a conscious effort to drive a wedge between blacks and labor. (Once he placed racial preferences on the table, Nixon adroitly extricated himself, and by 1972 was campaigning against racial quotas.)

The ironies were compounded by the Supreme Court. In the 1974 case *DeFunis* v. *Odegaard,* in which a system of racial preferences in law school admissions was at issue, it was the Court's liberal giant, William O. Douglas, who argued that racial preferences were unconstitutional, and suggested instead that preferences be based on disadvantage. Four years later, in the *Bakke* case, the great proponent of affirmative action as a means to achieve "diversity" was Nixon appointee Lewis F. Powell Jr. Somewhere along the line, the right wing embraced Douglas and Critical Race Theory embraced Powell.

Today, the left pushes racial preferences, even for the most advantaged minorities, in order to promote diversity and provide role models for disadvantaged blacks—an argument which, if it came from Ronald Reagan, the left would rightly dismiss as trickle-down social theory. Today, when William Julius Wilson argues the opposite of the Moynihan report—that the problems facing the black community are rooted more in class than race—it is Wilson who is excoriated by civil rights groups. The left can barely utter the word "class," instead resorting to euphemisms such as "income groups," "wage earners" and "people who play by the rules."

For all of this, the left has paid a tremendous price. On a political level, with a few notable exceptions, the history of the past twenty-five years is a history of white, working-class Robert Kennedy Democrats turning first into Wallace Democrats, then into Nixon and Reagan Democrats and ultimately into today's Angry White Males. Time and again, the white working class votes its race rather than its class, and Republicans win. The failure of the left to embrace class also helps turn poor blacks, for whom racial preferences are, in Stephen Carter's words, "stunningly irrelevant," toward Louis Farrakhan.

On the merits, the left has committed itself to a goal—equality of group results—which seems highly radical, when it is in fact rather unambi-

tious. To the extent that affirmative action, at its ultimate moment of success, merely creates a self-perpetuating black elite along with a white one, its goal is modest—certainly more conservative than real equality of opportunity, which gives blacks and whites and other Americans of all economic strata a fair chance at success.

The priority given to race over class has inevitably exacerbated white racism. Today, both liberals and conservatives conflate race and class because it serves both of their purposes to do so. Every year, when SAT scores are released, the breakdown by race shows enormous gaps between blacks on the one hand and whites and Asians on the other. The NAACP cites these figures as evidence that we need to do more. Charles Murray cites the same statistics as evidence of intractable racial differences. We rarely see a breakdown of scores by class, which would show enormous gaps between rich and poor, gaps that would help explain the differences in scores by race.

On the legal front, it once made some strategic sense to emphasize race over class. But when states moved to the remedial phase—and began trying to address past discrimination—the racial focus became a liability. The strict scrutiny that struck down Jim Crow is now used, to varying degrees, to curtail racial preferences. Class, on the other hand, is not one of the suspect categories under the Fourteenth Amendment, which leaves class-based remedies much less assailable.

If class-based affirmative action is a theory that liberals should take seriously, how would it work in practice? In this magazine, Michael Kinsley has asked, "Does Clarence Thomas, the sharecropper's kid, get more or fewer preference points than the unemployed miner's son from Appalachia?" Most conservative proponents of class-based affirmative action have failed to explain their idea with any degree of specificity. Either they're insincere—offering the alternative only for tactical reasons—or they're stumped.

The former is more likely. While the questions of implementation are serious and difficult, they are not impossible to answer. At the university level, admissions committees deal every day with precisely the type of apples-and-oranges question that Kinsley poses. Should a law school admit an applicant with a 3.2 GPA from Yale or a 3.3 from Georgetown? How do you compare those two if one applicant worked for the Peace Corps but the other had slightly higher LSATs?

In fact, a number of universities already give preferences for disadvantaged students in addition to racial minorities. Since 1989 Berkeley has granted special consideration to applicants "from socioeconomically disadvantaged backgrounds . . . regardless of race or ethnicity." Temple University Law School has, since the 1970s, given preference to "applicants who have overcome exceptional and continuous economic depravation." And at Hastings College of Law, 20 percent of the class is set aside for disadvantaged students through the Legal Equal Opportunity Program. Even the U.C. Davis medical program challenged by Allan Bakke was limited to "disadvantaged" minorities, a system which Davis apparently did not find impossible to administer.

Similar class-based preference programs could be provided by public employers and federal contractors for high school graduates not pursuing college, on the theory that at that age their class-based handicaps hide their true potential and are not at all of their own making. In public contracting, government agencies could follow the model of New York City's old class-based program, which provided preferences based not on the ethnicity or gender of the contractor, but to small firms located in New York City which did part of their business in depressed areas or employed economically disadvantaged workers.

The definition of class or disadvantage may vary according to context, but if, for example, the government chose to require class-based affirmative action from universities receiving federal funds, it is possible to devise an enforceable set of objective standards for deprivation. If the aim of class-based affirmative action is to provide a system of genuine equality of opportunity, a leg up to promising students who have done well despite the odds, we have a wealth of sociological data to devise an obstacles test. While some might balk at the very idea of reducing disadvantage to a number, we currently reduce intellectual promise to numbers — SATs and GPAs — and adding a number for disadvantage into the calculus just makes deciding who gets ahead and who does not a little fairer.

There are three basic ways to proceed: with a simple, moderate or complex definition. The simple method is to ask college applicants their family's income and measure disadvantage by that factor alone, on the theory that income is a good proxy for a whole host of economic disadvantages (such as bad schools or a difficult learning environment). This oversimplified approach is essentially the tack we've taken with respect to compensatory race-based affirmative action. For example, most affirmative action programs ask applicants to check a racial box and sweep all the ambiguities under the rug. Even though African Americans have, as Justice Thurgood Marshall said in *Bakke,* suffered a history "different in kind, not just degree, from that of other ethnic groups," universities don't calibrate preferences based on comparative group disadvantage (and, in the Davis system challenged by Bakke, two-thirds of the preferences went to Mexican-Americans and Asians, not blacks). We also ignore the question of when an individual's family immigrated in order to determine whether the family was even theoretically subject to the official discrimination in this country on which preferences are predicated.

"Diversity" was supposed to solve all this by saying we don't care about compensation, only viewpoint. But, again, if universities are genuinely seeking diversity of viewpoints, they should inquire whether a minority applicant really does have the "minority viewpoint" being sought. Derrick Bell's famous statement — "the ends of diversity are not served by people who look black and think white" — is at once repellent and a relevant critique of the assumption that all minority members think alike. In theory, we need some assurance from the applicant that he or she will in fact interact with students of different backgrounds, lest the cosmetic diversity of the freshman yearbook be lost to the reality of ethnic theme houses.

The second way to proceed, the moderately complicated calculus of class, would look at what sociologists believe to be the Big Three determinants of life chances: parental income, education and occupation. Parents' education, which is highly correlated with a child's academic achievement, can be measured in number of years. And while ranking occupations might seem hopelessly complex, various attempts to do so objectively have yielded remarkably consistent results—from the Barr Scale of the early 1920s to Alba Edwards' Census rankings of the 1940s to the Duncan Scores of the 1960s.

The third alternative, the complex calculus of disadvantage, would count all the factors mentioned, but might also look at net worth, the quality of secondary education, neighborhood influences and family structure. An applicant's family wealth is readily available from financial aid forms, and provides a long-term view of relative disadvantage, to supplement the "snapshot" picture that income provides. We also know that schooling opportunities are crucial to a student's life chances, even controlling for home environment. Some data suggest that a disadvantaged student at a middle-class school does better on average than a middle-class student at a school with high concentrations of poverty. Objective figures are available to measure secondary school quality—from per student expenditure, to the percentage of students receiving free or reduced-price lunches, to a school's median score on standardized achievement tests. Neighborhood influences, measured by the concentration of poverty within Census tracts or zip codes, could also be factored in, since numerous studies have found that living in a low-income community can adversely affect an individual's life chances above and beyond family income. Finally, everyone

from Dan Quayle to Donna Shalala agrees that children growing up in single-parent homes have a tougher time. This factor could be taken into account as well.

The point is not that this list is the perfect one, but that it *is* possible to devise a series of fairly objective and verifiable factors that measure the degree to which a teenager's true potential has been hidden. (As it happens, the complex definition is the one that disproportionately benefits African Americans. Even among similar income groups, blacks are more likely than whites to live in concentrated poverty, go to bad schools, and live in single-parent homes.) It's just not true that a system of class preferences is inherently harder to administer than a system based on race. Race only seems simpler because we have ignored the ambiguities. And racial preferences are just as easy to ridicule. To paraphrase Kinsley, does a new Indian immigrant get fewer or more points than a third-generation Latino whose mother is Anglo?

Who should benefit? Mickey Kaus, in "Class Is In" (TRB, March 27), argued that class preferences should be reserved for the underclass. But the injuries of class extend beyond the poorest. The offspring of the working poor and the working class lack advantages, too, and indeed SAT scores correlate lockstep with income at every increment. Unless you believe in genetic inferiority, these statistics suggest unfairness is not confined to the underclass. As a practical matter, a teenager who emerges from the underclass has little chance of surviving at an elite college. At Berkeley, administrators found that using a definition of disadvantaged, under which neither parent attended a four-year college and the family could not afford to pay $1,000 in education expenses, failed to bring in enough students who were likely to pass.

Still, there are several serious objections to class-based preferences that must be addressed.

1. We're not ready to be color-blind because racial discrimination continues to afflict our society. Ron

Brown said affirmative action "continues to be needed not to redress grievances of the past, but the current discrimination that continues to exist." This is a relatively new theory, which conveniently elides the fact that preferences were supposed to be temporary. It also stands logic on its head. While racial discrimination undoubtedly still exists, the Civil Rights Act of 1964 was meant to address prospective discrimination. Affirmative action—discrimination in itself—makes sense only to the extent that there is a current-day legacy of *past* discrimination which new prospective laws cannot reach back and remedy.

In the contexts of education and employment, the Civil Rights Act already contains powerful tools to address intentional and unintentional discrimination. The Civil Rights Act of 1991 reaffirmed the need to address unintentional discrimination—by requiring employers to justify employment practices that are statistically more likely to hurt minorities—but it did so without crossing the line to required preferences. This principle also applies to Title VI of the Civil Rights Act, so that if, for example, it can be shown that the SAT produces an unjustified disparate impact, a university can be barred from using it. In addition, "soft" forms of affirmative action, which require employers and universities to broaden the net and interview people from all races are good ways of ensuring positions are not filled by word of mouth, through wealthy white networks.

We have weaker tools to deal with discrimination in other areas of life—say, taxi drivers who refuse to pick up black businessmen—but how does a preference in education or employment remedy that wrong? By contrast, there is nothing illegal about bad schools, bad housing and grossly stunted opportunities for the poor. A class preference is perfectly appropriate.

2. *Class preferences will be just as stigmatizing as racial preferences.* Kinsley argues that "any debilitating self-doubt that exists because of affirmative action is not going to be mitigated by being told you got into Harvard because of your 'socioeconomic disadvantage' rather than your race."

But class preferences are different from racial preferences in at least two important respects. First, stigma—in one's own eyes and the eyes of others—is bound up with the question of whether an admissions criterion is accepted as legitimate. Students with good grades aren't seen as getting in "just because they're smart." And there appears to be a societal consensus—from Douglas to Scalia—that kids from poor backgrounds deserve a leg up. Such a consensus has never existed for class-blind racial preferences.

Second, there is no myth of inferiority in this country about the abilities of poor people comparable to that about African Americans. Now, if racial preferences are purely a matter of compensatory justice, then the question of whether preferences exacerbate white racism is not relevant. But today racial preferences are often justified by social utility (bringing different racial groups together helps dispel stereotypes) in which case the social consequences are highly relevant. The general argument made by proponents of racial preferences—that policies need to be grounded in social reality, not ahistorical theory—cuts in favor of the class category. Why? Precisely because there is no stubborn historical myth for it to reinforce.

Kaus makes a related argument when he says that class preferences "will still reward those who play the victim." But if objective criteria are used to define the disadvantaged, there is no way to "play" the victim. Poor and working-class teenagers are the victims of class inequality not of their own making. Preferences, unlike, say, a welfare check, tell poor teenagers not that they are helpless victims, but that we think their long-run potential is great, and we're going to give them a chance—if they work their tails off—to prove themselves.

3. *Class preferences continue to treat people as members of groups as opposed to individuals.* Yes. But so do university admissions policies that summarily reject students below a certain SAT level. It's hard to know what treating people as individuals means. (Perhaps if university admissions committees interviewed the teachers of

each applicant back to kindergarten to get a better picture of their academic potential, we'd be treating them more as individuals.) The question is not whether we treat people as members of groups—that's inevitable—but whether the group is a relevant one. And in measuring disadvantage (and hidden potential) class is surely a much better proxy than race.

4. *Class-based affirmative action will not yield a diverse student body in elite colleges.* Actually, there is reason to believe that class preferences will disproportionately benefit people of color in most contexts—since minorities are disproportionately poor. In the university context, however, class-based preferences were rejected during the 1970s in part because of fear that they would produce inadequate numbers of minority students. The problem is that when you control for income, African American students do worse than white and Asian students on the SAT—due in part to differences in culture and linguistic patterns, and in part to the way income alone as a measurement hides other class-based differences among ethnic groups.

The concern is a serious and complicated one. Briefly, there are four responses. First, even Murray and Richard Herrnstein agree that the residual racial gap in scores has declined significantly in the past two decades, so the concern, though real, is not as great as it once was. Second, if we use the sophisticated definition of class discussed earlier—which reflects the relative disadvantage of blacks vis-à-vis whites of the same income level—the racial gap should close further. Third, we can improve racial diversity by getting rid of unjustified preferences—for alumni kids or students from underrepresented geographic regions—which disproportionately hurt people of color. Finally, if the goal is to provide genuine equal opportunity, not equality of group result, and if we are satisfied that a meritocratic system which corrects for class inequality is the best possible approximation of that equality, then we have achieved our goal.

5. *Class-based affirmative action will cause as much resentment among those left out as race-based affirmative action.* Kinsley argues that the rejected applicant in the infamous Jesse Helms commercial from 1990 would feel just as angry for losing out on a class-based as a race-based preference, since both involve "making up for past injustice." The difference, of course, is that class preferences go to the actual victims of class injury, mooting the whole question of intergenerational justice. In the racial context, this was called "victim specificity." Even the Reagan administration was in favor of compensating actual victims of racial discrimination.

The larger point implicit in Kinsley's question is a more serious one: that any preference system, whether race- or class-based, is "still a form of zero-sum social engineering." Why should liberals push for class preferences at all? Why not just provide more funding for education, safer schools, better nutrition? The answer is that liberals should do these things; but we cannot hold our breath for it to happen. In 1993, when all the planets were aligned—a populist Democratic president, Democratic control of both Houses of Congress—they produced what *The New York Times* called "A BUDGET WORTHY OF MR. BUSH." Cheaper alternatives, such as preferences, must supplement more expensive strategies of social spending. Besides, to the extent that class preferences help change the focus of public discourse from race to class, they help reforge the coalition needed to sustain the social programs liberals want.

Class preferences could restore the successful formula on which the early civil rights movement rested: morally unassailable underpinnings and a relatively inexpensive agenda. It's crucial to remember that Martin Luther King Jr. called for special consideration based on class, not race. After laying out a forceful argument for the special debt owed to blacks, King rejected the call for a Negro Bill of Rights in favor of a Bill of Rights for the Disadvantaged. It was King's insight that there were nonracial ways to remedy racial wrongs, and that the injuries of class de-serve attention along with the injuries of race.

None of this is to argue that King would have opposed affirmative action if the alternative were to do nothing. For Jesse Helms to invoke King's color-blind rhetoric now that it is in the interests of white people to do so is the worst kind of hypocrisy. Some form of compensation is neces-sary, and I think affirmative action, though deeply flawed, is better than nothing.

But the opportunity to save affirmative action of any kind may soon pass. If the Supreme Court continues to narrow the instances in which racial preferences are justified, if California voters put an end to affirmative action in their state, and if Congress begins to roll back racial preferences in legislation which President Clinton finds hard to veto — or President Phil Gramm signs with gusto — conservatives will have less and less rea-son to bargain. Now is the time to call their bluff.

21
No, Poverty Has Not Disappeared

HERBERT J. GANS

As Kahlenberg noted in the third reading (20) in this chapter, we tend to blame the cause of poverty mostly on personal attributes rather than the re-strictions of reality. Gans indicates another reason for the persistence of poverty: poor people are fun-damentally needed.

As you read, ask yourself the following questions:

1. The author believes that poverty will continue until functional alternatives are developed. What are these alternatives?

2. Do you believe that the alternatives suggested or your own will eliminate poverty? Why?

GLOSSARY

Functional analysis A means of analysis that exam-ines the objective consequences of an action, a law, or the like.

Latent function A consequence that is not readily apparent.

Dysfunction An objective consequence that hinders the fulfillment of a goal.

Negative income tax Receipt of tax dollars when in-come falls below a set figure.

Family assistance plan Various programs for aiding needy families.

Vicarious Participating in another person's experi-ence through the imagination.

SOME TWENTY YEARS AGO Robert K. Merton ap-plied the notion of functional analysis[1] to explain the continuing though maligned existence of the urban political machine: if it continued to exist, perhaps it fulfilled latent — unintended or un-recognized — positive functions. Clearly it did. Merton pointed out how the political machine provided central authority to get things done when a decentralized local government could not act, humanized the services of the imper-

Reprinted from Social Policy, *July–August 1971, pp. 20–24, published by Social Policy Corporation, New York, NY 10036. Copyright 1971 by Social Policy Corporation.*

sonal bureaucracy for fearful citizens, offered concrete help (rather than abstract law or justice) to the poor, and otherwise performed services needed or demanded by many people but considered unconventional or even illegal by formal public agencies.

Today, poverty is more maligned than the political machine ever was; yet it, too, is a persistent social phenomenon. Consequently, there may be some merit in applying functional analysis to poverty, in asking whether it also has positive functions that explain its persistence.

Merton defined functions as "those observed consequences [of a phenomenon] which make for the adaptation or adjustment of a given [social] system." I shall use a slightly different definition; instead of identifying functions for an entire social system, I shall identify them for the interest groups, socioeconomic classes, and other population aggregates with shared values that "inhabit" a social system. I suspect that in a modern heterogeneous society, few phenomena are functional or dysfunctional for the society as a whole, and that most result in benefits to some groups and costs to others. Nor are any phenomena indispensable; in most instances, one can suggest what Merton calls "functional alternatives" or equivalents for them, in other words, other social patterns or policies that achieve the same positive functions but avoid the dysfunctions.[2]

Associating poverty with positive functions seems at first glance to be unimaginable. Of course, the slumlord and the loan shark are commonly known to profit from the existence of poverty, but they are viewed as evil men, so their activities are classified among the dysfunctions of poverty. However, what is less often recognized, at least by the conventional wisdom, is that poverty also makes possible the existence or expansion of respectable professions and occupations, for example, penology, criminology, social work, and public health. More recently, the poor have provided jobs for professional and paraprofessional "poverty warriors," and for journalists and social scientists, this author included, who have supplied the information demanded by the revival of public interest in poverty.

Clearly, then, poverty and the poor may well satisfy a number of positive functions for many nonpoor groups in American society. I shall describe thirteen such functions—economic, social, and political—that seem to me most significant.

The Functions of Poverty

First, the existence of poverty ensures that society's "dirty work" will be done. Every society has such work: physically dirty or dangerous, temporary, dead-end and underpaid, undignified and menial jobs. Society can fill these jobs by paying higher wages than for "clean" work, or it can force people who have no other choice to do the dirty work—and at low wages. In America, poverty functions to provide a low-wage labor pool that is willing—or, rather, unable to be *un*-willing—to perform dirty work at low cost. Indeed, this function of the poor is so important that in some Southern states, welfare payments have been cut off during the summer months when the poor are needed to work in the fields. Moreover, much of the debate about the Negative Income Tax and the Family Assistance Plan has concerned their impact on the work incentive, by which is actually meant the incentive of the poor to do the needed dirty work if the wages therefrom are no larger than the income grant. Many economic activities that involve dirty work depend on the poor for their existence: restaurants, hospitals, parts of the garment industry, and "truck farming," among others, could not persist in their present form without the poor.

Second, because the poor are required to work at low wages, they subsidize a variety of economic activities that benefit the affluent. For example, domestics subsidize the upper middle and upper classes, making life easier for their employers and freeing affluent women for a variety of professional, cultural, civic, and partying activities. Similarly, because the poor pay a higher proportion of their income in property and sales taxes, among others, they subsidize many state and local governmental services that benefit more affluent groups. In addition, the poor support innovation in medical practice as patients in teaching and research hospitals and as guinea pigs in medical experiments.

Third, poverty creates jobs for a number of occupations and professions that serve or "service" the poor, or protect the rest of society from them. As already noted, penology would be minuscule without the poor, as would the police. Other activities and groups that flourish because of the existence of poverty are the numbers game, the sale of heroin and cheap wines and liquors, pentecostal ministers, faith healers, prostitutes, pawn shops, and the peacetime army, which recruits its enlisted men mainly from among the poor.

Fourth, the poor buy goods others do not want and thus prolong the economic usefulness of such goods — day-old bread, fruit and vegetables that would otherwise have to be thrown out, second-hand clothes, and deteriorating automobiles and buildings. They also provide incomes for doctors, lawyers, teachers, and others who are too old, poorly trained, or incompetent to attract more affluent clients.

In addition to economic functions, the poor perform a number of social functions.

Fifth, the poor can be identified and punished as alleged or real deviants in order to uphold the legitimacy of conventional norms. To justify the desirability of hard work, thrift, honesty, and monogamy, for example, the defenders of these norms must be able to find people who can be accused of being lazy, spendthrift, dishonest, and promiscuous. Although there is some evidence that the poor are about as moral and law-abiding as anyone else, they are more likely than middle-class transgressors to be caught and punished when they participate in deviant acts. Moreover, they lack the political and cultural power to correct the stereotypes that other people hold of them and thus continue to be thought of as lazy, spendthrift, and so on, by those who need living proof that moral deviance does not pay.

Sixth, and conversely, the poor offer vicarious participation to the rest of the population in the uninhibited sexual, alcoholic, and narcotic behavior in which they are alleged to participate and which, being freed from the constraints of affluence, they are often thought to enjoy more than the middle classes. Thus many people, some social scientists included, believe that the poor not only are more given to uninhibited behavior (which may be true, although it is often motivated by despair more than lack of inhibition) but derive more pleasure from it than affluent people (which research by Lee Rainwater, Walter Miller, and others shows to be patently untrue). However, whether the poor actually have more sex and enjoy it more is irrelevant; so long as middle-class people believe this to be true, they can participate in it vicariously when instances are reported in factual or fictional form.

Seventh, the poor also serve a direct cultural function when culture created by or for them is adopted by the more affluent. The rich often collect artifacts from extinct folk cultures of poor people; and almost all Americans listen to the blues, Negro spirituals, and country music, which originated among the Southern poor. Recently they have enjoyed the rock styles that were born, like the Beatles, in the slums; and in the last year, poetry written by ghetto children has become popular in literary circles. The poor also serve as culture heroes, particularly, of course, to the left, but the hobo, the cowboy, the hipster, and the mythical prostitute with a heart of gold perform this function for a variety of groups.

Eighth, poverty helps to guarantee the status of those who are not poor. In every hierarchical society someone has to be at the bottom; but in American society, in which social mobility is an important goal for many and people need to know where they stand, the poor function as a reliable and relatively permanent measuring rod for status comparisons. This is particularly true for the working class, whose politics is influenced by the need to maintain status distinctions between themselves and the poor, much as the aristocracy must find ways of distinguishing itself from the *nouveaux riches*.

Ninth, the poor also aid the upward mobility of groups just above them in the class hierarchy. Thus a goodly number of Americans have entered the middle class through the profits earned from the provision of goods and services in the slums, including illegal or nonrespectable ones that upper-class and upper-middle-class businessmen shun because of their low prestige. As a result, members of almost every immigrant group have financed their upward mobility by providing slum housing, entertainment, gambling, narcotics, etc., to later arrivals—most recently to Blacks and Puerto Ricans.

Tenth, the poor help to keep the aristocracy busy, thus justifying its continued existence. "Society" uses the poor as clients of settlement houses and beneficiaries of charity affairs; indeed, the aristocracy must have the poor to demonstrate its superiority over other elites who devote themselves to earning money.

Eleventh, the poor, being powerless, can be made to absorb the costs of change and growth in American society. During the nineteenth century, they did the backbreaking work that built the cities; today, they are pushed out of their neighborhoods to make room for "progress." Urban renewal projects to hold middle-class taxpayers in the city and expressways to enable suburbanites to commute downtown have typically been located in poor neighborhoods, since no other group will allow itself to be displaced. For the same reason, universities, hospitals, and civic centers also expand into land occupied by the poor. The major costs of the industrialization of agriculture have been borne by the poor, who are pushed off the land without recompense; and they have paid a large share of the human cost of the growth of American power overseas, for they have provided many of the foot soldiers for Vietnam and other wars.

Twelfth, the poor facilitate and stabilize the American political process. Because they vote and participate in politics less than other groups, the political system is often free to ignore them. Moreover, since they can rarely support Republicans, they often provide Democrats with a captive constituency that has no other place to go. As a result, the Democrats can count on their votes, and be more responsive to voters—for example, the white working class—who might otherwise switch to the Republicans.

Thirteenth, the role of the poor in upholding conventional norms (see the fifth point, above)

also has a significant political function. An economy based on the ideology of laissez faire requires a deprived population that is allegedly unwilling to work or that can be considered inferior because it must accept charity or welfare in order to survive. Not only does the alleged moral deviancy of the poor reduce the moral pressure on the present political economy to eliminate poverty but socialist alternatives can be made to look quite unattractive if those who will benefit most from them can be described as lazy, spendthrift, dishonest, and promiscuous.

The Alternatives

I have described thirteen of the more important functions poverty and the poor satisfy in American society, enough to support the functionalist thesis that poverty, like any other social phenomenon, survives in part because it is useful to society or some of its parts. This analysis is not intended to suggest that because it is often functional, poverty *should* exist, or that it *must* exist. For one thing, poverty has many more dysfunctions than functions; for another, it is possible to suggest functional alternatives.

For example, society's dirty work could be done without poverty, either by automation or by paying "dirty workers" decent wages. Nor is it necessary for the poor to subsidize the many activities they support through their low-wage jobs. This would, however, drive up the costs of these activities, which would result in higher prices to their customers and clients. Similarly, many of the professionals who flourish because of the poor could be given other roles. Social workers could provide counseling to the affluent, as they prefer to do anyway; and the police could devote themselves to traffic and organized crime. Other roles would have to be found for badly trained or incompetent professionals now relegated to serving the poor, and someone else would have to pay their salaries. Fewer penologists would be employable, however. And pentecostal religion could probably not survive without the poor—nor would parts of the second- and third-hand-goods market. And in many cities, "used" housing that no one else wants would then have to be torn down at public expense.

Alternatives for the cultural functions of the poor could be found more easily and cheaply. Indeed, entertainers, hippies, and adolescents are already serving as the deviants needed to uphold traditional morality and as devotees of orgies to "staff" the fantasies of vicarious participation.

The status functions of the poor are another matter. In a hierarchical society, some people must be defined as inferior to everyone else with respect to a variety of attributes, but they need not be poor in the absolute sense. One could conceive of a society in which the "lower class," though last in the pecking order, received 75 percent of the median income, rather than 15–40 percent, as is now the case. Needless to say, this would require considerable income redistribution.

The contribution the poor make to the upward mobility of the groups that provide them with goods and services could also be maintained without the poor's having such low incomes. However, it is true that if the poor were more affluent, they would have access to enough capital to take over the provider role, thus competing with, and perhaps rejecting, the "outsiders." (Indeed, owing in part to antipoverty programs, this is already happening in a number of ghettos, where white storeowners are being replaced by Blacks.) Similarly, if the poor were more affluent, they would make less willing clients for upper-class philanthropy, although some would still use settlement houses to achieve upward mobility, as they do now. Thus "society" could continue to run its philanthropic activities.

The political functions of the poor would be more difficult to replace. With increased affluence the poor would probably obtain more political power and be more active politically. With higher incomes and more political power, the poor would be likely to resist paying the costs of growth and change. Of course, it is possible to imagine urban renewal and highway projects that properly reimbursed the displaced people, but such projects would then become considerably more expensive, and many might never be built. This, in turn, would reduce the comfort and convenience of those who now benefit from urban renewal and expressways. Finally, hippies could serve also as more deviants to justify the existing political economy—as they already do. Presumably, however, if poverty were eliminated, there would be fewer attacks on that economy.

In sum, then, many of the functions served by the poor could be replaced if poverty were eliminated, but almost always at higher costs to others, particularly more affluent others. Consequently, a functional analysis must conclude that poverty persists not only because it fulfills a number of positive functions but also because many of the functional alternatives to poverty would be quite dysfunctional for the affluent members of society. A functional analysis thus ultimately arrives at much the same conclusion as radical sociology, except that radical thinkers treat as manifest what I describe as latent; that social phenomena that are functional for affluent or powerful groups and dysfunctional for poor or powerless ones persist; that when the elimination of such phenomena through functional alternatives would generate dysfunctions for the affluent or powerful, they will continue to persist; and that phenomena like poverty can be eliminated only when they become dysfunctional for the affluent or powerful, or when the powerless can obtain enough power to change society.

NOTES

1. "Manifest and Latent Functions," in *Social Theory and Social Structure* (Glencoe, Ill.: Free Press, 1949), p. 71.

2. I shall henceforth abbreviate positive functions as functions and negative functions as dysfunctions. I shall also describe functions and dysfunctions, in the planner's terminology, as benefits and costs.

QUESTIONS FOR DISCUSSION

For further discussion of this topic, see the Wadsworth Sociology Resource Center, "Virtual Society," ***http://sociology.wadsworth.com,*** under *Sociological Footprints,* by Cargan and Ballantine. You can respond to the discussion questions there or enter your own comments in the online chat forum.

SUGGESTED READINGS AND SOCIOLOGY INTERNET RESOURCES

See the Wadsworth Sociology Resource Center, "Virtual Society," *http://sociology.wadsworth. com,* for additional links, suggestions for further reading, and learning tools related to this chapter.

Either from the "Virtual Society" website or directly from your web browser, you may access InfoTrac College Edition, an online university library that includes over 700 popular and scholarly journals in which you can find articles related to the topics in this chapter.

Part III

MAJOR PATTERNS
OF SOCIETY

A N INSTITUTION IS A formal relationship organized around common values to meet basic needs within the society. When a behavior pattern becomes fixed and expected, it can be said to have become part of an institution. Although long-standing normative patterns are difficult to change, new behavior patterns are occurring constantly because of innovations in the material culture (for example, the invention of the automobile) or challenges to expected behavior (the availability of the automobile removed parental supervision of dating from the home). As the new behavior is adopted, it becomes part of the institutionalized, normative expectation. Then, through the socialization function of that institution, the new behavior is passed on to the rest of the society.

All societies have certain functional prerequisites that are necessary for survival. To fulfill these prerequisites, societies have developed major institutions. They are economics, education, the family, politics, and religion. Although their forms may differ greatly, each of these institutions can be found in every society. Before we begin our discussion of major social institutions, we should review the prerequisites that social institutions fulfill. According to Talcott Parsons, there are two: (1) the social system must be relatively compatible with both the individual members of the society and the cultural system as a whole, and (2) the social system requires the support of the other systems around it.

It is the second of Parsons's prerequisites that leads us to expect social institutions to be interdependent and interrelated. What happens in one institution will affect all others. Suppose, for example, a recession in the economy occurs. Family members may lose jobs, churches may receive fewer and smaller donations, and politicians may have a hard time getting re-elected.

These ideas of Parsons's indicate that institutions are essential in helping us develop and deal with our wants and needs but that institutions can also be dysfunctional—the means and ends can mean distortion and negative interrelations, with negative results. The effects of socialization mean that transforming our institutions will be difficult and that the interrelationships among institutions will make it harder for them to accomplish their original ends.

Each of us is involved at all times with the major institutions of our society. The effects of family, education, religion, economics, and politics on our lives are constant. Illustrations of such effects are seen in the ways family morals or beliefs and our school experiences affect our thinking—and even the nonbeliever will be affected by religious beliefs embodied in laws or reform efforts. Similarly, the economy, with its cycles of inflation and unemployment, has an effect on prices and taxes, and politics affects our lives through the passage of laws and expenditure of taxes.

As you read through the various selections in Part III, think about the changes that our major social institutions are undergoing and need to undergo—for that is the major theme of each chapter in this section.

CHAPTER 6

Families

DIVERSITY AND CHANGE

OUR INTRODUCTION TO THE WORLD COMES through a primary group: the family. It is the institution with which we have the most contact and the one from which we traditionally receive the most emotional support because it combines emotional intimacy and sexual life. We depend on families for our early nurturance and socialization, and we could not survive without them. Society depends on families to carry out certain vital functions, such as socialization of the young and regulation of sexual activity.

Despite our personal closeness to this institution, a number of myths are held about it. For example, the image commonly held of marriage and the family is singular, whereas, as the chapter title suggests, there are many types of marriages (first marriage, remarriage, and at least two types of multimate marriages) and of families (newlywed, parental, dual-earner, dual-parent, single-parent, blended, and postparental). In fact, the image or model of father working, mother at home, and two to three children represents only about 15 percent of all families. A more recent myth about marriage has come about because of the increasing delay in age at first marriage in the United States. The myth that this is a rejection of marriage ignores the fact that the United States is one of the most marrying societies in the world, with over 90 percent of Americans marrying at least once. A final myth to be examined blames women's liberation for being antifamily and being responsible for the growing divorce rate. But as noted, most women want to marry and have children, and the divorce rate may be seen as both a reflection of the greater individual freedom for both sexes and as a replacement for death in ending marriage. These changes have led writers to indicate that the family is in a "postnuclear" trend.

The family is also being strained by the pressure of changes in other institutions, since all institutions are interrelated. The need for reforms in education (Chapter 7) is calling for more family involvement and financial support at a time when both are in short supply because of economic situations and because both sexes need to work and to work longer hours. Yet neither economic nor political institutions have come to the aid of the nuclear family—with flex-time, job sharing, family leave, part-time work, day care. The need is for institutional policy changes. For example, it would be relatively less costly to turn schools into full-time day care centers, because the buildings already exist and they must be heated or cooled anyway. Similarly, the impact on families by religious groups that seek to outlaw abortion and prevent sexual education may mean more unwanted children and children whose families are economically unable to care for them. In light of these efforts—or nonefforts—can the family still fulfill the functions it has in the past performed for both the individual and society? What projections can be made for the future of the family? The readings in this chapter address these and other issues related to the family.

Often ignored in the examination of the family are the reasons for getting married—an important influence on the future family. In this first reading, South examines these influences by asking whether we marry for love or money.

The second reading by Elkind comes to grips with this question by noting that this seemingly "staid" institution has been, and is, constantly changing. The reference to a permeable family is, perhaps, a continuation of the issue of why we get married.

The third reading in this chapter shifts the emphasis from changes affecting marriage and family to the nature of the marriage relationship and how it is changing—that it is not necessarily in contradiction to what South (Reading 22) claims people look for in marriage.

The final reading also notes the changes affecting the family and the importance of family enrichment programs to help it continue to carry out its important functions.

As you read this chapter, consider the significance of the family for all of us and consider what happens when this institution does not fulfill its functions for us and society.

22

For Love or Money?
Sociodemographic Determinants
of the Expected Benefits from Marriage

SCOTT J. SOUTH

This reading examines the myth of romantic love by asking about the benefits one expects from marriage. Not surprisingly, it depends on many factors such as race, education, marital experience, etc. What is surprising, though, are the expectations associated with each group.

As you read, ask yourself the following questions:

1. *What background and social resources determine one's outlook on marriage?*
2. *Considering the costs and benefits of marriage, which items do you think would aid your selection?*

GLOSSARY

Hypothesis Educated guess that guides research.

Sociodemographic determinants Variables such as age, sex, ethnic background that help determine an individual's status in society.

Theoretical framework Fundamental assumptions that guide theory and research.

IT IS SURPRISING, THEREFORE, that the perceived benefits of marriage have received little scrutiny. Most empirical studies of marriage analyze only the actual transition to marriage, with little or no concern for individuals' perceptions of the costs and benefits of making that transition.

An analysis of the expected benefits from marriage may prove particularly useful for explaining group differences in marriage rates. Both marital timing and the probability of ultimately marrying differ significantly by sex (Goldscheider and Waite 1986), by race and ethnicity (Bennett, Bloom, and Craig 1989; Schoen and Owens 1990; Teachman, Polonko, and Leigh 1987), and by age (Rodgers and Thornton 1985). For the most part, these sociodemographic differentials in marital behavior cannot be explained by group differences in other social and demographic attributes (Marini 1978; Michael and Tuma 1985; Testa, Astone, Krogh, and Neckerman 1989).

This chapter uses data from over 2000 unmarried respondents in the National Survey of Families and Households to examine sociodemographic differences in the anticipated benefits from marriage. Theories of marital entry, especially those emphasizing the structural characteristics of marriage markets, are reviewed to derive hypotheses relating age, race and ethnicity, sex, and other sociodemographic variables to the perceived benefits from marriage. The hypotheses link sociodemographic background to the expected improvement in overall happiness from marriage as well as to specific areas of benefit such as improvement in standard of living, emotional security, sex life, and relationships with friends.

Theoretical Framework

Much like theories of marital formation (Espenshade 1985), explanations for group differences

From Scott J. South and Stewart E. Tolnoy (eds.), 1992. The Changing American Family: Sociological and Demographic Perspectives. *Boulder, CO: WestviewPress. © 1992 by WestviewPress. Reprinted by permission of WestviewPress.*

in the expected benefits from marriage can be subsumed under two complementary rubrics: (1) factors related to the attractiveness of alternatives to marriage, especially as reflected by women's economic independence; and (2) the quantity and quality of potential spouses available to an individual. . . .

. . . Thus, the expected benefits from marriage are likely to increase along with the adoption of traditional sex roles and the absence of conflicting concurrent activities such as school attendance.

A growing explanation for subgroup differences in marital behavior—one with clear implications for the study of the expected benefits from marriage—emphasizes group variation in marriage opportunities. . . .

Another slant on the marriage squeeze stresses the quality rather than the quantity of potential spouses (Wilson 1987). From this perspective, a deficit of men with desirable economic characteristics is thought to lie behind women's growing disinclination to marry. . . .

These two explanations for marital formation—the attractiveness of alternatives to marriage and spouse availability—can be used to generate hypotheses linking the key variables of age, race, sex, and socioeconomic status to the expected benefits from marriage. First, marriage-market theories suggest that age will increase the expected benefits from marriage for men but reduce them for women. Because men tend to marry women younger than themselves, the supply of available spouses increases with age for men but decreases for women (Oppenheimer 1988). Given the greater likelihood that men will find an attractive mate, it seems probable that among men the anticipated happiness from marriage will increase with age; among women, it will decline.

One possible variation in this hypothesized interaction of sex and age involves the expected benefit of marriage to a woman's standard of living. The economic gains to marriage for older women are likely to be greater than those for younger women because the former are more

likely to marry older men, whose earnings are higher than younger men's. But this effect should vary further by race because the sex differential in income for never-married persons increases with age for whites but decreases for blacks (Espenshade 1985). Although the sex differential in earnings is generally lower among blacks than whites (Goldscheider and Waite 1986), it is especially low among older unmarried blacks. Hence the expected improvement in economic circumstances should decline with age among black women but increase with age among white women. . . .

Race and ethnic differences in the expected benefits to marriage are also apt to exist. Blacks' marriage rates are considerably lower than whites' (Bennett et al. 1989; Teachman et al. 1987), with the rates for Mexican-Americans generally resembling those of non-Hispanic whites (Schoen and Owens 1990). To the extent that actual marriage rates reflect group differences in the perceived benefits from marriage, significant racial and ethnic differences in expected marital payoffs are likely to exist. . . .

As noted earlier, socioeconomic status is also likely to affect the expected benefits from marriage. Resources such as high income and education, steady employment, and home ownership make marriage less likely to improve one's standard of living. Social and economic resources can also be used to acquire many other advantages outside of marriage. . . .

Finally, the formerly married may view marriage differently than do the never married. Although some divorced persons may have soured on the institution, high remarriage rates suggest that the formerly married see as much benefit and likely more benefits to marriage than do the never married.

Data and Methods

The data source for this analysis is the National Survey of Families and Households (NSFH), a national probability sample of 13,017 adults interviewed between March 1987 and May 1988

(Sweet, Bumpass, and Call 1988). The key questions used here are part of the self-administered questionnaire completed by 2214 unmarried, noncohabiting persons between the ages of 19 and 35. (Missing data on some items leaves 2095 respondents available for this analysis.) The NSFH oversamples minority groups, thus facilitating comparisons among blacks, Hispanics, and non-Hispanic whites. Sample weights are used to achieve the proper representation of respondents in the U.S. population.

The dependent variables capture a variety of possible perceived benefits from marriage. Respondents were directed, "For each of the following areas, please circle how you think your life might be different if you were married now." The nine areas of benefit were "overall happiness"; "standard of living"; "economic security"; "economic independence"; "freedom to do what you want"; "emotional security"; "sex life"; "friendships with others"; and "relations with parents." The five possible responses to each item were "much worse"; "somewhat worse"; "same"; "somewhat better"; and "much better." A principal components analysis and an analysis of the correlates of these items indicated that three of them—"standard of living," "economic security," and "economic independence"—tapped a very similar domain; they were combined by taking a simple average, the resulting index hereafter referred to as standard of living.

The items used to measure the expected benefits from marriage have several favorable features. First, the questions require that respondents *compare* their current situation with their perception of their lives upon marriage.[1] As noted earlier, such a comparison is central to several popular theories of marital behavior. Also, the use of several items allows respondents to weigh the pros and cons of marriage for different domains of life, some of which are expected to relate differently to sociodemographic background. Of course, respondents are limited to the benefits listed in the questionnaire. However, most of the life circumstances that could benefit from marriage are included in the survey. Moreover, because the pri-

mary focus is on subgroup differences rather than estimates for the total population, the failure to include all possible perceived benefits from marriage is not a serious flaw.

Perhaps the most serious disadvantage of these items is that they were understandably not asked of the married population. If married persons differ from the unmarried (or if the former were different before they married) in their expected benefits from marriage, then the possibility of sample selection bias exists (Berk 1983). . . . For each respondent, the predicted probability of exclusion from the sample (i.e., the hazard rate) was then calculated and included as an independent variable in the substantive equations described further on. This correction for possible selection bias had uniformly little effect on the coefficients for variables already in the model. Thus, although the possibility cannot be completely discounted, it does not appear that sample selection bias will severely vitiate the results. . . .

Because the theoretical framework suggests that age, race, and sex interact in their effects on the expected benefits from marriage, appropriate product terms were constructed and included in the equations. Initially, an equation with the three-way interaction of age, race, and sex, along with the lower-order two-way interactions (age by race, age by sex, race by sex) and all other variables in their original metric, was estimated. If the three-way interactions were not statistically significant, the equation was reestimated with only the two-way interactions. If none of the two-way interactions was significant, then the additive model containing no interaction terms was estimated.

Results

Table 1 presents some descriptive statistics on the expected benefits from marriage. With the exception of freedom to do what they want, on average the respondents expect marriage to improve their lives. The overall means are above 3, indicating that the mythical average respondent anticipates that most life domains would be somewhat better or much better upon marriage. Although having means above 3, fewer than half of the respondents expect improvement in friendships with others or relations with parents.[2] Most respondents expect no change in these areas upon marriage. If means are used as the yardstick, the respondents expect their sex lives to benefit most from marriage, followed by their overall happiness and emotional security.

The expected benefits to marriage are moderately related to sex and race/ethnicity. Only for relations with parents is the association between these variables not statistically significant by conventional standards, and even here it is of borderline significance. The strongest relationships are between sex/race and expected improvement in standard of living (eta = .25) and freedom (eta = .19). Both variables are characterized by strong sex differences, with females anticipating greater improvement (or less deterioration) than males. Although more modest, race differences are apparent as well. Compared to their white counterparts, blacks and Hispanics are more likely to perceive improvements in both areas. Sociodemographic differentials in the expected improvement in overall happiness are not large but indicate that black females anticipate the greatest benefit, black and Hispanic males the least. Of course, these patterns are useful for descriptive purposes but not for analytical ones because they obtain without controlling for other factors. Moreover, interactions among age, race, and sex cannot be easily discerned from Table 1. . . .

With the sole exception of the expected improvements to one's sex life, some statistically significant interaction between age, race, and sex is apparent. For three of the dependent variables (standard of living, freedom, and friendships with others), one of the three-way interactions of age by race by sex is statistically significant. For three others (overall happiness, emotional security, and relations with parents), at least one of the two-way interaction terms is significant. The nature of these effects is described more fully later; for now it should be emphasized that, by

themselves, the component variables involved in these interactions have no clear interpretation.

Two of the variables measuring socioeconomic resources, education and earnings, as well as school enrollment, have fairly consistent negative effects on the expected benefits from marriage. At least one of these characteristics is significantly and inversely related to each of the endogenous variables. Presumably, the availability of desirable alternatives to marriage that accompany these attributes reduces the perceived attractiveness of marriage. By contrast, two other indicators of resources—weeks unemployed (a reverse indicator) and home ownership—have weak and nonsignificant effects on the expected benefits from marriage.

The formerly married differ significantly from the never married on the expected improvements upon marriage in standard of living and freedom. In both cases, the formerly married view marriage in a more favorable light. Having been previously married does not appear to shape substantially, either positively or negatively, marital expectations in other domains. The two contextual variables, the percent of the county population that is urban and the percent in poverty, have inconsistent and relatively weak associations with the perceived benefits from marriage. Respondents in more urban counties tend to anticipate less improvement in their sex lives but better relations with their parents. The former association is perhaps due to higher rates of premarital sexual relations in more urban areas; marriage might lead to a smaller increase in the frequency of intercourse than in less urban areas. The positive association between urban county percent and expected improvements in parental relations is more difficult to explain; it is conceivable that respondents in urban counties anticipate a more marked shift upon marriage toward a family-oriented social network. Contrary to expectations, the county poverty rate is positively related to the expected improvement in standard of living. Perhaps the surrounding poverty makes individuals in poor counties view objectively smaller financial gains as having greater impact on their personal standard of living. Alterna-

TABLE 1 Descriptive Statistics for the Expected Benefits from Marriage by Race/Ethnicity and Sex

Expected Benefit	White Males	White Females	Black Males	Black Females	Hispanic Males	Hispanic Females	Total	P (ETA)
Overall happiness								
Mean	3.78	3.84	3.55	3.94	3.56	3.97	3.79	.00
Standard deviation	1.03	1.01	1.05	1.03	1.20	1.02	1.04	(.11)
% expecting improvement	64.70	69.00	53.10	70.10	61.50	75.30	66.00	
Standard of living								
Mean	3.01	3.40	3.20	3.75	3.12	3.46	3.25	.00
Standard deviation	.98	.93	.95	.83	1.07	.94	.98	(.25)
% expecting improvement	48.40	65.50	54.40	81.50	53.10	65.60	58.50	
Freedom								
Mean	2.47	2.78	2.60	3.04	2.64	3.06	2.67	.00
Standard deviation	1.05	1.02	1.14	1.18	1.07	1.23	1.09	(.19)
% expecting improvement	11.60	18.20	18.80	28.60	18.00	29.10	16.90	
Emotional security								
Mean	3.57	3.74	3.32	3.77	3.48	3.62	3.62	.00
Standard deviation	.95	.98	1.04	1.00	1.02	1.10	.99	(.13)
% expecting improvement	52.60	62.20	38.60	58.70	49.70	50.90	55.00	
Sex life								
Mean	4.19	4.23	3.79	4.04	4.10	4.08	4.15	.00
Standard deviation	.94	.92	.98	1.02	.83	1.05	.95	(.13)
% expecting improvement	76.30	77.10	61.30	67.80	80.60	74.40	74.80	
Friendships with others								
Mean	3.22	3.09	3.05	3.27	3.22	3.11	3.17	.00
Standard deviation	.83	.74	.83	.86	.76	1.00	.81	(.09)
% expecting improvement	28.70	17.90	20.10	25.90	26.90	27.80	24.20	
Relations with parents								
Mean	3.24	3.20	3.32	3.37	3.33	3.34	3.26	.06
Standard deviation	.80	.76	.81	.81	.89	.98	.80	(.07)
% expecting improvement	25.80	23.00	31.70	27.30	34.10	32.30	26.20	
N (weighted)	871	655	161	208	111	89	2095	

tively, perhaps the poverty rate for the total population is an inadequate indicator of the economic circumstances of potential spouses. . . .

Discussion and Conclusion

By and large, the findings are consistent with the hypotheses drawn from the theoretical framework. Individuals with greater personal resources tend to see fewer benefits to marriage than persons from less advantaged backgrounds. Presumably, individuals with greater resources have available more attractive alternatives to marriage, especially in the labor market. The perceived benefits to marriage increase with age for males; among females they either decrease or increase more slowly. These differences are consistent with an improvement in the quantity and quality of potential wives as men age and with a deterioration of the pool of eligible husbands for women as they grow older. Given the deficit of employed young black males, young black women appear to perceive greater benefits in marriage than their position in the marriage market warrants. However, the often precipitous decline with age in their expectations does accord with the hypotheses.

One striking result is that young black men and women have quite different visions of how marriage will affect their lives. Although these differences tend to converge and sometimes reverse themselves at older ages, at the youngest ages black men see much less benefit in marriage than black women. The low rates of marriage among black women appear to be as much a function of black males' reticence to marry as of black females' inability to find attractive mates (cf. Wilson 1987). These sex differences at the younger ages are also apparent among whites but in more muted form. It seems unlikely that market differentials can completely explain these sex and race differences, because black women are usually considered the most disadvantaged group in the marriage market. One possible explanation for young black women perceiving more benefits from marriage than others could be differential selection out of the unmarried population. If, because of a more severe marriage squeeze among black women than black men or whites, fewer black women who wish to marry are able to do so, then the remaining pool of unmarried black women might contain a relatively larger number who see substantial benefit to marriage. However, this explanation does not appear readily capable of explaining the different effect of age among black women than among others.

The larger sex difference among blacks than whites in the expected benefits from marriage might also be partly attributable to the experience of growing up in fatherless families. The higher proportion of female-headed families among blacks, coupled with the severe financial hardships endured by female-headed families (McLanahan 1985), might lead young black women to value marriage very highly. This might be especially true

for unmarried mothers. Relatedly, the deficit of black fathers to serve as role models for young black men may make it difficult for the latter to see the benefits of marriage. Again, however, it is not certain that this explanation can account for group differences in the effect of age on the expected benefits from marriage. Just as no single explanation can account for black-white differences in marital formation (Cherlin 1981), there is no simple interpretation of these patterns. Theories of the marriage market appear capable of explaining some of the sociodemographic determinants of the expected benefits from marriage, but the complexity of the differentials demands supplementary explanations.

Although racial differences in the expected benefits to marriage are fairly regular, differences between Anglos and Hispanics are weaker and less consistent. No support is found for the hypothesis that Hispanic men and women see greater benefits to marriage than do non-Hispanics or blacks.

These results also suggest that increased attention be paid to the role of age in the marital formation process. The age differences observed here in the cross-section could conceivably result from one of two processes. Individuals could change their minds about the expected benefits from marriage as they grow older (i.e., an aging effect), or older individuals could hold different attitudes than younger ones throughout their lives (i.e., a cohort effect).[3] It is impossible with these data to choose unequivocally between these alternatives, but both aging and cohort explanations can be accommodated by a theoretical framework emphasizing disequilibria in marriage markets. For example, imbalances in the number of marriageable men and women can be created by cohort fluctuations in fertility combined with the traditional age difference between spouses; by increasing sex differentials in mortality throughout the prime marriage years; and by changes in the volume of "competition" from other ages for spouses, as when older men court young brides, a situation that both increases competition among younger men and leaves older women facing a dwindling supply of older grooms. Clearly, the secular rise in age at marriage and the increase in remarriages warrant further analysis of marital attitudes and behavior at the older ages.

Finally, further analyses of the expected benefits from marriage seem justified. Not only are the anticipated costs and gains of marriage critical for evaluating theories of marital timing, but social changes that might alter the ledger of costs and benefits (and allegedly undermine the fundamental basis of marriage) are apt to continue in the future. Women's growing economic independence is the most cited change, but the deterioration of male earnings (Oppenheimer 1988), a weakened normative commitment to marry (Thornton 1989), and changes in the marriage squeeze may also affect perceptions of marital costs and benefits. The consequences of these perceptions might also be worth exploring insofar as unmet expectations can influence marital quality and stability.

NOTES

1. As with similar attitudinal variables, establishing the validity of these items is not a simple task. The questions appear straightforward enough to acquire face validity. Construct validity was assessed by correlating these items with a question asking respondents how much they agree or disagree (on a 5-point Likert-type scale) to the statement, I would like to get married someday. Presumably, individuals who see more benefits to marriage would express greater desire to marry, although clearly these variables are not intended to measure the same theoretical construct. The correlations between the desire to marry and the seven perceived benefits to marriage are all positive and statistically significant. Perhaps because of relatively little variance in the former variable (almost 80% of respondents agreed or strongly agreed) and random measurement error, the correlations are not overly strong, ranging from .09 (for relations with parents) to .30 (for overall happiness). That the strongest correlation is for overall happiness rather than the specific life domains, however, argues further for the validity of the dependent variables. It is perhaps worth stressing that the expected benefits from marriage are not intended to measure the desire to marry or the ultimate probability of marrying.

Of course, when some individuals answer these questions, they have a specific potential spouse in mind; others are considering a hypothetical partner. The theories under consideration do not appear to deem this an important distinction, nor is there reason to expect important differences between these two groups of respondents. Nonetheless, additional analyses controlled for whether respondents reported having a steady boyfriend or girlfriend, with no appreciable changes in the results.

2. Including a dummy variable for individuals whose earnings are imputed has no appreciable impact on the findings.

3. For the composite variable, standard of living, the percentage expecting improvement is the percentage of respondents whose average on the three items is greater than 3. For all other variables, this is the percentage of respondents expecting this area of their lives to be "somewhat better" or "much better" upon marriage.

4. In theory, the observed age differentials could also be caused by a selection effect if individuals are selected out of the unmarried population on the basis of their expectations about the benefits from marriage. However, the correction for sample selection bias left the effects of age unchanged, suggesting that other interpretations are required. It is also worth noting that if selection effects are operating, they must be very complex, because they would have to operate differently for the various race/sex groups.

REFERENCES

Bennett, Neil G., David E. Bloom, and Patricia H. Craig, 1989. "The Divergence of Black and White Marriage Patterns." *American Journal of Sociology* 95:692–722.

Berk, Richard A., 1983. "An Introduction to Sample Selection Bias in Sociological Data." *American Sociological Review* 48:386–398.

Cherlin, Andrew J., 1981. *Marriage, Divorce, Remarriage.* Cambridge, MA: Harvard University Press.

Espenshade, Thomas J., 1985. "Marriage Trends in America: Estimates, Implications, and Causes." *Population and Development Review* 11:193–245.

Goldscheider, Frances Kobrin, and Linda J. Waite, 1986. "Sex Differences in the Entry into Marriage." *American Journal of Sociology* 92:91–109.

Marini, Margaret Mooney, 1978. "The Transition to Adulthood: Sex Differences in Educational Attainment and Age of Marriage." *American Sociological Review* 43:483–507.

McLanahan, Sara, 1985. "Family Structure and the Reproduction of Poverty." *American Journal of Sociology* 90:873–901.

Michael, Robert T., and Nancy Brandon Tuma, 1985. "Entry into Marriage and Parenthood by Young

Men and Women: The Influence of Family Background." *Demography* 22:515–544.

Oppenheimer, Valerie Kincade, 1988. "A Theory of Marriage Timing." *American Journal of Sociology* 94:563–591.

Rodgers, Willard L., and Arland Thornton, 1985. "Changing Patterns of First Marriage in the United States." *Demography* 22:265–279.

Schoen, Robert, and Dawn Owens, 1990. "A Further Look at First Unions and First Marriages." Paper presented at the Conference on Demographic Perspectives on the American Family: Patterns and Prospects, April 6–7, Albany, NY.

Sweet, James, Larry Bumpass, and Vaughn Call, 1988. "The Design and Content of the National Survey of Families and Households." Working Paper NSFH-1, Center for Demography and Ecology, University of Wisconsin–Madison.

Teachman, Jay D., Karen A. Polonko, and Geoffrey K. Leigh, 1987. "Marital Timing: Race and Sex Comparisons." *Social Forces* 66:239–268.

Testa, Mark, Nan Marie Astone, Marilyn Krogh, and Kathryn M. Neckerman, 1989. "Employment and Marriage among Inner-City Fathers." *Annals of the American Academy of Political and Social Sciences* 501:79–91.

Thornton, Arland, 1989. "Changing Attitudes Toward Family Issues in the United States." *Journal of Marriage and the Family* 51:873–893.

Wilson, William Julius, 1987. *The Truly Disadvantaged.* Chicago: University of Chicago Press.

23
The Family in the Postmodern World

DAVID ELKIND

Having married, one begins to form a family. But as Elkind notes, the modern nuclear family has moved on to become a postmodern, permeable family. There has been a change in sentiments, values, and perceptions about family life, and this has created a new imbalance.

As you read, ask yourself the following questions:

1. *What social, historical, and cultural events have transformed the sentiments and values of the family?*
2. *Considering the changes affecting the family, what do you see as a picture of your future family?*

GLOSSARY

Postmodern world Contemporary society; post-industrial society.

Postmodern permeable family Encompasses many different family kinship arrangements.

Traditional nuclear family Two parents, one working, one home with the children.

ALTHOUGH WE HAVE NOT given up our beliefs in progress, universality, and lawfulness, we have adjusted and modified them in keeping with postmodern experience and discoveries. The tectonic shift in our basic ideas about ourselves

Reprinted from National Forum: The Phi Kappa Phi Journal, *Vol. 75, No. 3 (Summer 1995).*

and our world has affected all facets of our society, from science to the arts, from philosophy to architecture.

A common motif is that of *pastiche,* a mixture of seemingly disparate elements. In architecture, for example, postmodern buildings are a creative mixture of different architectural styles. Postmodern buildings often combine architectural styles of antiquity such as open courtyards, atria, and Greek columns with modern glass and steel. Such buildings celebrate difference rather than progress, particulars rather than universals, and irregularity rather than lawfulness.

The Modern Nuclear Family

The social, historical, and cultural events that have transformed the other institutions of American society have had an equal effect upon the basic sentiments, values, and perceptions of the American family. The nuclear family—two parents, one working, the other staying home to rear the children—clearly embodied the ideas of modernity. This family was regarded as the end result of an evolutionary process of variation and of natural selection. The nuclear family survived because it was the fittest for rearing children and thus for perpetuating the species. It also was regarded as the universal form that eventually would be found in all societies and cultures. Finally, the nuclear family was thought to be the most lawful kinship system; any family kinship system that was not nuclear was regarded as irregular, unlawful, and by extension, immoral.

SENTIMENTS OF THE NUCLEAR FAMILY

Accompanying this modern view of the nuclear family were the sentiments that enlivened it. The first of these was the sentiment, as described by Edward Shorter in *The Making of the Modern Family,* of *romantic love.* Beginning with nineteenth-century individualism, the belief arose that for each of us there is one other individual who was created as our perfect mate. Once we encountered that person, we would know it instantly

and proceed to spend the rest of our lives forever "happily-ever-aftering." An essential condition of this romantic ideal was that a young woman would "save" herself for her fated partner. In this romantic context, virginity was a valuable commodity that could be exchanged for a lifelong commitment to the relationship. Romantic love worked to keep couples together even when they were unhappy. While this ideal was unfortunate for parents in unrewarding relationships, it often benefited children because parents stayed together and usually did not blame the children for the failure of the marriage.

A second sentiment of the nuclear family, according to Shorter, was that of *maternal love.* This sentiment grew out of a belief in a maternal "instinct" such that women had a biological urge to stay home and take care of their children. Women were thought to be totally fulfilled if they were able to follow the dictates of this biological imperative. Although the evidence for instincts in humans was, and is, hard to find, the lack of evidence did not prevent the belief in the maternal instinct from gaining wide popular acceptance. Unfortunately, for women at any rate, husbands came to believe that the maternal instinct extended to them as well as to their offspring. Husbands expected their wives to look after them in the same manner that they looked after the children, which placed a heavy burden upon wives and mothers.

A third sentiment of the nuclear family, described by Shorter, was that of *domesticity.* According to this sentiment, the family was the emotional center of one's life, and all other emotional attachments had to be subordinated to it. The home and hearth were where you obtained your emotional support and nurturance. The family was, as Christopher Lasch described it in the title of his book, *A Haven in a Heartless World.* Domesticity also provided creative outlets for women to the extent that quilting, needlepoint, clothes-making, canning, and so on, were carried on in the home. Unfortunately, these creative outlets were closed to women around the turn of the century when the home economics movement and its corporate symbol, "Betty Crocker," convinced homemakers that machine-made and store-bought products were cheaper and more efficient than those made at home.

THE NUCLEAR FAMILY VALUE

In keeping with these romantic, presumably biologically grounded sentiments was a corresponding family value, namely, that of *togetherness*. Togetherness meant that family members were loyal to one another and supported one another. It required that they share holidays, remember birthdays, and call and/or write when they were far from home. The value of togetherness was often best exemplified at family mealtimes. Meals were the occasions when everyone got together to provide a family forum for sharing the day's events, for arguing with, as well as for consoling, one another. Family members, though following different paths during the day, nonetheless felt reunited at mealtimes. Having meals together took precedence over personal interests and individual pursuits.

NUCLEAR FAMILY PERCEPTIONS

Nuclear family sentiments and values were accompanied by complementary perceptions of parents, children, and adolescents that I have described in my book *Ties That Stress*. Parents were perceived as being *intuitively knowledgeable* about children. In part, this perception grew out of the fact that many modern parents had grown up in extended families and had experience with children at different age levels. It also derived from the belief in a parental instinct according to which not only the mother but the father as well had inborn parenting skills. Parenting was not something you learned how to do; it was something you just did when the children arrived.

Professionals who wrote for nuclear parents accepted this intuitive perception. Famed English pediatrician Donald Winicott told parents in his book entitled *Babies and Their Mothers* that "You Are a Good Enough Parent" and you do not have to be a college graduate to do an excellent job of parenting. And Benjamin Spock in his book *Infant and Child Care* encouraged parents to use their common sense and to follow their feelings.

Children in the modern family were perceived as *innocent*, as in need of adult care, protection, and guidance. The perception of children as innocent, however, did not become widespread until the end of the nineteenth century. Before that time children were often looked at from a religious perspective and were thought to be endowed with original sin—which had to be rooted out of them by whatever dreadful means necessary. But towards the end of the last century, children came to be seen as mischievous rather than evil, and childhood was looked upon as a very special, magical time that needed to be cherished and protected. This perception was reinforced by legislation prohibiting child labor and ensuring school attendance and also was mirrored by a new literature for children and adolescents such as *Peter Pan, The Wind in the Willows, Tom Sawyer,* and *Huckleberry Finn.* Schools were child-centered and often followed Dewey's progressive educational philosophy of learning by doing.

Adolescents, in turn, were regarded as *immature* and as passing through a period of emotional *Sturm and Drang* (storm and stress). To successfully make the passage to adulthood, they needed firm limits, clear values, and moral leadership. Without such limits, values, and leadership, young people drifted into states that bordered on mental illness. Holden Caulfield, the hero of J. D. Salinger's *The Catcher in the Rye*, is an example of such a rudderless modern adolescent. To assist parents in providing the limits, values, and leadership deemed necessary for young people to make the transition to adulthood, teachers at school and at church provided programs and clubs for young men and women. In addition, malt shops and soda fountains as well as large movie theaters provided young people with their own space to socialize with their friends.

In the modern nuclear family, therefore, the needs of children and youth were better served than were the needs of parents, particularly mothers. This phenomenon is what I have called, in

Ties That Stress, the *old family imbalance.* Whenever there is an institutionalized need imbalance, those who experience it undergo inordinate stress. In the 1950s and 1960s, for example, women consumed millions of pounds of tranquilizers. The old imbalance was one contributor to the Women's Movement, while need imbalances of other kinds helped give rise to the Civil Rights Movement and demands from other groups, such as the handicapped and ethnic minorities, for a more balanced societal pattern of need satisfaction.

These events and many others ushered in a new era of social egalitarianism that extended to the family and transformed its sentiments, value, and perceptions.

The Postmodern Permeable Family

After mid-century, postmodern ideas began to transform the family. Mothers moved into the work force, divorce became easier and more common, the privacy of the home was invaded by television, and children and adolescents began to be hurried to grow up fast. Gradually a new family form emerged that is now more common than the nuclear family. The postmodern *permeable* family encompasses many different family kinship arrangements, nuclear families, two-parent working families, single-parent families, adoptive families, remarried families, as well as gay and lesbian parent families. We recognize today that children can be effectively raised in many different family systems and that it is the emotional climate of the family, rather than its kinship structure, that primarily determines a child's emotional well-being and healthy development. The permeable family differs from the nuclear family not only in its structure but also in its sentiments, values, and perceptions.

PERMEABLE FAMILY SENTIMENTS

The 1960s were witness to what Shorter has called the second sexual revolution. It was ushered in partly by the introduction of new contraceptive methods (the pill), partly by the sexual experimentation that took place during World War II both at home and abroad, and partly by the decline in the moral authority of the government after the Vietnam War and Watergate. In effect, the second sexual revolution amounted to the social acceptance of premarital sex. This acceptance effectively destroyed the sentiment of romantic love inasmuch as now young people could have a succession of sexual partners before marriage. Thanks to the second sexual revolution, virginity has lost its value as a commodity; it can no longer be "exchanged" for anything.

As a result of this second sexual revolution, a new family sentiment, which I call *consensual love,* now predominates in the permeable family. Consensual love is based upon mutual agreement and decision-making. It assumes that the participants are equal and that no participant is more dependent or independent than the other. In addition, the couple recognizes that the relationship may not be permanent and that either party may decide that it is no longer a rewarding or satisfying arrangement. The legal system recognizes this new egalitarian marital relationship with the introduction of "no fault" divorce. Unlike earlier marital laws which insisted that marital break-up was the fault of one partner or the other, "no fault" recognizes that break-ups are rarely one-sided.

Secondly, as a result of the large numbers of mothers entering the work force over the last few decades, the modern sentiment of maternal love has largely been given up as well. In its place is the sentiment of what might be called *shared parenting,* the recognition that parenting can be shared with father, with other relatives, and with nonparental caregivers. Most children can thrive in high-quality, out-of-home care. The tragedy of the postmodern family is, however, that high quality child care is accessible to only a small proportion of those parents and children who need it.

Finally, the openness of the family to the larger world also has undermined the sentiment of domesticity. The privacy that was the hallmark of the nuclear family has been invaded by television, which brings the outside world of crime and punishment, of shopping and traveling, of manufacturing and building immediately into our homes. Moreover, we also are made privy, thanks to a growing number of talk programs, to the many permutations of human relations that can, and apparently do, occur within our homes. We have become (and this is our new family sentiment) *urbane*—knowledgeable about the world in its many glories, passions, and depravities.

THE PERMEABLE FAMILY VALUE

The value of the postmodern permeable family is that of *autonomy,* the importance of individual choice and personal life journey. Autonomy is the natural corollary to the sentiments of consensual love, shared parenting, and urbanity— all of which support individual choice and freedom. Perhaps the value of autonomy is most evident in the transformation of family mealtime. That icon of togetherness is no longer held sacred. Today, soccer practice, a business meeting, or a music lesson takes precedence over the imperative to share a meal together.

PERMEABLE FAMILY PERCEPTIONS

Permeable family parents also are perceived in a new way. Because many contemporary parents have grown up in small families, they may have limited experience with children. They look to experts for advice, and they receive it in abundance. Parenting is no longer a matter of unlearned intuition, but rather a matter of learned *technique.* The postmodern advice to parents is now of a different order from that given to modern parents. Whereas modern writers emphasized intuition and limited their advice to how children grow and develop, postmodern writers see parents as in need of technique and give them advice on how to rear their children. Unfortunately, this advice is often given without reference to child growth and development, as if one size fits all. While how-to techniques are easier for parents, they are harder on children who no longer experience tailor-made parenting practices.

In the postmodern era, children have come to be seen as *competent,* ready, and able to deal with all of life's vicissitudes. This new perception of childhood competence does not grow out of any new revolutionary findings about child growth and development. It does derive from the inability of postmodern parents to protect their children in the way that modern parents were able to protect their offspring. Even the most well-intentioned and committed parents cannot make the schools less competitive, the streets less violent, and the media less vulgar. Parents have no control over an economy that necessitates both parents working to maintain a standard of living above the poverty line. Parents have to believe that children can cope with the world today or, as parents, they will go mad. And children are more competent than we gave them credit for being in the modern world. Nonetheless, they may be less competent then we would like or possibly need them to be.

Immaturity is no longer a fitting sequel to the perception of childhood competence. Our new, postmodern perception of adolescents is that of *sophistication*. We now look at adolescents as sophisticated in matters of sex, drugs, media, and computer technology. The social acceptance of premarital sexual activity has now extended down to adolescents, with the number of teenagers who are sexually active tripling since the 1960s. Partly in response to increased work demands, partly in response to the new perception of adolescents as knowledgeable and competent, adults provide many fewer organized activities for teenagers than were provided in the modern era. Young people's needs for limit-setting, guidance, and value-modeling are not being met. This lack contributes to the new imbalance and to the many stress-related problems of contemporary youth.

The postmodern permeable family has corrected some of the injustices and inequities inherent in the sentiments and perceptions of the nuclear family. But the need imbalance that worked such hardships on women in the modern family has now shifted to children and youth in the postmodern permeable family. The needs of children and adolescents for protection and security, guidance, limit-setting, and monitoring are being weighted less heavily than are the needs of the adults in our society. This is what I have called the *new imbalance,* and it is the new imbalance that stresses children and accounts for dramatic increase in all manner of stress-related problems among young people.

A New Family Balance

We have today what has been called the "New Morbidity," which refers to the fact that a higher percentage of young people now die from stress-related causes (for example, substance abuse-related automobile accidents, suicide, anorexia)

as once did from diseases such as polio and tuberculosis. The solution is not to return to the old imbalance of the nuclear family but rather to move towards a new, vital family form that provides a more nearly equal balance between the needs of children and youth and those of parents and adults.

This transformation will happen only if the institutions most closely associated with the family — the schools, the media, the legal system — change direction as well. At the most basic level we must reinvent adulthood. We must stop focusing on our own personal concerns and recommit ourselves to ensuring the health and well-being of future generations.

24
Social Structure, Families, and Children

LEONARD BEEGHLEY

Beeghley examines how the transformation in the social structure of the United States has affected the family and children, and the implications for understanding family problems. He shows the changing relationships of children in families. This discussion may have you wondering why couples have children. Why?

As you read, ask yourself the following questions:

1. *How have children been affected by the structural changes affecting the family?*
2. *Considering the problems with children, would you still have any? Why?*

GLOSSARY

Progress Social impact of increasing scientific knowledge.

Social structure Relatively stable pattern of social behavior.

FAMILIES ARE THE CENTER of children's lives. In families, children receive their initial socialization (Brim and Wheeler, 1966). That is, they develop psychological traits and behavioral patterns that remain forever. They learn many of the social skills necessary for participating in the larger

Written for the sixth edition. Copyright by Leonard Beeghley, Department of Sociology, University of Florida, Gainesville, FL 32611.

society. They acquire much of the knowledge and wisdom they will need as adults. They obtain the material goods—food, clothing, and shelter—necessary for living. Hence, the most important people in children's lives are their parents and siblings. Yet, historically, families have been buffeted by changes over which they had little control.

In the nineteenth century, most families struggled against nature, using muscle power to produce nearly everything they needed (Bell, 1976). Nearly all work occurred in and around the home, which constituted a productive enterprise. Every member of the family, including children, performed manual labor to ensure the group's survival. Families were large. Divorce was rare. Most education occurred at home. Effective medical treatment was minimal, which meant that pregnancy and births could not be regulated. In this context, parents functioned relatively autonomously from the larger society, rearing children as best they could.

Today, in contrast, few families produce what they need for themselves. The home has become a center for consumption, not production. Most productive work is paid and occurs away from the house. Wives as well as husbands are employed. Children are usually exempt from paid work because the knowledge needed and the psychological sophistication necessary to be effective have become so great. Families are now small. Divorce is common. Formal education is required and nearly always occurs away from home. Medical science has advanced so that pregnancy and birth now can be regulated. Despite these changes, primary responsibility for ensuring children's well-being continues to fall on the family. It remains the center of their lives.

This quite different context means that a greater range of choices exists in every sphere of life and that parents can no longer function autonomously in rearing children. Today, families mediate between children, government, and other social institutions. In so doing, however, they confront problems that did not exist in the past. This makes the socialization process more difficult. Research shows that children mature into well-adjusted adults in loving, nurturing, and stable environments (Maccoby, 1980). This insight remains true regardless of the changes just described. Thus, the problem faced by the United States and other nations involves making sure that such environments occur as often as possible.

This reading has two purposes. First, I describe how the transformation of the social structure in the United States has altered the family and show how these changes affect children. The following issues are examined: family size, mother's employment, family breakdown, and teenage pregnancy and childbearing. Second, in the brief concluding section I discuss the implications this analysis has for understanding U.S. society and the problems of families and children.[1]

Family Size

Families have become smaller over time. This change began nearly two centuries ago and the downward trend has been steady, except for the "baby boom" after World War II. Thus, in the early 1800s the average woman had seven to eight live children. Today, she has about two (Ryder, 1980; Cherlin, 1981). This pattern of declining fertility is not unique to the United States. Other Western societies have experienced the same phenomenon. How can it be explained?

Social Structure and Family Size

One reason for smaller families lies in the increasing value people place on children, childhood, and parenting. Although this description seems counterintuitive, Ariès shows that childhood has not always been significant and children have not always been valued (1962). Remember, life was short in the past. The average person who survived infancy died by age 35. Children became "little adults" by age 6 or 7. They dressed like adults. They worked like adults. They filled responsibility like adults. This situation changed over time, however, as the family drew in on itself and became conjugal (Goode,

1963). That is, families now live rather privately, nurturing one another. In this context, children are valued for their own sake. Yet as children have become important, family size has had to be reduced to allow parents to attend to the needs of each child. It is very difficult to provide a nurturing environment for many children. Furthermore, as the number of children fell, norms about appropriate care and attention devoted to each child became stricter. Children have become "priceless," an intrinsic value (Zelizer, 1985). This development enhanced the idea that families should be child centered and that parenting carries distinctive rewards. Today, the loss of a child to accident or disease constitutes a great tragedy. Childhood is now a special time. Hence, couples bear fewer children in order to nurture them properly.

Second, with industrialization children are no longer economic assets. By "industrialization," I mean the transformation of the economy as new forms of energy are substituted for muscle power, leading to advances in productivity. Prior to the twentieth century, societies barely produced enough material goods so that everyone could survive (Bell, 1976; Lenski and Lenski, 1987). Most people subsisted by means of hard manual labor, scratching a living from the soil. In such a context, children produced more than they cost. From a very young age, 5–6 years old, children, as "little adults," became productive workers—like everyone else. In addition, they functioned much like pension plans today, as a means for parents' security in old age. Hence, an economic incentive existed to have many children. Such a strategy was magnified, of course, by high levels of infant death. This situation still characterizes many developing societies. It resembled the United States in the last century. Today, however, children cost a great deal. Put bluntly, they are an economic drain on the family. Hence, family size has fallen.

Third, advances in nutrition, sanitation, medical treatment, and birth control allow more children to survive and parents to regulate whether and when to bear them (McKeown, 1976). Children, as special objects of love, are healthy today.

Hence, parents know that children are likely to grow to maturity. These advances reduce the desire to bear many children.

The reasons for the long-term fall in family size are deeply embedded in modern life. It is hard to imagine public policies that would convince couples to have seven to nine children. This is because the value placed on children and industrialization combine with technological advances in other areas to provide a wider range of choices available to people. Today, couples can choose to have as many children as they wish. Most bear few.

Family Size and Children

Small families present new problems, of course. For example, fewer relatives exist to inhibit child abuse and serve as caretakers for the young. In addition, as is noted later, parents' employment often limits time they spend with children. Nonetheless, small families can be good for children, because the ratio of adult caretakers to children is now very low. In the past, the ratio usually ranged from two adults to seven to nine children, assuming that both parents survived. Today, it typically ranges from two adults to one to three children, assuming both parents are present. It ranges from one adult to one to two children if only one parent is present. In principle, such ratios imply beneficial outcomes for children, suggesting they can be closely supervised and protected and that their psychological needs can be met.

Research shows that this is just what happens. Small families are advantageous for children. Those reared in such contexts attain higher grades and remain in school longer (Blake, 1989). As a result, they display higher levels of occupational achievement (Hout, 1988). What happens, of course, is that children in small families are more likely to have a broad array of intellectually and culturally enriching experiences. These range from music and dance lessons to being read to by their parents. Of course, these advantages do not always occur. For the moment, however, I simply want to make clear that small families can have positive effects on children.

Mothers' Employment

In every society, people work. Historically, the imperative to labor applied to everyone — young and old, mothers and fathers. That situation changed in the latter part of the nineteenth century. Most men began working for pay outside the home. Most women did not. Over time, however, an increasing proportion of women have joined the paid labor force. Thus, the labor force participation rate of men and women has been converging over the years. Since about 1870, that of men has remained relatively stable, at 75 to 80 percent. In contrast, the proportion of women employed rose steadily over this same period, from 13 percent to 56 percent (U.S. Census Bureau, 1975, p. 139; U.S. Census Bureau, 1991, p. 47). This trend has occurred regardless of marital status. In addition, a similar pattern has occurred among mothers: Those with young children, less than 6 years of age, now work outside the home at the same rate as all married women. Moreover, mothers of school-age children exhibit the highest level of employment, 68 percent. This long-term process constitutes a return to older forms of family life, altered to fit a new situation. How can it be explained?

SOCIAL STRUCTURE AND MOTHERS' EMPLOYMENT

One reason for this trend lies in the increased opportunities created by industrialization, especially when coupled with capitalism (Berger, 1986). The use of machines in the process of production affected the female labor force participation rate by expanding the number of jobs, especially white-collar jobs. In this context, most adults can physically perform most jobs today. The average person works with (or against) other people, not nature. Hence, women's smaller size and the possibility of pregnancy and childbirth no longer impede employment. In this historically unique situation, women have opportunities never before available.

Second, the rewards associated with paid work attract women as well as men: prestige, income, and self-respect. In modern societies, people obtain prestige in light of their occupation. I refer not only to how well they perform a job, but what it is. One does not need to be a world-class physician. Rather, just being a physician brings a woman more prestige than, say, being a housewife. Paid work, in short, is more highly valued than unpaid work in the United States. In addition, employed women have an income and, it follows, the potential for independence if they wish. Those who are unemployed remain economically dependent, like children. Furthermore, a job constitutes a primary source of self-respect in modern societies. Employed people display a more positive self-concept than those who are not (Schlosman and Verba, 1979). Hence, in a context where rewards exist for joining the labor force, increasing numbers of women have sought employment.

Third, medical progress makes it possible for women and men to prevent disease and pregnancy. Thus, sexually transmitted diseases can be avoided or cured, pregnancy can be regulated, and lactation (breastfeeding) is unnecessary to the health of babies.[2] As a result, couples can choose whether, when, and under what conditions to bear children. This ability facilitates mothers' employment.

Finally, values have changed: The notion that women ought to be equal to men has become prevalent (Cott, 1987). Historically, women have been productive workers in every known society. This is because household tasks were also productive acts, required if families were to survive. In addition, as noted earlier, bearing children was also a productive act. Although a brief moment occurred in the late nineteenth and early twentieth centuries in which middle-class families cultivated an ideal of the stay-at-home mother who developed the "domestic arts" (keeping house), cared for children, and volunteered her time for charity, this ideal was unique (Rothman, 1984). It has given way to paid work, partly because women who do not earn an income cannot hope to be equal to men in an industrial society.

The increasing rate of mothers' employment reflects long-term structural changes in the United States. Once again, it is hard to imagine public policies that would convince very many families that wives ought to remain at home. Because of industrialization, the rewards associated with paid work, the ability to regulate pregnancy, and the value placed on equality, women have more choices than were available in the past. These factors should not ignore, perhaps, the prime reason for mothers working: financial need.

MOTHERS' EMPLOYMENT AND CHILDREN

It turns out, however, that mothers' employment is not bad for children (Hayes and Kamerman, 1983). The children gain economic advantages. They also gain more diffuse (but real) benefits that follow from an employed woman's increased self-respect and personal development. Moreover, the children of employed women display levels of educational achievement similar to those of unemployed women (Blake, 1989). Education, of course, is the key to occupational attainment.

Negative consequences do exist, however. Children sometimes lose time with their parents when they are employed (Kamerman and Kahn, 1988; Nock and Kingston, 1988). This is a generalized family problem. When both parents work for pay, time becomes a serious problem—time with children, time to keep house, time to relax, time to reflect. All are in short supply. These difficulties are most acute, of course, for single parents. The question that must be asked is whether strategies exist or can be developed to mitigate this problem.

Family Breakdown and Divorce

Divorce rates rose steadily over the last century until recently. Thus, of all marriages begun in 1880, only 8 percent eventually ended in divorce. In comparison, demographers project that half of all marriages begun in the 1980s will eventually end in divorce (Preston and McDonald, 1979; Weed, 1980; Martin and Bumpass, 1989).

Divorce frequently involves children (National Center for Health Statistics, 1991). Thus, the number of children affected by divorce in the United States rose from 900,000 in 1970 to 1.2 million in 1980, declining slightly to 1.1 million in 1987. It follows that children's living arrangements have changed. The proportion living with one parent rose from 12 to 23 percent between 1970 and 1985 (National Center for Health Statistics, 1991). Why has the divorce rate gone up?

SOCIAL STRUCTURE AND DIVORCE

In considering this issue, it is important to remember that families break down in all societies (Beeghley and Dwyer, 1989). Husbands and wives become estranged from one another and stop communicating. The household becomes tense, marred by arguments and fights. Sometimes spouses coexist in a sort of familial cold war. Violence occurs in the form of spouse or child abuse. Regardless of these events, in the past people got married and stayed married. Today, however, family breakdown often leads to divorce and to children living with one parent. This is because people today expect their marriages to be based on mutual respect, personal devotion to one another, and emotional and sexual satisfaction. These marital characteristics are not only vital to the spouses, but they also provide the context in which children can be nurtured.

One reason for the divorce trend lies in the process of industrialization. This process not only reduced the economic value of children and revolutionized the workplace, but also transformed the household. In the nineteenth century, the household served mainly as a productive center in which family members were economically interdependent. Today, however, the household only serves as a consumption center. Family and home have become a refuge from the external world, rather than built into it. Hence the ties between husbands and wives have changed.

Second, industrialization brings urbanization, because modern methods of production and distribution require that people live in close proximity to one another. Urbanization allows married people to compare their relationship with that of others. In addition, urbanization allows people to meet others. In the nineteenth century, rural families lived in relative isolation because transportation and communication were so difficult. Urban living, in contrast, not only increases social contact, but also enhances people's autonomy and individual freedom. In such a context, most people are strangers and norms demand they be nonintrusive with one another (Simmel, 1908; Lofland, 1973). This fact allows for personal anonymity. Hence, urban life provides an environment in which those mired in unhappy marriages can observe that this unhappiness is not inevitable and can meet others.

Third, increasing female labor force participation changes both women's and men's options and raises the possibility of sexual equality. In a social context where a small proportion of women worked for pay, few sought divorce, simply because they could not be economically independent. Similarly, few men wanted a divorce, simply because (if they did not flee to the frontier) they were obligated to continue supporting, more or less indefinitely, both their previous family and any new one they formed. This situation alters as women earn an income, feel competent to do so, and recognize the possibility for independence. Women also have greater bargaining power because a steady income provides a means to support oneself should the need arise (Scanzoni, 1972).

Fourth, the declining fertility rate reduces the complications inherent in divorce. These "complications," of course, involve children. The fewer children, the less complex and difficult a divorce becomes, both economically and psychologically.

Fifth, the new value placed on gender equality implies that men and women are equal status partners who join together, or not, as they wish (Pearson and Hendrix, 1979; Beeghley, 1983b). Moreover, changing gender role norms, such as those dictating that women should be treated equally on the job, increase their independence.

Sixth, religious diversity diluted the political impact of organized religion. No religion encourages divorce. The dilemma facing each faith group has been how to respond, religiously and morally, to a situation in which increasing numbers of people want to dissolve their marriages (O'Neill, 1967). Some faiths find theological justification for it under certain circumstances. Moreover, I would speculate that the religious diversity typical of the United States is significant in other ways as well. For example, the acceptance of divorce by one denomination may have influenced other faith groups with similar theological orientations to allow it as well. Furthermore, in a religiously diverse society, people can switch to other faiths relatively easily. Such switching occurs a great deal in the United States (Beeghley and others, 1990), probably more than in other societies. Thus, if they wish, people belonging to faith groups opposed to divorce can switch to one of those allowing it and continue, with some change in beliefs, to be religious.

Finally, the diffuse governmental responsibility for determining divorce law facilitates its liberalization. Divorce law is not set by Congress but by each individual state. Historically, forty-nine states allowed it, based on such acts as adultery, desertion, cruelty, chronic drunkenness, or criminal conviction (Blake, 1962; O'Neill, 1967). In addition, so-called divorce colonies existed in North Dakota, South Dakota, Florida, and Nevada, where divorces were (relatively) easier to obtain. In 1970, California introduced no-fault divorce, which was emulated in forty-eight other states within a decade. Couples' legal right to dissolve their marriages reduces the impediments to divorce.

Although the ability to divorce and, hence, its rate, could be restricted, the structural pressures favoring it remain strong. Thus, the United States is likely to choose to live with this phenomenon. An obvious problem follows: children grow to adulthood in their families and, from their angle divorce may or may not be a good thing.

Divorce and Children

As nearly everyone has observed, people often become distressed when a divorce occurs and it sometimes takes time for both children and adults to recover. Research shows that these observations are basically correct. It is important to distinguish between family breakdown and divorce, however, in order to place the problems of children in perspective.

Families in which the marital relationship has broken down are not nurturing environments. Hence, children in such families usually display psychological and behavioral problems before a divorce (Cherlin and others, 1991). These difficulties typically become more severe when the final rupture occurs (Wallerstein and Blakeslee, 1989). Young children, who are confused by the situation, often blame themselves for the breakup of the family. Older children frequently become angry. Their grades drop. They often act out by fighting, skipping school, running away, and the like. Such behavior reflects children's frustration and their realization that they can do nothing to re-establish their former lifestyle. This situation is frequently compounded as home life becomes chaotic, with more pickup meals than previously, erratic bedtimes, and less emotional support from the custodial parent. Yet, over the long term, most children settle down and return to a normal process of maturation (Wallerstein and Blakeslee, 1989). A minority, however, display long-term psychological problems in the aftermath of divorce. Whether this latter fact reflects the marital dissolution or the dysfunctional marriage that preceded it always remains a question.

Probably the establishment of a stable, nurturing environment is the key factor determining whether children do reasonably well after a divorce. The ability to create such an atmosphere varies idiosyncratically depending on such factors as custody arrangements, the presence of grandparents or other kin, remarriage of the parents, educational practices, neighborhood experiences, religious influences, and economic resources. This last factor is especially important, since the standard of living of women and their children typically declines significantly after a divorce, while that of men increases (Weitzman, 1985). One author provides some perspective on what this difference means by noting that men's child support payments are often less than their car payments after a divorce (Glendon, 1987). Other disruptions also occur. For example, in one-third of all cases, the family house is sold to facilitate the "equal" division of property by the spouses (Weitzman, 1985). The disruptive effect of this change on children is especially great. Not only must they cope with the disappearance of one parent, but they must also adjust to a new (and usually less attractive) environment. Children find security in their own bedroom, the contours of their yard, the faces in the neighborhood and school. When these things are taken away, along with one of the parents, the effects can be devastating. Children learn that others cannot be counted on and that life is not very predictable. These lessons can have long-term negative consequences for individuals and, sometimes, for the society.

The impact of this disruption, of course, varies by social class. Children from poor and working-class backgrounds sometimes endure more instability in their lives simply because of their class of origin. This situation worsens when divorce occurs. For example, many women and their children become impoverished after a divorce and do not emerge from this situation until remarriage occurs (Weitzman, 1985). This is one reason why half of all poor families are headed by single women (Duncan, 1984; Beeghley, 1983a).

As with the other issues dealt with here, a high divorce rate may be essential to modern societies. It is difficult to imagine states opting to eliminate divorce. Rather, the question is how to regulate it and whether the consequences for children can be minimized.

Teenage Pregnancy and Childbearing

Rates of sexual intercourse among young people have risen steadily over the course of this century (Beeghley and Sellers, 1986).[3] Today, most unmarried people become sexually active by the end of adolescence: about 80 percent of males and 70 percent of females (Hayes, 1987). By "sexually active," I mean a person has had intercourse at least once. It is important to understand, however, that the United States is not unique. A similar historical process has occurred in other nations similar to our own. Thus, countries like England, France, Germany, Netherlands, and Sweden display rates of intercourse similar to or greater than our own (Jones and others, 1986).

Pregnancy rates and their consequences, however, differ dramatically from one nation to another. Some illustrative data are presented below (Jones and others, 1986, pp. 25–30). The figures refer to rates per 1000 young women, 18 to 19 years old.

	United States	England	Netherlands
Pregnancy	129	75	25
Births	72	54	17
Abortions	57	21	8

Thus, while the rate of intercourse is similar, its consequences vary. How can these differences be explained? I am going to proceed in two steps: offering first an account of the increasing rate of sexual intercourse and then dealing with differences in pregnancy and its consequences.

SOCIAL STRUCTURE AND SEXUAL INTERCOURSE

One reason for the high rate of sexual activity is that the average gap between puberty and marriage has become wider over time (Beeghley and Sellers, 1986). Over the past century, the average age of the onset of female puberty declined from about 16 to about 12. Although this decline is biological, it has social causes: better nutrition, improved medical care, and reduced child labor have produced healthier and stronger children who mature earlier. Over this same period, the average age of marriage for young women has increased from about 20–21 to 23–24, primarily because they remain in school longer. Thus, the gap between female puberty and marriage has changed from about five to six years to eleven to twelve years. Similar changes have occurred among boys, although they mature and marry later. In the current context, many people who are biologically capable will act on their strong desires before marriage.

Second, interaction among teenagers has increased. When most adolescents lived in rural areas, contact with the opposite sex was more difficult than today. Urbanization changed that. It provided young people with more independence and contact than ever before. The impact of this factor is greater because of advances in communication, such as the car and telephone. They are increased still further by school attendance requirements, a twentieth-century phenomenon. Many teenagers, of course, continue to college, where they live in dorms and apartments with no parental supervision. Sex sometimes follows from opportunities to meet and make friends.

Third, medical progress now makes it possible to prevent pregnancy. It is not accidental that the birth rate among teenage girls began falling around 1960, when the birth control pill was invented. The young have learned, from their parents, in school, and other sources, how to prevent pregnancy. Because of court rulings, congressional legislation, and state policies, young people now have access to contraception and abortion services. In effect, the combination of technological advance with governmental and private efforts has produced falling adolescent birth rates over the last thirty years.

The historical changes underlying teenage sexual activity indicate how the social structure in which young people grow up has changed. While young people could choose to have intercourse outside of marriage in the past, restrictions in years between maturation and puberty, opportunities to meet, and sexual safety made this choice difficult. Today, it is far easier.

SOCIAL STRUCTURE AND ADOLESCENT PREGNANCY

Paradoxically, the data shown earlier reveal that while sex need not result in pregnancy, it often does in the United States. Young women in England, Netherlands, and other Western nations display pregnancy rates one-half to one-third that in the United States. The main reason for this difference is the comparatively low level of contraceptive use by teenagers in this country. The most practical and effective method of birth control currently available for adolescents is the birth control pill. Other methods may become available, such as RU-486 and the "morning after" pill (Trussell and Stewart, 1992). All data show that widespread use of the pill leads to a drop in the pregnancy rate and, hence, in abortions and births. U.S. teenagers, however, have less access to these methods than those in other Western societies (Jones and others, 1986; Trussell and others, 1992). Hence, the relative lack of attention to preventing pregnancy leads to high numbers of adolescent abortions and births.

ADOLESCENT PREGNANCY, CHILDBEARING, AND CHILDREN

Western societies have developed relatively clear indicators of the passage into adulthood: leaving home, completing education, gaining economic independence, and marrying. The first three criteria are usually seen as prerequisites for marriage and children, mainly because they signify the end of the initial socialization process. They indicate that individuals have developed the necessary psychological traits, social skills, and knowledge to participate in a modern society as full-fledged adults. They are ready to become parents. But sometimes children become pregnant, bear children, and form families before they are ready. Giving birth transforms a child's life.

When children have children, negative consequences follow for both. The first set of problems involves the health of teenage mothers and their children (Hayes, 1987). Such mothers have a higher probability of maternal mortality (death) and morbidity (illness). Their children are more likely to be premature and suffer low birth weight. These phenomena, in turn, produce an increased likelihood of mental retardation, long-term illness, and many other difficulties (McCormick, 1992).

The second set of problems involves the living standards of young mothers and their children (Hayes, 1987; Furstenberg, Brooks-Gunn, and Morgan, 1987). Teenage mothers achieve lower

education, lower earnings, and lower occupational prestige than those who do not bear children. Such mothers and their children are also more likely to live below the poverty line and receive welfare. Moreover, while the impact of these negative results declines over time, most young mothers never catch up to their peers who did not give birth. In contrast, while abortion is rarely a desirable choice, adolescents who terminate pregnancy are more likely to graduate and less likely to become pregnant again compared to those who give birth (Zabin, Hirsch, and Emerson, 1989).

The third set of problems involves family structure and size (Hayes, 1987). Teenage mothers are more likely to marry than other adolescents. They are also more likely to divorce than those who delay childbearing. They display higher total fertility (number of children), closer spacing of births, more nonmarital births, and more unintended births.

The final set of problems involves the children of adolescent mothers (Hayes, 1987). In addition to the health issues mentioned earlier, such children are more likely to be neglected and abused by their young parents. They are much more likely to be injured and hospitalized before the age of 5. They display less intellectual development and lower academic achievement. They are more likely to have behavioral problems in school and other contexts. Most of these difficulties reflect the immaturity of young parents and their poverty. In addition, however, it is probable that the large families characteristic of teenage mothers have negative consequences for children independently of other factors. In sum, teenage childbearing negatively affects the long-term future of both the mother and the child.

All these problems have been phrased as personal losses. It is important to remember, however, that these losses generate many costs to the society. For example, it has been estimated that families started by teenage mothers absorb 53 percent of the cost of welfare programs, such as Aid for Families with Dependent Children, Food Stamps, and Medicaid (Burt, 1986). More generally, the social and economic costs of allowing high rates of teenage childbearing are staggering. Such people are far less happy and far less productive citizens.

Concluding Remarks

In this reading, I have described how fundamental transformations in the nature of society have altered family life. These changes illustrate a general trend: the history of the last century is one of increasing choice in human affairs. The ability to choose means that people have greater control over their lives. They decide on their own goals and construct their own way to happiness. An understanding of this trend leads to a surprising insight: most issues defined as social problems, such as those considered here, constitute by-products of progress. I am using the term *progress* to describe the social impact of increasing scientific knowledge. Science is a peculiar way of understanding the world. It emphasizes that each event has a cause that can be discovered and verified by observation. Its successful application to the practical problems human beings face has transformed our world over the past several centuries, producing many unexpected changes. For example, industrialization, which represents applied knowledge, led to the transformation of the economy, which led in turn to similar family size and increased opportunities for women. Yet problems result as people have adapted to changing conditions produced by progress.

Many persons find it difficult to accept the idea that social problems are the paradoxical results of improving life. The conventional assumption is that social problems represent the breakdown of society, not its advancement. The word *decline,* in fact, is often used in connection with common sense interpretations of changes in family life. We hear, then, of the decline of the nuclear family and morality. The result is a call for a return to the old verities (or values). If mothers would just stay home and have lots of babies, if parents would just stay together, and if teenagers would just stay out of each other's arms, it is said, problems of the family would disappear. Yet this plea suggests a Mom-and-apple-pie vision of reality. It is as if the United States could go backward in time to a different, half-forgotten, mythical past and thereby solve the nation's problems. While this orientation is, perhaps, well intended, it will not work and, besides, no one really wants to do it. For example, a century ago the family was a center for productive activity and spouses were economically dependent on one another. In this context, the divorce rate was low. Today, however, thanks to technological advances and other changes, the family is a center for consumption and spouses are less economically dependent on each other. In this rather different context, the divorce rate is high. To alter this situation, the United States would merely have to go back to a relatively preindustrial and rural lifestyle. My use of the word "merely" in this context is sarcasm, of course, since almost no one wants to live in the past. Being rustic may be fun on vacation, but after a few days nearly everyone wants to use a bathroom and take a shower. Such amenities, which reflect industrialization, were unavailable just a short time ago. Although it sounds odd, the hidden reality is that the old verities people believe in and act on—among them a belief in hard work, wealth acquisition, individualism, and scientific advance—frequently cause the very situations they define as social problems.

So the trick is to develop strategies that take into account the needs of adults and children as we move forward into the twenty-first century (Scanzoni, 1991). I am going to close this reading by making some suggestions. You should understand that they are speculative and deliberately provocative. My goal is to stimulate you to think creatively about public policy in the context of a realistic consideration of the alternatives people face.

The problem of combining employment and parenting needs to be thought through. The issue is how to maximize the ability of parents in small families to care for children. In modern economies, people not only must be employed to make ends meet, but they also want to work. It makes them economically autonomous, gives them opportunities for self-fulfillment, and provides contact with the larger society. But child care becomes a perennial problem. It seems to me that employers, especially large ones, should provide childcare on site. This is not just for the sake of parents and children. It is my impression that employees who know their children are well cared for are more productive. Older children should remain at school (if not in school) all day long. When holidays or teacher workdays occur during the normal academic year, many parents still have to work. For example, except for Christmas Day, the Christmas "vacation" does not occur for most people. Every year, then, parents face a two-week crisis. The schools ought to be the location for child care on such occasions. My point is that children who are in stable, nurturing environments at home, at child care facilities, and at school become more well-adjusted adults. All citizens, even those who choose not to parent, have a stake in producing such people.

Problems of parenting lead to a more general concern with family organization. It seems to me that we need to teach men as well as women to view householding and parenting, if they choose to have children, as joint tasks. In the past, men remained aloof from the home. Children, hence, were denied the nurturance of fathers and fathers were denied the opportunity to nurture others. Women, in contrast, were prevented from competing and achieving outside the family. Such artificial limits waste human resources.

In addition to interaction within the family, the timing of its formation needs to be considered. The problem here is to reduce the chance of estrangement and divorce, and to ensure that children are raised in stable environments. Probably the best single predictor of divorce is age at marriage: The younger the spouses, the greater the odds of divorce. Hence, young people should be taught to delay marriage until, say, age 25 at the earliest. Furthermore, they should delay having children for several years after marriage. This probably means tolerating, perhaps formalizing, cohabitation (without children) before marriage. While such strategies may reduce the number of divorces, divorce will occur. In my opinion, the interests of children should be kept paramount

when families break apart. When childbearing is a choice, as in modern societies, it is also a responsibility. People should not walk away from it.

Finally, the special problems of adolescents need to be dealt with. I would prefer that adolescents delay becoming sexually active. This is because the older they are, the less likely they will become involved in exploitive relationships and more likely they will protect themselves from pregnancy and disease. This is not, however, realistic public policy. In policy (rather than theological) terms, the problem has to be avoiding the consequences of sex rather than the act itself. Hence, we should make birth control and, if necessary, abortions, easily available. Preferably free. Data from other nations show clearly that pregnancies can be prevented even when sexual activity is high. It is an act of wisdom to seize on what works rather than, Quixote-like, to ask the young to "Just say no."

On that note, I would like to end by remarking on one of the paradoxes of sociology: It leads people to think daringly but act prudently (Berger, 1963). Just as the transformative changes of the last century have led to increased choice in human affairs, so have attempts at regulating social life. Such efforts do not always succeed (Sieber, 1981). Experience remains limited. Wisdom is rare. So I advise you to think carefully about some of the changes I have suggested. Consider them nondogmatically. And be careful.

NOTES

1. The theoretical underpinning for this analysis comes from Durkheim (1982 [1895]) and Merton (1968). For further explanation, see Beeghley (1989, pp. 20–22) and Beeghley and Dwyer (1989).

2. The appearance of the human immunodeficiency virus (HIV) obviously complicates this situation. Even in this case, however, prevention is relatively easy and it is reasonable to expect a method for managing the virus—that is, preventing the onset of AIDS (acquired immune deficiency syndrome)—will become available in the future (Mitsuya, Yarchoan, and Broder, 1990).

3. In presenting data, I focus mostly on girls. This is because they usually bear the consequences of pregnancy and childbirth. It is important to remember,

however, that young boys are intimately involved in this problem (Marsiglio, 1988).

REFERENCES

Ariès, Philippe, 1962. *Centuries of Childhood: A Social History of Family Life*. New York: Vintage.

Beeghley, Leonard, 1983a. *Living Poorly in America*. New York: Praeger.

Beeghley, Leonard, 1983b. Spencer's analysis of the evolution of the family and the status of women. *Sociological Perspectives* 26: 299–323.

Beeghley, Leonard, 1989. *The Structure of Stratification in the United States*. Boston, MA: Allyn & Bacon.

Beeghley, Leonard, John K. Cochran, and E. Wilbur Bock, 1990. The religious switcher and alcohol use: An application of reference group theory. *Sociological Forum* 5: 261–278.

Beeghley, Leonard, and Jeffrey W. Dwyer, 1989. Social structure and the rate of divorce. In James A. Holstein and Gale Miller (eds.), *Perspectives on Social Problems*. Vol. 1. Greenwich, CT: Jai Press.

Beeghley, Leonard, and Christine Sellers, 1986. Adolescents and sex: A structural theory of premarital sex in the United States. *Deviant Behavior* 7 (2): 313–336.

Bell, Daniel, 1976. *The Coming of Post-Industrial Society*. New York: Basic Books.

Berger, Peter M., 1963. *Invitation to Sociology*. Garden City, NY: Doubleday.

Berger, Peter M., 1986. *The Capitalist Revolution*. New York: Basic Books.

Blake, Judith, 1989. *Family Size and Achievement*. Berkeley: University of California Press.

Blake, Nelson M., 1962. *The Road to Reno*. New York: Macmillan.

Brim, Orville G., and Stanton Wheeler, 1966. *Socialization After Childhood*. New York: Wiley.

Burt, Martha R., 1986. Estimating the cost of teenage childbearing. *Family Planning Perspectives* 18: 221–226.

Cherlin, Andrew, 1981. *Marriage, Divorce, and Remarriage*. Cambridge, MA: Harvard University Press.

Cott, Nancy, 1987, *The Grounding of Modern Feminism*. New Haven, CT: Yale University Press.

Duncan, Greg, 1984. *Years of Poverty, Years of Plenty*. Ann Arbor: Institute for Social Research, University of Michigan.

Durkheim, Emile, 1982 (1895). *The Rules of Sociological Method*. New York: Free Press.

Furstenberg, Frank F., J. Brooks-Gunn, and S. Philip Morgan, 1987. *Adolescent Mothers in Later Life*. New York: Cambridge University Press.

Glendon, Mary Ann, 1987. *Abortion and Divorce in Western Law*. Cambridge, MA: Harvard University Press.

Goode, William J., 1963. *World Revolution and Family Patterns*. New York: Free Press.

Hayes, Cheryl D. (ed.), 1987. *Risking the Future: Adolescent Sexuality, Pregnancy, and Childbearing*. Vol. 1. Washington, DC: National Academy Press.

Hayes, C. D., and S. B. Kamerman, 1983. *Children of Working Mothers: Experiences and Outcomes*. Washington, DC: National Academy Press.

Hout, Michael, 1988. More universalism, less structural mobility: The American occupational structure in the 1980s. *American Journal of Sociology* 93: 1358–1400.

Jones, Elise, Jacqueline D. Forrest, Noreen Goldman, Stanley Henshaw, Richard Lincoln, Jeannie I. Rosoff, Charles F. Westoff, and Deirdre Wulf, 1986. *Teenage Pregnancy in Industrial Countries*. New Haven, CT: Yale University Press.

Kamerman, Sheila, and Alfred J. Kahn, 1988. *Mothers Alone: Strategies for a Time of Change*. Dover, MA: Auburn House.

Lenski, Gerhard, and Jean Lenski, 1987. *Human Societies*. 5th ed. New York: McGraw-Hill.

Lofland, Lyn, 1973. *A World of Strangers*. New York: Basic Books.

Maccoby, Eleanor, 1980. *Social Development: Psychological Growth and the Parent-Child Relationship*. New York: Harcourt, Brace Jovanovich.

Marsiglio, William, 1988. Adolescent male sexuality and heterosexual masculinity: A conceptual model and review. *Journal of Adolescent Research* 3/4: 285–303.

Martin, Teresa Castro, and Larry L. Bumpass, 1989. Recent trends in marital disruption. *Demography* 26: 37–51.

McCormick, M. C., 1992. The health and developmental status of very-low-birth-weight children at school age. *Journal of the American Medical Association* 267: 2204–2210.

McKeown, Thomas, 1976. *The Modern Rise of Population*. New York: Academic Press.

Merton, Robert K., 1968. *Social Theory and Social Structure*. New York: Free Press.

Mitsuya, H., R. Yarchoan, and S. Broder, 1990. Molecular targets for AIDS therapy. *Science* 249: 1533–1543.

National Center for Health Statistics, 1991. Annual summary of births, marriages, divorces, and deaths: United States, 1990. *Monthly Vital Statistics Report* 39 (August): 28. Washington, DC: U.S. Government Printing Office.

Nock, Steven L., and Paul W. Kingston, 1988. Time with children: The impact of couples' work-time commitments. *Social Forces* 67: 59–85.

O'Neill, William L., 1967. *Divorce in the Progressive Era.* New Haven, CT: Yale University Press.

Pearson, Willie, and Llewellyn Hendrix., 1979. Divorce and the status of women. *Journal of Marriage and Family* 41: 375–385.

Preston, Samuel H., and John McDonald, 1979. The incidence of divorce within cohorts of American marriages contracted since the Civil War. *Demography* 16: 1–25.

Rothman, Ellen K., 1984. *Hands and hearts: A history of courtship in America.* New York: Basic Books.

Ryder, Norman B., 1980. Components of temporal variations in American fertility. In R. W. Hiorns (ed.), *Demographic Patterns in Developed Societies.* London: Taylor & Francis.

Scanzoni, John, 1972. *Sexual Bargaining: Power and Politics in the American Marriage.* Englewood Cliffs, NJ: Prentice-Hall.

Scanzoni, John, 1991. On balancing the family interests of children and adults. In Elaine Anderson and Richard Hula (eds.), *Reconstructing Family Policy.* New York: Greenwood Press.

Schlosman, Kay Lehman, and Sidney Verba, 1979. *Injury to Insult: Unemployment, Class, and Political Response.* Cambridge, MA: Harvard University Press.

Simmel, Georg, 1908. The metropolis and mental life. In *The Sociology of Georg Simmel.* New York: Free Press, 1950.

Trussell, James, and Felicia Stewart, 1992. The effectiveness of postcoital hormonal contraception. *Family Planning Perspectives* 24: 262–265.

Trussell, James, Felicia Stewart, Felicia Guest, and Robert A. Hatcher, 1992. Emergency contraceptive pills: A simple proposal to reduce unintended pregnancies. *Family Planning Perspectives* 24: 269–273.

U.S. Census Bureau, 1975. *Historical Statistics of the U.S.: Colonial Times to 1970.* Washington, DC: U.S. Government Printing Office.

U.S. Census Bureau, 1991. *Statistical Abstract of the United States.* Washington, DC: Government Printing Office.

Wallerstein, Judith S., and Sandra Blakeslee, 1989. *Second Chances: Men, Women, and Children a Decade After Divorce.* New York: Ticknor & Fields.

Weed, James, 1980. *National Estimates of Marital Dissolution and Survivorship: United States.* Washington, DC: U.S. Department of Health and Human Services.

Weitzman, Lenore, 1985. *The Divorce Revolution.* New York: Free Press.

Zabin, Laurie S., Marilyn B. Hirsch, and Mark R. Emerson, 1989. When urban adolescents choose abortion: Effects on education, psychological status, and subsequent pregnancy. *Family Planning Perspectives* 21: 248–255.

Zelizer, Viviana A., 1985. *Pricing the Priceless Child: The Changing Social Value of Children.* New York: Basic Books.

25

Family Enrichment: Programmes to Foster Healthy Family Development

CLAUDIA S. ARP, DAVID H. ARP, AND VERA MACE

With today's pace of rapid social change, the question becomes whether the authors' family enrichment programs can work. What do you think? If they do not, what are some of the possible outcomes?

As you read, ask yourself the following questions:

1. *In what ways would the programs mentioned help to foster healthy family development?*
2. *What other means can you think of that will add to the development of healthy families?*

GLOSSARY

Family Universal social institution uniting individuals in a group that bears and raises children.

Proactive services Programs that deal with remedial needs and potential problems before they become serious troubles.

Introduction

THROUGHOUT HISTORY, STRONG MARRIAGES and families have been the foundation of well-being. Nicholas Stinnett wrote:

"As we look back in history we see that the quality of family life is important to the strength of nations. There is a pattern in the rise and fall of great societies. . . . When these societies were at the peak of their power and prosperity, the family was strong and highly valued. When family life became weak in these societies, when the family was not valued—when goals became extremely individualistic—the society began to deteriorate and eventually fell."[1]

History is a lesson that the continuance of the family must be safeguarded and strengthened in the world today.

Why are families needed? Unlike other forms of life, human children have to be nurtured and cared for over an extended period or they will die. The continued existence of the human race is a primary function of families.[2, 3]

Family patterns have changed and adapted through cultural development over the ages and continue to change. The primary need to love and nurture the human child exists so that the generations may continue despite the ongoing changes in the world.

The need for the family in human society is as great as ever and may be even greater because of the increasing complexity of life. The functions and opportunities of the family are not limited to individual self-contained units: they have ever-increasing possibilities of expanding. The healthier and stronger a family unit, the more likely that such units may constitute a strong nation and, in turn, strong nations make a stronger world.

With the recurrence of profound changes, families are facing difficult challenges. Families are at risk. With such realities as poverty, war, mass migration, changes in the workforce, domestic violence, drug addiction and crime, the need is for programmes focused on families ranging from the educational and preventive to therapeutic. This spectrum can be graphically demonstrated as follows:

United Nations, 1993.

Education/prevention . . Intervention/therapeutic
(Strength-centered) . . . . (Problem-centered)

Prevention programmes are described in terms such as "enrichment," "wellness," and "self-improvement." Since the family is the basic social unit and thus affects the direction of societies, helping families meet their own needs is critical. Family enrichment programmes can provide competencies to enable families to be appropriately self-sufficient. Enrichment programmes empower individuals within families and the family as a unit; this enables families to contribute to the communities and nations they live in, strengthening governmental and non-governmental organizations.

Proactive services give priority to the prevention of serious trouble rather than to meeting remedial needs. Catalano and Dooley define proactive services as those that prevent the occurrence of risk factors, and reactive services as those that improve the responses to risk factors.[4] There are several reasons to emphasize prevention programmes. For instance, prevention is cheaper. Widespread family wellness programmes could reduce the economic burdens of some nations. Prevention is broad-based. In recent years an increase of skill-oriented prevention programmes have overlapped with remedial programmes, reaching many of those people who were overlooked in the past. Prevention is also easier. Working with families who are basically healthy is not as complicated as working with unhealthy (dysfunctional) families. Working with families before they are in crises conserves energy and time that can be used to promote growth. By increasing prevention-oriented services, disruptive and destructive patterns can be changed and constructive help offered to families.[5] . . .

Prevention may be seen as the equivalent of the prenatal care of future families and immunization and vaccination against disease.

Present marriages need to be strengthened and couples must be prepared for healthy marriages to improve family life in the future. Abuse prevention, family planning, and preparation-for-parenthood programmes can help bring needed change in future families. Family-life education programmes can help families raise emotionally healthy children who are well prepared to be partners and parents in the twenty-first century. . . .

Basic Elements of Functional Families

"A functional family is the healthy soil out of which individuals can become mature human beings."[6]

Family enrichment or wellness programmes can best be developed and implemented if the functions of families are used as a foundation. The primary functions of families might be summarized as set out below.

The first function of the family is to pass on human values from generation to generation. Some civilizations believed that people gained immortality by passing on their essential selves to their children. But a child is not merely the propagation of parental personality. Children do not belong to parents. . . . The task of parents is to accept children for who they are; to love them; to care for them; and to strive to surround them with wholesome and positive influences. In this way, positive human values such as honesty, caring, and tolerance will be passed on to the next generation.[7]

The second function of the family is to provide a place for all its members, from birth to death. To fulfil this function, skills for achieving and maintaining human relationships must be learned. Human relationships need to be based upon an acceptance of the commonalities of all while valuing diversity within the family, which includes respect for each family member's unique personality. Extending this concept beyond the individual family can lead to improved relations between races, faiths, and nations.

The third function of the family is to provide a place where its members can learn how to achieve and maintain human relationships and how to work together for the common good. Ultimately, this includes working for world peace.

The fourth function of the family is to find ways of sharing and passing on the message of a better world so that all persons will want it and will work for it. Families are the primary vehicle for the preparation of ideas and human goals across generations.

These are noble family functions that can be better fulfilled through family enrichment.

QUALITIES OF STRONG FAMILIES

Much of what has been written about families has focused on what is wrong with the family. From an enrichment-oriented preventive focus, the emphasis has been upon pathology. It is critical, however, to build upon what is right about families and to provide tools for families to be strengthened. Research on family strengths provides, therefore, a valuable point of beginning for family enrichment programmes.[8]

Stinnett's research project spanned more than a decade and included over 3,000 families in every state of the United States of America and 20 other countries in Africa, Europe and Latin America. The top six qualities of strong families identified through Stinnett's research[9, 10] are summarized below.

Appreciation Every individual has a basic need to be appreciated. Strong families have the habit of looking for each other's good qualities and expressing their appreciation. Every person has strengths and positive qualities; the family is the primary place where the affirmation of strengths should occur. Healthy families concentrate on the positive.

Spending Time Together Strong families genuinely enjoy being together. This requires pre-planning and structure. Time together doesn't just happen; strong families make it happen. One interesting pattern emerged from Stinnett's research: there was a high frequency of families participating in outdoor activities. A possible explanation is that when families are together outdoors, distractions are fewer and they can concentrate on each other. Also physical exercise (especially outdoors) contributes to health and to personal feelings of well-being; both can strengthen the family.

Commitment Strong families are deeply committed to the family group and to promoting each other's happiness and welfare. While there has been little research on family commitment, Yankelovich observed that society was in the process of leaving behind an excessive self-centered orientation and of moving towards a new ethic of commitment emphasizing new rules of living that supported self-fulfilment through deeper personal relationships.[11] Mace[12] writes that commitment produces behavioural change. When life becomes overwhelming and family members are pulled in many different directions, strong families take the initiative in restructuring their lifestyle. They may eliminate the discretionary activities that do not benefit the family. Such families also set realistic family goals that will enhance the quality of their family relationships.[13]

Good Communication Patterns Strong families demonstrate good communication patterns. This quality is closely related to the previous ones. Good communication requires an investment of time.[12]

Not only do strong families talk to each other, but they also listen well, thus communicating a mutual respect for one another. They are also willing to deal with conflict and are committed to finding solutions that are best for everyone. They are able to express their feelings and do so in non-combative, constructive ways. They are able to make creative use of conflict. This kind of broad-based communication is integral to the life of strong families.

High Degree of Religious Orientation Strong families tend to have a high degree of religious orientation. The research undertaken during the past 40 years shows a positive relationship of religion to marital happiness and family well-being. Mace[13] and Stinnett point out that this quality of strong families goes beyond attending the church, synagogue, or mosque, or participating in religious activities. Stinnett found that an awareness of a higher power gave many families a sense of purpose and strength. This awareness also helped them to be more patient with each other, more forgiving, quicker to get over anger, more positive, and more supportive in their relationships.[14]

Ability to Deal with Crises in a Positive Manner
Strong families are able to deal with crises in a constructive way. They are able to deal with problems while remaining supportive of one another. They have problem-solving skills and are able to see some positive elements even in the darkest of situations. They are able to identify the crisis and to attack the problem and not each other.[15]

These characteristics of strong families coincided with the findings of other researchers who had examined healthy families.[16-19] What is needed to develop these qualities?

DEVELOPING HEALTHY FAMILIES

Individuals need to be prepared for marriage and parenthood. Both marriage and parenting are complex and challenging responsibilities that affect not only individuals and family members but also society as a whole. Bradshaw[6] proposed that families should be viewed as systems, with the marital partnership the chief component. If that is functioning well, the children will have an opportunity of growing up as happy, healthy people. He summarized the needs of families within the system: a sense of worth, a sense of physical security or productivity, a sense of intimacy and relatedness, a sense of unified structure, a sense of responsibility, a sense of being challenged and stimulated, a sense of joy, and an opportunity for affirmation and spiritual grounding. A child needs a mother and father who are committed to each other in a basically healthy relationship. Thus, family enrichment programmes consider the couple, the children, and the family as an interrelated unit.

Although a healthy family has many components, a significant one is the positive role models provided by parents. Positive parental role models help children work towards their own maturity. Child development specialists generally agree that the ideal setting for the healthy development of a child is two loving parents, mutually fulfilled in their marriage, who can provide their children with warm, non-possessive love. In such a family, a child identifies positively with the parent of his or her own sex, and learns by observa-

tion how to relate positively to the parent of the opposite sex.[20]

Many researchers believe the evidence for the emphasis upon strong marriages is compelling. Popenoe states:

"Social science research is almost never conclusive. There are always methodological difficulties and stones left unturned. Yet in the decades of work as a social scientist, I know of few other bodies of data in which the weight of evidence is so decisively on one side of the issue: for children, two parent-families are preferable to single-parent and step-families!"*

Popenoe's research strongly suggests the need for developing well-balanced programmes designed to strengthen existing and future family relationships.

Children who grow up in happy, intact, functional families unconsciously learn the roles they will later need in marriage and parenthood. They also gain a deep, satisfying sense of self-worth and a respect for others who share their lives. Such experiences help them to mature as self-confident adults capable of relating positively and creatively to other people, not only in the small world of the home but also in the community.

Family wellness is characterized by maturity in which parental roles model the following:

How to be a man or woman;
How be a husband or wife;
How to be a father or mother;
How to achieve intimate relationships with others;
How to be a functional human being.

Parents can best model healthy roles when they are part of a functional, strong family.[6]

REALITIES: TODAY'S FAMILIES AT RISK

Never have families and individuals had to live in such a complex, problematic world with so many choices to make, choices that often lead to feelings of guilt, fear, despair, and helplessness. Families

---

*As quoted in Barbara Dafoe Whitehead, "Dan Quayle was right," *The Atlantic Monthly,* (April 1993), p. 82.

are at risk. Families are more transient and mobile and can no longer count on the extended family for support. The fragmentation and dissolution of extended families and the increase of families headed by single parents are all issues in today's world. Family life in the 1990s is unlike that of any other period in history and is changing more rapidly than ever before. These changes are accelerated by the influences of recently developed technological advances including transportation, telecommunications, and computers.

Environmental issues, overpopulation, and economic recession along with war, crime and violence, political upheaval, famine, and the resettlement of refugees are some of the serious global problems concerning caregivers.

Acknowledging that these problems exist and have a direct and indirect impact on families is important. Poor relationships within the family are known to be closely related to many problems in society such as juvenile delinquency and domestic abuse. This interrelationship of what happens within the families, communities, and the world needs to be considered in the development and implementation of family enrichment programmes. . . .

FAMILY ENRICHMENT

Family enrichment programmes have existed for many years. Those who have developed and implemented programmes in the area of family enrichment, and written on the subject, agree that certain basic skills are derived from family functions and strengths. These skills are described below.

Learning to Communicate Positive communication means expressing oneself clearly as well as listening well. Communication can enhance close relationships because thoughts and feelings are expressed and clarified. Communication includes the expression of caring and affection. The expression of thoughts and feelings in positive ways consistent with actions can be learned.[20, 21]

Accentuating the Positive Observing and sharing positive actions, making thoughtful gestures, and acknowledging accomplishments makes others feel validated. Parents need to be encouraged to accentuate the positive. Simple actions can be used to emphasize positive qualities and actions. For example, a brief note to a child acknowledging something he or she did well could be put into the child's lunch box for school.

Expressing Feelings Human beings have a range of feelings that are normal and need to be expressed and validated. Expressing feelings in constructive ways is crucial in healthy relationships. Sadness, joy, anger, and happiness all exist and are neither good nor bad; they exist. The way feelings are expressed can lead to close positive relationships where growth occurs or to non-constructive, often destructive, relationships.

Members of a family usually live so close together that unless they can share their feelings with each other, misunderstandings and tensions tend to develop and they become angry. What should families do with anger? Some people bottle it up and allow it to simmer. Others explode and get relief, often at the expense of someone else. Such an expression of anger is healthier than suppressing it. But there is an even better way: acknowledge anger; regard it as your emotional state; and ask the person with whom you're angry to help you handle it. This approach can lead to constructive results.[20]

Finding ways of expressing feelings that work well for each family member and the family unit is important. For example, some individuals prefer to express feelings in writing or drawing, others orally; some want time to reflect while others do not.

The following recommendations foster positive communication:

A. Avoid "you" statements;
B. Avoid "why" questions;
C. Express feelings. Begin sentences with "I feel . . ." or "I believe. . . ." Share your own feelings and let the statement reflect back on you. In this way, you will avoid attacking the other person;
D. Allow time for resolving issues. To have meaningful family relationships, it is necessary to bring the issues and the feelings

about the issues out into the open, and to work at resolving them. This needs to be done again and again. Acknowledging and dealing with differences is necessary in order for each member to live together in a modern democratic family.

Learning to Cooperate Parents with appropriate expectations and limits provide a foundation for cooperation. Children need limits with an external structure that will help them develop inner control. Their unbridled impulses can lead them into trouble and scary situations if they are not supported by adults. Children can best learn self-discipline with the cooperation of wise parents.[20]

In previous generations, many parents exercised discipline in an authoritarian manner. Most modern parents reject autocratic rule but have never been taught the more difficult cooperative method of discipline. As a result, many give up in despair and become "permissive," which often means indulgent, allowing children to do whatever they wish. These young people develop little inner control, intensifying their alienation from their parents and their parents' way of life. Parents who value autonomy and self-discipline set appropriate limits.

The cooperative method of parenting includes good communication and interaction between parent and child. This no longer means demanding unquestioning obedience as it once did when the job of parenthood was seen as just an effort to subdue the child's will.

Working cooperatively within the family develops the potential for cooperation and collaboration beyond the home.

Learning to Negotiate Healthy families have learned to deal with conflict in positive ways. This is certainly not to suggest that conflict is absent in their homes. The issue is not whether conflict occurs but how conflicts are resolved within families. How can parents teach their children to deal with conflict in constructive ways? In their behaviour, parents need to provide model ways of resolving conflict. Helping their children to resolve their differences is difficult for parents if they are not willing or able to work

out their own. Parents must be encouraged to apply these skills as an educative responsibility to their children.

Many parents are plagued with doubts about whether they are doing the right thing. Healthy families talk things over with their children, especially when in disagreement. When parents feel that they must say no to their children's requests or set limits the children disagree with, careful explanations should be provided so their children understand the reasons for the decisions. Children must feel they deserve an explanation for parental decisions that affect them.

Parents must learn to negotiate if they want to influence their children positively. As children grow older, outside influences loom larger. For instance, because families are smaller than in the past, children often seek close companions outside the home, especially when siblings are far apart in age. In increasingly pluralistic societies, values children have learned at home are more likely to be contradicted by different values their companions have acquired in their own families. At quite an early age, children sense this conflict of loyalties.

The conflict grows acute for adolescents who in gaining independence may find themselves more and more alienated from their parents. At this time of family life, support groups are especially helpful for parents and adolescents as well as for the family.

Learning to Release Children into Adulthood Adolescence is, by definition, the state or process of growing up. Parents need to prepare their teenagers to make their own decisions, which means gradually releasing decision-making power into their hands.[22]

Parents of older children may support the increasing autonomy of their teenage children by expressing the process in a manner such as the following:

"My job is not to give you orders, but to help you make your own decisions and stand on your own feet. I'm also responsible for you and want to protect you from the dangerous situations you may get into because of your lack of experience.

I'll give you freedom and responsibility a little at a time, as we both feel you're ready for it. If you can handle it, then you've earned the right to a little more freedom. If you can't, then we'll wait until you've gained enough new experience to try again."

Once this kind of agreement has been established between parent and child, the way is open for cooperation.

Passing the torch to the next generation is not easy, but parents can do some things to facilitate the transition from adolescence to responsible adulthood. Parents can do a better job of letting go if they understand the goals of adolescence.

In an adolescent-parent study that involved 8,165 young adolescents and 10,467 parents, Merton and Irene Stromm[23] identified seven goals that most adolescents intuitively seek to achieve during their teenage years:

A. Achievement: the satisfaction of achieving a standard of excellence in some area of endeavour;

B. Friendships: the broadening of one's social base by having learned to make friendships and maintain them;

C. Feelings: the self-understanding gained through having learned to share one's feelings with another person;

D. Identity: the sense of knowing "who I am," of being recognized as a significant person;

E. Responsibility: the confidence of knowing "I can stand alone and make responsible decisions";

F. Maturity: the transformation from a child into an adult;

G. Sexuality: the acceptance of responsibility for one's new role as a sexual being.

Parents can help achieve these goals through helping their adolescents understand themselves and the world around them. Parents need to encourage responsible citizenship and the equality of all, stressing the worth of the individual.

Basically the goal of parenting can be summarized in one phrase: "To prepare your child to function independently as an adult."[22]

State of the Contemporary Family . . .

The current situation in the United States illustrates how families are at risk and how family enrichment can be used in intervention as well as prevention programmes for families.

The high rate of marital breakdown and the growing trend of serial monogamy means that fewer and fewer children have the opportunity of growing up in stable and predictable family settings. Large numbers of children spend at least some of their lives in one-parent families. For example, before they reach the age of 18, two out of three children born in the United States in 1993 could expect to live in a single-parent household.[24]

The following statistics are illustrative:

One out of four households in the United States is headed by a single parent with one or more children;

Nine out of ten (90 per cent) single-parent families are headed by the mother;

The United States has more than 14 million single parents;

The Census Bureau of the United States estimates that more than six out of every ten children born in the mid-1990s will live in a single-parent home before they reach their eighteenth birthday.* . . .

The current situation is not encouraging. Children in single-parent families are two to three times as likely to have emotional or behavioural problems. They are two and a half times as likely to give birth out of wedlock. According to an article in The Atlantic Monthly, the out-of-wedlock birth rate jumped from 5 per cent in 1960 to 27 per cent in 1990.[25]

Often the parent who is parenting alone is experiencing considerable emotional stress. The trauma of divorce often means that a child has to deal with parting company with one parent

---

*Marriage and divorce statistics are from the Barna Research Group and taken from research studies the Group has conducted over the past decade and from data accumulated and reported by the National Center for Health Statistics in Washington, D.C. in a statistical abstract of the United States for 1991.

and sometimes adjusting to a step-parent, as well as to any other children brought into the new family circle created by a new marriage. Such changes can be traumatic experiences and result in a loss of security. While children suffer if their parents' marriages break up, they also suffer in homes where their parents are together but in conflict. . . .

By any measure, the disruptive and damaging effects of familial breakdown underscores the need to give families both intervention and preventive support. In an initial crisis as a result of divorce, death, or an unwanted pregnancy, remedial supports are needed. Help is also needed with ongoing issues faced by families at risk as well as in developing strong marriages, educating for parenthood, and implementing early intervention when families are at risk.

INTERVENTION

Intervention-based programmes are needed to deal with family crises such as a divorce, domestic violence, or the death of one's partner. This is comparable to the hospital emergency room or intensive-care treatment. Offering this kind of support can help the affected parties stabilize their lives and ensure that basic protection and material needs are met.

Intervention programmes such as divorce-recovery workshops, emergency financial aid, and, where abuse is an issue, the provision and protection of safe houses or shelters of safety are necessary and required.

In any intervention programme, the goal is to sustain and move the individual or family towards self-sufficiency. If this goal is achieved, the economic cost to the State of such interventions is more than returned. Empowered healthy families have a potential to contribute to society that is not present in families with problems.

ENRICHMENT

Examples of the community services that fall under enrichment programmes are providing parenting support groups and providing family-life education and short-term shelter services that help families in need with tasks and everyday necessities. Community services may include financial planning and food and baby-sitting cooperatives offered in a variety of ways.

Supportive networks, strategies on how to stay in touch with grandparents and other relatives, child care, tips for economical shopping, and even short-term financial support are all areas that may need to be provided for families at risk. Questionnaire research and other methods are indispensable in identifying current family needs.

While intervention and family enrichment programmes are vital, ultimately preventative measures must be emphasized so that families are able to avoid some of those problems requiring intervention. . . .

Challenge to Action

In this period of profound technological advances and unprecedented change, the family supports of the past are insufficient for meeting the familial needs of the present. Humankind is able to send people to the moon, to travel faster than the speed of sound, to fax information halfway across the world in seconds, and to access think-tanks with computers, but lacks the interpersonal competencies needed to build loving relationships. Action on both the personal, local, national, and international levels to make family wellness a priority needs to be encouraged.

Community-based, family-life education and enrichment programmes supported by policies and by human and economic resources can contribute in major ways to the development of healthy families, and to building the smallest unit in democracy at the heart of society. Action for programmes and services to foster the healthy development of families must be a priority supported by individuals, families, governments, and non-governmental organizations. The future depends upon humankind taking responsibility immediately for this change in priorities from an emphasis on therapy to promoting healthy development. . . .

REFERENCES

1. Nicholas Stinnett, "Strong families," in *Marriage and Family in a Changing Society*, J. Hemslin and others, eds. (New York, The Free Press, 1992), pp. 496–506.

2. David and Vera Mace, *Close Companions* (New York, Continuum, 1982).

3. Vera Mace, *365 Meditations for Women* (Nashville, Tennessee, Abingdon Press, 1989), pp. 126–127.

4. R. Catalano and P. Dooley, "Economic change in primary prevention," in *Prevention in Mental Health: Research, Policy and Practice*, D. Mace and others, eds. (Beverly Hills, California, Sage, 1983), pp. 21–40.

5. Luciano L'Abate, "Prevention as a profession" in *Prevention in Family Services*, D. Mace and others, eds. (Beverly Hills, California, Sage, 1983), pp. 49–61.

6. John Bradshaw, *Bradshaw on: The Family* (Deerfield Beach, Florida, Health Communications, 1988).

7. David and Vera Mace, *In the Presence of God* (Philadelphia, the Westminster Press, 1985), p. 9.

8. University of Nebraska-Lincoln, Department of Human Development and Family and Conferences and Institutes, *Building Family Strengths: A Manual for Families* (March, 1986).

9. Nicholas Stinnett, "Strong families: a portrait" in *Prevention in Family Services*, D. Mace and others, eds. (Beverly Hills, California, Sage, 1983), pp. 27–38.

10. Nicholas Stinnett and John DeFrain, *Secrets of Strong Families* (Boston, Little, Brown and Company, 1986).

11. D. Yankelovich, New Rules: *Searching for Fulfillment in a World Turned Upside Down* (New York, Random House, 1981).

12. David Mace, *Prevention in Family Services* (Beverly Hills, California, Sage, 1983).

13. Claudia Arp, *Beating the Winter Blues* (Nashville, Tennessee, Thomas Nelson, 1991).

14. James M. Hemslin, *Marriage and Family in a Changing Society* (New York, The Free Press, 1992).

15. Dave and Claudia Arp, *60 One-Minute Family Builder Series* (Nashville, Tennessee, Thomas Nelson, 1993).

16. H. A. Otto, "The personal and family strength research projects: some implications for the therapist," *Mental Hygiene* No. 48 (1964), pp. 349–450.

17. J. M. Lewis and others, *No Single Thread: Psychological Health in Family Systems* (New York, Brunner/Mazel, 1976).

18. J. M. Lewis, How's Your Family? (New York, Brunner/Mazel, 1979).

19. P. T. Nelson and B. Banonis, "Family concerns and strengths identified at Delaware's White House Conference on Families" In *Family Strengths* 3: Roots of Well-Being (Lincoln, Nebraska, University of Nebraska Press, 1981).

20. David and Vera Mace, *Men, Women and God* (Atlanta, Georgia, John Knox Press, 1976).

21. David Mace, *Love and Anger in Marriage* (Grand Rapids, Michigan, Zondervan, 1982).

22. Claudia Arp, *Almost 13* (Nashville, Tennessee, Thomas Nelson, 1986).

23. Merton and Irene Strammen, *Five Cries of Parents* (New York, Harper & Row, 1985), p. 6.

24. George Barna, *The Future of the American Family* (Chicago, Illinois, Moody Press, 1993).

25. Barbara Dafoe Whitehead, "Dan Quayle was right," *The Atlantic Monthly* (April 1993), pp. 47–84.

QUESTIONS FOR DISCUSSION

For further discussion of this topic, see the Wadsworth Sociology Resource Center, "Virtual Society," *http://sociology.wadsworth.com,* under *Sociological Footprints,* by Cargan and Ballantine. You can respond to the discussion questions there or enter your own comments in the online chat forum.

SUGGESTED READINGS AND SOCIOLOGY INTERNET RESOURCES

See the Wadsworth Sociology Resource Center, "Virtual Society," *http://sociology.wadsworth. com,* for additional links, suggestions for further reading, and learning tools related to this chapter.

Either from the "Virtual Society" website or directly from your web browser, you may access InfoTrac College Edition, an online university library that includes over 700 popular and scholarly journals in which you can find articles related to the topics in this chapter.

CHAPTER 7

Education

INSTITUTION IN A CROSSFIRE

MUCH OF OUR TIME UP TO the age of 18, and often well beyond, is spent learning the roles necessary for survival in society. All societies are concerned with socializing their young to develop skills for knowledge deemed necessary and inculcating loyalty to the social system. In some societies, this education takes place informally through imitation of elders; in others, such as the United States, formal schools have become key mechanisms for the transmission of knowledge and culture. Schools function to perpetuate dominant societal values, prepare young people for roles in society, and allocate societal positions to them.

It is within these functions that controversy arises. For instance, whose culture should be transmitted? Vocal pressure groups argue for curricula to be "inclusive," to represent all groups. Such voices influence textbook content and selection, courses offered, and course content. A comparison of your course catalog today with one of 10 to 20 years ago would probably reflect some of these changes.

Sociologists interested in the institution of education focus on the structure of the system, including the functions performed, the roles played by the various groups and individuals involved, the processes operating within the system, and the pressures from other parts of society that influence the system. Of primary concern is the role education plays in providing the opportunities we have in society; thus, many sociologists of education study questions of equality of educational opportunity. Many argue that opportunity is not equal for all who pursue education, and that we lose many bright students in the process of education. Several of our readings focus on this important issue of equal opportunity in education.

Likewise, questions of who gets what training and positions creates controversy. Education in most countries is viewed as key to a better life and opportunities. Competition for entrance into the best schools and universities is keen.

Some assume that if we change certain aspects of the system we can remedy the problems, whereas others see a need to change the larger societal structure, not just education, to reduce inequality in society. Issues of equality of educational opportunity are relevant to most minority groups. The question is whether the dominant group in society has an educational advantage over minority groups, and therefore an advantage in the job market and in long-range earning capacity.

In the first reading, Sally Lubeck observes two types of early childhood educational classrooms—one African-American and working class, and the other white and middle class—to learn how cultural values and attitudes are transmitted to children. In the excerpt from the conclusion of her book *Sandbox Society,* she shows how the socialization experience of African-American children in school is more collective or group oriented than that of white children, and she points out some implications of this finding.

Moving from nursery school to kindergarten, Harry Gracey discusses how children initially learn the rules of society by learning the rules of the classroom—routine, obedience, and discipline. Thus, schools are important tools in preparing children for their later roles in the work world.

The next reading by Ballantine outlines a number of current issues in the debate over equal opportunity, from tracking and testing to teacher expectations and home environment. Included in this reading is an insert by Robert Lake, an impassioned plea by a Native American father to his son's teacher, expressing the frustrations faced by this minority group when dealing with the education system of the dominant society.

Jonathan Kozol dramatically illustrates the unequal opportunity that can exist in schools in these three excerpts from his book *Savage Inequalities.* The first excerpt describes a school in East St. Louis in which the defects of the physical plant lead to the need for periodic school closings. The second selection describes a poor school in New York City in which children face large classes, lack of facilities, and violence. The third school, located in a wealthy suburb of Chicago, provides an example of a contrasting situation, proving the savage inequalities.

As you read this chapter, consider the following questions: What are some causes of inequality in education and society? Some consequences? What solutions do the readings propose?

26
Sandbox Society: Summary Analysis

SALLY LUBECK

Clear examples of differences in educational opportunity are described in the next two selections. The first focuses on preschool differences, illustrating that inequality begins early. If children are to receive equal educational opportunity, schools must provide it. Yet educational experiences of children from different racial and ethnic backgrounds and social classes vary widely. Lubeck gives dramatic testimony to this fact in her findings from observations at two very different nursery schools.

Sandbox Society contains detailed description of ways in which teachers in two early education classrooms constructed different social contexts through the organization of time, space, activities and materials, and social interaction. The study involved a year of fieldwork in which the author was a participant observer in the two programs. The final chapter (reprinted below) provides a summative analysis and, as such, omits the many examples on which the analysis itself is based.

As you read her account, think about the following:

1. *How do different structures and experiences in educational settings influence children's futures?*
2. *Have your own educational experiences influenced what you feel you can do with your life or the opportunities available to you?*

GLOSSARY

School learning environment Atmosphere and setting of a school.
Bureaucratic structure Hierarchy of positions in organizations.
Cultural transmission theory How children learn the rules and norms of their society.

USING TWO EARLY EDUCATION classrooms as "windows" through which to observe the child-rearing practices of two sets of women, the present study has attempted to delineate how cultural values and attitudes are transmitted to children. In these settings, which are approximately the same size and located in the same community in middle America, the teachers have been shown to construct different orders based on their differing life experiences. At the Irving Head Start Center, teachers and students are black and from working-class backgrounds; at the Harmony preschool, teachers and students are white and middle-class.

Both the black and the white women in the respective settings have clear ideas about how children learn and about how they should behave. They convey their own life orientations and expectations to children by creating total environments that reinforce values that give their own lives meaning. The learning environments thus become different means of reaching different ends. The Head Start teachers work closely together and reinforce collective values; the preschool teachers work alone with children much of the time and encourage values of individualism and self-expression.

The present study has had two major dimensions: first, to compare the child-rearing strategies of women in two early education settings and to demonstrate how they differ and, secondly, to explain how these differences arise within different social contexts. In both cases, the teachers live in families very like those of the children they teach, and, in both cases, they structure an environment that is consonant with their experiences outside of school.

From *Sally Lubeck,* Sandbox Society: Early Education in Black and White America *(London: Falmer Press, 1985), pp. 133–147. Reprinted with permission.*

Environmental Differences

From slavery to Selma the history of American blacks has been characterized by prejudice, repression, and inequity, and black Americans, like their African forebears, have adapted to a hostile environment by binding closely with one another. Many writers have noted the importance of "kin" in the black community. Black women frequently rear children cooperatively rather than separately.

In the lives of the women discussed here differences in family structure are distinguishing characteristics. The three Harmony teachers live in nuclear families in single-family dwellings. All three women are married, are able to share household chores, and take responsibility themselves for specific child-rearing tasks. The Head Start teachers, by contrast, both live in extended families, one in a household with her mother (who is bedridden), her father, her own two daughters, her sister, brother-in-law, and their daughter, one in a rented house with her son, daughter, and granddaughter. Though each makes less than $5,000 per year, they provide major financial support for their families.

Differences in School Environments

THE HARMONY PRESCHOOL

The preschool teachers appear to exercise some degree of control over their work situations; all work part-time by choice in a nonsectarian preschool located within a community church. Within the school setting, the teachers experience some constraints. For example, wall-to-wall carpeting was installed in the room which also seconds as a church meeting room. However, they had a platform made so they could still have paint, water, and sand in the classroom. And they have free rein in terms of the curriculum they will provide, the scheduling of activities, and the general organization of the program.

The pastor's office overlooks the A-frame, high-ceilinged room. However, I never observed either him or his secretary enter the room while class was in session, and telephone messages were discreetly dropped into the room on a line. Likewise, parents seemed to hover in the corridor outside and only entered when the children were singing songs or when the teachers otherwise indicated that activities were winding down. The classroom, like the traditional middle-class home, was the domain of women whose main role and function was the care and educating of the young.

THE IRVING HEAD START CENTER

A powerful bureaucracy structures the administration of local Head Start programs, and personnel at each level are responsible for reporting on the people under them. The supervisor visits unexpectedly or calls to say that she drove by at 7:30 A.M. and did not see the teachers' cars; in turn, she reports her findings. The teachers resent this intrusion, also her imposition of values (such as the advocacy of "play"), for she is perceived as an "outsider," someone who never works in the classroom herself. Yet the teachers' job likewise is defined by demands of the system, and they too must go into parents' homes and "rate" them. Their ambivalent relationship with parents, in part, is created by this extension of the monitoring role.

The bureaucratic structure creates an atmosphere of vigilance and extraneous control that has a kind of "ripple effect." This appears to bind the teachers together, uniting them against those who neither understand nor participate in the reality which they share.

However, the teachers are inevitably bound to this greater system even within the classroom. The telephone rings frequently, and it is not uncommon for someone to arrive at the door unexpectedly. Be it a parent, the bus driver, the supervisor, someone from the central office, a member of the ancillary staff, or another teacher in the building, one teacher or the other must interrupt what she is doing. Differences in "lag" time are indicative of the fact that the Head Start teachers are subjected to more external disruption.

Not only do the Head Start teachers experience less personal control over their situations than the preschool teachers, they also have more extensive responsibilities both at work and at home. Ironically, the social service demands of the program appear to create a situation not unlike that which black women have been shown to experience in the society at large (Stack, 1975).

Adaptations

In both settings, the teachers' beliefs and behaviors appear to be influenced by and continuous with their experience outside of school. Whereas the preschool teachers recreate a setting quite like that found in child-centered interactions, the Head Start teachers appear to re-enact patterns of interaction that have been shown to prevail within extended family networks (Hill, 1972; Stack, 1975; and Nobles, 1974), working closely together, sharing tasks, decisions, and resources, and sharing also the perception of those outside as hostile to their interests and efforts.

The preschool women work alone with children much of the morning, encouraging children also to work alone, making "unique" products, and fostering language that will enable the individual child to differentiate his or her experience from that of others. The Head Start women work together most of the morning, and, in their structuring of events, provide group-oriented activities that encourage children to do what others do.

Adaptations Conveyed Through Interactions with Children

The structuring of time and space, the choice of activities and utilization of materials, and patterns of adult-child interaction have been shown to be implicit forms through which adult values are transmitted to children. The two programs have been shown to differ systematically. It is as though the dynamic in the preschool was a sort of centrifugal force, working children away from the group, while the Head Start center dynamic was centripetal, encouraging children to cohere and to be part of the group. In their focus on individual children and in their structuring of an environment that maximizes individual choice and action, the preschool teachers encourage children to be uniquely different from others. "Free time" provides time for children to select activities of interest. Time is a continuum through which both children and activities change. Because children "develop" (or change) over time, the teachers provide different materials for children of different ages and separate the children into three different age groups for "developmentally appropriate" activities.

For much of the morning individuals move in different directions, at different rates of speed. Likewise time can be broken out into "units." And the classroom space is highly differentiated, providing different activities in different areas of the room. Transformational materials provide opportunities for children to have unique experiences, to make unique products, to impose their own order onto things. Teachers spend most of their time in the classroom with the children, and they are generally responsive to their requests. Children frequently initiate conversations with the teachers, and they call them by their first names. The classroom thus would seem to reinforce values of individuality and autonomy and to promote positive feelings toward change.

The Head Start teachers appear to structure time, space, and activity so as to reinforce values of collectivism, authority, and traditional (repetitive) modes of interaction that reinforce the group experience. Children spend most of their time together in group activities, and social knowledge conveyed through verbal exchange appears more important than the manipulation of things.

But the adults likewise spend much of their time together, engaged in a proliferation of tasks that are integral to the social service demands of the program. A separate "teacher space" is set up to provide a place for these duties to be accomplished. The "peer-centered" nature of the classroom is evident not only in the bond between the women who share tasks and problems, but also in the "smoke-screen" maintained by the peer group that gives the illusion of conforming to authority.

By structuring time, space, and activity so that children do what others do, while also conforming to the directives of the teachers, the teachers thus seem to socialize children to adapt to the reality which they themselves experience.

Toward a Theory of Cultural Transmission

Though understanding how children become group members and how group norms are reproduced and perpetuated in schools is a timely concern, a theory of education as cultural transmission has yet to be developed (Wilcox, 1982). Based on observation of the enculturation processes in two classrooms in middle America, the following models isolate constellations of factors that help to explain how adaptations to specific environments are transmitted to children.

A theoretical model is a specific form of a theory, highlighting key relationships among factors which are thought to be significant. In these cases, the models isolate factors that appear to be adaptive responses to differing environments and postulate how these alternative adult orientations might have subsequent outcomes in the behavioral and cognitive styles of children. Extending anthropologists' belief that a culture is a self-perpetuating system (Cohen, 1971), the models provide a framework for understanding how child-rearing practices, as expressed in different environments, may serve to perpetuate social and cognitive characteristics within a population.

ADULT-INDIVIDUALISTIC

Importance of Experimentation; Knowledge Gained from the Manipulation of Things Piaget proposes that children gradually learn to abstract from their experience in the concrete world by acting on objects during the early years of life. He called the young child a "little scientist" who constantly strives to understand the natural world. In the preschool the teachers assume that "acting on objects" is important. They provide materials that can be transformed into many things, and, through experimentation, children "discover"

possibilities inherent in those materials. In a seemingly significant way, teachers assume that, once children have experienced something, they will know it. ("It might be fun to make an island in the sandbox and pour water all around it to make an ocean.")

Language to Differentiate Experience and Modulate Choice The emphasis on the individual child, the changing nature of the curriculum, and the transformational nature of materials provide opportunities for increasingly differentiated experience. As Bernstein (1971) hypothesizes: "The greater the differentiation of the child's experience, the greater his ability to differentiate and elaborate objects in his environment" (p. 28). Children are encouraged to enumerate characteristics of an object. A pineapple is "brown" and "round," "sweet" and "prickly." Levels of intensity in the verbal directives of adults require children to decode what is being allowed and/or encouraged.

Distinction Between "Parts" and "Wholes" The room is divided into different areas, and children are allowed to move from section to section; the curriculum is divided into "units" and even the group is a whole composed of parts ("Who is missing?"), composed of children who are frequently called by name and whose names appear on boards on the wall. The nature of the curriculum is such that superordinate classifications (fruit, number, etc.) are constantly being broken down into their component parts.

Linear Projection of Social Space Children spend much of the morning moving in different directions at different rates of speed.

Construction of Time as a Continuum The teachers believe that children "develop" in different ways and at different rates of speed. By providing an environment that offers choices, by working closely with the children, and by implementing a curriculum that changes over time, they strive to facilitate the "growth" of the individual child.

Interaction Between Adults and Children The preschool teachers are able to work with the children most of the morning. They sit with groups of children in specific areas of the room for much of the morning "free play" time. They help children clean up, sit with their age groups of children at snack, lead three ropes of children to the park. For most of the morning time, space and activity are constructed so as to maximize adult-child interaction.

CHILD-INDIVIDUALISTIC

"Seeks Help — Seeks Attention — Seeks Dominance" The preschool children generally are able to maximize their interactions with the teachers, gaining access to their attempts to individualize and also gaining attention for their individual efforts. Fieldnotes indicate that more interactions take place between children and teachers than between the children themselves. A group of older boys who frequently generated their own pretend play were the exception.

Verbal Communication Space is usually shared with adults, and conversation is encouraged. Frequently, verbal exchange is related to what the child has done to materials.

Abstract Thinking Cohen (1969/76) presents the argument that abstract thought has three major components: breadth of knowledge, analytic abstraction, and the ability to extract information from an embedding context. Although it was not possible to "get inside" children's heads, their observable behavior could be seen to lay the groundwork for such an orientation: in their diversity of experience and use of a differentiating vocabulary, in the multiple opportunities they had for experiencing concrete components of classifications (number, fruit, color, etc.) and in the opportunities provided in which they could "pull out" of a material an aspect they wished to elaborate. Though such experiences seem only tangentially related to the requirements of public schooling, the teachers firmly believed that these experiences were what were appropriate for young children. Interestingly, when I returned to the classroom late in the year, the head teacher greeted me with pride:

"Oh, I'm so glad you've come to see us. We're so proud of our children. Three of them are reading already."

ADULT-COLLECTIVE

Manipulation of Objects Considered Unimportant: Knowledge Transferred from Adults to Children The Irving Head Start teachers believe that children learn by listening and by repeating after the teacher. Blocks and other manipulatives are considered playthings, things to keep the children occupied during "free time," while the teachers attend to other responsibilities. Activities frequently require children to follow directions, to perform the same actions, and to arrive at the same product. To this extent, the objects serve as media through which the adults' goals can be realized, rather than as objects of interest in themselves. The focus appears to be on people, not on things.

Language to Reinforce the Group Orientation and Authority of Adults The adults in the center share decisions, resources, and responsibilities. In interaction with children, it is common for the teachers to give verbal directives to the group, often to aggregates of children ("Boys," "Boys and girls," "Girls" . . .) These directions tend to be quite specific, stipulating a particular place or way in which children are to wait or stand, or a particular way in which a task is to be accomplished (gluing feathers on a band, coloring, etc.). During group time the teacher will often ask the same question of every child in the group, seemingly stressing the equality of group members. Teachers sit or stand above the children. Since they are called by their last names, the authority of adults would likewise seem to be reinforced.

Holistic Worldview Although the present study has not addressed this topic directly, a holistic worldview would seem to be indicated in the construction of time as a "container," in the spatial centering of the group for much of each day, in the convergent responses expected from the children, as well as in the aggregating of groups of children according to sex.

Spatial Centering Children spend most of their time at the center "grouped." Most ride to school in the same van. They eat breakfast together, waiting on the red rug until all are finished. During "free play" three to four children are assigned to each area and are to remain there. After cleanup, the children have a formal "group" time: They rest together, and then eat lunch together. In almost every time frame the group is "centered" on a specific activity.

Construction of Time as a Series of Discrete Moments Time appeared to be constructed as a series of "containers," in which the same activity was repeated day after day. Time was also divided between adult responsibilities and time for the children.

Separation of Adult Responsibilities and Children Because the teachers are responsible for many tasks besides purely educational ones, time and space are provided so they can attend to "adult" responsibilities (providing meals, cleaning up, filling the refrigerator, ordering supplies, doing paperwork, keeping records, speaking with supervisors and ancillary personnel, preparing materials, etc.). In "teacher space" many of these "adult" tasks are carried out. In "children" space, the teachers see to it that the children have breakfast and a rest time, and they interact with the group in formal question-and-answer type lessons that are repeated each day over time.

CHILD-COLLECTIVE

"Offers Help—Offers Support—Suggests Responsibility" The first day that I entered the center I was struck by the realization that older children were supervising younger ones while mothers filled out forms and spoke with the teacher and social worker. Although Whiting and Whiting (1975) cites this as a primary personality characteristic of children in kin-based traditional societies, such behavior was not much in evidence in the day-to-day observations of the present study. To some extent, an educational setting with children of the same age places constraints on such behavior, since children more typically "care" for those younger than themselves.

Generalized Verbal Communication At the center children are with their peers more than with adults. Children frequently use generalized terms such as "thing" and "stuff" and are not encouraged through interaction with adults (and through changing and highly differentiated objects) to use more particularized vocabularies. In adult-child interactions the focus on convergence, similarity, and agreement appears to be at odds with values of divergence, differentiation, and distinction implied by a particularized language.

Relational Thinking Cohen (1969/76) argues that relational thinking is characteristic of persons reared in "shared function" environments, in family and friendship groups in which functions such as child care, leadership, and control of funds are shared rather than being assigned to status roles. Such experiences, she argues, lead to a particular form of conceptual patterning. The relational style "requires a descriptive mode of abstraction and is self-centered in its orientation to reality; only the global characteristics of a stimulus have meaning to its users, and these only in reference to some total context" (p. 301). She notes that, in addition to incompatible selection and classification rule sets, three other characteristics distinguish relationship thinkers from abstract thinkers: the perception of time as a series of discrete moments, of self in the center of social space, and in specific causality rather than multiple causality. Such conceptual patterning would seem reflective of the patterns of social organization observed in the center.

Conclusion

To suggest a way of looking is an appropriate research enterprise, but to suggest a way of doing things differently is to venture upon the thin ice of social policy. Evidence from the present study lends support to the contention that poor black children in America do not "lack stimulation," as the formulators of so many social programs have assumed. Rather they appear to be methodically socialized into the preeminent values of a society in which the needs of the group prevail over the needs of the individual. To the extent that values and attitudes are culturally mediated, to that extent poor black children can be said to be "different" from children of white, middle-income families, though, of course, such a generalization would not hold in every instance.

Clearly the collective orientation of Afro-Americans has both historical roots and contemporary efficacy. The Head Start women operate under real constraints both distal and proximal. Yet to lodge the "problem" in institutional structures is also too simple. For in their ability to work cooperatively, to share resources and responsibilities, to see humor in adversity and to accomplish many tasks which include the care of children, the Head Start women achieve something increasingly beyond the grasp of upwardly mobile America — a oneness with others and a common purpose. Far from being helpless pawns in a system, they actively construct a meaningful and consistent order that "works" within — and even in opposition to — the constraints of the bureaucratic order and the society at large.

Ultimately it must be realized that differences among people within American society are deeply grounded in a social system that is premised on an unequal distribution of resources, a society that is product, more than people, oriented, one in which verbal facility, abstraction, and self-achievement are prized to the exclusion of other values.

If policy is to be effected, a first step may be to acknowledge that individuals are not "right" or "wrong," that people are not "culturally deprived," and that social groups do not stretch out along a line from "primitive" to "advanced." Rather we live in a world of multiple orders and multiple meanings, and, if we are to fulfill our most basic needs, both to achieve — and to belong, to change — and to maintain, to grow both individually and together, we must find ways to learn from each other about these seemingly contrary forces. In the social sciences we stand, like Newton, on a beach of sand, looking first at this pebble or this shell, while a sea of truth lies undiscovered before us.

REFERENCES

Bernstein, B., 1971. *Class, Codes and Control*, Vol. 1, London, Routledge and Kegan Paul.

Cohen, R., 1969. 'Conceptual styles, culture conflict and non-verbal tests of intelligence,' *American Anthropologist*, 71, 5, October and in Roberts, J. and Akinsanya, S. (Eds.) (1976) *Schooling in the Cultural Context: Anthropological Studies of Education*, New York, David McKay.

Cohen, Y., 1971. 'The shaping of men's minds: adaptations to imperatives of culture' in Wax, M., Diamond, S. and Gearing, F. (Eds.) *Anthropological Perspectives on Education*, New York, Basic Books, pp. 19–49.

Hill, R., 1972. *The Strengths of Black Families*, New York, Emerson Hall.

Nobles, W., 1974. 'Africanity: Its role in black families,' *Black Scholar*, 5, 9, June, pp. 10–17.

Stack, C., 1975. *All our Kin: Strategies for Survival in a Black Community*, New York, Harper and Row.

Whiting, B., and Whiting, J., 1975. *Children of Six Cultures*, Cambridge, Harvard University Press.

Wilcox, K., 1982. 'Differential socializations in the classroom: Implications for equal opportunity' and 'Ethnography as a methodology and its applications to the study of schooling: a review' in Spindler, G. (Ed.) *Doing the Ethnography of Schooling: Educational Anthropology in Action*, New York, Holt, Rinehart and Winston, pp. 268–309 and 456–88.

27

Learning the Student Role: Kindergarten as Academic Boot Camp

HARRY L. GRACEY

Schools are agents of socialization, preparing students for life in society In preparing children, they teach expectations and demand conformity to society's norms. They help teach children the attitudes and behaviors appropriate to the society. Gracey points out that this process of teaching children to "fit in" begins early. His focus is the routines children in kindergarten are expected to follow.

As you read the article, think about the following:

1. How does the kindergarten routine encourage conformity and help prepare children for later life?

2. How did your schooling prepare you for what you are doing and plan to do?

GLOSSARY

Educational institution/system The structure in society that provides systematic socialization of young.
Student role Behavior and attitudes regarded by educators as appropriate to children in schools.

EDUCATION MUST BE CONSIDERED one of the major institutions of social life today. Along with the family and organized religion, however, it is a "secondary institution," one in which people are prepared for life in society as it is presently organized. The main dimensions of modern life, that is, the nature of society as a whole, is determined principally by the "Primary institutions," which today are the economy, the political system, and the military establishment. Education has been defined by sociologists, classical and contemporary, as an institution which serves society by socializing people into it through a formalized, standardized procedure. At the beginning of this century Emile Durkheim told student teachers at the University of Paris that education "consists of a methodical socialization of the younger generation." He went on to add:

> It is the influence exercised by adult generations on those that are not ready for social life. Its object is to arouse and to develop in the child a certain number of physical, intellectual, and moral states that are demanded of him by the political society as a whole and by the special milieu for which he is specifically destined. . . . To the egotistic and asocial being that has just been born, [society] must, as rapidly as possible, add another, capable of leading a moral and social life. Such is the work of education.[1]

The education process, Durkheim said, "is above all the means by which society perpetually recreates the conditions of its very existence."[2] The contemporary educational sociologist, Wilbur Brookover, offers a similar formulation in his recent textbook definition of education:

> Actually, therefore, in the broadest sense education is synonymous with socialization. It includes any social behavior that assists in the induction of the child into membership in the society or any behavior by which the society perpetuates itself through the next generation.[3]

The educational institution is, then, one of the ways in which society is perpetuated through the systematic socialization of the young, while

In Dennis Wrong and Harry L. Gracey (eds.), Reading in Introductory Sociology. *New York: Macmillan, 1967.*

the nature of the society which is being perpetuated—its organization and operation, its values, beliefs, and ways of living—are determined by the primary institutions. The educational system, like other secondary institutions, *serves* the society which is *created* by the operation of the economy, the political system, and the military establishment.

Schools, the social organizations of the educational institution, are today for the most part large bureaucracies run by specially trained and certified people. There are few places left in modern societies where formal teaching and learning is carried on in small, isolated groups, like the rural, one-room schoolhouses of the last century. Schools are large, formal organizations which tend to be parts of larger organizations, local community School Districts. These School Districts are bureaucratically organized and their operations are supervised by state and local governments. In this context, as Brookover says:

> The term education is used . . . to refer to a system of schools, in which specifically designated persons are expected to teach children and youth certain types of acceptable behavior. The school system becomes a . . . unit in the total social structure and is recognized by the members of the society as a separate social institution. Within this structure a portion of the total socialization process occurs.[4]

Education is the part of the socialization process which takes place in the schools; and these are, more and more today, bureaucracies within bureaucracies.

Kindergarten is generally conceived by educators as a year of preparation for school. It is thought of as a year in which small children, five or six years old, are prepared socially and emotionally for the academic learning which will take place over the next twelve years. It is expected that a foundation of behavior and attitudes will be laid in kindergarten on which the children can acquire the skills and knowledge they will be taught in the grades. A booklet prepared for parents by the staff of a suburban New York school system says that the kindergarten experience will stimulate the child's desire to learn and cultivate the skills he will need for learning in the rest of his school career. It claims that the child will find opportunities for physical growth, for satisfying his "need for self-expression," acquire some knowledge, and provide opportunities for creative activity. It concludes, "The most important benefit that your five-year-old will receive from kindergarten is the opportunity to live and grow happily and purposefully with others in a small society." The kindergarten teachers in one of the elementary schools in this community, one we shall call the Wilbur Wright School, said their goals were to see that the children "grew" in all ways: physically, of course, emotionally, socially, and academically. They said they wanted children to like school as a result of their kindergarten experiences and that they wanted them to learn to get along with others.

None of these goals, however, is unique to kindergarten; each of them is held to some extent by teachers in the other six grades at Wright School. And growth would occur, but differently, even if the child did not attend school. The children already know how to get along with others, in their families and their play groups. The unique job of the kindergarten in the educational division of labor seems rather to be teaching children the student role. The student role is the repertoire of behavior and attitudes regarded by educators as appropriate to children in school. Observation in the kindergartens of the Wilbur Wright School revealed a great variety of activities through which children are shown and then drilled in the behavior and attitudes defined as appropriate for school and thereby induced to learn the role of student. Observations of the kindergartens and interviews with the teachers both pointed to the teaching and learning of classroom routines as the main element of the student role. The teachers expended most of their efforts, for the first half of the year at least, in training the children to follow the routines which teachers created. The children were, in a very real sense, *drilled* in tasks and activities created by the teachers for their own purposes and beginning and ending quite arbitrarily (from the child's point of view) at the command of the teacher. One teacher remarked that she hated September, because during the first month "everything has to be done rigidly, and repeatedly, until they know exactly what they're supposed to do." However, "by January," she said, "they know exactly what to do [during the day] and I don't have to be after them all the time." Classroom routines were introduced gradually from the beginning of the year in all the kindergartens, and the children were drilled in them as long as was necessary to achieve regular compliance. By the end of the school year, the successful kindergarten teacher has a well-organized group of children. They follow classroom routines automatically, having learned all the command signals and the expected responses to them. They have, in our terms, learned the student role. The following observation shows one such classroom operating at optimum organization on an afternoon late in May. It is the class of an experienced and respected kindergarten teacher.

An Afternoon in Kindergarten

At about 12:20 in the afternoon on a day in the last week of May, Edith Kerr leaves the teachers' room where she has been having lunch and walks to her classroom at the far end of the primary wing of Wright School. A group of five- and six-year-olds peers at her through the glass doors leading from the hall cloakroom to the play area outside. Entering her room, she straightens some material in the "book corner" of the room, arranges music on the piano, takes colored paper from her closet and places it on one of the shelves under the window. Her room is divided into a number of activity areas through the arrangement of furniture and play equipment. Two easels and a paint table near the door create a kind of passageway inside the room. A wedge-shaped area just inside the front door is made into a teacher's area by the placing of "her" things there: her desk, file, and piano. To the left is the book corner, marked off from the rest of the room by a puppet stage and a movable chalk-

board. In it are a display rack of picture books, a record player, and a stack of children's records. To the right of the entrance are the sink and clean-up area. Four large round tables with six chairs at each for the children are placed near the walls about halfway down the length of the room, two on each side, leaving a large open area in the center for group games, block building, and toy truck driving. Windows stretch down the length of both walls, starting about three feet from the floor and extending almost to the high ceilings. Under the windows are long shelves on which are kept all the toys, games, blocks, paper, paints, and other equipment of the kindergarten. The left rear corner of the room is a play store with shelves, merchandise, and cash register; the right rear corner is a play kitchen with stove, sink, ironing board, and bassinette with baby dolls in it. This area is partly shielded from the rest of the room by a large standing display rack for posters and children's art work. A sandbox is found against the back wall between these two areas. The room is light, brightly colored and filled with things adults feel five- and six-year-olds will find interesting and pleasing.

At 12:25 Edith opens the outside door and admits the waiting children. They hang their sweaters on hooks outside the door and then go to the center of the room and arrange themselves in a semi-circle on the floor, facing the teacher's chair, which she has placed in the center of the floor. Edith follows them in and sits in her chair checking attendance while waiting for the bell to ring. When she has finished attendance, which she takes by sight, she asks the children what the date is, what day and month it is, how many children are enrolled in the class, how many are present, and how many are absent.

The bell rings at 12:30 and the teacher puts away her attendance book. She introduces a visitor, who is sitting against the wall taking notes, as someone who wants to learn about schools and children. She then goes to the back of the room and takes down a large chart labeled "Helping Hands." Bringing it to the center of the room, she tells the children it is time to change jobs. Each child is assigned some task on the chart by placing his name, lettered on a pa-per "hand," next to a picture signifying the task—e.g., a broom, a blackboard, a milk bot-tle, a flag, and a Bible. She asks the children who wants each of the jobs and rearranges their "hands" accordingly. Returning to her chair, Edith announces, "One person should tell us what happened to Mark." A girl raises her hand, and when called on says, "Mark fell and hit his head and had to go to the hospital." The teacher adds that Mark's mother had written saying he was in the hospital.

During this time the children have been in-teracting among themselves, in their semi-circle. Children have whispered to their neighbors, poked one another, made general comments to the group, waved to friends on the other side of the circle. None of this has been disruptive, and the teacher has ignored it for the most part. The children seem to know just how much of each kind of interaction is permitted—they may greet in a soft voice someone who sits next to them, for example, but may not shout greetings to a friend who sits across the circle, so they confine themselves to waving and remain well within un-derstood limits.

At 12:35 two children arrive. Edith asks them why they are late and then sends them to join the circle on the floor. The other children vie with each other to tell the newcomers what happened to Mark. When this leads to a general disorder Edith asks, "Who has serious time?" The children become quiet and a girl raises her hand. Edith nods and the child gets a Bible and hands it to Edith. She reads the Twenty-third Psalm while the children sit quietly. Edith helps the child in charge begin reciting the Lord's Prayer; the other children follow along for the first unit of sounds, and then trail off as Edith finishes for them. Everyone stands and faces the American flag hung to the right of the door. Edith leads the pledge to the flag, with the children again fol-lowing the familiar sounds as far as they remem-ber them. Edith then asks the girl in charge what song she wants and the child replies, "My Coun-try." Edith goes to the piano and plays "Amer-ica," singing as the children follow her words.

Edith returns to her chair in the center of the room and the children sit again in the semi-circle on the floor. It is 12:40 when she tells the chil-dren, "Let's have boys' sharing time first." She calls the name of the first boy sitting on the end of the circle, and he comes up to her with a toy helicopter. He turns and holds it up for the other children to see. He says, "It's a heli-copter." Edith asks, "What is it used for?" and he replies, "For the army. Carry men. For the war." Other children join in, "For shooting sub-marines." "To bring back men from space when they are in the ocean." Edith sends the boy back to the circle and asks the next boy if he has some-thing. He replies "No" and she passes on to the next. He says "Yes" and brings a bird's nest to her. He holds it for the class to see, and the teacher asks, "What kind of bird made the nest?" The boy replies, "My friend says a rain bird made it." Edith asks what the nest is made of and different children reply, "mud," leaves," and "sticks." There is also a bit of moss woven into the nest, and Edith tries to describe it to the chil-dren. They, however, are more interested in see-ing if anything is inside it, and Edith lets the boy carry it around the semi-circle showing the chil-dren its insides. Edith tells the children of some baby robins in a nest in her yard, and some of the children tell about baby birds they have seen. Some children are asking about a small object in the nest which they say looks like an egg, but all have seen the nest now and Edith calls on the next boy. A number of children say, "I know what Michael has, but I'm not telling." Michael brings a book to the teacher and then goes back to his place in the circle of children. Edith reads the last page of the book to the class. Some children tell of books which they have at home. Edith calls the next boy, and three children call out, "I know what David has." "He always has the same thing." "It's a bang-bang." David goes to his table and gets a box which he brings to Edith. He opens it and shows the teacher a scale-model of an old-fashioned dueling pistol. When David does not turn around to the class, Edith tells him, "Show it to the children" and he does. One child says, "Mr. Johnson [the principal] said no guns."

Edith replies, "Yes, how many of you know that?" Most of the children in the circle raise their hands. She continues, "That you aren't supposed to bring guns to school?" She calls the next boy on the circle and he brings two large toy soldiers to her which the children enthusiastically identify as being from "Babes in Toyland." The next boy brings an American flag to Edith and shows it to the class. She asks him what the stars and stripes stand for and admonishes him to treat it carefully. "Why should you treat it carefully?" she asks the boy. "Because it's our flag," he replies. She congratulates him, saying, "That's right."

"Show and Tell" lasted twenty minutes and during the last ten one girl in particular announced that she knew what each child called upon had to show. Edith asked her to be quiet each time she spoke out, but she was not content, continuing to offer her comment at each "show." Four children from other classes had come into the room to bring something from another teacher or to ask for something from Edith. Those with requests were asked to return later if the item wasn't readily available.

Edith now asks if any of the children told their mothers about their trip to the local zoo the previous day. Many children raise their hands. As Edith calls on them, they tell what they liked in the zoo. Some children cannot wait to be called on, and they call out things to the teacher, who asks them to be quiet. After a few of the animals are mentioned, one child says, "I liked the spooky house," and the others chime in to agree with him, some pantomiming fear and horror. Edith is puzzled, and asks what this was. When half the children try to tell her at once, she raises her hand for quiet, then calls on individual children. One says, "The house with nobody in it"; another, "The dark little house." Edith asks where it was in the zoo, but the children cannot describe its location in any way which she can understand. Edith makes some jokes but they involve adult abstractions which the children cannot grasp. The children have become quite noisy now, speaking out to make both relevant and irrelevant comments, and three little girls have become particularly assertive.

Edith gets up from her seat at 1:10 and goes to the book corner, where she puts a record on the player. As it begins a story about the trip to the zoo, she returns to the circle and asks the children to go sit at the tables. She divides them among the tables in such a way as to indicate that they don't have regular seats. When the children are all seated at the four tables, five or six to a table, the teacher asks, "Who wants to be the first one?" One of the noisy girls comes to the center of the room. The voice on the record is giving directions for imitating an ostrich and the girl follows them, walking around the center of the room holding her ankles with her hands. Edith replays the record, and all the children, table by table, imitate ostriches down the center of the room and back. Edith removes her shoes and shows that she can be an ostrich too. This is apparently a familiar game, for a number of children are calling out, "Can we have the crab?" Edith asks one of the children to do a crab "so we can all remember how," and then plays the part of the record with music for imitating crabs by. The children from the first table line up across the room, hands and feet on the floor and faces pointing toward the ceiling. After they have "walked" down the room and back in this posture they sit at their table and the children of the next table play "crab." The children love this; they run from their tables, dance about on the floor waiting for their turns and are generally exuberant. Children ask for the "inch worm," and the game is played again with the children squirming down the floor. As a conclusion Edith shows them a new animal imitation, the "lame dog." The children all hobble down the floor on three "legs," table by table to the accompaniment of the record.

At 1:30 Edith has the children line up in the center of the room: she says, "Table one, line up in front of me," and children ask, "What are we going to do?" Then she moves a few steps to the side and says, "Table two over here; line up next to table one," and more children ask, "What for?" She does this for table three and table four, and each time the children ask, "Why, what are we going to do?" When the children are lined up in four lines of five each, spaced so that they are not touching one another, Edith puts on a new record and leads the class in calisthenics, to the accompaniment of the record. The children just jump around every which way in their places instead of doing the exercises, and by the time the record is finished, Edith, the only one following it, seems exhausted. She is apparently adopting the President's new "Physical Fitness" program for her classroom.

At 1:35 Edith pulls her chair to the easels and calls the children to sit on the floor in front of her, table by table. When they are all seated she asks, "What are you going to do for worktime today?" Different children raise their hands and tell Edith what they are going to draw. Most are going to make pictures of animals they saw in the zoo. Edith asks if they want to make pictures to send to Mark in the hospital, and the children agree to this. Edith gives drawing paper to the children, calling them to her one by one. After getting a piece of paper, the children go to the crayon box on the righthand shelves, select a number of colors, and go to the tables, where they begin drawing. Edith is again trying to quiet the perpetually talking girls. She keeps two of them standing by her so they won't disrupt the others. She asks them "Why do you feel you have to talk all the time?" and then scolds them for not listening to her. Then she sends them to their tables to draw.

Most of the children are drawing at their tables, sitting or kneeling in their chairs. They are all working very industriously and, engrossed in their work, very quietly. Three girls have chosen to paint at the easels, and having donned their smocks, they are busily mixing colors and intently applying them to their pictures. If the children at the tables are primitives and neo-realists in their animal depictions, these girls at the easels are the class abstract-expressionists, with their broad-stroked, colorful paintings.

Edith asks of the children generally, "What color should I make the cover of Mark's book?" Brown and green are suggested by some children "because Mark likes them." The other children are puzzled as to just what is going on and

ask, "What book?" or "What does she mean?" Edith explains what she thought was clear to them already, that they are all going to put their pictures together in a "book" to be sent to Mark. She goes to a small table in the play-kitchen corner and tells the children to bring her their pictures when they are finished and she will write their message for Mark on them.

By 1:50 most children have finished their pictures and given them to Edith. She talks with some of them as she ties the bundle of pictures together—answering questions, listening, carrying on conversations. The children are playing in various parts of the room with toys, games, and blocks which they have taken off the shelves. They also move from table to table examining each other's pictures, offering compliments and suggestions. Three girls at the table are cutting up colored paper for a collage. Another girl is walking about the room in a pair of high heels with a woman's purse over her arm. Three boys are playing in the center of the room with the large block set, with which they are building walk-ways and walking on them. Edith is very much concerned about their safety and comes over a number of times to fuss over them. Two or three other boys are pushing trucks around the center of the room, and mild altercations occur when they drive through the block constructions. Some boys and girls are playing at the toy store, two girls are serving "tea" in the play kitchen and one is washing a doll baby. Two boys have elected to clean the room, and with large sponges they wash the movable blackboard, the puppet stage, and then begin on the tables. They run into resistance from the children who are working with construction toys on the tables and do not want to dismantle their structures. The class is like a room full of bees, each intent on pursuing some activity, occasionally bumping into one another, but just veering off in another direction without serious altercation. At 2:05 the custodian arrives pushing a cart loaded with half-pint milk containers. He places a tray of cartons on the counter next to the sink, then leaves. His coming and going is unnoticed in the room (as, incidentally, is the presence of the observer, who is completely ignored by the children for the entire afternoon).

At 2:15 Edith walks to the entrance of the room, switches off the lights, and sits at the piano and plays. The children begin spontaneously singing the song, which is "Clean up, clean up. Everybody clean up." Edith walks around the room supervising the clean-up. Some children put their toys, the blocks, puzzles, games, and so on back on their shelves under the windows. The children making a collage keep right on working. A child from another class comes in to borrow the 45-rpm adapter for the record player. At more urging from Edith the rest of the children shelve their toys and work. The children are sitting around their tables now, and Edith asks, "What record would you like to hear while you have your milk?" There is some confusion and no general consensus, so Edith drops the subject and begins to call the children, table by table, to come get their milk. "Table one," she says, and the five children come to the sink, wash their hands and dry them, pick up a carton of milk and a straw, and take it back to their table. Two talking girls wander about the room interfering with the children getting their milk and Edith calls out to them to "settle down." As the children sit, many of them call out to Edith the name of the record they want to hear. When all the children are seated at tables with milk, Edith plays one of these records called "Bozo and the Birds" and shows the children pictures in a book which go with the record. The record recites, and the book shows the adventures of a clown, Bozo, as he walks through a woods meeting many different kinds of birds who, of course, display the characteristics of many kinds of people or, more accurately, different stereotypes. As children finish their milk, they take blankets or pads from the shelves under the windows and lie on them in the center of the room, where Edith sits on her chair showing the pictures. By 2:30 half the class is lying on the floor on their blankets, the record is still playing, and the teacher is turning the pages of the book. The child who came in previously returns the

45-rpm adaptor, and one of the kindergartners tells Edith what the boy's name is and where he lives.

The record ends at 2:40. Edith says, "Children, down on your blankets." All the class is lying on blankets now. Edith refuses to answer the various questions individual children put to her because, she tells them, "it's rest time now." Instead she talks very softly about what they will do tomorrow. They are going to work with clay, she says. The children lie quietly and listen. One of the boys raises his hand and when called on tells Edith, "The animals in the zoo looked so hungry yesterday." Edith asks the children what they think about this and a number try to volunteer opinions, but Edith accepts only those offered in a "rest-time tone," that is, softly and quietly. After a brief discussion of animal feeding, Edith calls the names of the two children on milk detail and has them collect empty milk cartons from the tables and return them to the tray. She asks the two children on clean-up detail to clean up the room. Then she gets up from her chair and goes to the door to turn on the lights. At this signal, the children all get up from the floor and return their blankets and pads to the shelf. It is raining (the reason for no outside play this afternoon) and cars driven by mothers clog the school drive and line up along the street. One of the talkative little girls comes over to Edith and pointing out the window says, "Mrs. Kerr, see my mother in the new Cadillac?"

At 2:50 Edith sits at the piano and plays. The children sit on the floor in the center of the room and sing. They have a repertoire of songs about animals, including one in which each child sings a refrain alone. They know these by heart and sing along through the ringing of the 2:55 bell. When the song is finished, Edith gets up and coming to the group says, "Okay, rhyming words to get your coats today." The children raise their hands and as Edith calls on them, they tell her two rhyming words, after which they are allowed to go into the hall to get their coats and sweaters. They return to the room with these and sit at their tables. At 2:59 Edith says, "When you have your coats on, you may line up at the door." Half of the children go to the door and stand in a long line. When the three o'clock bell rings, Edith returns to the piano and plays. The children sing a song called "Goodbye," after which Edith sends them out.

Training for Learning and for Life

The day in kindergarten at Wright School illustrates both the content of the student role as it has been learned by these children and the processes by which the teacher has brought about this learning, or "taught" them the student role. The children have learned to go through routines and to follow orders with unquestioning obedience, even when these make no sense to them. They have been disciplined to do as they are told by an authoritative person without significant protest. Edith has developed this discipline in the children by creating and enforcing a rigid social structure in the classroom through which she effectively controls the behavior of most of the children for most of the school day. The "living with others in a small society" which the school pamphlet tells parents is the most important thing the children will learn in kindergarten can be seen now in its operational meaning, which is learning to live by the routines imposed by the school. This learning appears to be the principle content of the student role.

Children who submit to school-imposed discipline and come to identify with it, so that being a "good student" comes to be an important part of their developing identities, *become* the good students by the school's definitions. Those who submit to the routines of the school but do not come to identify with them will be adequate students who find the more important part of their identities elsewhere, such as in the play group outside school. Children who refuse to submit to the school routines are rebels, who become known as "bad students" and often "problem children" in the school, for they do not learn the academic curriculum and their behavior is often disruptive in the classroom. Today schools engage clinical psychologists in part to help teachers deal with such children.

In looking at Edith's kindergarten at Wright School, it is interesting to ask how the children learn this role of student—come to accept school-imposed routines—and what, exactly, it involves in terms of behavior and attitudes. The most prominent features of the classroom are its physical and social structures. The room is carefully furnished and arranged in ways adults feel will interest children. The play store and play kitchen in the back of the room, for example, imply that children are interested in mimicking these activities of the adult world. The only space left for the children to create something of their own is the empty center of the room, and the materials at their disposal are the blocks, whose use causes anxiety on the part of the teacher. The room, being carefully organized physically by the adults, leaves little room for the creation of physical organization on the part of the children.

The social structure created by Edith is a far more powerful and subtle force for fitting the children to the student role. This structure is established by the very rigid and tightly controlled set of rituals and routines through which the children are put during the day. There is first the rigid "locating procedure" in which the children are asked to find themselves in terms of the month, date, day of the week, and the number of the class who are present and absent. This puts them solidly in the real world as defined by adults. The day is then divided into six periods whose activities are for the most part determined by the teacher. In Edith's kindergarten the children went through Serious Time, which opens the school day, Sharing Time, Play Time (which in clear weather would be spent outside), Work Time, Clean-up Time, after which they have their milk, and Rest Time after which they go home. The teacher has programmed activities for each of these Times.

Occasionally the class is allowed limited discretion to choose between proffered activities, such as stories or records, but original ideas for activities are never solicited from them. Opportunity for free individual action is open only once in the day, during the part of Work Time left after the general class assignment has been completed (on the day reported the class assignment was drawing animal pictures for the absent Mark). Spontaneous interests or observations from the children are never developed by the teacher. It seems that her schedule just does not allow room for developing such unplanned events. During Sharing Time, for example, the child who brought a bird's nest told Edith, in reply to her question of what kind of bird made it, "My friend says it's a rain bird." Edith does not think to ask about this bird, probably because the answer is "childish," that is, not given in accepted adult categories of birds. The children then express great interest in an object in the nest, but the teacher ignores this interest, probably because the object is uninteresting to her. The soldiers from "Babes in Toyland" strike a responsive note in the children, but this is not used for a discussion of any kind. The soldiers are treated in the same way as objects which bring little interest from the children. Finally, at the end of Sharing Time the child-world of perception literally erupts in the class with the recollection of "the spooky house" at the zoo. Apparently this made more of an impression on the children than did any of the animals, but Edith is unable to make any sense of it for herself. The tightly imposed order of the class begins to break down as the children discover a universe of discourse of their own and begin talking excitedly with one another. The teacher is effectively excluded from this child's world of perception and for a moment she fails to dominate the classroom situation. She reasserts control, however, by taking the children to the next activity she has planned for the day. It seems never to have occurred to Edith that there might be a meaningful learning experience for the children in re-creating the "spooky house" in the classroom. It seems fair to say that this would have offered an exercise in spontaneous self-expression and an opportunity for real creativity on the part of the children. Instead, they are taken through a canned animal imitation procedure, an activity which they apparently enjoy, but

which is also imposed upon them rather than created by them.

While children's perceptions of the world and opportunities for genuine spontaneity and creativity are being systematically eliminated from the kindergarten, unquestioned obedience to authority and role learning of meaningless material are being encouraged. When the children are called to line up in the center of the room they ask "Why?" and "What for?" as they are in the very process of complying. They have learned to go smoothly through a programmed day, regardless of whether parts of the program make any sense to them or not. Here the student role involves what might be called "doing what you're told and never mind why." Activities which might "make sense" to the children are effectively ruled out, and they are forced or induced to participate in activities which may be "senseless," such as calisthenics.

At the same time the children are being taught by rote meaningless sounds in the ritual oaths and songs, such as the Lord's Prayer, the Pledge to the Flag, and "America." As they go through the grades children learn more and more of the sounds of these ritual oaths, but the fact that they have often learned meaningless sounds rather than meaningful statements is shown when they are asked to write these out in the sixth grade; they write them as groups of sounds rather than as a series of words, according to the sixth grade teachers at Wright School. Probably much learning in the elementary grades is of this character, that is, having no intrinsic meaning to the children, but rather being tasks inexplicably required of them by authoritative adults. Listening to sixth grade children read social studies reports, for example, in which they have copied material from encyclopedias about a particular country, an observer often gets the feeling that he is watching an activity which has no intrinsic meaning for the child. The child who reads, "Switzerland grows wheat and cows and grass and makes a lot of cheese" knows the dictionary meaning of each of these words but may very well have no conception at all of this "thing" called Switzerland. He is simply carrying out a task assigned by the teacher *because* it is assigned, and this may be its only "meaning" for him.

Another type of learning which takes place in kindergarten is seen in children who take advantage of the "holes" in the adult social structure to create activities of their own, during Work Time or out-of-doors during Play Time. Here the children are learning to carve out a small world of their own within the world created by adults. They very quickly learn that if they keep within permissible limits of noise and action they can play much as they please. Small groups of children formed during the year in Edith's kindergarten who played together at these times, developing semi-independent little groups in which they created their own worlds in the interstices of the adult-imposed physical and social world. These groups remind the sociological observer very much of the so-called "informal groups" which adults develop in factories and offices of large bureaucracies.[5] Here, too, within authoritatively imposed social organizations people find "holes" to create little subworlds which support informal, friendly, unofficial behavior. Forming and participating in such groups seems to be as much part of the student role as it is of the role of bureaucrat.

The kindergarten has been conceived of here as the year in which children are prepared for their schooling by learning the role of student. In the classrooms of the rest of the school

grades, the children will be asked to submit to systems and routines imposed by the teachers and the curriculum. The days will be much like those of kindergarten, except that academic subjects will be substituted for the activities of the kindergarten. Once out of the school system, young adults will more than likely find themselves working in large-scale bureaucratic organizations, perhaps on the assembly line in the factory, perhaps in the paper routines of the white collar occupations, where they will be required to submit to rigid routines imposed by "the company" which may make little sense to them. Those who can operate well in this situation will be successful bureaucratic functionaries. Kindergarten, therefore, can be seen as preparing children not only for participation in the bureaucratic organization of large modern school systems, but also for the large-scale occupational bureaucracies of modern society.

NOTES

1. Emile Durkheim, *Sociology and Education* (New York: The Free Press, 1956), pp. 71–72.

2. *Ibid.,* p. 123.

3. Wilbur Brookover, *The Sociology of Education* (New York: American Book Company, 1957), p. 4.

4. *Ibid.,* p. 6.

5. See, for example, Peter M. Blau, *Bureaucracy in Modern Society* (New York: Random House, 1956), Chapter 3.

28
Schools and Equal Opportunity

JEANNE H. BALLANTINE

Schools are in the front line because they are located in communities, accessible to the local population. Therefore, they are often the target of frustration over lack of equal opportunity. Parents count on the schools to prepare their children for opportunities, yet not all children have equal opportunity in schools. The following reading outlines some barriers to equal educational opportunity and some reasons why it is difficult to achieve.

As you read, consider the following questions:

1. What are some practices that affect equal opportunity in schools?

2. What solutions are suggested, and what solutions have been tried in your community to bring about equal opportunity in schools?

GLOSSARY

Equal opportunity When all people, regardless of cultural, racial, gender, or social "background," have an equal chance of achieving a high socioeconomic status in society.

Ability grouping Placing students in groups based on some measure of their ability in the subject area.

"At risk" students Students who may drop out of school because of background factors such as home environment and peer groups.

IMAGES OF A LAND of opportunity have drawn people from around the world to the United States in search of new lives. Hopes that children would have a chance to improve their lot have hinged on schools and teachers. Yet the hopes of many groups for equal opportunity through education have not materialized.

Equal opportunity exists when all people, including those without status, wealth, or membership in a privileged group, have an equal chance of achieving a high socioeconomic status

in society, regardless of their sex, minority status, or social class. This requires removing obstacles to individual achievement such as prejudice, ignorance, and treatable impairments.[1] Minority groups whose racial, ethnic, religious, and linguistic culture and behavior patterns differ from the dominant U.S. middle-class patterns most often have problems in schools. However, in 1995, minority children made up 35 percent of elementary and secondary school enrollment: African-American 16.8 percent; Hispanic 13.5 percent; and Asian, Pacific Islander, and Native American 4.8 percent. In 1971, 58.5 percent of black and 48.3 percent of Hispanic high school students graduated, compared to 81.7 percent of white students. By 1994, those figures had risen to 84.1 percent blacks, 60.3 percent Hispanic, and 91.1 percent white students. In 1996, 15 percent of blacks, 30 percent of Hispanics, and 9.3 percent of whites between 18 and 19 years of age had dropped out of school. Though the retention figures in high schools are slowly improving, we cannot afford to ignore these young people, both for their sake and for the economic and social well-being of the country.[2]

Practices That Affect Equal Opportunity in Schools

How do schools inhibit equal opportunity, and how can schools provide equal opportunity to all young people regardless of race, ethnic, gender, or religious background? Research in the sociology of education has provided a great deal of information about problems faced by children and their teachers in schools. Understanding these problems is the first step in alleviating them. Several classroom interaction patterns, school and systemwide structures and decision-making patterns, and policy issues that affect equal opportunity are discussed. Although the actions in these areas overlap, they have been grouped for convenience.

Classroom Interactions that Affect Equal Opportunity

TEACHER EXPECTATIONS

As teachers come into contact with children, they are affected by the manner of dress, name, physical appearance, race, sex, language usage and accent, parents' occupations, and the children's responses to teachers. Most teachers come from middle-class backgrounds, and as is true of most individuals, gravitate toward those who are most similar. These factors may affect teacher behaviors and expectations toward students, and may lead to higher expectations for some than for others. Children pick up these subtle cues, may take on the belief of the teacher, and begin to behave accordingly; achievement in school reflects self-concept. The result may be a prevailing climate of hopelessness—teachers discouraged about the children's ability to learn and children believing they cannot learn. Consider the following example:

AN INDIAN FATHER'S PLEA
by Robert Lake (Medicine Grizzlybear)

Dear Teacher,
I would like to introduce you to my son, Wind-Wolf. He is probably what you would consider a typical Indian kid. He was born and raised on the reservation. He has black hair, dark brown eyes, and an olive complexion. And, like so many Indian children his age, he is shy and quiet in the classroom. He is 5 years old, in kindergarten, and I can't understand why you have already labeled him a "slow learner."

He has already been through quite an education compared with his peers in Western society. He was bonded to his mother and to the Mother Earth in a traditional native childbirth ceremony. And he has been continuously cared for by his mother, father, sisters, cousins, aunts, uncles, grandparents, and extended tribal family since this ceremony.

The traditional Indian baby basket became his "turtle's shell" and served as the first seat for his classroom. It is the same kind of basket our people have used for thousands of years. It is specially designed to provide the child with the kind of knowledge and experience he will need to survive in his culture and environment.

Wind-Wolf was strapped in snugly with a deliberate restriction on his arms and legs. Although Western society may argue this hinders motor-skill development and abstract reasoning, we believe it forces the child to first develop his intuitive faculties, rational intellect, symbolic thinking, and five senses. Wind-Wolf was with his mother constantly, closely bonded physically, as she carried him on her back or held him while breast-feeding. She carried him everywhere she went, and every night he slept with both parents. Because of this, Wind-Wolf's educational setting was not only a "secure" environment, but it was also very colorful, complicated, sensitive, and diverse.

As he grew older, Wind-Wolf began to crawl out of the baby basket, develop his motor skills, and explore the world around him. When frightened or sleepy he could always return to the basket, as a turtle withdraws into its shell. Such an inward journey allows one to reflect in privacy on what he has learned and to carry the new knowledge deeply into the unconscious and the soul. Shapes, sizes, colors, texture, sound, smell, feeling, taste, and the learning process are therefore functionally integrated—the physical and spiritual, matter and energy, and conscious and unconscious, individual and social.

It takes a long time to absorb and reflect on these kinds of experiences, so maybe that is why you think my Indian child is a slow learner. His aunts and grandmothers taught him to count and know his numbers while they sorted materials for making abstract designs in native baskets. And he was taught to learn mathematics by counting the sticks we use in our traditional native hand game. So he may be slow in grasping the methods and tools you use in your classroom, ones quite familiar to his white peers, but I hope you will be patient with him. It takes time to adjust to a new cultural system and learn new things.

He is not culturally "disadvantaged," but he is culturally "different."*

DIFFERENCES IN GENDER TREATMENT

The classroom provides very different experiences for girls and boys. These differences in achievement levels, academic and social interests, and discipline stem from several sources: cultural traditions, stereotypes about the proper roles of women and men, and norms that stem from these beliefs. However, the evidence of cultural influences makes it difficult to sift out what part is biological. Gender variations in learning styles, math and science ability, and achievement show great differences across cultures, making it virtually impossible to generalize about biological influences.

A major focus of study is achievement differences by gender in math and science. Although some have suggested that biological differences are at the root, Asian-American women and women in some other societies outperform men in science and math. Differences in science and math achievement can be attributed primarily to degree of parental support, teacher expectations, student study habits, and values related to women excelling in math—all cultural, not biological, factors.[3]

Studies of school and classroom experiences that differ by gender have focused on several topics:

1. Boys are called on and spoken to more than girls by teachers, although this includes negative disciplinary interaction.
2. Girls tend to form intimate "chumships" while boys relate through groups organized around activities such as sports.[4] Some have drawn attention to similarities between sports courts and board rooms in the interaction skills needed to be successful in each. Even conversational styles differ.[5]
3. Teacher behaviors and expectations often differ depending on student gender. Do group-

ings reflect gender differences? Are boys asked to do the "heavy" tasks? Do boys play outside more than girls? Are boys more competitive than girls? Evidence indicates the answer to these questions is often yes.
4. Girls achieve higher grades throughout their public school education, and do better in reading, writing, and literature. However, girls fall behind boys in math and science during high school, and they are seldom tracked toward subjects that can lead to higher-paying careers in engineering and technology.

School Structures That Affect Equal Opportunity

ABILITY GROUPING OR TRACKING

One of the most hotly debated practices in school systems is ability grouping, the practice of grouping students according to their estimated ability. The practice is most common in math and reading, but sometimes occurs in science. Does it help or hinder the majority of students? The answers are not simple, because many variables enter into the discussion. How is ability measured? Are we talking about low- or high-ability students? How are placements made? Are we considering one subject area or all subjects? Can students move from one track to another, or is the system rigid—once placed, that's it?

Ability grouping often begins in elementary school, where placement correlates with children's language skills, appearance, gender, and other socioeconomic variables.[6] Once placed in a group, change is uncommon unless it is downward. By eighth grade, for instance, students' science grouping affects their future science curriculum.[7] Some children come from families that are more sophisticated in dealing with the schools, and can influence the placement of their children in classrooms.[8]

Poor and minority students fall disproportionately at the bottom of the placements. In some schools, placements reflect school needs—filling study halls, boosting classes with low enroll-

*From Robert Lake, "An Indian Father's Plea," *Teacher Magazine* 2 (September 1990): 48–53. Reprinted with permission from *Teacher Magazine*.

ments, assigning students to remedial classes to keep federal funding, and meeting staff preferences for course assignments.[9]

The key point is that ability grouping does not always represent students' actual ability; many factors filter into placement. If grouping is used, social scientists suggest flexible placements that are reviewed frequently and that differ by subject areas. Thus, a student may be in a higher math group than reading group. Teachers and counselors alike should consider factors in addition to test scores in placements. With these cautions, ability grouping, used for limited numbers of courses, can be effective. However, when group placement locks students into courses and tracks such as college preparatory or vocational, it can impair achievement and equal opportunity.

"AT RISK" STUDENTS AND DROPOUTS

One in four students in the United States won't graduate from high school. Where ethnic diversity is greatest, retention rates are lowest. Table 1 shows dropout rates for various groups.[10] Drop-out rates for whites and Hispanics have leveled off since 1988, and have dropped slightly for African-Americans.[11] The very schools that would become dumping grounds under choice systems have the highest dropout rates, in part because of the high sense of alienation, powerlessness, and isolation among students and teachers. Because dropouts remain unemployed at a much higher rate than the population at large, this group costs society in lost human resources, welfare costs, and sometimes prison costs.

What causes students to drop out? "Many youngsters arrive in school homeless, sick, hungry, and destitute, plagued by problems that often make staying and succeeding in school virtually impossible."[12] Many schools are so poor and crowded that they cannot begin to offer support to students who need help. Counseling, advising, help with family problems, peer issues, courses, and career planning are all lacking in poor schools. There is also often no supervision away from the school. Drugs and alcohol, bilingual language problems, and jobs that take a

TABLE 1 High School Dropout Rates, 1972 vs. 1993

	1972	*1993*
Total	6.1%	4.5%
White	5.3%	3.9%
Black	9.5%	5.8%
Hispanic	11.2%	6.7%

Source: From page 186 of National Center for Education Statistics, "The Condition of Education," 1995; U.S. Department of Commerce, Bureau of the Census, October 1993 Current Population Surveys; U.S. Department of Education, National Center for Education Statistics, *Dropout Rates in the United States: 1993.* Washington, D.C.: 1994.

great deal of a student's time can be risk factors, but probably the most serious problem is peer group influence. Many children are faced with peer pressures to join gangs, and problems in "socially deprived" neighborhoods can have a negative effect on educational attainment.

Many schools suspend or expel problem students; this action not only reinforces to the students the message that some students do not fit, but also puts the excluded students farther behind in their studies. Many "at-risk" students are trouble makers, and certain discipline techniques accentuate the problems.

Policy Issues Affecting Equal Opportunity

TESTING

Arguments for and against the use of tests for placement in school classes and in higher education focus on the issue of equal opportunity. Many functionalists argue that standardized tests give everyone an equal opportunity by testing them with the same instrument. However, others—including many conflict theorists—argue that tests are biased in several ways against poor and minority test takers. Let us focus on the arguments against tests used for school placement.

One argument relates to the question "What are tests measuring?" IQ tests, for instance, have been used for placement, but there are questions about the nature of intelligence and what the tests actually measure. Several scientists have challenged the notion of "one" intelligence, proposing instead that there may be multiple intelligences not measurable with our current instruments,[13] including problem solving, critical thinking, and other areas.

Another question centers on cultural bias in tests. Can we really devise a culture-free aptitude or achievement test that doesn't disadvantage students from certain backgrounds? Questions may rely on content that is more familiar to one subcultural group than another, and some groups receive more preparation for the tests both in their classrooms and in private courses to improve scores. Since testing is likely to continue as a means of placement, it behooves us to be aware of the controversies and take precautions in our own schools and classrooms.

DESEGREGATION AND INTEGRATION

In 1954 the U.S. Supreme Court ruled that "Separate is not equal," meaning that segregated schools do not provide equal educational opportunity for students. Ornstein and Levine note that

> *Desegregation* refers to enrollment patterns wherein students of different racial groups attend the same schools, and students are not separated in racially isolated schools or classrooms. *Integration* refers to situations in which students of different racial groups not only attend school together, but effective steps have been taken . . . to overcome the disadvantages of minority students and develop positive interracial relationships.[14]

Since the 1964 Civil Rights Act required school districts to take action to desegregate, courts at every level have been interpreting rulings for their districts. A 1992 U.S. Supreme Court ruling (*Freeman v. Pitts:* Case No. 89-1290) gives back to local districts that were under court-ordered desegregation programs some control over aspects of their operation such as assignment of students to classes and schools. Some fear this could resegregate districts.

A group of social scientists presented a summary statement of social science research on the importance and value of desegregation efforts over the past 20 years. The findings fall into four categories:

1. The desegregation of a school district can positively influence residential integration in the community.
2. Desegregation is associated with moderate academic gains for minority group students and does no harm to white students.
3. Desegregation plans work best when they cover as many grades as possible, when they encompass as large a geographic area as possible, and when they stick to clearly defined goals over the long haul.
4. Effective desegregation is linked to other types of educational reform.[15]

Desegregation attempts include locally and federally sponsored programs. Magnet schools provide a way of distributing students on the basis of subject concentration areas. In the fall of 1992, there were approximately 5000 magnet schools nationwide. Schools specializing in the arts or science and math attract students from a variety of backgrounds and in some cities have proven to be an effective way to desegregate schools by bringing together a diverse group of students. However, for most students this option is not available.

The federal government is involved in funding programs from preschool (such as Head Start) to high school (Upward Bound). These programs give students from poor backgrounds extra help to prepare them for first grade and for college. The most positive results are seen in the improved elementary school achievement of Head Start children. However, one major concern of integration attempts is that students entering integrated schools face social and academic isolation.

CHOICE AND "AMERICA 2000"

Each new president, each new secretary of education, each new "crisis in education" brings forth new plans to solve school problems. Most of these plans present little that hasn't been thought of and tried in the past, but memories are short and "new" plans are presented as the salvation for schools. Progressive education, back-to-basics, alternative schools, competency-based education, accountability, and choice movements have all been touted as the answer to our educational problems.

One plan presented as the solution to the educational system's problems is "choice"[16] — the idea that parents should select the school with the philosophy and curriculum best suited to their children's needs. This system would put schools in competition for the money each child brings, the theory being that competition would enhance efforts of school personnel to perform and produce quality education. Ideally, parents would become more involved in their children's education, a factor we know increases achievement.

On the surface, the idea sounds exciting — give parents a say in their children's education and create competition to improve schools. However, critics — including many conflict theorists — have several problems with the proposals. Related to equal opportunity and choice, urban public schools (already in serious trouble) would become the dumping ground for students not enrolled in other public or private schools. This situation would create further divisions in society by supporting private schools and the best public schools, which could become more selective about their student bodies and could siphon off the best teachers; urban public schools would sink further, according to critics.

Some have predicted the demise of public school education and desegregated schools under choice plans. One critic put it this way: "'choice' plans of the kind the White House (under the Bush administration) has proposed threaten to compound the present fact of racial segregation with the added injury of caste discrimination, further isolating those who . . . have been consigned to places *nobody* would choose if he had any choice at all."[17]

The bottom line seems to be that little research supports the prediction that education for low-income youth would improve under a choice plan. In fact, it could perpetuate or increase the gap between wealthy and poor youth, because high-income youth would continue to receive better educations.

Solutions and Recommendations

What are we to learn from this discussion of equal opportunity? That there are many activities we as individuals in the classroom and in interaction with students can carry out to help ensure that each child gets the education to which she or he is entitled, and that children are prepared for fruitful, productive lives in society. Awareness of problem areas is the first step in moving toward change.

Positive teacher attitudes, approaches, encouragement, and expectations for learning are key to success. We know that some inner-city schools have turned achievement levels around through rigorous expectations[18] and belief in children's abilities to achieve. Understanding the dynamics of classroom interaction can make teachers aware of actions related to gender differences in classroom behaviors and achievement.

Individual classroom teachers need to be aware of the ways in which students are grouped and group themselves in classrooms. Avoiding rigid groupings and encouraging integrated activities can improve school climate and achievement levels for minority students and women.

As long as tests are used for placement of students, other means of corroborating test scores must be used. What are the teacher observations of the students' abilities in classroom and other activities? Do the tests accurately reflect the students' abilities and interests?

Experts advise early identification of and intervention with "at-risk" students. Some programs that are receiving attention include accelerated academics for students who are bored, after-school and Saturday programs, alternative programs with different learning environments for alienated students, laws to deny drivers' licenses to students who are doing poorly in school or who drop out before 18, and programs to get parents involved. Finding alternatives to suspensions and expulsions is a key element in keeping students in school. For instance, in-school suspensions, tutoring, and rigid work requirements have proven successful in some schools.

Activities in the classroom that recognize the strengths of "at-risk" students are helpful in providing a positive sense of self. Assigning students special tasks and responsibilities gives them a stake in their education.

In addition to the in-school recommendations, factors outside the school influence equal opportunity for students. Mounting evidence supports the positive effects of parental involvement on student achievement.[19] If parents participate in school-sponsored activities, attend school meetings, help children with their scheduling, and oversee homework, children get the message that education matters and that their parents are supportive. Some schools that have sponsored programs to involve parents have had good success.

Early childhood education experiences such as Head Start have proven successful in giving children from less advantaged backgrounds a positive start in school. Some public schools are also offering early childhood classes.[20] For children of working parents and for those living in poverty, preschool offers a home away from home and early socialization in what, ideally, is a stimulating environment.[21] One of the few federal programs to continue to receive funding support is the early childhood education program Head Start; research shows its effectiveness for children from poverty backgrounds. The program deals with the health, educational, and emotional development of the children enrolled.

School partnerships with corporate America are taking a number of forms.[22] Some corporations promise to pay for the college expenses of those inner-city students who stay in school and graduate. Others are sponsoring model schools. Still others are operating schools on corporate premises or bringing students in for courses and training. Foundations also provide grants to school districts for projects to improve education. Although some fear the influence of the corporate world on schools, others note that local districts can seek the type of involvement they wish from corporations in their areas.

The term "multicultural curriculum" refers to teaching history, literature, and other subjects in ways that accurately reflect the different cultural strands in our society and world.[23] Integration of multicultural materials into existing curricula involves inclusion of reading materials by and about minority groups (including women), history that gives a fair and integrated portrayal of all groups, and other themes that promote understanding of all segments of multicultural society in the United States and the world. This movement is raising awareness of different interests and influencing the choices schools and teachers are making about classroom materials. However, some academics are concerned that traditional curricula are being eliminated to make room for new content.[24] The main importance of the debate is its relevance for the way we select classroom materials, and the impact this selection has on equal opportunity.

The United States can provide equal opportunity to all its citizens, but reforms to enhance equal opportunity such as careful selection of texts, fair use of testing and tracking, and other efforts must be launched. These efforts must affect individual students and teachers, groups such as classrooms, and the society at large. The task cannot be successfully completed unless efforts for change work at all levels.

NOTES

1. John W. Gardner, *Excellence* (New York: Harper & Row, 1984), p. 46; James S. Coleman, *Equality and Achievement in Education* (Boulder, CO: Westview Press, 1990).

2. U.S. Department of Education, *Digest of Education Statistics 1997.* (Washington, D.C: National Center for Education Statistics, 1998): NCES 98-015.

3. Anna Bellisari, Male superiority in mathematical aptitude: An artifact, *Human Organization* 48, no. 3 (Fall 1989): 273–279; David P. Baker (Catholic University of America) and Deborah Perkins Jones (U.S. Department of State), Creating gender differences: A cross-national assessment of gender inequality and sex differences in mathematics performance, unpublished

manuscript, 1991; Elizabeth L. Useem, Student selection into course sequences in mathematics: The impact of parental involvement and school policies, *Journal of Research on Adolescence* 1, no. 1 (1991).

4. Carol Tavris, Boys trample girls' turf, *Los Angeles Times,* May 7, 1990, B5.

5. Deborah Tannen, *You Just Don't Understand: Women and Men in Conversation* (New York: Ballantine Books, 1991); Teachers' classroom strategies should recognize that men and women use language differently, *Chronicle of Higher Education* 37, no. 40 (June 19, 1991): B2–3.

6. Jeannie Oakes, Multiplying inequalities: The effects of race, social class, and tracking on opportunities to learn mathematics and science (Santa Monica: Rand Corporation, 1990).

7. Kathryn S. Schiller and David Stevenson, Sequences of opportunities for learning mathematics, paper presented to the American Sociological Association, Pittsburgh, Pennsylvania, August 1992.

8. Sally B. Kilgore, The organizational context of tracking in schools, *American Sociological Review* 56, no. 2 (April 1991): 40; Samuel R. Lucas, Secondary school tracking in the United States: Existence, extension, and equity, paper presented to the American Sociological Association, Pittsburgh, Pennsylvania, August 1992.

9. Carolyn Riehl, Gary Natriello, and Aaron M. Pallas, Losing track: The dynamics of student assignment processes in high school, paper presented to American Sociological Association, Pittsburgh, Pennsylvania, August 1992.

10. U.S. Department of Education, National Center for Educational Statistics, Dropout rates in the United States, 1993. (Washington, D.C.: U.S. Department of Education, 1994).

11. U.S. Department of Education, *The Condition of Education* (Washington, DC: U.S. Department of Education, 1992), p. 59.

12. Catherine L. Garner and Stephen W. Raudenbush, Neighborhood effects on educational attainment: A multilevel analysis, *Sociology of Education* 64, no. 4 (October 1991): 251–262.

13. Howard Gardner, The theory of multiple intelligences, *Annual Dyslexia* 37 (1987): 19–35.

14. Allan C. Ornstein and Daniel U. Levine, *An Introduction to the Foundations of Education,* 3d ed. (Boston: Houghton Mifflin, 1985), p. 398.

15. Gary A. Orfield and others, Status of school desegregation: The next generation, Report to the National School Board Association (Alexandria, VA: National School Board Association, 1992).

16. America 2000: An education strategy (Washington, DC: U.S. Department of Education, 1991).

17. Jonathan Kozol, *Savage Inequalities: Children in America's Schools* (New York: Crown, 1991), p. 63.

18. T. L. Good, Teacher expectations and student perceptions: A decade of research, *Educational Leadership* 38 (1980–1981): 415–422; Hancock, Lynnell, In defiance of Darwin. *Newsweek,* October 24, 1994, p. 35.

19. Joyce L. Epstein, School/family/community Partnerships. *Phi Delta Kappan,* May 1995, 701–712.

20. Ann Mitchell, Michelle Seligson, and Fern Marx, *Early Childhood Programs and the Public Schools: Between Promise and Practice* (Dover, MA: Auburn House, 1990).

21. Gill Barrett, *Disaffection from School: An Early Years Perspective* (New York: Falmer Press, 1989).

22. *Chronicle of Higher Education* 38, no. 13 (November 20, 1991): 15.

23. Diane Ravitch, Multiculturalism yes, particularism no, *Chronicle of Higher Education* 37, no. 15 (December 12, 1990): A13.

24. D'Souza, Dinesh, *Illiberal Education: The Politics of Race and Sex on Campus* (New York: Free Press, 1991).

29

Savage Inequalities: Children in America's Schools

JONATHAN KOZOL

In addition to differences in structure and curricula in schools, rich and poor neighborhoods have different schools. The following excerpts from Jonathan Kozol's Savage Inequalities *illustrate the gap between schools in three different communities.*

As you read, consider the following:

1. In what ways do the schools in these three communities differ?

2. Are you aware of differences in schools in your area?

GLOSSARY

Savage inequalities Differences between schools for children from poor families in inner cities or poor suburbs compared to children from rich communities.

THE PROBLEMS OF THE streets in urban areas, as teachers often note, frequently spill over into public schools. In the public schools of East St. Louis this is literally the case.

"Martin Luther King Junior High School," notes the *Post-Dispatch* in a story published in the early spring of 1989, "was evacuated Friday afternoon after sewage flowed into the kitchen. . . . The kitchen was closed and students were sent home." On Monday, the paper continued, "East St. Louis Senior High School was awash in sewage for the second time this year." The school had to be shut because of "fumes and backed-up toilets." Sewage flowed into the basement, through the floor, then up into the kitchen and the students' bathrooms. The backup, we read, "occurred in the food preparation areas."

School is resumed the following morning at the high school, but a few days later the overflow recurs. This time the entire system is affected, since the meals distributed to every student in the city are prepared in the two schools that have been flooded. School is called off for all 16,500 students in the district. The sewage backup, caused by the failure of two pumping stations, forces officials at the high school to shut down the furnaces.

At Martin Luther King, the parking lot and gym are also flooded. "It's a disaster," says a legislator. "The streets are underwater; gaseous fumes are being emitted from the pipes under the schools," she says, "making people ill."

In the same week, the schools announce the layoff of 280 teachers, 166 cooks and cafeteria workers, 25 teacher aides, 16 custodians, and 18 painters, electricians, engineers and plumbers. The president of the teachers' union says the cuts, which will bring the size of kindergarten and primary classes up to 30 students, and the size of fourth to twelfth grade classes up to 35, will have "an unimaginable impact" on the students. "If you have a high school teacher with five classes each day and between 150 and 175 students . . . , it's going to have a devastating effect." The school system, it is also noted, has been using more than 70 "permanent substitute teachers," who are paid only $10,000 yearly, as a way of saving money.

Governor Thompson, however, tells the press that he will not pour money into East St. Louis to solve long-term problems. East St. Louis residents, he says, must help themselves. "There is money in the community," the governor insists. "It's just not being spent for what it should be spent for."

New York: Crown Publishers, Inc., 1991 (excerpts pp. 23–26, 65–66, 88–90).

The governor, while acknowledging that East St. Louis faces economic problems, nonetheless refers dismissively to those who live in East St. Louis. "What in the community," he asks, "is being done right?" He takes the opportunity of a visit to the area to announce a fiscal grant for sewer improvement to a relatively wealthy town nearby.

In East St. Louis, meanwhile, teachers are running out of chalk and paper, and their paychecks are arriving two weeks late. The city warns its teachers to expect a cut of half their pay until the fiscal crisis has been eased.

The threatened teacher layoffs are mandated by the Illinois Board of Education, which, because of the city's fiscal crisis, has been given supervisory control of the school budget. Two weeks later the state superintendent partially relents. In a tone very different from that of the governor, he notes that East St. Louis does not have the means to solve its education problems on its own. "There is no natural way," he says, that "East St. Louis can bring itself out of this situation." Several cuts will be required in any case—one quarter of the system's teachers, 75 teacher aides, and several dozen others will be given notice—but, the state board notes, sports and music programs will not be affected.

East St. Louis, says the chairman of the state board, "is simply the worst possible place I can imagine to have a child brought up. . . . The community is in desperate circumstances." Sports and music, he observes, are, for many children here, "the only avenues of success." Sadly enough, no matter how it ratifies the stereotype, this is the truth; and there is a poignant aspect to the fact that, even with class size soaring and one quarter of the system's teachers being given their dismissal, the state board of education demonstrates its genuine but skewed compassion by attempting to leave sports and music untouched by the overall austerity.

Even sports facilities, however, are degrading by comparison with those found and expected at most high schools in America. The football field at East St. Louis High is missing almost everything—including goalposts. There are a couple of metal pipes—no crossbar, just the pipes. Bob Shannon, the football coach, who has to use his personal funds to purchase footballs and has had to cut and rake the football field himself, has dreams of having goalposts someday. He'd also like to let his students have new uniforms. The ones they wear are nine years old and held together somehow by a patchwork of repairs. Keeping them clean is a problem, too. The school cannot afford a washing machine. The uniforms are carted to a corner laundromat with fifteen dollars' worth of quarters.

Other football teams that come to play, according to the coach, are shocked to see the field and locker rooms. They want to play without a halftime break and get away. The coach reports that he's been missing paychecks, but he's trying nonetheless to raise some money to help out a member of the team whose mother has just died of cancer.

"The days of the tight money have arrived," he says. "It don't look like Moses will be coming to this school."

He tells me he has been in East St. Louis 19 years and has been the football coach for 14 years. "I was born," he says, "in Natchez, Mississippi. I stood on the courthouse steps of Natchez with Charles Evers. I was a teen-age boy when Michael Schwerner and the other boys were murdered. I've been in the struggle all along. In Mississippi, it was the fight for legal rights. This time, it's a struggle for survival.

"In certain ways," he says, "it's harder now because in those days it was a clear enemy you had to face, a man in a hood and not a statistician. No one could persuade you that you were to blame. Now the choices seem like they are left to you and, if you make the wrong choice, you are made to understand you are to blame. . . .

"Night-time in this city, hot and smoky in the summer, there are dealers standin' out on every street. Of the kids I see here, maybe 55 percent will graduate from school. Of that number, maybe one in four will go to college. How many will stay? That is a bigger question.

"The basic essentials are simply missing here. When we go to wealthier schools I look at the faces of my boys. They don't say a lot. They have their faces to the windows, lookin' out. I can't tell what they are thinking. I am hopin' they are saying, 'This is something I will give my kids someday.'"

Tall and trim, his black hair graying slightly, he is 45 years old.

"No, my wife and I don't live here. We live in a town called Ferguson, Missouri. I was born in poverty and raised in poverty. I feel that I owe it to myself to live where they pick up the garbage."

In the visitors' locker room, he shows me lockers with no locks. The weight room stinks of sweat and water-rot. "See, this ceiling is in danger of collapsing. See, this room don't have no heat in winter. But we got to come here anyway. We wear our coats while working out. I tell the boys, 'We got to get it done. Our fans don't know that we do not have heat.'"

He tells me he arrives at school at 7:45 A.M. and leaves at 6:00 P.M.—except in football season, when he leaves at 8:00 P.M. "This is my life. It isn't all I dreamed of and I tell myself sometimes that I might have accomplished more. But growing up in poverty rules out some avenues. You do the best you can." . . .

On the following morning I visit P.S. 79, an elementary school in the same district. "We work under difficult circumstances," says the principal, James Carter, who is black. "The school was built to hold one thousand students. We have 1,550. We are badly overcrowded. We need smaller classes but, to do this, we would need more space. I can't add five teachers. I would have no place to put them."

Some experts, I observe, believe that class size isn't a real issue. He dismisses this abruptly. "It doesn't take a genius to discover that you learn more in a smaller class. I have to bus some 60 kindergarten children elsewhere, since I have no space for them. When they return next year, where do I put them?

"I can't set up a computer lab. I have no room. I had to put a class into the library. I have no librarian. There are two gymnasiums upstairs but they cannot be used for sports. We hold more classes there. It's unfair to measure us against the suburbs. They have 17 to 20 children in a class. Average class size in this school is 30.

"The school is 29 percent black, 70 percent Hispanic. Few of these kids get Head Start. There is no space in the district. Of 200 kindergarten children, 50 maybe get some kind of preschool."

I ask him how much difference preschool makes.

"Those who get it do appreciably better. I can't overestimate its impact but, as I have said, we have no space."

The school tracks children by ability, he says. "There are five to seven levels in each grade. The highest level is equivalent to 'gifted' but it's not a full-scale gifted program. We don't have the funds. We have no science room. The science teachers carry their equipment with them."

We sit and talk within the nurse's room. The window is broken. There are two holes in the ceiling. About a quarter of the ceiling has been patched and covered with a plastic garbage bag.

"Ideal class size for these kids would be 15 to 20. Will these children ever get what white kids in the suburbs take for granted? I don't think so. If you ask me why, I'd have to speak of race and social class. I don't think the powers that be in New York City understand, or want to understand, that if they do not give these children a sufficient education to lead healthy and productive lives, we will be their victims later on. We'll pay the price someday—in violence, in economic costs. I despair of making this appeal in any terms but these. You cannot issue an appeal to conscience in New York today. The fair-play argument won't be accepted. So you speak of violence and hope that it will scare the city into action."

While we talk, three children who look six or seven years old come to the door and ask to see the nurse, who isn't in the school today. One of the children, a Puerto Rican girl, looks haggard. "I have a pain in my tooth," she says. The principal says, "The nurse is out. Why don't you call your mother?" The child says, "My mother doesn't have a phone." The principal sighs. "Then go back to your class." When she leaves, the principal is angry. "It's amazing to me that these children ever make it with the obstacles they face. Many *do* care and they *do* try, but there's a feeling of despair. The parents of these children want the same things for their children that the parents in the suburbs want. Drugs are not the cause of this. They are the symptom. Nonetheless, they're used by people in the suburbs and rich people in Manhattan as another reason to keep children of poor people at a distance."

I ask him, "Will white children and black children ever go to school together in New York?"

"I don't see it," he replies. "I just don't think it's going to happen. It's a dream. I simply do not see white folks in Riverdale agreeing to cross-bus with kids like these. A few, maybe. Very few. I don't think I'll live to see it happen."

I ask him whether race is the decisive factor. Many experts, I observe, believe that wealth is more important in determining these inequalities.

"This," he says — and sweeps his hand around him at the room, the garbage bag, the ceiling — "would not happen to white children."

In a kindergarten class the children sit cross-legged on a carpet in a space between two walls of books. Their 26 faces are turned up to watch their teacher, an elderly black woman. A little boy who sits beside me is involved in trying to tie bows in his shoelaces. The children sing a song: "Lift Every Voice." On the wall are these handwritten words: "Beautiful, also, are the souls of my people."

In a very small room on the fourth floor, 52 people in two classes do their best to teach and learn. Both are first grade classes. One, I am informed, is "low ability." The other is bilingual.

"The room is barely large enough for one class," says the principal.

The room is 25 by 50 feet. There are 26 first graders and two adults on the left, 22 others and two adults on the right. On the wall there is the picture of a small white child, circled by a Valentine and a Gainsborough painting of a child in a formal dress. . . .

Children who go to school in towns like Glencoe and Winnetka do not need to steal words from a dictionary. Most of them learn to read by second or third grade. By the time they get to sixth or seventh grade, many are reading at the level of the seniors in the best Chicago high schools. By the time they enter ninth grade at New Trier High, they are in a world of academic possibilities that far exceed the hopes and dreams of most schoolchildren in Chicago.

"Our goal is for students to be successful," says the New Trier principal. With 93 percent of seniors going on to four-year colleges — many to schools like Harvard, Princeton, Berkeley, Brown, and Yale — this goal is largely realized.

New Trier's physical setting might well make the students of Du Sable High School envious. The *Washington Post* describes a neighborhood of "circular driveways, chirping birds and white-columned homes." It is, says a student, "a maple land of beauty and civility." While Du Sable is sited on one crowded city block, New Trier students have the use of 27 acres. While Du Sable's science students have to settle for makeshift equipment, New Trier's students have superior labs and up-to-date technology. One wing of the school, a physical education center that includes three separate gyms, also contains a fencing room, a wrestling room, and studios for dance instruction. In all, the school has seven gyms as well as an Olympic pool.

The youngsters, according to a profile of the school in *Town and Country* magazine, "make good use of the huge, well-equipped building, which is immaculately maintained by a custodial staff of 48."

It is impossible to read this without thinking of a school like Goudy, where there are no science labs, no music or art classes and no playground — and where the two bathrooms, lacking toilet paper, fill the building with their stench.

"This is a school with a lot of choices," says one student at New Trier; and this hardly seems an overstatement if one studies the curriculum. Courses in music, art and drama are so varied and abundant that students can virtually major in these subjects in addition to their academic programs. The modern and classical language department offers Latin (four years) and six other foreign languages. Elective courses include the literature of Nobel winners, aeronautics, criminal justice, and computer languages. In a senior literature class, students are reading Nietzsche, Darwin, Plato, Freud, and Goethe. The school also operates a television station with a broadcast license from the FCC, which broadcasts on four channels to three counties.

Average class size is 24 children; classes for slower learners hold 15. This may be compared to Goudy—where a remedial class holds 39 children and a "gifted" class has 36.

Every freshman at New Trier is assigned a faculty adviser who remains assigned to him or her through graduation. Each of the faculty advisers—they are given a reduced class schedule to allow them time for this—gives counseling to about two dozen children. . . .

The ambience among the students at New Trier, of whom only 1.3 percent are black, says *Town and Country*, is "wholesome and refreshing, a sort of throwback to the Fifties." It is, we are told, "a preppy kind of place." In a cheerful photo of the faculty and students, one cannot discern a single nonwhite face.

New Trier's "temperate climate" is "aided by the homogeneity of its students," *Town and Country* notes. ". . . Almost all are of European extraction and harbor similar values."

"Eighty to 90 percent of the kids here," says a counselor, "are good, healthy, red-blooded Americans."

The wealth of New Trier's geographical district provides $340,000 worth of taxable property for each child; Chicago's property wealth affords only one-fifth this much. Nonetheless, *Town and Country* gives New Trier's parents credit for a "willingness to pay enough . . . in taxes" to make this one of the state's best-funded schools. New Trier, according to the magazine, is "a striking example of what is possible when citizens want to achieve the best for their children." Families move here "seeking the best," and their children "make good use" of what they're given.

QUESTIONS FOR DISCUSSION

For further discussion of this topic, see the Wadsworth Sociology Resource Center, "Virtual Society," *http://sociology.wadsworth.com,* under *Sociological Footprints,* by Cargan and Ballantine. You can respond to the discussion questions there or enter your own comments in the online chat forum.

SUGGESTED READINGS AND SOCIOLOGY INTERNET RESOURCES

See the Wadsworth Sociology Resource Center, "Virtual Society," *http://sociology.wadsworth. com,* for additional links, suggestions for further reading, and learning tools related to this chapter.

Either from the "Virtual Society" website or directly from your web browser, you may access InfoTrac College Edition, an online university library that includes over 700 popular and scholarly journals in which you can find articles related to the topics in this chapter.

CHAPTER 8

Religion

THE SUPERNATURAL AND SOCIETY

RELIGION HAS BEEN DEFINED BY EARLY sociologist Emile Durkheim as a more or less coherent system of beliefs (monotheism, polytheism) and practices (fasts, feasts) that concerns a supernatural order of beings (gods, goddesses, angels), places (heaven, hell, purgatory), and forces (mana). This rather simplistic definition explains *what* religion is but does not explain *why* it is a universal phenomenon.

As Durkheim notes in the first reading of this chapter, religion allows for the transcendence of human existence, and this perspective gives people a means of dealing with conditions of uncertainty. It helps people to overcome their feeling of power-lessness to control the conditions that affect their lives, and it gives them a way to cope with unfulfilled psychological and economic needs. The result of these beliefs for religious believers is a loyalty that transcends national loyalty; that is, the religious claim will have priority. In addition, those in organized religious congregations tend to participate in larger numbers and are more generous with their time and money to these organizations than any other organization that transcends the family and its cir-cle of friends.

As Durkheim notes in the first reading, religion is society. Therefore it is not sur-prising to find its impact in other social institutions. In addition, as noted in the sec-tion introduction, all institutions are interdependent and interrelated. Billingsley and Caldwell, in the second reading, indicate this relationship between the three institutions of religion, the family, and education in the African-American community.

For society, religion provides a means for social control; it helps support some of the primary societal rules. However, as a glance around the world would reveal, reli-gious beliefs can also be dysfunctional: religious beliefs can lead to discrimination and even to killing those who are not of the same faith. As the abortion issue in the United States shows, religious beliefs can lead to political conflict, and vice versa.

The functions performed by religion do not mean that religions will not undergo changes. On the contrary, it may mean that new beliefs arise to fulfill these important personal functions. This also means a need for change in the more traditional religions. It would appear that change is also an important ingredient in this normally staid institution. In the third reading, Peter Berger sheds light on the nature of religious functions by pointing out that they can change. He notes, for instance, that the oth-erworldliness of many fundamentalist religions has given way to a greater concern with values in current life.

The U.S. Constitution may insist on a separation of church and state; that is, there is a right to religious expression but no religion is guaranteed favored governmental dominance. This assurance, however, does not mean that religion is not involved in the political institution. The conflict over abortion and other moral issues as well as conflict in other societies for religious hegemony alerts us to the relationship between the religious and political institutions. This is seen in the final reading on the many other beliefs and goals of the antiabortion movement.

30

The Elementary Forms of the Religious Life

EMILE DURKHEIM

The author claims that religious beliefs and rituals are real and reflect the societies in which they exist. Note his reasons for this claim. If you agree with his thesis, ask yourself what effect it has on your personal religious beliefs.

As you read, ask yourself the following questions:

1. Why does the author say that religious beliefs and rituals are real and reflect society?

2. How do your religious beliefs and rituals reflect American society?

GLOSSARY

Profane Not devoted to religious purposes; secular.
Sui generis In itself.

THE THEORISTS WHO HAVE undertaken to explain religion in rational terms have generally seen in it before all else a system of ideas, corresponding to some determined object. This object has been conceived in a multitude of ways: nature, the infinite, the unknowable, the ideal, etc.; but these differences matter but little. In any case, it was the conceptions and beliefs which were considered as the essential elements of religion. As for the rites, from this point of view they appear to be only an external translation, contingent and material, of these internal states which alone pass as having any intrinsic value. This conception is so commonly held that generally the disputes of which religion is the theme turn about the question whether it can conciliate itself with science or not, that is to say, whether or not there is a place beside our scientific knowledge for another form of thought which would be specifically religious.

But the believers, the men who lead the religious life and have a direct sensation of what it really is, object to this way of regarding it, saying that it does not correspond to their daily experience. In fact, they feel that the real function of religion is not to make us think, to enrich our knowledge, nor to add to the conceptions which we owe to science others of another origin and another character, but rather, it is to make us act, to aid us to live. The believer who has communicated with his god is not merely a man who sees new truths of which the unbeliever is ignorant; he is a man who is *stronger.* He feels within him more force, either to endure the trials of existence, or to conquer them. It is as though he were raised above the miseries of the world, because he is raised above his condition as a mere man; he believes that he is saved from evil, under whatever form he may conceive this evil. The first article in every creed is the belief in salvation by faith. But it is hard to see how a mere idea could have this efficacy. An idea is in reality only a part of ourselves; then how could it confer upon us powers superior to those which we have of our own nature? Howsoever rich it might be in affective virtues, it could add nothing to our natural vitality; for it could only release the motive powers which are within us, neither creating them nor increasing them. From the mere fact that we consider an object worthy of being loved and sought after, it does not follow that we feel ourselves stronger afterwards; it is also necessary that this object set free energies superior to these which we ordinarily have at our command and also that we have some means of making these

The Elementary Forms of Religious Life: A Study in Religious Sociology. Emile Durkheim, Routledge, 1915. Translated from the French by Joseph Ward Swain. London: Allen & Urwin, 1915. Reprinted with the permission of Routledge.

enter into us and unite themselves to our interior lives. Now for that, it is not enough that we think of them; it is also indispensable that we place ourselves within their sphere of action, and that we set ourselves where we may best feel their influence; in a word, it is necessary that we act, and that we repeat the acts thus necessary every time we feel the need of renewing their effects. From this point of view, it is readily seen how that group of regularly repeated acts which form the cult get their importance. In fact, whoever has really practised a religion knows very well that it is the cult which gives rise to these impressions of joy, of interior peace, of serenity, of enthusiasm which are, for the believer, an experimental proof of his beliefs. The cult is not simply a system of signs by which the faith is outwardly translated; it is a collection of the means by which this is created and recreated periodically. Whether it consists in material acts or mental operations, it is always this which is efficacious.

Our entire study rests upon this postulate that the unanimous sentiment of the believers of all times cannot be purely illusory. Together with a recent apologist of the faith[1] we admit that these religious beliefs rest upon a specific experience whose demonstrative value is, in one sense, not one bit inferior to that of scientific experiments, though different from them. We, too, think that "a tree is known by its fruits," and that fertility is the best proof of what the roots are worth. But from the fact that a "religious experience," if we choose to call it this, does exist and that it has a certain foundation — and, by the way, is there any experience which has none? — it does not follow that the reality which is its foundation conforms objectively to the idea which believers have of it. The very fact that the fashion in which it has been conceived has varied infinitely in different times is enough to prove that none of these conceptions express it adequately. If a scientist states it as an axiom that the sensations of heat and light which we feel correspond to some objective cause, he does not conclude that this is what it appears to the senses to be. Likewise, even if the impressions which the faithful feel are

not imaginary, still they are in no way privileged intuitions; there is no reason for believing that they inform us better upon the nature of their object than do ordinary sensations upon the nature of bodies and their properties. In order to discover what this object consists of, we must submit them to an examination and elaboration analogous to that which has substituted for the sensuous idea of the world another which is scientific and conceptual.

This is precisely what we have tried to do, and we have seen that this reality, which mythologies have represented under so many different forms, but which is the universal and eternal objective cause of these sensations *sui generis* out of which religious experience is made, is society. We have shown what moral forces it develops and how it awakens this sentiment of a refuge, of a shield, and of a guardian support which attaches the believer to his cult. It is that which raises him outside himself; it is even that which made him. For that which makes a man is the totality of the intellectual property which constitutes civilization, and civilization is the work of society. Thus is explained the preponderating role of the cult in all religions, whichever they may be. This is because society cannot make its influence felt unless it is in action, and it is not in action unless the individuals who compose it are assembled together and act in common. It is by common action that it takes consciousness of itself and realizes its position; it is before all else an active cooperation. The collective ideas and sentiments are even possible only owing to these exterior movements which symbolize them, as we have established. Then it is action which dominates the religious life, because of the mere fact that it is society which is its source. . . .

Religious forces are therefore human forces, moral forces. It is true that since collective sentiments can become conscious of themselves only by fixing themselves upon external objects, they have not been able to take form without adopting some of their characteristics from other things: They have thus acquired a sort of physical nature; in this way they have come to mix themselves with the life of the material world, and then have considered themselves capable of explaining what passes there. But when they are considered only from this point of view and in this role, only their most superficial aspect is seen. In reality, the essential elements of which these collective sentiments are made have been borrowed by the understanding. It ordinarily seems that they should have a human character only when they are conceived under human forms;[2] but even the most impersonal and the most anonymous are nothing else than objectified statements. . . .

Some reply that men have a natural faculty for idealizing, that is to say, of substituting for the real world another different one, to which they transport themselves by thought. But that is merely changing the terms of the problem; it is not resolving it or even advancing it. This systematic idealization is an essential characteristic of religions. Explaining them by an innate power of idealization is simply replacing one word by another which is the equivalent of the first; it is as if they said that men have made religions because they have a religious nature. Animals know only one world, the one which they perceive by experience, internal as well as external. Men alone have the faculty of conceiving the ideal, of adding something to the real. Now where does this singular privilege come from? Before making it an initial fact or a mysterious virtue which escapes science, we must be sure that it does not depend upon empirically determinable conditions.

The explanation of religion which we have proposed has precisely this advantage, that it gives an answer to this question. For our definition of the sacred is that it is something added to and above the real: Now the ideal answers to this same definition; we cannot explain one without explaining the other. In fact, we have seen that if collective life awakens religious thought on reaching a certain degree of intensity, it is because it brings about a state of effervescence which changes the conditions of psychic activity. Vital energies are overexcited, passions more active, sensations stronger; there are even some which are produced only at this moment. A man does not recognize himself; he feels himself transformed and consequently he transforms the environment which surrounds him. In order to account for the very particular impressions which he receives, he attributes to the things with which he is in most direct contact properties which they have not, exceptional powers and virtues which the objects of everyday experience do not possess. In a word, above the real world where his profane life passes he has placed another which, in one sense, does not exist except in thought, but to which he attributes a higher sort of dignity than to the first. Thus, from a double point of view it is an ideal world.

The formation of the ideal world is therefore not an irreducible fact which escapes science; it depends upon conditions which observation can touch; it is a natural product of social life. For a society to become conscious of itself and maintain at the necessary degree of intensity the sentiments which it thus attains, it must assemble and concentrate itself. Now this concentration brings about an exaltation of the mental life which takes form in a group of ideal conceptions where is portrayed the new life thus awakened; they correspond to this new set of psychical forces which is added to those which we have at our disposition for the daily tasks of existence. A society can neither create itself nor recreate itself without at the same time creating an ideal. This creation is not a sort of work of supererogation for it, by which it would complete itself, being already formed; it is the act by which it is periodically made and remade. Therefore when some oppose the ideal society to the real society, like two antagonists which would lead us in opposite directions, they materialize and oppose abstractions. The ideal society is not outside of the real society; it is a part of it. Far from being divided between them as between two poles which mutually repel each other, we cannot hold to one without holding to the other. For a society is not made up merely of the mass of individuals who compose it, the ground which they occupy, the things which they use and the movements which they perform, but above all is the idea which it forms of itself. It is undoubtedly true that it hesitates over the manner in which it ought to conceive itself; it feels itself drawn in divergent directions. But these conflicts which break forth are not between the ideal and reality, but between two different ideals, that of yesterday and that of today, that which has the authority of tradition and that which has the hope of the future. There is surely a place for investigating whence these ideals evolve; but whatever solution may be given to this problem, it still remains that all passes in the world of the ideal.

Thus the collective ideal which religion expresses is far from being due to a vague innate power of the individual, but it is rather at the school of collective life that the individual has learned to idealize. It is in assimilating the ideals elaborated by society that he has become capable of conceiving the ideal. It is society which, by leading him within its sphere of action, has made him acquire the need of raising himself above the world of experience and has at the same time furnished him with the means of conceiving another. For society has constructed this new world in constructing itself, since it is society which this expresses. Thus both with the individual and in the group, the faculty of idealizing has nothing mysterious about it. It is not a sort of luxury which a man could get along without, but a condition of his very existence. He could not be a social being, that is to say, he could not be a man, if he had not acquired it. It is true that in incarnating themselves in individuals, collective ideals tend to individualize themselves. Each understands them after his own fashion and marks them with his own stamp; he suppresses certain elements and adds others. Thus the personal ideal disengages itself from the social ideal in proportion as the individual personality develops itself and becomes an autonomous source of action. But if we wish to understand this aptitude, so singular in appearance, of living outside of reality, it is enough to connect it with the social conditions upon which it depends.

Therefore it is necessary to avoid seeing in this theory of religion a simple restatement of historical materialism: That would be misunderstanding our thought to an extreme degree. In showing that religion is something essentially so-

cial, we do not mean to say that it confines itself to translating into another language the material forms of society and its immediate vital necessities. It is true that we take it as evident that social life depends upon its material foundation and bears its mark, just as the mental life of an individual depends upon his nervous system and in fact his whole organism. But collective consciousness is something more than a mere epiphenomenon of its morphological basis, just as individual consciousness is something more than a simple efflorescence of the nervous system. In order that the former may appear, a synthesis *sui generis* of particular consciousnesses is required. Now this synthesis has the effect of disengaging a whole world of sentiments, ideas, and images which, once born, obey laws all their own. They attract each other, repel each other, unite, divide themselves, and multiply, though these combinations are not commanded and necessitated by the conditions of the underlying reality. The life thus brought into being even enjoys so great an independence that it sometimes indulges in manifestations with no purpose or utility of any sort, for the mere pleasure of affirming itself. We have shown that this is often precisely the case with ritual activity and mythological thought.[3] . . .

That is what the conflict between science and religion really amounts to. It is said that science denies religion in principle. But religion exists; it is a system of given facts; in a word, it is a reality. How could science deny this reality? Also, insofar as religion is action, and insofar as it is a means of making men live, science could not take its place, for even if this expresses life, it does not create it; it may well seek to explain the faith, but by that very act it presupposes it. Thus there is no conflict except upon one limited point. Of the two functions which religion originally fulfilled, there is one, and only one, which tends to escape it more and more: That is its speculative function. That which science refuses to grant to religion is not its right to exist, but its right to dogmatize upon the nature of things and the special competence which it claims for itself for knowing man and the world. As a matter of fact, it does not know itself. It does not even know what it is made of, nor to what need it answers. It is itself a subject for science, so far is it from being able to make the law for science! And from another point of view, since there is no proper subject for religious speculation outside that reality to which scientific reflection is applied, it is evident that this former cannot play the same role in the future that it has played in the past.

However, it seems destined to transform itself rather than to disappear. . . .

NOTES

1. James, William. 1902. *The Varieties of Religious Experience*. New York: Longmans, Green.

2. It is for this reason that Frazer and even Preuss set impersonal religious forces outside of, or at least on the threshold of religion, to attach them to magic.

3. On this same question, see our article, "Representations individuelles et representations collectives." In *Revue du Metaphysique*, May, 1898.

31

The Church, the Family, and the School in the African American Community

ANDREW BILLINGSLEY AND CLEOPATRA HOWARD CALDWELL

The fact that our social institutions are interrelated—that is, affect each other—is sometimes forgotten. In this reading, the authors show how the church as a social institution interacts with and influences two other important social institutions.

As you read, ask yourself the following questions:

1. In what ways does the church interact with the family and the school?

2. Do you approve of the church's interactions with the family and the school? Why?

GLOSSARY

Social institution Organized sphere of social life; subsystem of society to meet human needs.

THIS READING FOCUSES ON the church as social institution and examines how it interacts with and influences two other important social institutions in society (as identified by Moberg, 1962)—

the family and the school. Historically, the church, the family, and the school are the three most critical institutions whose interactions have been responsible for the viability of the African American community (Roberts, 1980). The strengths of these three institutions are due in large measure to their function as expressions of the most basic values of the African American cultural heritage. These values include spirituality, high achievement aspirations, and commitment to family as enduring, flexible and adaptive functional mechanisms for survival (Billingsley, in press). . . .

In this reading we discuss the church's role in assisting families in need and in supporting educational institutions. The data presented herein are based on an ongoing, nationwide, multi-year study of the family-oriented community outreach programs that Black churches sponsor to assist those in need. Preliminary findings from a

From the Journal of Negro Education, *Vol. 60, No. 3 (1991). Copyright © 1992, Howard University.*

representative sample of 315 Black churches in the northeastern region of the country will be presented. . . .

The Black Church

The Black church continues to hold the allegiance of large numbers of African Americans and exerts great influence over their behavior. . . .

Why is the Black church so important? If the church is the institutionalized expression of the religious life of a people, as many sociologists generally believe, then the Black church is a powerful institution. Spirituality, according to Hill (1971), is one of the most distinctive features of African American culture. This is partially reflected in overt religious expression. A reanalysis of the data from the University of Michigan's National Survey of Black Americans (Billingsley, in press) shows that:

- 84% of African American adults considered themselves to be religious;
- 80% considered it very important to send their children to church;
- 78% indicated that they pray often;
- 76% said that the church was a very important institution in their early childhood socialization;
- 77% reported that the church was still very important;
- 71% attend church at least once a month; and
- Nearly 70% were members of a church.

In the African American community the church is more than a religious institution. Lincoln (1989) describes the multiple functions of this institution as follows:

> Beyond its purely religious function, as critical as that has been, the Black church in its historical role as lyceum, conservatory, forum, social service center, political academy, and financial institution, has been and is for Black America the mother of our culture, the champion of our freedom, the hallmark of our civilization. (p. 3)

Moreover, utilizing data from the National Survey of Black Americans, Taylor and Chatters (1988) demonstrate empirically that the church is a major source of social support in the African American community. For example, they found that 80% of the elderly surveyed received tangible support from church members.

The African American Family

The family is also a strong and functional institution in the African American community. As Hill (1971) explains, African American families are sustained by five major sources of strength including, notably, a strong religious orientation, flexibility of family roles, and a strong achievement orientation. Despite the strains on contemporary families these strengths remain evident. The majority of African Americans are part of family units. Indeed, as late as 1988, 70% of all African American households were family households (U.S. Census, 1988), albeit those households represented a highly diverse set of structures. Some of these were nuclear families, some were extended families, and some were augmented families. Further, some nuclear families were married-couple families without children, some were married couples with children, and some were single-parent families. Some Black households were headed by males and some were headed by females.

All of these variations are families in the sense that they are comprised of persons related to each other by blood, marriage, formal adoption, informal adoption, or by simple appropriation. The overwhelming majority of children, who constitute the major basis of family formation, live in family units. As late as 1988, 94% of African American children lived in families while less than 1% lived in nonfamily households (U.S. Census, 1988).

This tells us that the family is still a very important institution in African American communities. It also tells us that the family is a diverse, flexible and highly adaptive institution. Even the structure that is under the greatest assault — the married-couple family — is not an insignificant aspect of African American life (Billingsley, in press). Indeed, a recent study by Tucker and

Mitchell-Kernan (in press) suggests that those who have pronounced the death of married-couple families in the African American community may be premature. Despite the rapid rise in the alternatives to marriage cited above, the marriage relation continues to be a major expression of African American life. In 1983, for example, there were about 3.5 million Black married-couple families in the nation. Five years later, in 1988, the number had increased to 3.7 million. Most of these married-couple families, contrary to another myth, lived in central cities and not in suburbs (U.S. Census, 1988).

The School

The school is also a highly valued institution for many African American people as expressed through high parental educational aspirations for their children (Willie, Garibaldi & Reed, 1990). For nearly half a century African American parents, their children, leaders, and organizations have been in the forefront of efforts to desegregate schools so that African American children could receive a quality education. In the process they have endured enormous hardships and sacrifices for limited gains.

A recent national study by Orfield (1987) shows that African Americans continue to provide strong support for desegregation. In Orfield's study 66% of African Americans, as compared to 58% of Hispanics and only 36% of Whites, still support bussing as a means of desegregation. As Orfield notes, "School desegregation is far from the panacea for unequal education, but no urban community has yet been able to produce segregated schools that are equal" (p. 7).

The importance of the school is not a modern phenomenon in the African American community. In a study covering the period 1865 to 1954, Johnson (1986) reports that African Americans historically have placed high value on education and on the schooling of their children:

Black parents, and the Black community from the era of the Reconstruction period in Ameri-

can history to the *Brown* decision (1954) have been interested in the education of Black children; have placed high value on their children's education; have maintained a belief that education is a method of benefiting from the opportunity of the American society; and that interest has been translated into action by providing schooling for Black children throughout the period covered by this study. (p. 199)

Research on Black Churches, Families, and Schools

Despite the paucity of empirical investigations, three large-scale studies do exist that have been particularly informative regarding the role of the Black church as a supportive institution for families and schools. The first is a study conducted by Dr. Benjamin E. Mays and his colleague Joseph Nicholson in the late 1920s and published in 1933 as *The Negro's Church.* . . .

While the Mays and Nicholson study focused primarily on the religious aspects of the church, a number of relevant findings emerged from it that are essential to one's understanding of the church as a supportive, socializing agent in the African American community. First, they found that community outreach was a prominent feature of the urban churches of that time. Support for education was at the top of these outreach priorities. Other programs identified in descending order of prevalence among the 609 urban churches included: relief for the poor, recreational work, gymnasium classes, feeding the unemployed, benevolent societies, free health clinics, cooperation with YWCA and YMCA programs, Girl Scouts, Boy Scouts, kindergarten, and day nurseries. Second, they found that the minister was the key to the programs and operations of the church, both religious and nonreligious. In part, outreach programs depended on and grew as a result of the orientation of the minister. A third finding was what they termed the "genius" or "soul" of the Black church. As they concluded:

. . . there is in the genius or the "soul" of the Negro church something that gives it life and vitality that makes it stand out significantly above its buildings, creeds, rituals and doctrines—something that makes it a unique institution. (p. 278)

... Lincoln and Mamiya (1990) launched a national survey of Black churches in the late 1970s. . . .

A number of findings from Lincoln and Mamiya's study are relevant to the present study. Among them are the following:

- By 1979 community outreach had become a major activity of Black churches. Altogether, 71% of the churches indicated that they had participated with other organizations or sponsored community outreach programs, while 29% reported they had not. The overwhelming majority of outreach programs were financed by the churches.

- The churches provided strong support for education at all levels. For example, the vast majority of rural churches provided support for Black colleges either through special offerings or annual contributions.

- Government-sponsored programs played a minor but important role in the outreach of Black churches. About 8% of the churches had participated with government-sponsored programs—mostly Head Start, child care centers, and programs for the elderly.

Lincoln and Mamiya conclude that the relevance of the Black church reaches far beyond its membership. "But mere numbers aside," they add, "the impact of the Black church on the spiritual, social, economic, educational, and political interests that structure life in America . . . can scarcely be overlooked" (p. xii).

A third study that provides some guidance was conducted by George et al. (1989) of the American Association for the Advancement of Science (AAAS). . . . A number of findings from this study help to substantiate the important role of the Black church in the community as it interacts with families and schools. George et al. found, for example, that Black churches, now as in the past, are extensively involved with their communities outside the walls of the church. Moreover, much of this involvement has an educational focus.

Nearly two-thirds of the churches in this study (214 out of 378 churches) responded that they were actually conducting nonreligious educational programs in the community. Notably, a majority of the churches' educational program participants (60%) were not members of the churches. These Black churches offered a wide variety of types of educational programs, with the most common being tutoring (57%), day care (56%) and field trips (56%). When not offering educational programs directly, 22 churches provided scholarships for their members to attend college and other schools. Additionally, some churches operated their own schools ($N = 45$).

The findings from these three studies show that the Black church, the family, and educational institutions can interact to provide a major resource for strengthening African American communities. The results of the present study of family-oriented community outreach programs sponsored by Black churches further illuminates the potential of this interaction.

The Black Church Family Project

The Black Church Family Project was designed to examine the nature and scope of the family-oriented community outreach programs sponsored by a nationally representative sample of Black churches. . . .

Prior to conducting our comprehensive national study of church sponsored family-oriented community outreach programs, we undertook a massive effort to determine the state of current knowledge regarding the number of Black churches in this country as well as the distribution of these churches by selected denominations. Findings from this undertaking follow.

DISTRIBUTION OF BLACK CHURCHES BY DENOMINATION

Although we used a region-by-region approach to develop the sampling frame and have initially focused most of our efforts on the Northeast, we have obtained information on various churches from around the nation, with assistance from a variety of sources including denominational headquarters, the National Urban League, mayors' offices, Afro-American Studies departments at various colleges, and local funeral directors. This information, coupled with available publications (Jacquet, 1989; Melton, 1988), has allowed us to estimate conservatively that there are between 65,000 and 75,000 Black congregations in the nation. We have also been advised that this number would increase dramatically if "storefront churches"—small, independent churches not associated with major denominations that are located in renovated store sites or houses—were fully taken into consideration. . . .

CHARACTERISTICS OF THE CHURCHES

Several characteristics of the Black churches in our study are relevant to their community outreach missions. For instance, the Black church is rather well known for its independence and self-sufficiency. These characteristics, referred to by Mays and Nicholson (1933), are dramatically reflected in our study by the finding that most congregations own their church building. Over 90% of the church buildings are owned by their congregations, while nearly 70% have paid off their church's mortgage. No other segment of the African American community represents such ownership and independence.

A heterogeneous social-class structure is also a distinctive characteristic of most Black churches. That is, a majority (55.9%) of the churches in our sample have congregations composed of both working-class and middle-class members. Less than 10% have mostly middle-class members, while another 30% have mostly working-class members.

The tendency for Black churches to serve as centers of community activities was documented in our study. Better than 43% of the churches with outreach programs allow their buildings to be used by nonchurch groups, which suggests that the Black church can be considered a community institution. Moreover, 57.5% of the ministers are active in community groups.

OUTREACH PROGRAMS

When we examined the number of churches that supported community outreach programs, we found that nearly 70% operated 1 or more programs, nearly 60% operated 2 or more, and nearly

one-half operated 3 or more programs. The number of programs operated per church ranged from 1 to 15, with a median of 4. A total of 900 programs were sponsored by the 216 churches that sponsored such programs.

A wide range of activity is reflected in the types of programs offered by these churches. . . .

Nearly 4% of the churches sponsor programs for the children of the communities in which they are located. These programs include child development centers, day care centers, child health, and other programs for children. Other types of programs offered for the young are youth services and youth development programs that often have an educational component. Altogether, 16% of the churches conduct such youth programs. Some of these programs illustrate the church's informal and sometimes indirect role in providing educational support for youth in areas such as teen parenting skills, teen pregnancy prevention, employment opportunities, health education, drug use prevention, AIDS awareness, cultural awareness, mentoring, and recreational programs.

Formal educational programs and assistance for children and youth constituted about 11% of the total number of programs offered by the 216 churches. These ranged from Head Start-type programs for children to after-school academic support programs, to full-scale elementary and secondary schools as well as college preparatory and college support programs. Moreover, a number of those churches not offering educational programs directly provide scholarships for college attendance.

When all programs aimed primarily at children and youth are considered, they account for almost one-third of the total number of programs offered. Children and youth programs, therefore, constitute a major investment for Black churches with community outreach programs for nonchurch members.

By far, the largest category of programs offered is family support and family assistance programs. Forty-two percent of all programs offered are for family-oriented community outreach programs. These include family support programs such as basic assistance with food, clothing, and shelter as well as emergency financial aid.

Individual services for adults constitute 9% of the total. These include services to incarcerated men and women, and counseling programs focusing on drug abuse, AIDS, and other health issues. Moving further up the age range, services and aid to the elderly represent 8% of the total. Finally, community service and community development programs comprise 10% of the total number of programs offered by these churches.

COLLABORATION WITH OTHER CHURCHES AND COMMUNITY AGENCIES

In addition to conducting community outreach programs independently, many of the churches in our sample have established elaborate and extensive networks of collaboration with other churches and community agencies to carry out this work. Indeed, our evidence confirms the existence of what Lincoln (1990) has termed "the dialectic between the communal and the privatistic" orientations of the Black church. According to Lincoln the communal orientation reflects the historical tendency of Black churches to embrace "all aspects of the lives of their members, including political, economic, educational, and social concerns." The privatistic orientation, on the other hand, implies "a withdrawal from the concerns of the larger community to focus on meeting only the religious needs of its adherents" (p. 13).

When we asked the churches' representative (the senior minister in 72% of the cases) for their own personal orientation toward the mission of the church, they overwhelmingly (83%) indicated that both the religious needs of church members and the concerns of the larger community were important. In practice, however, we found a sharper split between the two orientations. Consistent with Lincoln's "privatistic orientation" nearly one-third of the churches in our study operated no community outreach programs at all. The two-thirds who did operate outreach programs provide support for Lincoln's "communal orientation" as being the dominant perspective among contemporary Black churches, at least in the northeastern region of the nation.

We have gathered more specific information about the nature of this communal orientation as practiced by these churches. For example, 65% of the churches have cooperative arrangements with other churches in the community to jointly conduct outreach programs. Further, we found that about 73% of the churches ($N = 156$) with outreach programs have collaborative relations with secular agencies in the community as part of their outreach efforts. . . .

[S]chools are second only to police departments as agencies with which a number of churches have established collaborative working relations in sponsoring community outreach programs. Specifically, about 76% of the churches that have worked with agencies have relationships with the schools, while about 84% have working relationships with police departments. In third place are welfare departments followed by hospitals, local prisons, health departments, and housing departments. All these agencies reflect the most pressing needs and current predicament of African American people to which the contemporary Black church offers active response and assistance. Much of this activity on the part of the church involves helping families make more effective contact with these community agencies—proving once again that the school, the family, and the church have enormous needs for collaboration and substantial potential for mutual enhancements. . . .

Conclusion

Nonetheless, it is in the public school system that the Black church will for some time make its greatest impact on the education of African American children and youth. This is the arena to which most African American families will continue to turn for the education of their children. The church can use its considerable and demonstrated influence to help African American families, and the African American community generally, extract from the public schools better performance in the education of African American children. This explains why efforts like Project Spirit hold such promise. The commitment to and dependence upon public schools by African American parents has caused them to place a great deal of hope in the promise of school desegregation, compensatory education, and Head Start. Thus far, however, all of these approaches have exhibited major shortcomings.

Indeed, the direction the Black community as a whole must take in this regard has been spelled out in a recent policy statement issued by a group of Black intellectuals under the leadership of the esteemed historian John Hope Franklin and sociologist Sarah Lawrence Lightfoot (Committee on Policy for Racial Justice, 1989). "What we must demand," they write, "is this: that the schools shift their focus from the supposed deficiencies of the black child — from the alleged inadequacies of black family life — to the barriers that stand in the way of academic success" (p. 2). As our ongoing research reveals, what better, more powerful, independent, numerous and resourceful ally can Black families count on to help them establish such relationships with schools than the Black church?

REFERENCES

Billingsley, A. (in press). *Climbing Jacob's ladder: The future of African American families.* New York: Simon & Schuster/Touchstone Books.

Blackwell, J. E., 1985. *The Black community: Diversity and unity,* 2nd ed. New York: Harper & Row.

Committee on Policy for Racial Justice. (1989). *Visions of a better way.* Washington, DC: Joint Center for Political Studies.

Drake, S. C., and H. R. Clayton, 1945. *Black metropolis.* New York: Harper & Row.

DuBois, W. E. B., January 1898. The study of the Negro problem. *Annals, 1,* 1–23.

Frazier, E. F., 1974. *The Negro church in America.* New York: Schocken Books.

George, Y., V. Richardson, M. Lakes-Matyas, and F. Blake, 1989. *Saving Black minds: Black churches and education.* Washington, DC: American Association for the Advancement of Science.

Hess, I., 1985. *Sampling for social research surveys, 1947–1980.* Ann Arbor, MI: Institute for Social Research, University of Michigan.

Hill, R., 1971. *The strengths of Black families.* New York: Emerson Hall.

Jacquet, C. H. (Ed.), 1989. *Yearbook of American and Canadian churches 1989.* Nashville, TN: Abingdon Press.

Johnson, C. S., 1934. *Shadow of the plantation.* Chicago: University of Chicago Press.

Johnson, J., 1986. *An historical review of the role Black parents and the Black community played in providing school for Black children in the South: 1865–1954.* Unpublished doctoral dissertation, University of Massachusetts, School of Education.

Lewis, H., 1957. *Black ways of Kent.* New York: Van Rees Press.

Lincoln, C. E., April 1989. *The Black church and Black self-determination.* Paper presented at the Association of Black Foundation Executives, Kansas City, MO.

Lincoln, C. E., and L. H. Mamiya, 1990. *The Black church in the African American experience.* Durham, NC: Duke University Press.

Mays, B. E., and J. W. Nicholson, 1933. *The Negro's church.* New York: Russess.

Melton, J. G. (Ed.), 1988. *Encyclopedia of American religions,* 3rd ed. Detroit, MI: Gale Research.

Moberg, D. O., 1962. *The church as a social institution.* Englewood Cliffs, NJ: Prentice-Hall, Inc.

Neighbors, H. and J. S. Jackson, 1984. The use of informal and formal help: Four patterns of illness behavior in the Black community. *American Journal of Community Psychology, 12,* 629–644.

Orfield, G., 1987. School desegregation needed now. *Focus, 15,* 5–7.

Roberts, J. D., 1980. *Roots of a Black future: Family and church.* Philadelphia: The Westminster Press.

Taylor, R., and L. Chatters, 1988. Church members as a source of informal social support. *Review of Religious Research, 30,* 432–438.

Tucker, B., and C. Mitchell-Kernan, (in press). Sex ratio imbalance among Afro-Americans: Conceptual methodological issues. In R. L. Jones (Ed.), *Advances in Black psychology.* Berkeley, CA: Cobb & Henry.

U.S. Bureau of the Census, Department of Commerce, 1988. *Current population reports: Population characteristics,* Series P-20, Nos. 388 and 437. Washington, DC: U. S. Government Printing Office.

Willie, C. V., A. M. Garibaldi, and W. L. Reed, (Eds.), 1990. *The education of African Americans,* Vol. 3. Boston: University of Massachusetts Press.

32
Religion in Post-Protestant America

PETER L. BERGER

*The U.S. Constitution states that there will be no
law establishing a particular religion. The results
are not only religious pluralism but also, as Berger
notes, moral pluralism. This factor may result in
conflict because religious pluralism may entail political efforts to enforce particular moral beliefs.*

As you read, ask yourself the following questions:

1. *What is the price paid by religion for pluralistic relativism?*

2. *How have the various debates—church-state
separation, secularization-strict conservatism—
affected your beliefs? In what ways?*

GLOSSARY

Ecclesiastical Formally allied with the state.
First Amendment Free exercise of religion and establishment of religion.

IN JUNE 1985 THE U.S. Supreme Court over-
ruled an Alabama statute authorizing public
schools in that state to observe a one-minute
silence "for meditation or voluntary prayer."
The reasoning behind the decision was that the
statute represented an establishment of religion
and thereby violated the First Amendment to the
United States Constitution, which states that
"Congress shall make no law respecting an establishment of religion, or prohibiting the free
exercise thereof."

I was abroad a few weeks after this decision and
was put in the position of trying to explain it to a
group of by no means unfriendly Europeans. It
was not an easy task. Their first puzzlement: how
is it that an act of a state legislature can be over-
ruled on the basis of an amendment specifically
referring to acts of Congress?—was relatively easy

to dispel. I explained that the Fourteenth Amendment had "nationalized" all the rights spelled out
in the first ten amendments to the federal Constitution. Fortunately, no one asked me to explain
just what the Fourteenth Amendment was in the
first place, or I would have had to discourse at
length on the way in which an act intended to
outlaw slavery came to bear such a grave additional burden. But there was no way I could deal
with their second puzzlement: how a minute of silence in a classroom of noisy schoolchildren could
possibly endanger the religious liberty of Americans. We are not an easy people to understand.

Sidney Mead, the eminent American church
historian, has described the United States as "a
nation with the soul of a church." The phrase is
apt and serves well to illuminate various aspects
of the American national character, such as its
deep-rooted sense of historic mission and its in-
veterate moralism.

If the nation is churchlike, then surely the
Supreme Court is its most obviously ecclesiastical institution, endlessly engaged in interpreting
and reinterpreting the sacred text on which the
nation-church is supposed to be based. As Muslims call Jews and Christians people of the Book,
the United States may be called a polity of the
Book. No wonder we have more lawyers, both
in absolute numbers and per capita, than any
other country on earth. And at the pinnacle of
this hierarchy of clerks are those whose business
is constitutional exegesis in the federal courts, a
business with striking similarities to the scholas-
ticisms spawned by those religious traditions
(notably Judaism, Christianity, and Islam) that
derive from a revelation in a holy book.

Reprinted from Commentary, *May 1986, by permission; all rights reserved.*

Scholastic interpretations evolve from generation to generation; after a while, it becomes a little difficult to relate the latest exegetical exercise to the original meaning of the texts at issue. Some look altogether askance at the effort; these are the judicial activists, who, like Roman Catholic theorists of the Church, put their faith less in the original texts than in the institutional process by which they are transmitted. Others, our strict constructionists, recapitulate the classical Protestant effort to get back to the texts — though this often involves them in rather complicated intellectual acrobatics, for no amount of determination can change the fact that present-day America is a very different place from the America in which the Constitution was written.

As several recent books make clear, the men who drafted the First Amendment not only did not have in mind what later exegetes have imputed to them, but different ones among them had quite different things in mind. In view of the historical evidence, for example, it is hard to sustain the later interpretation that the two clauses of the First Amendment, the "establishment" clause and the "free-exercise" clause, implied a balanced theory of church-state relations. In any case, the main concern of those who drafted the amendment was not to formulate a positive statement about the rights of religion but rather to make sure that the new national institutions (some members of Congress balked at the very word "national") would not usurp rights and powers won in the several states. What is more, the present wording was arrived at fairly quickly after bargaining over several discarded alternatives, as Congress was anxious to get on to other matters.

This is how Thomas Curry puts it in *The First Freedoms: Church and State in America to the Passage of the First Amendment*[1]:

> The passage of the First Amendment constituted a symbolic act, a declaration for the future, an assurance to those nervous about the federal government that it was not going to reverse any of the guarantees for religious liberty won by the revolutionary states. . . . Congress approached the subject in a somewhat hasty

and absentminded manner. To examine the two clauses of the amendment as a carefully worded analysis of church-state relations would be to overburden them.

That is a historian's judgment. It goes without saying that lawyers and constitutional theorists are not likely to be much restrained by it: overburdening texts is their vocation.

From Curry, and from William Lee Miller[2] and A. James Reichley[3] as well, we get a vivid picture of the bewildering variety of the colonies in the matter of religious arrangements (as indeed in everything else). There was New England, dominated by the powerful Puritan presence — although with a sharp divergence between theocratic Massachusetts and latitudinarian Rhode Island. There was Virginia with its Anglican establishment, which was latitudinarian in its own way, permitting the sort of mellow deism that characterized most of the great Virginians of the period, notably Thomas Jefferson, James Madison, and, last but not least, George Washington. Maryland and Pennsylvania were two different experiments in religious liberty, the first a rather short-lived attempt at Catholic libertarianism, the second, rooted in Quaker ideology, leading to what was surely the most motley collection of religious eccentrics in 18th-century Western civilization. And New York, with its tradition of Dutch mercantilism, demonstrated that religious liberty could be the fruit as much of commercial pragmatism as of lofty ideas.

It is one of the miracles of American history that the representatives of these discrepant political entities could agree on anything; possibly the First Amendment was one of their less miraculous agreements, since no one desired to see the new nation institutionalize church-state relations along the lines of the English establishment. In this, as is not unusual for a revolutionary coalition, they were mainly united in what they were against.

Miller devotes a large part of his book to two representative if wildly divergent individuals, James Madison and Roger Williams. He makes a good case for seeing these men as prototypes of the two principal traditions of religious liberty in the American republic. Madison, of course, was one of the great Virginians—a gentleman of ample means, a friend of Jefferson, a relaxed deist comfortable in the Anglican church, an advocate of "republican virtue" in the spirit of the American Enlightenment. Williams, born amid the murderous fanaticism of English Protestant dissent, was a cantankerous Puritan, no less so for eventually concluding that religious liberty was a desideratum in this world. Williams arrived at this conclusion not out of any Enlightenment ideas about the nobility of man's quest for truth but out of his deeply Calvinistic conviction that, all existing churches being hopelessly corrupt, none must be given the power to coerce.

Out of these twin roots, Miller writes, grew a set of peculiar political and ecclesiastical attitudes. Enlightenment skepticism brought the idea that since no certainties were available in the area of religion, tolerance was the only reasonable position. Protestant sectarianism contributed a strong sense of theological certitude which, however, included the conviction that, since the state and any state-established church would always misuse its power, the church that was faithful to the truth of the gospel must be fiercely independent.

These discrepant ideas combined, at a particular juncture of history, to create a unique fabric of church-state relations. To be sure, circumstances helped a great deal: even in the absence of these ideas, it would have been difficult to replicate on American soil any of the religious establishments of Europe. There were simply too many different groups, too many religio-political interests, for any one to establish dominance. Willy-nilly, they had to learn to live with one another. But whether the ideas were conveniently at hand to legitimate pressing pragmatic concerns, or whether the ideas had history-changing efficacy of their own (or whether, more likely, both things were true), the arrangement so created is with us today. And so, indeed, are the two traditions out of which it grew. Thus, regularly, lawyers for the American Civil Liberties Union with nary a religious certitude in their bones

find themselves on the same side of church-state issues—i.e., the side of strict separation—with born-again Baptists who, if one is to believe them, have never doubted their orthodox faith by one iota.

Another thing made clear by these authors is that within the institutional framework of liberty brought about by this paradoxical conspiracy of skeptics and believers, there has been an unending intrusion of religiously basic values into the public life of the nation. The Constitution, as it came to be interpreted, might well have set up a wall of separation between church and state, but it most certainly did not set up a wall between religion and politics. Reichley traces such interventions right through the 1984 presidential election, and he clearly expects them to continue, possibly with renewed force, in the coming years.

Indeed, the "nation with the soul of a church" has always had a deeply religious tendency to turn political campaigns into great crusades. Two high-points were the anti-slavery and the temperance campaigns. The first achieved permanent victory (though at the price of a bloody civil war), the latter only a brief and Pyrrhic victory. Both demonstrated the enormous capacity of churches and church-related groups to mobilize Americans for a political campaign defined as a struggle between virtue and vice. The format of these two classical crusades continues to reproduce itself in this country whenever political agendas are infused with high religious pathos. There is the same unwillingness to compromise or to weigh costs against benefits (prophets do not calculate), the same tendency to diabolize the opposition, and the same curious alliance of clergy and activist women that in other contexts the historian Ann Douglas has dubbed the "feminization of American culture."

In recent decades we have seen three such religiopolitical eruptions, each characterized by this now-typical format: the civil rights movement, the antiwar movement with its various Left-leaning offshoots, and most recently, the powerful phenomenon that Richard John Neuhaus has aptly called the "bourgeois insurgency," once again generating miscellaneous offshoots but with the anti-abortion movement at its core.

These movements have mobilized different strata of the American population and the last two are in sharp antagonism to each other. Yet all three stand in "apostolic succession" to all the earlier crusades of moral fervor sustained by religious certitude. Perhaps a final irony is that, today, each movement defines its opponent as an illegitimate intrusion of religion into politics. Thus the Reverend Jesse Jackson and the Reverend Jerry Falwell accuse each other of misusing their religious status for political ends, and the followers of each believe (sincerely, one may stipulate) that any increase in the power of the opposing movement will lead to an age of inquisitorial intolerance and oppression.

To be sure, religious liberty is not an American monopoly. It is a common feature of all democracies and one of the important legitimations of the democratic form of government. Yet the United States can validly claim to have most successfully institutionalized religious liberty. Given the heterogeneity of the American population, this has been a remarkable achievement, all the more so when one compares the civic peace that has characterized relations among religious groups here with the record of many other societies. In this achievement, as speakers on many patriotic occasions regularly and correctly point out, present-day America demonstrates the wisdom of the generation that produced the Constitution. Here, too, we see a remarkable continuity between past and present.

But it is just as obvious that the social environment of religion in America has changed profoundly since the time of the First Amendment. The most important change can be stated quite simply: the Protestant social and cultural establishment has ended. Indeed, with just a little license one can speak today of a post-Protestant America.

The term "Protestant establishment" may at first sound grating. Nevertheless, although the Constitution effectively prevented the legal establishment of Protestantism, first in the new

nation and then in the states, it did not and could not change the fact that American society was crucially shaped by Protestantism. Its social order, its ethos, even its manners bore an unmistakably Protestant cast, readily recognized as such especially by observers who were not themselves Protestant (Alexis de Tocqueville, for example).

Although to some extent this remains the case today, the social and cultural domination of the country by a cohesive Protestant elite has been severely weakened. Other groups have successfully made their way, different elites have been formed, and both society and culture have become vastly more pluralistic than could have been imagined even at the turn of this century. The Protestant character of American civilization has become more of an echo, a nuanced survival of what used to be a robust sociocultural reality. John Murray Cuddihy has aptly described this as a change from a Protestant ethic to a Protestant etiquette: the fervid moral intensity of the earlier period has become a mood, a morally neutral civility—that "Protestant smile," now widely diffused among Americans from every conceivable ethnic and religious background, which continues to impress newcomers to these shores as either heartwarming ("Americans are so friendly") or hypocritical ("they don't really mean it").

The major reason for the change has been the pattern of immigration, bringing masses of non-Protestant people, especially Catholics and Jews, into society. Today, the influx of Latin Americans continues the gradual "Catholicization" of American society (though one should note that Latin American Catholicism differs in important ways from that of Southern and Eastern Europe, not to mention the very particular Catholicism of the Irish). A new and potentially significant factor is the larger immigration from Asia. Although religious statistics in the United States are not models of exactitude, it has been reasonably claimed that Hawaii is the first state of the union in which Christians are a minority. Anyone driving out of Honolulu on the Pali Highway, past edifices and shrines of Asian cults of every description, can easily visualize an America far more pluralistic than the one described in 1955 by Will Herberg in his perceptive book, *Protestant-Catholic-Jew.* (Public-service advertisements on Honolulu buses: "Attend the church, synagogue, or shrine of your choice!")

There may be another, more subtle reason for the demise of the Protestant establishment. Those at its core came to lose their own belief in it. There were cultural as well as political failures of nerve. E. Digby Baltzell, the only American sociologist who has studied the Protestant upper class with care and without destructive intent, has traced this sociocultural decline in several excellent books; the recent history of the Episcopal church, once the elite denomination *par excellence,* offers the best insight into it. One glance at the weekly program of the Cathedral of St. John the Divine in New York City provides wondrous examples of the Protestant transformation from self-confidence to a nervous and guilt-prone search for some way, any way, of sociocultural survival.

As long as the Protestant establishment was a reality, it provided a reliable symbolic center despite (perhaps even because of) the legal disestablishment of religion. Its inner decline, combined with the increasingly rigorous interpretation by the courts of church-state separation, has led to the situation characterized by Richard John Neuhaus as the "naked public square": that is, a public life increasingly swept clean of the religious symbols that used to legitimate it and make it plausible.

This is a problem in and of itself. Society, deprived of plausible legitimating symbols, increasingly becomes a merely contractual arrangement. Although this may be fine as long as there are no serious problems confronting a society, when such problems do arise it becomes more and more difficult to motivate people to make sacrifices for any collective purpose. Emile Durkheim, in his last great work, *The Elementary Forms of the Religious Life,* argued that a society will not survive unless people are prepared, if necessary, to die for it. Human beings are not prepared to

die for a contract based on the pragmatic accommodation of interests. When, a few years ago, the Carter administration reintroduced draft registration (not the draft itself, just registration for a possible draft), a picture that appeared in many newspapers showed a demonstration of Princeton students, one of whom carried a poster with the inscription, "Nothing is worth dying for." To say that this is a problem is an understatement at a moment when American democracy faces the greatest military threat ever to its survival.

But there is more to the matter. As Neuhaus has observed, the public square will not remain "naked": other contents inevitably come to fill it. In our case today, the contents are those of "secular humanism." The phrase is used pejoratively by, for example, the followers of Jerry Falwell, but in fact it quite aptly catches the gist of the values, antiseptically free of religious referents and emphatically humanistic, that have been elevated to the status of a largely tacit and unacknowledged new cultural establishment. It is precisely such an ethos of "secular humanism" that is embodied, for example, in Lawrence Kohlberg's program of "moral education" and in the various pedagogic systems based on "values clarification." It is the ethos of a new, post-Protestant establishment (many if not most of whose members are themselves of elite Protestant ancestry).

A key problem of our time is the relation between this new elite and the rest of the American people. In recent years there has been an impressive accumulation of new data about American religion (as in the survey on American values undertaken by the Connecticut Mutual Life Insurance Company, or the superb *All Faithful People,* by Theodore Caplow et al.). These data show that the majority of Americans are as furiously religious as ever — and very probably *more* religious than they were when Tocqueville marveled at this quality of American life.

The most dramatic expression of American religious turbulence today has been the upsurge of evangelicalism in its various forms. At least since the mid-1970's — Jimmy Carter's campaign for the Presidency can be taken as a convenient marker—traditional religious and moral beliefs, which had been assumed to survive, if at all, in the hinterlands of society, suddenly erupted into the center of public life. Even those safely located within the cultural enclaves of "secular humanism" have been forced to take notice of this veritable explosion of religious fervor, first in bewilderment, then (especially as significant elements of the evangelical world became mobilized in the service of Right-of-Center political causes) with growing alarm.

The evangelical renascence, though, is only the most spectacular expression of the perduring religiosity of most of the American people. With the exception of the so-called "mainline" Protestant denominations (a misnomer if ever there was one, as these denominations are increasingly marginal to the nation's religious life), it has manifested itself elsewhere as well. Spurred on by Rome, especially under the present papacy, powerful neotraditionalist impulses have made themselves felt in the Catholic community. There has occurred an increasingly robust religious revival within all three denominations of American Judaism. Least noticed but of great potential importance have been the increasing vitality and "Americanization" of Eastern Orthodox Christianity, now no longer an ethnic enclave. There has been the spectacular increase of the Mormons. Whatever else most Americans are, they are *not* "secular humanists."

These developments have created serious intellectual difficulties for those (like myself) who thought that modernization and secularization were inexorably linked phenomena. To say the least, the United States appears to be an exception to this linkage. Indeed, if one compares sociological data on religion worldwide, America sticks out as a remarkable exception. The most modernized societies, notably in Western Europe and in Japan, are in fact high on any secularization scale, be it subjective (recording what people say they believe) or objective (recording what people actually do in terms of religious practice). The prototypes of secular modernity

are the Scandinavian countries, with Sweden in the unchallenged lead. At the other end of the list would be societies like India.

When one looks at the United States as a whole, in terms of the subjective and objective indicators of religiosity, one sees a remarkable resemblance to India (the content, needless to say, being different). Yet if one moves around in the cultural centers of America, one breathes a remarkably Scandinavian air. (Could this be why there are so many Volvos on Ivy League campuses?) And thereby hangs a tale of great significance: when it comes to religion, America is an India, with a little Sweden superimposed.

In more conventional social-scientific language we might say that secularization in America today tends to be class-specific. It is most diffused in the college-educated upper middle class—or, more pointedly, in those strata that Irving Kristol and others have called the New Class, people who derive their livelihood from the production and distribution of symbolic knowledge. It is least diffused in the old middle class, the lower middle class, and the "respectable" working class. This means that the general drama of secularization and counter-secularization has been drawn into a major conflict between classes. In America today and very likely in other Western democracies as well, the New Class is increasingly resented by other strata in society, and not the least factor in this resentment is the hegemony exercised by the New Class over the institutions of elite culture, including large segments of the educational system and the media.

The evangelical upsurge may at least in part be understood as a rebellion of other groups against the culture of the secularized New Class, and especially against the coercive imposition of that culture on children in the public schools. Thus it makes sense that education has been one of the major battlefields in this conflict. The "Indians" are rising against the "Swedes," and neotraditionalist religious symbols are weapons in this uprising. (To say this, of course, in no way negates the sincerely religious motives of those engaged in the various evangelical campaigns.)

There is also a specifically religious dynamic here, in which counter-secularization takes a more or less fundamentalist form. This dynamic can be observed in many parts of the world, not least in the Muslim societies. In the United States the ironic fact is that an overall social climate of tolerance and relativism has kept on being disturbed by seemingly irrational eruptions of unbridled fanaticism. The various "revolutions" and "liberations" of the last two decades provide a number of secular analogues. How can it be that all these nice people, many of them long-time smilers of the Protestant smile, have suddenly come out of the closet with their various agendas of absolute claims, non-negotiable demands, and the determination to pursue their purposes by any means necessary? How can Protestant niceness be so quickly replaced by rage?

This alternation becomes more understandable if one reflects on the psychic costs of pluralistic relativism. The latter is experienced by many as a great liberation, affording the opportunity to emerge from one's own narrow subculture into a broad milieu of cosmopolitanism, and in addition freeing one from all sorts of taboos and "repressions." But the price of liberation is the loss, or at least the weakening, of all certitude — social, moral, religious. Pan-tolerant individuals are thus peculiarly conversion-prone, and conversions can be sudden and violent. The same individuals may quite frequently come to experience as oppressive that which was once liberating; their rage is a reaction to the intuition that they have been had. In this light, the emergence of various movements with claims to absolute certitude is no longer surprising, although the choice an individual makes among the movements available in the ideological marketplace will remain a matter of biographical chance or social location.

But to return to the church-state issues with which we began: Reichley gives a helpful overview of the different positions in play today. The two most important are held by those who advocate a strict separation and by those who would somehow accommodate the religious character of the American people. Both positions (there are, of course, many gradations in each) face serious difficulties.

The doctrine of strict separation flies in the face of a social reality that remains persistently, stubbornly religious. The unending turmoil engendered by the Supreme Court decision on prayer in the public schools is the best example of this collision: whatever may have been the careful legal reasoning behind the decision, the fact remains that millions of Americans have perceived it as a solemn declaration that, their most dearly held beliefs and values having been disavowed by the highest authority of the Republic, their country has become officially godless. The political difficulties this has led to are evident.

But there is a deeper difficulty. As I have already noted, every society, and a democratically governed society more than others, requires legitimation. It requires the belief that it is morally justified. And in the nature of the case, such legitimation cannot be invented *ex nihilo*, it must be credible in terms of beliefs and values that people actually hold. When a people is as religious as the American people, it is going to be very difficult to purge the official legitimations of society of all religious symbols and still have them remain credible. The doctrine of strict separation (not just between the state and specific denominations, but between the state and religion as such) has thus contributed to a crisis of legitimation.

The accommodationist position, however, faces serious difficulties as well. These are all grounded in the equally stubborn reality of American pluralism. Just *which* religious symbols are to be accommodated in the "public square"? The answer to this question has usually been rendered in terms of the so called Judeo-Christian tradition. One may leave aside for the moment how this answer will sit with Hawaiian Buddhists and with agnostics (not all of them members of the New Class) who have experienced their emancipation from various Christian or Jewish subcultures as a personal liberation; taken together these groups still constitute a minority of the American population, and perhaps they, or most of them, might settle for a few Judeo-Christian symbols as an acceptable price for civic peace. But there is the more intractable fact that those who do identify with that great tradition, in its various denominational forms, are themselves deeply divided as to its moral meaning.

America has been very successful in dealing with its religious pluralism. The *moral* pluralism that has now emerged will be much more difficult to deal with. Currently the most dramatic expression of this is the abortion issue. What symbolic accommodation can be struck between two groups, each containing roughly 50 percent of the American people, one of whom perceives abortion as involving the fundamental right of a woman over her own body and the other as being an act of homicide? How can such an accommodation be struck when each side makes use of religious symbols and religious language to legitimate its position?

On the other hand, American society has long experience with settling seemingly irreconcilable positions. And the American polity has developed a genius for institutionalizing such settlements. Barring national disasters of catastrophic scope, it is not unreasonable to expect that new formulas can be found to institutionalize new balances, new compromises. Religious institutions will certainly have a major role to play in such a new settlement. But they will be in a better position to play that role if they draw back from their respective positions of partisan advocacy (most of it very class-specific indeed) and return to what has been their most creative function, that of mediation. Churches as mediating institutions are more likely to contribute to the solution of these problems than churches as partisan armies.

NOTES

1. Oxford University Press.
2. *The First Liberty: Religion and the American Public*, Knopf.
3. *Religion in American Public Life*, Brookings.

33

Motivation and Ideology: What Drives the Anti-Abortion Movement . . .

DALLAS A. BLANCHARD

The anti-abortion movement involves a large number of beliefs and activities affecting all aspects of life. The question, then, is how one feels about this whole range of issues.

As you read, ask yourself the following questions:

1. *What are the determining factors in the anti-abortion movement?*
2. *In what ways do you agree/disagree with the anti-abortion movement?*

GLOSSARY

Social movement Organized activities to encourage or oppose some aspect of change (in this case oppose abortion).

Fundamentalism Adherence to traditional religious or cultural norms such as respect for authority.

Why and How People Join the Anti-Abortion Movement

RESEARCHERS HAVE POSITED A variety of explanations for what motivates people to join the anti-abortion movement. As with any other social movement, the anti-abortion movement has within it various sub-groups, or organizations, each of which attracts different kinds of participants and expects different levels of participation. It might in fact be more appropriate to speak of anti-abortion *movements.*

Those opposing abortion are not unified. Some organizations have a single-issue orientation, opposing abortion alone, while others take what they consider to be a "pro-life" stance on many issues, opposing abortion as well as eu-

thanasia, capital punishment, and the use of nuclear and chemical arms. . . .

Researchers have identified a number of pathways for joining the anti-abortion movement. Luker, in her 1984 study of the early California movement, found that activists in the initial stages of the movement found their way to it through professional associations. The earliest opponents of abortion liberalization were primarily physicians and attorneys who disagreed with their professional associations' endorsement of abortion reform. It is my hypothesis that membership in organizations that concentrate on the education of the public or religious constituencies and on political lobbying is orchestrated primarily through professional networks. With the passage of the California reform bill and the increase in abortion rates several years later, many recruits to the movement fell into the category Luker refers to as "self-selected"; that is, they were not recruited through existing networks but sought out or sometimes formed organizations through which to express their opposition.

Himmelstein (1984), in summarizing the research on the anti-abortion movement available in the 10 years following *Roe v. Wade*, concluded that religious networks were the primary source of recruitment. Religious networks appear to be more crucial in the recruitment of persons into high-profile and/or violence prone groups (Blanchard and Prewitt 1992)—of which Operation Rescue is an example—than into the earlier, milder activist groups (although such net-

Blanchard, Dallas A. The Anti-Abortion Movement and the Rise of the Religious Right: From Polite to Fiery Protest. *NY: Twayne, 1994.*

works are generally important throughout the movement). Such networks were also important, apparently, in recruitment into local Right to Life Committees, sponsored by the National Right to Life Committee and the Catholic church. The National Right to Life Committee, for example, is 72 percent Roman Catholic (Granberg 1981). It appears that the earliest anti-abortion organizations were essentially Catholic and dependent on church networks for their members; the recruitment of Protestants later on has also been dependent on religious networks (Cuneo 1989, Maxwell 1992).

Other avenues for participation in the anti-abortion movement opened up through association with other issues. Feminists for Life, for example, was founded by women involved in the feminist movement. Sojourners, a socially conscious evangelical group concerned with issues such as poverty and racism, has an anti-abortion position. Some anti-nuclear and anti-death-penalty groups have also been the basis for the organization of anti-abortion efforts.

Clearly, pre-existing networks and organizational memberships are crucial in initial enlistment into the movement. Hall (1993) maintains that individual mobilization into a social movement requires the conditions of attitudinal, network, and biographical availability. My conclusions regarding the anti-abortion movement support this contention. Indeed, biographical availability — the interaction of social class, occupation, familial status, sex, and age — is particularly related to the type of organization with which and the level of activism at which an individual will engage.

General social movement theory places the motivation to join the anti-abortion movement into four basic categories: status defense; anti-feminism; moral commitment; and cultural fundamentalism, or defense.

The earliest explanation for the movement was that participants were members of the working class attempting to shore up, or defend, their declining social status. Clarke, in his 1987 study of English anti-abortionists finds this explanation to be inadequate, as do Wood and Hughes in

their 1984 investigation of an anti-pornography movement group.

Petchesky (1984) concludes that the movement is basically anti-feminist — against the changing status of women. From this position, the primary goal of the movement is to "keep women in their place" and, in particular, to make them suffer for sexual "libertinism." Statements by some anti-abortion activists support this theory. Cuneo (1989), for example, finds what he calls "sexual puritans" on the fringe of the anti-abortion movement in Toronto. Abortion opponent and long-time right-wing activist Phyllis Schlaffley states this position: "It's very healthy for a young girl to be deterred from promiscuity by fear of contracting a painful, incurable disease, or cervical cancer, or sterility, or the likelihood of giving birth to a dead, blind, or brain-damaged baby (even ten years later when she may be happily married)" (Planned Parenthood pamphlet, no title, n.d. [1990]). . . . A number of researchers have concluded that sexual moralism is the strongest predictor of anti-abortion attitudes.

The theory of moral commitment proposes that movement participants are motivated by concern for the human status of the fetus. It is probably as close as any explanation comes to "pure altruism." Although there is a growing body of research on altruism, researchers on the abortion issue have tended to ignore this as a possible draw to the movement, while movement participants almost exclusively claim this position: that since the fetus is incapable of defending itself, they must act on its behalf.

In examining and categorizing the motivations of participants in the anti-abortion movement in Toronto, Cuneo (1989:85ff.) found only one category — civil rights — that might be considered altruistic. The people in this category tend to be nonreligious and embarrassed by the activities of religious activists; they feel that fetuses have a right to exist but cannot speak for themselves. Cuneo's other primary categories of motivation are characterized by concerns related to the "traditional" family, the status of women in the family, and religion. He also finds an activist fringe composed of what he calls religious seekers; sexual therapeutics, "plagued by guilt and fear of female sexual power" (115); and punitive puritans, who want to punish women for sexual transgressions. All of Cueno's categories of participant, with the exception of the civil rights category, seek to maintain traditional male/female hierarchies and statuses.

The theory of cultural fundamentalism, or defense, proposes that the anti-abortion movement is largely an expression of the desire to return to what its proponents perceive to be "traditional culture." This theory incorporates elements of the status defense and anti-feminist theories.

It is important to note that a number of researchers at different points in time (Cuneo 1989; Ginsburg 1990; Luker 1984; Maxwell 1991, 1992) have indicated that (1) there have been changes over time in who gets recruited into the movement and why, (2) different motivations tend to bring different kinds of people into different types of activism, and (3) even particular movement organizations draw different kinds of people with quite different motivations. At this point in the history of the anti-abortion movement, the dominant motivation, particularly in the more activist organizations such as Operation Rescue, appears to be cultural fundamentalism. Closely informing cultural fundamentalism are the tenets of religious fundamentalism, usually associated with certain Protestant denominations but also evident in the Catholic and Mormon faiths. . . .

Religious and Cultural Fundamentalism Defined

Cultural fundamentalism is in large part a protest against cultural change: against the rising status of women; against the greater acceptance of "deviant" life-styles such as homosexuality; against the loss of prayer and Bible reading in the schools; and against the increase in sexual openness and freedom. Wood and Hughes (1984) describe cultural fundamentalism as "adherence to traditional norms, respect for family and religious authority, asceticism and control of impulse. Above all, it is an unflinching and thoroughgoing

moralistic outlook on the world; moralism provides a common orientation and common discourse for concerns with the use of alcohol and pornography, the rights of homo-sexuals, 'pro-family' and 'decency' issues." The theologies of Protestants and Catholics active in the anti-abortion movement—many of whom could also be termed fundamentalists—reflect these concerns.

Protestant fundamentalism arose in the 1880s as a response to the use by Protestant scholars of the relatively new linguistic techniques of text and form criticism in their study of the Bible. These scholars determined that the Pentateuch was a compilation of at least five separate documents written from differing religious perspectives. Similar techniques applied to the New Testament questioned the traditionally assigned authorship of many of its books. The "mainline" denominations of the time (Episcopalian, Congregational, Methodist, Unitarian, and some Presbyterian) began to teach these new insights in their seminaries. With these approaches tended to go a general acceptance of other new and expanding scientific findings, such as evolution.

By 1900 the urbanization and industrialization of the United States were well under way, with their attendant social dislocations. One response to this upheaval was the Social Gospel movement, which strove to enact humanitarian laws regulating such things as child labor, unions, old-age benefits, and guaranteed living wages. With the Social Gospel movement and the acceptance of new scientific discoveries tended to go an optimism about the perfectibility of human nature and society.

Fundamentalism solidified its positions in opposition to these trends as well as in response to social change and the loss of the religious consensus, which came with the influx of European Jewish and Catholic immigrants to America. While fundamentalism was a growing movement prior to 1900, its trumpet was significantly sounded with the publication in 1910 of the first volume of a 12-volume series titled *The Fundamentals*. Prior to the 1920s and 1930s, it was primarily a movement of the North; growth in the South came mostly after 1950.

There are divisions within Protestant fundamentalism, but there is a general common basis of belief. Customary beliefs include a personal experience of salvation; verbal inspiration and literal interpretation of scriptures as worded in the King James Bible; the divinity of Jesus; the literal, physical resurrection of Jesus; special creation of the world in six days, as opposed to the theory of evolution; the virgin birth of Jesus; and the substitutionary atonement (Jesus' death on the cross as a substitution for each of the "saved"). The heart of Protestant fundamentalism is the literal interpretation of the Bible, secondary to this is the belief in substitutionary atonement.

Fundamentalists have also shown a strong tendency toward separatism—separation from the secular world as much as possible and separation from "apostates" (liberals and nonfundamentalist denominations). They are also united in their views on "traditional" family issues: opposing abortion, divorce, the Equal Rights Amendment, and civil rights for homosexuals. While their separatism was expressed prior to the 1970s in extreme hostilities toward Roman Catholicism, fundamentalists and Catholics share a large set of family values, which has led to their pragmatic cooperation in the anti-abortion movement.

Another tie between some Protestant fundamentalists and some Catholics lies in the Charismatic Renewal Movement. This movement, which coalesced in the 1920s (Harrell 1975), emphasized glossolalia (speaking in tongues) and faith healing. Charismatics were denied admission to the World's Christian Fundamental Association in 1928 because of these emphases. By the 1960s charismaticism began spreading to some mainline denominations and Catholic churches. While Presbyterians, United Methodists, Disciples of Christ, Episcopalians, and Lutherans sometimes reluctantly accepted this new charismatic movement, Southern Baptists and Churches of Christ adamantly opposed it, especially the speaking in tongues. But the Catholic charismatics have tended to be quite conservative both theologically and socially, giving them ideological ties with the charismatic

Protestant fundamentalists and fostering cooperation between the two in interdenominational charismatic conferences, a prelude to cooperation in the anti-abortion movement.

There is also a Catholic "fundamentalism," which may or may not be charismatic and which centers on church dogma rather than biblical literalism, the Protestant a priori dogma. Catholic fundamentalists accept virtually unquestioningly the teachings of the church. They share with Protestant fundamentalists the assumption that dogma precedes and supersedes analytical reason, while in liberal Catholic and Protestant thought and in the nation's law reason supersedes dogma. Similar to Catholic fundamentalism is the Mormon faith (or the Church of Jesus Christ of Latter Day Saints), which also takes church dogma at face value.

There are at least six basic commonalities to what can be called Protestant, Catholic, and Mormon fundamentalisms: (1) an attitude of certitude—that one may know the final truth, which includes antagonism to ambiguity; (2) an external source for that certitude—the Bible or church dogma; (3) a belief system that is at root dualistic; (4) an ethic based on the "traditional" family; (5) a justification for violence; and, therefore, (6) a rejection of modernism (secularization). (I am not alone in this position).

Taking those six commonalities point by point:

1. The certitude of fundamentalism rests on dependence on an external authority. That attitude correlates with authoritarianism, which includes obedience to an external authority, and, on that basis, the willingness to assert authority over others.
2. While the Protestant fundamentalists accept their particular interpretation of the King James Version of the Bible as the authoritative source, Mormon and Catholic fundamentalists tend to view church dogma as authoritative.
3. The dualism of Catholics, Protestants, and Mormons includes those of body/soul, body/mind, physical/spiritual. More basically, they see a distinction between God and

Satan, the forces of good and evil. In the fundamentalist worldview, Satan is limited and finite; he can be in only one place at one time. He has servants, however, demons who are constantly working his will, trying to deceive believers. A most important gift of the Spirit is the ability to distinguish between the activities of God and those of Satan and his demons.

4. The "traditional" family in the fundamentalist view of things has the father as head of the household, making the basic decisions, with the wife and children subject to his wishes. Obedience is stressed for both wives and children. Physical punishment is generally approved for use against both wives and children.

   This "traditional" family with the father as breadwinner and the mother as homemaker, together rearing a large family, is really not all that traditional. It arose on the family farm, prior to 1900, where large numbers of children were an economic asset. Even then, women were essential in the work of the farm. . . . In the urban environment, the "traditional" family structure was an option primarily for the middle and upper classes, and they limited their family size even prior to the development of efficient birth control methods. Throughout human history women have usually been breadwinners themselves, and the "traditional" family structure was not an option.

5. The justification for violence lies in the substitutionary theory of the atonement theology of both Protestants and Catholics. In this theory, the justice of God demands punishment for human sin. This God also supervises a literal hell, the images of which come more from Dante's *Inferno* than from the pages of the Bible. Fundamentalism, then, worships a violent God and offers a rationale for human violence (such as Old Testament demands for death when adultery, murder, and other sins are committed). The fundamentalist mindset espouses physical punish-

ment of children, the death penalty, and the use of nuclear weapons; fundamentalists are more frequently wife abusers, committers of incest, and child abusers.

6. Modernism entails a general acceptance of ambiguity contingency, probability (versus certitude), and a unitary view of the universe; that is, the view that there is no separation between body and soul, physical and spiritual, body and mind (when the body dies, the self is thought to die with it). Rejection of modernism and postmodernism is inherent in the rejection of a unitary worldview in favor of a dualistic worldview. The classic fundamentalist position embraces a return to religion as the central social institution, with education, the family, economics, and politics serving religious ends, fashioned after the social structure characteristic of medieval times.

Also characteristic of Catholic, Protestant, and Mormon fundamentalists are beliefs in individualism (which supports a naive capitalism); pietism; a chauvinistic Americanism (among some fundamentalists) that sees the United States as the New Israel and its inhabitants as God's new chosen people; and a general opposition to intellectualism, modern science, the tenets of the Social Gospel, and communism. (Some liberal Christians may share some of these views.) Amid this complex of beliefs and alongside the opposition to evolution, interestingly, is an underlying espousal of social Darwinism, the "survival of the fittest" ethos that presumes American society to be truly civilized, the pinnacle of social progress. This nineteenth-century American neo-colonialism dominates the contemporary political views held by the religious right. It is also inherent in their belief in individualism and opposition to social welfare programs.

Particular personality characteristics also correlate with the fundamentalist syndrome: authoritarianism, self-righteousness, prejudice against minorities, moral absolutism (a refusal to compromise on perceived moral issues), and anti-analytical, anti-critical thinking. Many fundamentalists refuse to accept ambiguity as a given in moral decision making and tend to arrive at simplistic solutions to complex problems. For example, many hold that the solution to changes in the contemporary family can be answered by fathers' reasserting their primacy, by forcing their children and wives into blind obedience. Or, they say, premarital sex can be prevented by promoting abstinence. One popular spokesman, Tim LaHaye, asserts that the antidote to sexual desire, especially on the part of teenagers, lies in censoring reading materials (LaHaye 1980). Strict parental discipline automatically engenders self-discipline in children, he asserts. The implication is that enforced other-directedness by parents produces inner-directed children, while the evidence indicates that they are more likely to exchange parental authoritarianism for that of another parental figure. To develop inner-direction under such circumstances requires, as a first step, rebellion against the rejection of parental authority—the opposite of parental intent.

One aspect of fundamentalism, particularly the Protestant variety, is its insistence on the subservient role of women. The wife is expected to be subject to the direction of her husband, children to their father. While Luker (1983) found that anti-abortionists in California supported this position and that proponents of choice generally favored equal status for women, recent research has shown that reasons for involvement in the anti-abortion movement vary by denomination. That is, some Catholics tend to be involved in the movement more from a "right to life" position, while Protestants and other Catholics are more concerned with sexual morality. The broader right to life position is consistent with the official Catholic position against the death penalty and nuclear arms, while Protestant fundamentalists generally support the death penalty and a strong military. Thus, Protestant fundamentalists, and some Catholic activists, appear to be more concerned with premarital sexual behavior than with the life of the fetus.

Protestants and Catholics (especially traditional, ethnic Catholics) however, are both concerned with the "proper," or subordinate, role of women and the dominant role of men. Wives should obey their husbands, and unmarried women should refrain from sexual intercourse. Abortion, for the Protestants in particular, is an indication of sexual licentiousness (see, for example, LaHaye 1980). Therefore, the total abolition of abortion would be a strong deterrent to such behavior helping to reestablish traditional morality in women. Contemporary, more liberal views of sexual morality cast the virgin female as deviant. The male virgin has long been regarded as deviant. The fundamentalist ethic appears to accept this traditional double standard with its relative silence on male virginity.

Another aspect of this gender role ethic lies in the home-related roles of females. Women are expected to remain at home, to bear children, and to care for them, while also serving the needs of their husbands. Again, this is also related to social class and the social role expectations of the lower and working classes, who tend to expect women to "stay in their place."

Luker's (1983) research reveals that some women in the anti-abortion movement are motivated by concern for maintaining their ability to rely on men (husbands) to support their social roles as mothers, while pro-choice women tend to want to maintain their independent status. Some of the men involved in anti-abortion violence are clearly acting out of a desire to maintain the dependent status of women and the dominant roles of men. Some of those violent males reveal an inability to establish "normal" relationships with women, which indicates that their violence may arise from a basic insecurity with the performance of normal male roles in relationships with women. This does not mean that these men do not have relationships with women. Indeed, it is in the context of relationships with women that dominance-related tendencies become more manifest. It is likely that insecurity-driven behaviors are characteristic of violent males generally, but psychiatric data are not available to confirm this, even for the population in question.

The Complex of Fundamentalist Issues

The values and beliefs inherent to religious and cultural fundamentalism are expressed in a number of issues other than abortion. Those issues bear some discussion here, particularly as they relate to the abortion question.

1. *Contraception.* Fundamentalists, Catholic, Protestant, and Mormon, generally oppose the use of contraceptives since they limit family size and the intentions of God in sexuality. They especially oppose sex education in the schools and the availability of contraceptives to minors without the approval of their parents. (See *Nightline,* 21 July 1989.) This is because control of women and sexuality are intertwined. If a girl has knowledge of birth control, she is potentially freed of the threat of pregnancy if she becomes sexually active. This frees her from parental control and discovery of illegitimate sexual intercourse.

2. *Prenatal testing, pregnancies from rape or incest, or those endangering a woman's life.* Since every pregnancy is divinely intended, opposition to prenatal testing arises from its use to abort severely defective fetuses and, in some cases, for sex selection. Abortion is wrong regardless of the origins of the pregnancy or the consequences of it.

3. *In vitro fertilization, artificial fertilization, surrogate motherhood.* These are opposed because they interfere with the "natural" fertilization process and because they may mean the destruction of some fertilized embryos.

4. *Homosexuality.* Homophobia is characteristic of fundamentalism, because homosexual behavior is viewed as being "unnatural" and is prohibited in the Bible.

5. *Uses of fetal tissue.* The use of fetal tissue in research and in the treatment of medical conditions such as Parkinson's disease is opposed, because it is thought to encourage abortion. (See *New York Times,* 16 August

1987; *Good Morning America,* 25 July 1991; *Face to Face with Connie Chung,* 25 November 1989; and *Nightline,* 6 January 1988.)

6. *Foreign relations issues.* Fundamentalists generally support aid to Israel and military funding (Diamond 1989). Indeed, as previously mentioned, they commonly view the United States as the New Israel. Protestant fundamentalists tend to be pre-millenialists, who maintain that biblical prophecies ordain that the reestablishment of the State of Israel will precede the Second Coming of Christ. Thus, they support aid to Israel to hasten the Second Coming, which actually, then, has an element of anti-Semitism to it, since Jews will not be among the saved.

7. *Euthanasia.* So-called right to life groups have frequently intervened in cases where relatives have sought to remove a patient from life-support systems. Most see a connection with abortion in that both abortion and removal of life-support interfere with God's decision as to when life should begin and end.

The most radical expression of cultural fundamentalism is that of Christian Reconstructionism, to which Randall Terry, former director of Operation Rescue, subscribes. The adherents of Christian Reconstructionism, while a distinct minority, have some congregations of up to 12,000 members and count among their number Methodists, Presbyterians, Lutherans, Baptists, Catholics, and former Jews. They are unabashed theocratists. They believe every area of life—law, politics, the arts, education, medicine, the media, business, and especially morality—should be governed in accordance with the tenets of Christian Reconstructionism. Some, such as Gary North, a prominent reconstructionist and son-in-law of Rousas John Rushdoony, considered the father of reconstructionism, would deny religious liberty—the freedom of religious expression—to "the enemies of God," whom the reconstructionists, of course, would identify.

The reconstructionists want to establish a "God-centered government," a Kingdom of God on Earth, instituting the Old Testament as the

Law of the Land. The goal of reconstructionism is to reestablish biblical, Jerusalemic society. Their program is quite specific. Those criminals which the Old Testament condemned to death would be executed, including homosexuals, sodomites, rapists, adulterers, and "incorrigible" youths. Jails would become primarily holding tanks for those awaiting execution or assignment as servants indentured to those whom they wronged as one form of restitution. The media would be censored extensively to reflect the views of the church. Public education and welfare would be abolished (only those who work should eat), and taxes would be limited to the tithe, 10 percent of income, regardless of income level, most of it paid to the church. Property, Social Security, and inheritance taxes would be eliminated. Church elders would serve as judges in courts overseeing moral issues, while "civil" courts would handle other issues. The country would return to the gold standard. Debts, including, for example, 30-year mortgages, would be limited to six years. In short, Christian Reconstructionists see democracy as being opposed to Christianity, as placing the rule of man above the rule of God. They also believe that "true" Christianity has its earthly rewards. They see it as the road to economic prosperity, with God blessing the faithful.

REFERENCES

Editor's Note: The original chapter from which this selection was taken has extensive footnotes and references that could not be listed in their entirety here. For more explanation and documentation of sources, please see Dallas A. Blanchard. 1994. *The Anti-Abortion Movement and the Rise of the Religious Right: From Polite to Fiery Protest.* New York: Twayne Publishers.

Blanchard, Dallas A. and Terry J. Prewitt. 1993. *Religious Violence and Abortion: The Gideon Project.* Gainesville: University Press of Florida.

Brinkerhoff, Merlin B. and Eugene Pupri. 1988. "Religious Involvement and Spousal Abuse: The Canadian Case." Paper presented at the Society of the Scientific Study of Religion.

Clarke, Alan. 1987. "Collective Action against Abortion Represents a Display of, and Concern for, Cultural Values, Rather than an Expression of Status Discontent." *British Journal of Sociology* 38:235–53.

Cuneo, Michael. 1989. *Catholics against the Church: Anti-Abortion Protest in Toronto, 1969–1985.* Toronto: University of Toronto Press.

Diamond, Sara. 1989. *Spiritual Warfare: The Politics of the Christian Right.* Boston: South End Press.

Ginsburg, Faye. 1990. *Contested Lives: The Abortion Debate in an American Community.* Berkeley: University of California Press.

Granberg, Donald. 1978. "Pro-Life or Reflection of Conservative Ideology? An Analysis of Opposition to Legalized Abortion." *Sociology and Social Research* 62:421–23.

————. 1981. "The Abortion Activists." *Family Planning Perspectives* 18:158–61.

Hall, Charles. 1993. "Social Networks and Availability Factors: Mobilizing Adherents for Social Movement Participation" (Ph.D. dissertation, Purdue University).

Himmelstein, Jerome L. 1984. "The Social Basis of Anti-Feminism: Religious Networks and Culture." *Journal for the Scientific Study of Religion* 25:1–25.

LaHaye, Tim. 1980. *The Battle for the Mind.* Old Tappan, NJ: Fleming H. Revell Co.

Luker, Kristin. 1984. *Abortion and the Politics of Motherhood.* Berkeley: University of California Press.

Maxwell, Carol. 1991. "Where's the Land of Happy?: Individual Meanings in Pro-Life Direct Action." Paper presented at the Society for the Scientific Study of Religion.

————. 1992. "Denomination, Meaning, and Persistence: Difference in Individual Motivation to Obstruct Abortion Practice." Paper presented at the Society for the Scientific Study of Religion.

Petchesky, Rosalind P. 1984. *Abortion and Woman's Choice: The State, Sexuality, and Reproductive Freedom.* New York: Longman.

Planned Parenthood Federation of America. 1990. Public Affairs Action Letter. (No title, no date).

Wood, M. and M. Hughes. 1984. "The Moral Basis of Moral Reform: Status Discontent vs. Culture and Socialization as Explanations of Anti-Pornography Social Movement Adherence." *American Sociological Review* 44:86–99.

QUESTIONS FOR DISCUSSION

For further discussion of this topic, see the Wadsworth Sociology Resource Center, "Virtual Society," *http://sociology.wadsworth.com,* under *Sociological Footprints,* by Cargan and Ballantine. You can respond to the discussion questions there or enter your own comments in the online chat forum.

SUGGESTED READINGS AND
SOCIOLOGY INTERNET RESOURCES

See the Wadsworth Sociology Resource Center, "Virtual Society," *http://sociology.wadsworth. com,* for additional links, suggestions for further reading, and learning tools related to this chapter.

Either from the "Virtual Society" website or directly from your web browser, you may access InfoTrac College Edition, an online university library that includes over 700 popular and scholarly journals in which you can find articles related to the topics in this chapter.

CHAPTER 9

Economics

NECESSITIES FOR SURVIVAL

EACH SOCIETY MUST HAVE A SYSTEM for the production and distribution of goods and services. The *economic* institution fulfills these functions. The economic institution is supported by and interrelated with all the other social institutions, but it is particularly related to the political institution. Consider three influential economic systems: feudalism, colonialism, and capitalism. The strong central government of ancient Rome came to an end with the fall of the Roman Empire. One result was the end of the central government's function as protector of monetary surety and safety and the beginning of a period in which people had to return to the land for survival and to submit to marauding ex-soldiers for protection. With the confiscation of the land and the election of a strong individual from their ranks as head-king, a feudal system of government became dominant. Later, under the colonial system, workers were needed for the large plantations and mines. However, the natives saw no reason to work for industry, since they had their own land for food and shelter. A political solution was found to meet the needs of this economic system: The best native lands were put into reserve, natives were required to pay in cash, and people arrested for various reasons were assigned to work on plantations and in mines. Under capitalism, economic organizations such as the corporation have become so powerful that its main operating thesis of laissez-faire (a free market) cannot, in most situations, operate without the need for government becoming more and more involved in the economic sphere. In addition, measures have been implemented to increase the purchasing power of the poor and the unemployed, to subsidize and regulate corporations. . . .

The theme of this chapter is an examination of the continuing changes occurring within capitalism. The first of those changes is the removal of savings and the wide use of credit cards accompanied by an easy spending philosophy, main ingredients in the changing economic system. In the first reading of Chapter 9, Ritzer notes the implications of this change.

Another change in capitalism is occurring due to changes in the labor market. The authors of the second reading believe that vast changes will have to be made if, as indicated, work is not central in our lives.

In the third reading, Joan Acker carries this theme of structural changes even further. She notes that the various structural trends in the economy and the workplace have not been as beneficial as some have claimed, especially for women. Acker indicates what steps are needed to improve womens' futures.

Wilson, in the final reading, carries this idea of a changing labor market to the next step. What can be done to lessen its negative impact on those most affected?

The message of this chapter is that people apparently accept individual creeds as the main cause of people's personal troubles but that this is a rather simplistic explanation for a complicated issue—an issue affected by such structural factors as the change to a postindustrial economy, political authorization of oligarchical control of major facets of the economy, and the need for having poor people perform various economic and social functions.

As you read this chapter, think about some current economic issues and how they might be addressed. Consider whether reforms proposed in the articles or your own ideas are possible in the current political climate. That is, what can be done about those public issues such as wage stagnation, unemployment, oligopoly, and the budget deficits of both the individual and the government, and how will these economic factors and their resolution be affected by other social institutions?

34

The Credit Card:
Private Troubles and Public Issues

GEORGE RITZER

The credit card is seen as a personal part of the individual's economy. Ritzer reveals that anything this ubiquitous has many ramifications for the society in general.

As you read, ask yourself the following questions:

1. According to the author, what are some good and bad things that can be said about credit cards?

2. How does your personal experience tie in with the author's claims about credit cards?

GLOSSARY

Machinations Crafty schemes.

THE CREDIT CARD HAS become an American icon. It is treasured, even worshipped, in the United States and, increasingly, throughout the rest of the world. The title of this book, *Expressing America,* therefore has a double meaning: The credit card expresses something about the essence of modern American society and, like an express train, is speeding across the world's landscape delivering American (and more generally consumer) culture. . . .

The credit card is not the first symbol of American culture to play such a role, nor will it be the last. Other important contemporary American icons include Coca-Cola, Levi's, Marlboro, Disney, and McDonald's. What they have in common is that, like credit cards, they are products at the very heart of American society, and they are highly valued by, and have had a profound effect on, many other societies throughout the world. However, the credit card is distinctive because it is a means that can be used to obtain those other icons, as well as virtually anything else available in the world's marketplaces. It is because of this greater versatility that the credit card may prove to be the most important American icon of all. If nothing else, it is likely to continue to exist long after other icons have become footnotes in the history of American culture. When the United States has an entirely new set of icons, the credit card will remain an important means for obtaining them. . . .

The Advantages of Credit Cards

. . . The most notable advantage of credit cards, at least at the societal level, is that they permit people to spend more than they have. Credit cards thereby allow the economy to function at a much higher (and faster) level than it might if it relied solely on cash and cash-based instruments.

Credit cards also have a number of specific advantages to consumers, especially in comparison to using cash for transactions:

- Credit cards increase our spending power, thereby allowing us to enjoy a more expansive, even luxurious lifestyle.
- Credit cards save us money by permitting us to take advantage of sales, something that might not be possible if we had to rely on cash on hand.
- Credit cards are convenient. They can be used 24 hours a day to charge expenditures by phone, mail, or home computer. Thus, we need no longer be inconvenienced by

In Expressing America. *Pine Forge Press, 1995.*

the fact that most shops and malls close overnight. Those whose mobility is limited or who are housebound can also still shop.

■ Credit cards can be used virtually anywhere in the world, whereas cash (and certainly checks) cannot so easily cross national borders. For example, we are able to travel from Paris to Rome on the spur of the moment in the middle of the night without worrying about whether we have, or will be able to obtain on arrival, Italian lire.

■ Credit cards smooth out consumption by allowing us to make purchases even when our incomes are low. If we happen to be laid off, we can continue to live the same lifestyle, at least for a time, with the anticipation that we will pay off our credit card balances when we are called back to work. We can make emergency purchases (of medicine, for example) even though we may have no cash on hand.

■ Credit cards allow us to do a better job of organizing our finances, because we are provided each month with a clear accounting of expenditures and of money due.

■ Credit cards may yield itemized invoices of tax-deductible expenses, giving us systematic records at tax time.

■ Credit cards allow us to refuse to pay a disputed bill while the credit card company investigates the transaction. Credit card receipts also help us in disputes with merchants.

■ Credit cards give us the option of paying our bills all at once or of stretching payments out over a length of time.

■ Credit cards are safer to carry than cash is and thus help to reduce cash-based crime. . . .

A Key Problem with Credit Cards

In the course of the twentieth century, the United States has gone from a nation that cherished savings to one that reveres spending, even spending beyond one's means. . . .

At the level of the national government, our addiction to spending is manifest in a once-unimaginable level of national debt, the enormous growth rate of that debt, and the widespread belief that the national debt cannot be significantly reduced, let alone eliminated. As a percentage of gross national product (GNP),* the federal debt declined rather steadily after World War II, reaching 33.3% in 1981. However, it then rose dramatically, reaching almost 73% of GNP in 1992. In dollar terms, the federal debt was just under $1 trillion in 1981, but by September 1993, it had more than quadrupled, to over $4.4 trillion. There is widespread fear that a huge and growing federal debt may bankrupt the nation and a near consensus that it will adversely affect future generations.

Our addiction to spending is also apparent among the aggregate of American citizens. Total personal savings was less in 1991 than in 1984, in spite of the fact that the population was much larger in 1991. Savings fell again in the early 1990s from about 5.2% of disposable income in late 1992 to approximately 4% in early 1994. A far smaller percentage of families (43.5%) had savings accounts in 1989 than did in 1983 (61.7%). And the citizens of many other nations have a far higher savings rate. At the same time, our indebtedness to banks, mortgage companies, credit card firms, and so on is increasing far more dramatically than similar indebtedness in other nations. . . .

Who Is to Blame?

THE INDIVIDUAL

In a society that is inclined to "psychologize" all problems, we are likely to blame individuals for not saving enough, for spending too much, and for not putting sufficient pressure on officials to restrain government expenditures. We also tend to "medicalize" these problems, blaming them

---

*While the term *GNP* is still used for historical purposes, it should be noted that the term *GDP* (gross domestic product) is now preferred.

on conditions that are thought to exist within the individual. . . . Although there are elements of truth to psychologistic and medicalistic perspectives, there is also a strong element of what sociologists term "blaming the victim." That is, although individuals bear some of the responsibility for not saving, for accumulating mounting debt, and for permitting their elected officials to spend far more than the government takes in, in the main individuals have been victimized by a social and financial system that discourages saving and encourages indebtedness.

Why are we so inclined to psychologize and medicalize problems like indebtedness? For one thing, American culture strongly emphasizes individualism. We tend to trace both success and failure to individual efforts, not larger social conditions. For another, large social and financial systems expend a great deal of time, energy, and money seeking, often successfully, to convince us that they are not responsible for society's problems. Individuals lack the ability and the resources to similarly "pass the buck." Of perhaps greatest importance, however, is the fact that individual, especially medical, problems appear to be amenable to treatment and even seem curable. In contrast, large-scale social problems (pollution, for example) seem far more intractable. It is for these, as well as many other reasons, that American society has a strong tendency to blame individuals for social problems.

THE GOVERNMENT

. . . Since the federal debt binge began in 1981, the government has also been responsible for creating a climate in which financial imprudence seems acceptable. After all, the public is led to feel, if it is acceptable for the government to live beyond its means, why can't individual citizens do the same? If the government can seemingly go on borrowing without facing the consequences of its debt, why can't individuals?

If the federal government truly wanted to address society's problems, it could clearly do far more both to encourage individual savings and to discourage individual debt. For example, the gov-

ernment could lower the taxes on income from savings accounts or even make such income tax-free. Or it could levy higher taxes on organizations and agencies that encourage individual indebtedness. The government could also do more to control and restrain the debt-creating and debt-increasing activities of the credit card industry.

BUSINESS

Although some of the blame for society's debt and savings problem must be placed on the federal government, the bulk of the responsibility belongs with those organizations and agencies associated with our consumer society that do all they can to get people to spend not only all of their income but also to plunge into debt in as many ways, and as deeply, as possible. We can begin with American business.

Those in manufacturing, retailing, advertising, and marketing (among others) devote their working hours and a large portion of their energies to figuring out ways of getting people to buy things. . . . One example is the dramatic proliferation of seductive catalogs that are mailed to our homes. Another is the advent and remarkable growth in popularity of the television home shopping networks. What these two developments have in common is their ability to allow us to spend our money quickly and efficiently without ever leaving our homes. Because the credit card is the preferred way to pay for goods purchased through these outlets, catalogs and home shopping networks also help us increase our level of indebtedness.

BANKS AND OTHER
FINANCIAL INSTITUTIONS

The historical mission of banks was to encourage savings and discourage debt. Today, however, banks and other financial institutions lead us away from savings and in the direction of debt. Saving is discouraged by, most importantly, the low interest rates paid by banks. It seems foolish to people to put their money in the bank at an interest rate of, say, 2.5% and then to pay taxes on the interest, thereby lowering the real rate of return to 2% or even less. This practice seems especially asinine when the inflation rate is, for example, 3% or 4%. Under such conditions, the saver's money is declining in value with each passing year. It seems obvious to most people that they are better off spending the money before it has a chance to lose any more value.

While banks are discouraging savings, they are in various ways encouraging debt. One good example is the high level of competition among the banks (and other financial institutions) to offer home equity lines of credit to consumers. As the name suggests, such lines of credit allow people to borrow against the equity they have built up in their homes. . . . Banks eagerly lend people money against this equity. Leaving the equity in the house is a kind of savings that appreciates with the value of the real estate, but borrowing against it allows people to buy more goods and services. In the process, however, they acquire a large new debt. And the house itself could be lost if one is unable to pay either the original mortgage or the home equity loan.

The credit card is yet another invention of the banks and other financial institutions to get people to save less and spend more. In the past, only the relatively well-to-do were able to get credit, and getting credit was a very cumbersome process (involving letters of credit, for example). Credit cards democratized credit, making it possible for the masses to obtain at least a minimal amount. Credit cards are also far easier to use than predecessors like letters of credit. Credit cards may thus be seen as convenient mechanisms whereby banks and other financial institutions can lend large numbers of people what collectively amounts to an enormous amount of money.

Normally, no collateral is needed to apply for a credit card. The money advanced by the credit card firms can be seen as borrowing against future earnings. However, because there is no collateral in the conventional sense, the credit card companies usually feel free to charge usurious interest rates.

Credit cards certainly allow the people who hold them to spend more than they otherwise would. . . . Some . . . overwhelmed by credit card debt, take out home equity lines of credit to pay it off. Then, with a clean slate, at least in the eyes of the credit card companies, such people are ready to begin charging again on their credit cards. Very soon many of them find themselves deeply in debt both to the bank that holds the home equity loan and to the credit card companies.

A representative of the credit card industry might say that no one forces people to take out home equity lines of credit or to obtain credit cards; people do so of their own volition and therefore are responsible for their financial predicament. Although this is certainly true at one level, at another level it is possible to view people as the victims of a financial (and economic) system that depends on them to go deeply into debt and itself grows wealthy as a result of that indebtedness. The newspapers, magazines, and broadcast media are full of advertisements offering various inducements to apply for a particular credit card or home equity loan rather than the ones offered by competitors. Many people are bombarded with mail offering all sorts of attractive benefits to those who sign up for yet another card or loan. More generally, one is made to feel foolish, even out of step, if one refuses to be an active part of the debtor society. Furthermore, it has become increasingly difficult to function in our society without a credit card. For example, people who do not have a record of credit card debt and payment find it difficult to get other kinds of credit, like home equity loans, car loans, or even mortgage loans.

An Indictment of the Financial System

The major blame for our society's lack of savings and our increasing indebtedness must be placed on the doorstep of large institutions. This book focuses on one of those institutions—the financial system, which is responsible for making credit card debt so easy and attractive that many of us have become deeply and perpetually indebted to the credit card firms. . . .

Case in Point: Getting Them Hooked While They're Young

Before moving on to a more specific discussion of the sociological perspective on the problems associated with credit cards, one more example of the way the credit card industry has created problems for people would be useful: the increasing effort by credit card firms to lure students into possessing their own credit cards. The over 9 million college students (of which 5.6 million are in school on a full-time basis) represent a huge and lucrative market for credit card companies. According to one estimate, about 82% of full-time college students now have credit cards. The number of undergraduates with credit cards increased by 37% between 1988 and 1990. The credit card companies have been aggressively targeting this population not only because of the immediate increase in business it offers but also because of the long-term income possibilities as the students move on to full-time jobs after graduation. To recruit college students, credit card firms are advertising heavily on campus, using on-campus booths to make their case and even hiring students to lure their peers into the credit card world. In addition, students have been offered a variety of inducements. I have in front of me a flyer aimed at a college-age audience. It proclaims that the cards have no annual fee, offer a comparatively low interest rate, and offer "special student benefits," including a 20% discount at retailers like MusicLand and Gold's Gym and a 5% discount on travel.

The credit card firms claim that the cards help teach students to be responsible with money (one professor calls it a "training-wheels operation"). The critics claim that the cards teach students to spend, often beyond their means, instead of saving. . . .

In running up credit card debt, it can be argued, college students are learning to live a lie. They are living at a level that they cannot afford at the time or perhaps even in the future. They may establish a pattern of consistently living beyond their means. However, they are merely postponing the day when they have to pay their debts. . . .

Not satisfied with the invasion of college campuses, credit card companies have been devoting increasing attention to high schools. One survey found that as of 1993, 32% of the country's high school students had their own credit cards and others had access to an adult's card. Strong efforts are underway to greatly increase that percentage. The president of a marketing firm noted, "It used to be that college was the big free-for-all for new customers. . . . But now, the big push is to get them between 16 and 18." Although adult approval is required for a person under 18 years of age to obtain a credit card, card companies have been pushing more aggressively to gain greater acceptance in this age group. . . .

The motivation behind all these programs is the industry view that about two-thirds of all people remain loyal to their first brand of card for 15 or more years. Thus the credit card companies are trying to get high school and college students accustomed to using their card instead of a competitor's. The larger fear is that the credit card companies are getting young people accustomed to buying on credit, thereby creating a whole new generation of debtors.

A Sociology of Credit Cards

. . . Sociologists have grown increasingly dissatisfied with having to choose between large-scale, macroscopic theories and small-scale, microscopic theories. Thus, there has been a growing interest in theories that integrate micro and macro concerns. In Europe, expanding interest in what is known there as agency-structure integration parallels the increasing American preoccupation with micro-macro integration.

MILLS: PERSONAL TROUBLES, PUBLIC ISSUES

Of more direct importance here is the now-famous distinction made by Mills in his 1959 work, *The Sociological Imagination,* between micro-level personal troubles and macro-level public issues. Personal troubles tend to be problems that affect an individual and those immediately around him or her. For example, a father who commits incest with his daughter is creating personal troubles for the daughter, other members of the family, and perhaps himself. However, that single father's actions are not going to create a public issue; that is, they are not likely to lead to a public outcry that society ought to abandon the family as a social institution. Public issues, in comparison, tend to be problems that affect large numbers of people and perhaps society as a whole. The disintegration of the nuclear family would be such a public issue. . . .

A useful parallel can be drawn between the credit card and cigarette industries. The practices of the cigarette industry create a variety of personal troubles, especially illness and early death. Furthermore, those practices have created a number of public issues (the cost to society of death and illness traceable to cigarette smoke), and thus many people have come to see cigarette industry practices themselves as public issues. Examples of industry practices that have become public issues are the aggressive marketing of cigarettes overseas . . . Similarly, the practices of the credit card industry help to create personal problems (such as indebtedness) and public issues (such as the relatively low national savings rate). Furthermore, some industry practices—such as the aggressive marketing of credit cards to teenagers—have themselves become public issues.

One of the premises of this book is that we need to begin adopting the same kind of critical outlook toward the credit card industry that we use in scrutinizing the cigarette industry. . . .

Mills's ideas give us remarkably contemporary theoretical tools for undertaking a critical analysis of the credit card industry and the problems it generates. . . .

MARX: CAPITALIST EXPLOITATION

In addition to Mills's general approach, there is the work of the German social theorist Karl Marx (1818–1881), especially his ideas on the exploitation that he saw as endemic to capitalist society. . . .

There have been many changes in the capitalist system, and a variety of issues have come to the fore that did not exist in Marx's day. As a result, a variety of neo-Marxian theories have arisen to deal with these capitalist realities. One that concerns us here is the increasing importance to capitalists of the market for goods and services. According to neo-Marxians, exploitation of the worker continues in the labor market, but capitalists also devote increasing attention to getting consumers to buy more goods and services. Higher profits can come from both cutting costs and selling more products.

The credit card industry plays a role by encouraging consumers to spend more money, in many cases far beyond their available cash, on the capitalists' goods and services. In a sense, the credit card companies have helped the capitalists to exploit consumers. Indeed, one could argue that modern capitalism has come to depend on a high level of consumer indebtedness. Capitalism could have progressed only so far by extracting cash from the consumers. It had to find a way to go further. . . .

SIMMEL: THE MONEY ECONOMY

. . . Simmel pointed to many problems associated with a money economy, but three are of special concern in this book:

- The first problem . . . is the "temptation to imprudence" associated with a money economy. Simmel argued that money, in comparison to its predecessors, such as barter, tends to tempt people into spending more and going into debt. My view is that credit cards are even more likely than money to make people imprudent. People using credit cards are not only likely to spend more but are also more likely to go deeply into debt. . . .

- Second, Simmel believed that money makes possible many types of "mean machinations" that were not possible, or were more difficult, in earlier economies. For example, bribes for political influence or payments for assassinations are more easily made with money than with barter. . . . Although bribes or assassinations are generally less likely to be paid for with a credit card than with cash, other types of mean machinations become more likely with credit cards. For example, some organizations associated with the credit card industry engage in fraudulent or deceptive practices in order to maximize their income from credit card users. . . .

- The third problem with a money economy that concerned Simmel was the issue of secrecy, especially the fact that a money economy makes payments of bribes and other types of secret transactions more possible. However, our main concern in this book is the increasing lack of secrecy and the invasion of privacy associated with the growth of the credit card industry. . . .

WEBER: RATIONALIZATION

. . . Weber defined rationalization as the process by which the modern world has come to be increasingly dominated by structures devoted to efficiency, predictability, calculability, and technological control. Those rational structures (for example, the capitalist marketplace and the bureaucracy) have had a progressively negative effect on individuals. Weber described a process by which more and more of us would come to be locked in an "iron cage of rationalization." . . . The credit card industry has also been an integral part of the rationalization process. By rationalizing the process by which consumer loans are made, the credit card industry has contributed to our society's dehumanization. . . .

GLOBALIZATION AND AMERICANIZATION

A sociology of credit cards requires a look at the relationship among the credit card industry, personal troubles, and public issues on a global scale. It is not just the United States, but also much of the rest of the world, that is being affected by the credit card industry and the social

problems it helps create. To some degree, this development is a result of globalization, a process that is at least partially autonomous of any single nation and that involves the reciprocal impact of many economies. In the main, however, American credit card companies dominate the global market. . . .

The central point is that, in many countries around the world, Americanization is a public issue that is causing personal troubles for their citizens. This book addresses the role played by the credit card industry in this process of Americanization and in the homogenization of life around the world, with the attendant loss of cultural and individual differences. . . .

Other Reasons for Devoting a Book to Credit Cards

SOMETHING NEW IN THE HISTORY OF MONEY

Money in all its forms, especially (given the interest of this book) in its cash form, is part of a historical process. It may seem hard to believe from today's vantage point, but at one time there was no money. Furthermore, some predict that there will come a time in which money, at least in the form of currency, will become less important if not disappear altogether, with the emergence of a "cashless society." . . .

More important for our purposes, money in the form of currency is being increasingly supplanted by the credit card. Instead of plunking down cash or even writing a check, more of us are saying "Charge it!" This apparently modest act is, in fact, a truly revolutionary development in the history of money. Furthermore, it is having a revolutionary impact on the nature of consumption, the economy, and the social world more generally. In fact, rather than simply being yet another step in the development of money, I am inclined to agree with the contention that credit cards are "an entirely new idea in value exchange." A variety of arguments can be marshaled in support of the idea that in credit cards we are seeing something entirely new in the history of economic exchange, especially relative to cash:

- Credit card companies are performing a function formerly limited to the federal government. That is, they create money . . . the Federal Reserve is no longer alone in this ability. The issuing of a new credit card with a $1,000 limit can be seen as creating $1,000. Thus, the credit industry is creating many billions of dollars each year and, among other things, creating inflationary pressures in the process.
- Credit cards to not have a cash or currency form. In fact, they are not even backed by money until a charge is actually made.
- With cash we are restricted to the amount on hand or in the bank, but with credit cards our ceiling is less clear. We are restricted only by the ever-changing limits of each of our credit cards as well as by the aggregate of the limits of all those cards.
- Although we can use our cash anytime we wish, the use of our credit card requires the authorization of another party.
- Unlike cash, which allows for total anonymity, one's name is printed on the front of the credit card and written on the back; a credit card may even have one's picture on it. Furthermore, credit card companies have a great deal of computerized information on us that is drawn on to approve transactions.
- Although cash is simple to produce and use over and over, credit cards require the backing of a complex, huge, and growing web of technologies.
- There is no direct cost to the consumer for using cash, but fees and interest may well accrue with credit card use.
- Although everyone, at least theoretically, has access to cash, some groups (the poor, the homeless, the unemployed) may be denied access to credit cards. Such restrictions sometimes occur unethically or illegally through the "redlining" of certain types of consumers or geographic areas.

- Because of their accordionlike limits, credit cards are more likely than cash to lead to consumerism, overspending, and indebtedness. . . .

A GROWING INDUSTRY

Another reason for focusing on credit cards is their astounding growth in recent years, which reflects their increasing importance in the social and economic worlds. There are now more than a billion credit cards of all types in the United States. Receivables for the industry as a whole in 1993 were up by almost 16% from the preceding year and by over 400% in a decade. The staggering proliferation of credit cards is also reflected in other indicators of use in the United States:

- Sixty-one percent of Americans now have at least one credit card.
- The average cardholder carries nine different cards.
- In 1992, consumers used the cards to make 5 billion transactions, with a total value of $420 billion.

There has been, among other things, growth in the number of people who have credit cards, the average number of credit cards held by each person, the amount of consumer debt attributable to credit card purchases, the number of facilities accepting credit cards, and the number of organizations issuing cards. . . . The average outstanding balance owed to Visa and MasterCard increased from less than $400 in the early 1980s to $970 in 1989 and to $1,096 in 1993. The amount of high-interest credit card debt owed by American consumers rose from $2.7 billion in 1969 to $50 billion in 1980 and was approaching $300 billion in 1994. . . .

A SYMBOL OF AMERICAN VALUES

A strong case can be made that the credit card is one of the leading symbols of 20th-century America or, as mentioned earlier, that the credit card is an American icon. Indeed, one observer calls the credit card "the twentieth century's symbol par excellence." Among other things, the credit card is emblematic of affluence, mobility, and the capacity to overcome obstacles in the pursuit of one's goals. Thus, those hundreds of millions of people who carry credit cards are also carrying with them these important symbols. And when they use a credit card, they are turning the symbols into material reality. . . .

Debunking Credit Card Myths

To most of us, credit cards appear to have near-magical powers, giving us greater access to a cornucopia of goods and services. They also seem to give us something for nothing. That is, without laying out any cash, we can leave the mall with an armload of purchases. Most of us like what we can acquire with credit cards, but some like credit cards so much that they accumulate as many as they can. Lots of credit cards, with higher

and higher spending limits, are important symbols of success. That most people adopt a highly positive view of credit cards is borne out by the proliferation of the cards throughout the United States and the world. . . .

A debunking sociology is aimed at revealing the spuriousness of various ideologies. As Berger puts it, "In such analyses the ideas by which men explain their actions are unmasked as self-deception, sales talk, the kind of sincerity . . . of a man who habitually believes his own propaganda." From this perspective, the credit card companies can be seen as purveyors of self-deceptive ideologies. They are, after all, in the business of selling their wares to the public, and they will say whatever is necessary to accomplish their goal. . . .

35
The Jobless Future?

STANLEY ARANOWITZ AND WILLIAM DIFAZIO

Headlines tell the story of mergers, transfers of production overseas, and the impact of technology on jobs. The heading is followed by a question mark — perhaps not a jobless future, but what about this lesser need for workers?

As you read, ask yourself the following questions:

1. What are some of the factors affecting the future of work?
2. How have the changes affecting the future of work affected you?

GLOSSARY

Scientific-technological revolution Transformation of the workplace from focus on agricultural to industrial to information/technological innovation.
"Workfare" Alternative program to welfare requiring many recipients of government funds to hold jobs.

Is Work a Need?

"MEN LIKE TO WORK. It's a funny thing, but they do. They may moan about it every Monday

In The Jobless Future: Sci-Tech and the Dogma of Work (*University of Minnesota Press, 1994*) *pp. 328–358—excerpts.*

morning and they may agitate for shorter hours and longer holidays, but they need to work for their self-respect."[1] . . .

It is not merely the comparative economic advantage of paid labor over an increasingly inferior "dole" that motivates "men" to take jobs. Nor does the meaning of work derive from its intrinsic interest; in principle, technology eliminates the workers' skills and, finally, the workers themselves. Instead, we are driven by the fact that the "self" is constituted, at least for most of us, by membership in the labor force, as a member of either the job bourgeoisie — the "professions" — or the working class.

Thus paid work is . . . a socially and psychologically constructed "need" shared by those who have been successfully habituated to think that the link between holding a job and having "dignity" is a given. Put bluntly, in this view the self is identical to its place in the paid labor force. No job, no (secure) self. . . . better to take pride in the fact that, as workers, we are able to provide for self and family without state aid or charity.

Even when one-third of the U.S. labor force was officially unemployed throughout the 1930s, and many workers were on short-time schedules, they still blamed themselves for their joblessness. There was no dignity for those who could not find jobs; the conventional wisdom, shaken for more than a decade but not displaced, was that there was "always" plenty of work for those who wanted it. . . . Individuals, not the economic and social system, are ultimately responsible for their fate; the market adjusts itself at a level approximating full employment, and any joblessness is "frictional" — that is, temporary — for responsible and able-bodied individuals. This key precept of the dominant ideology resumed its virtually uncontested hegemony after World War II, when official statistics recorded jobless rates of less than 6 percent until the early 1980s.

There are, of course, exceptions to the universal principle of paid labor as the sole path to male (and, increasingly, female) dignity, but these turn out to be only variations on the theme that work is a "need." One may retain "dignity"

if income has been "earned" through past usury or ownership of business. Unwork becomes dignified only if income is derived from retirement or disability. The implicit assumption is that the retired and the disabled would have remained in the paid labor force if they were able-bodied or younger. Retirement is still considered a reward for a lifetime of faithful paid work, although some research has contended that relatively few retirees in the United States prosper unless they have income acquired through labor or property in addition to their Social Security benefits. From the standpoint of the conventional ethic, paid labor is considered optional for women.

Inherited income is ambiguous. Even when heirs do not need to take paid work in order to live, it is always implied that they should be subject to the work ethic. Even in higher circles the "playboy" and "playgirl" are morally condemned; heirs who live on income that derives from trust funds and other repositories of the past labor of others but are unwilling to engage in socially useful activities face censure from peers. The socially responsible heir seeks redemption through performing good "works" in charities or other civic activities or may become a patron of the arts, science, or other types of knowledge. . . .

The Decline of "Work"

. . . With the decisive passing of craft, except in the crevices of the modern labor system, the main value of having a job (besides its economic function for individuals and households) is that it once provided a "community," which for many men replaced the traditional agricultural family. With the partial breakdown of the urban family, many women found the workplace a source of social solidarity as well. Moreover, contrary to popular depiction of craft labor as intrinsically satisfying in comparison to stupefying mass production, historical and ethnographhic evidence demonstrates that skilled workers were no less eager to be liberated from the workshop than assemblers and laborers . . . for it was not the object that provided satisfaction for the craftspersons, but the opportunity for a richer human interaction. This benefit of collective labor was not generally available to assembly-line and other less-qualified laborers. Production workers found after-hours conviviality in bars and social clubs.

But the culture of the factory and the large office is dying; for most workers, even those classified as "skilled," the old bonds are considerably loosened, even when they have not completely disappeared, for a very good reason: most craftspersons—in construction, in factories, car and instrument mechanics—know that the division of labor and the computer rationalize craft nearly as much as they do manual labor. Most auto and instrument mechanics, for example, rely on computer-mediated electronic tests for diagnostic purposes and work with parts that are engineered as modules. When Vic tells Robyn that the men like to work, what he forgets to mention is the reason for this affection. It certainly does not lie in work "satisfaction," that ambiguous sociological and psychological category that might issue from the substantial aspects of the task, but in the reality of the shop as a "dwelling place," a home that has little to do with the end product of their collective labors.

Achieving a home in the traditional industrial culture did not entail crafting a reified object that represented suprabiological dignity. . . . The markers of these bonds are less "tools," instruments of fabrication, than a shared discursive universe replete with rituals, linguistic codes, jokes, and worldviews—in a word, a culture. The shop or office may be regarded as a universe that visits exhaustion and frustration upon inhabitants but provides, at least for some workers, an irreplaceable network of relationships and, taken in its multiple significations, a discourse, which together constitute the class culture of the factory.

Contrary to the ideologically conditioned theory shared by sociologists, psychologists, and policy analysts that "nonwork" produces, and is produced by, social disorganization and is symbolic of irresponsibility and personal dysfunctionality, recipients of guaranteed annual income who are relieved of most obligations to engage in labor do not fall apart. The incidence of alcoholism, divorce, and other social ills associated with conditions of dysfunctionality does not increase among men who are not working. Nor do they tend to experience higher rates of mortality than those of comparable age who are engaged in full-time work. Given the opportunity to engage in active nonwork, they choose this option virtually every time.[2] . . .

Under current economic and social conditions, the major casualties of technological changes on the waterfront and, increasingly, in the auto, electronic, and communications industries, are the children and grandchildren who will never have the chance to work on the docks or in the factories and accumulate enough time to achieve dignified nonwork. The time of the new generations of never-to-be industrial workers is not free even though they are relieved of paid labor. Instead, it is suffused with anxiety that they may never again enter the cycle of labor and consumption that defined working lives in the Fordist era. . . .

The remaining repositories of "work" within the wage-labor system are, despite the ruthless transformations of virtually all of its products and producers into commodities, the diminishing instances of petty craft production (occupations that are frequently suffused with the uncertainties connected with self-employment), art, and the products of the relatively small proportion of those who produce knowledge of all kinds, even those, like teachers and writers, who through transmission (or translation) re-produce knowledge. This work retains its objectivity, depending on neither a knowing subject nor the immediacy of the labor process. . . .

The North American Free Trade Agreement (NAFTA) may be viewed as merely the conclusion of the first chapter in a long process of overcoming some aspects of the traditional unequal division of labor between north and south. In the next decade, U.S. wages and living standards are likely to continue to deteriorate. If labor organization emerges in Mexico and other parts of Latin America, wages there will rise, but not by enough to deter migration of U.S. plants, at least during the 1990s. In the near future, Texans and

Californians will cross the border in greater numbers every day to work in Mexico and Mexican workers will continue to migrate to certain jobs in the United States, approximating the situation at the already blurred U.S.-Canadian border. At the same time, as in Canada, Mexican industry is increasingly subject to U.S. investment; this will set a pattern for transnational investment in other countries of Latin America, particularly Brazil, which, along with Mexico, had before the current economic crisis succeeded to some extent in developing its own industrial base.

Deterritorialized production applies also to knowledge. By the early 1990s, for example, China and India were offering U.S., Japanese, and European capital access to highly qualified scientific and technical labor. U.S. computer corporations began to let contracts to software corporations in India. Du Pont and other chemical corporations were building petro-chemical complexes in Shanghai, employing Chinese engineers and chemists at eighty dollars a month. The fairly well developed Mexican bioengineering sector is actively negotiating with U.S. corporations to "share" discoveries and technical achievements. . . .

Just as the scientific-technological revolution has utterly transformed the workplace in all categories of labor, we are obliged to examine its consequences for the conception of work that undergirds cultural identity, the self, and our collective understanding of the norms by which the moral order imposes a mode of conduct upon us. . . .

On June 30, 1993, the *New York Times* reported that U.S. companies are cutting funds for scientific and technological research:

> Scientific research by private industry, the traditional powerhouse of innovations and technological leadership in the United States, is suffering deeper financial woes than previously disclosed, suggesting that America is slipping in the international race for discoveries that form the basis of new goods and services. The National Science Foundation reported in February that industrial research on research and development had begun to shrink after decades of growth.[3]

Of course, much of the previous growth was military, and was therefore driven by and dependent on public funds. But with recession, the tapering off of the cold war, and the enormous deficits accumulated by government and by corporations caught up in the swirl of the leveraged buyout mania of the 1980s, funds have dried up. . . . More to the point, the priorities of the federal scientific and technical bureaucracies, which are increasingly tied to the requirements of corporations, have restricted the *kind* of research they are willing to support. . . . In short, the commodification of basic science, combined with its increasingly technical character and declining funds, may in the future all but seal the fate of the United States as a major economic power.

For the plain truth is that overfunding and "useless" knowledge is the key to discovery. From the discoveries of Galileo to the "idle" ruminations of Frege, Gödel, Einstein, and Bohr, patronage, whether public or private, permitted unbounded dreaming that led to new ways of seeing and ultimately—but only ultimately—new modes of producing. When government and corporate policy makers insist on "dedicated" research as a condition of support, they announce that they have opted for failure rather than long-range innovation. This blatant act of research shooting itself in the foot is by no means intentional. Rather, it is the result of the logic of technoscience and the human capital paradigm according to which unsubordinated knowledge is perceived to threaten the social order either by draining economic resources or by proposing unpalatable jolts to the imagination. . . . For the social sciences and the humanities, cost reductions exacted a steep toll on research. . . .

The crisis in research of course has serious consequences for the U.S. national economy, but it augurs equally badly for hope that intellectual work will be possible for more than a tiny fraction of scientists and artists in the future. Its effects are even more far-reaching. For, in a higher education system already incurring severe criticism for the low number of U.S.-born scientific and technical majors and graduates at the undergraduate and graduate levels, the decline in basic research constitutes a disincentive for young people to enter the sciences. At leading universities, many if not most advanced-level physics, mathematics, and chemistry students are foreign born. . . .

Toward a New Labor Policy

. . . Our proposals are based on the presuppositions of this study: that economic growth grounded in technological innovation does not necessarily increase employment unless there is a sharp reduction in working hours, and even then may not be sufficient to sustain a level approaching full employment; and that since a considerable number of recently created jobs are part-time, poorly paid dead ends, there is a powerful argument that we have reached the moment when less work is entirely justified. In addition, our proposals assume the goal of assuring the *possibility* of the full development of individual and social capacities.

These statements further imply that—if our assertions that the world economy will not sustain full employment in the coming decades and the social safety net will remain full of holes are correct—we need to reconsider the pace of technological change and the effects of corporate reorganizations that have shed tens of thousands of employees in the past several years. Until measures such as a substantial reduction of working hours, a guaranteed income plan, a genuine national health scheme, and the revitalization of the progressive tax system have been introduced into law and union contracts, job-destroying technologies and mergers and acquisitions should be rigorously *evaluated* in terms of their implications for the well-being of communities and workers. In an era of uncontrolled growth amid economic stagnation, corporate efforts to make workers and communities pay the costs of falling profits are exacting heavy tolls and should be stopped.

We have used the general concept "evaluation" to connote the urgent need for social controls, perhaps in the form of re-regulation of business, over untrammeled labor-saving technological change and mergers that result in permanent reductions of labor forces. Needless to say, this proposal directly opposes the dominant strategy of U.S., European, and Japanese corporations and would assume a political situation in which national states were independent of these corporate interests. Unfortunately, for the most part this is the case among neither conservative nor social-democratic and social-liberal regimes. Free enterprise and free market ideology enjoy global hegemony in the current political and economic environment. Thus, even the suggestion that technological change and mergers may not be in the public interest flies in the face of the prevailing common sense. . . .

We wish to point out, however, that deregulation has been most consistently applied to corporate prerogatives: to reduce labor in production; to relocate, at will, factories and professional services; to eliminate workers through consolidation; to put labor in competition with itself by breaking and otherwise reducing the traditional protections provided by union contracts for decent wages and against working conditions that threaten health and safety; and to weaken employer- or government-financed health and pension benefits.

When it comes to regulating the poor, there is no absence of programs: workfare; more prisons for convicted drug dealers and users; armed guards in urban high schools. The largest corporations have never insisted upon competition for government contracts, nor have they hesitated to support tariff and trade restrictions when their particular interests are at stake. Nor have conservatives failed to temper their opposition to open borders to Latin American, Caribbean, and Asian immigrants, demanding draconian measures to regulate their flow. . . .

Finally, social ecology, which has emerged as a major paradigm of social and economic theory as much as it is a significant social movement, has taught us that untrammeled growth is by no means an unmixed blessing. At the most fundamental level, ecological thought is a powerful counterweight to the Western idea of progress.

We have learned that technologically driven growth has had disastrous consequences that can no longer be ignored. Hazardous waste, industrial pollution and its consequent global warming, life-threatening power sources (alternating-current electricity, nuclear energy), and increased radiation resulting from high-powered computer technologies are some of the most visible results of the rapid expansion of industrial and consumer societies.

To be sure, a theoretical model according to which human survival depends on maintaining a sustainable biosphere and stable ecosystems suggests that there may be enormous costs to uncontrolled economic growth, but no consensus has emerged, even among the most insistent critics of uncontrolled growth, concerning possible solutions. At one end of the spectrum are those who warn that unless growth is arrested, even reversed, the ecosystems that sustain life are in imminent jeopardy. This view proceeds from the indisputable fact that "development," one of the cherished names for capital accumulation and urbanization, has exacerbated what are called "natural" disasters—soil erosion, floods, global warming, the cancer epidemic that afflicts nearly a third of all people in industrial societies (by the year 2000, the figure may rise to 40 percent). Cancer, which many biologists argue is directly linked to living and working conditions, is rapidly becoming the major disease of industrial societies. Beyond industrial and commercial pollution, it is linked to the spreading contamination of food and water. This position argues for elimination of entire sectors of industry—especially nuclear energy, many branches of chemical production, the use of most fossil fuels—and conversion of the highly centralized electric power industry to locally based water, wind, and solar energy. Decentralizing power production suggests bioregional economies in which communities produce and distribute their own basic food and many other everyday products. In this economic arrangement, the scale of production is reduced. This regime would not entirely eliminate the division of labor and commodity exchanges, but would limit these activities in order to minimize disturbances to the ecosystem.

At the other end are the proponents of environmental protection through state and voluntary regulation by industrial corporations and developers. This group includes many social liberal governments and their professional retainers, who insist that the political and economic climate is permanently unfavorable to draconian ecological regulations. Growth can be selectively moderated by conservation measures such as creating national parks and wilderness areas, limiting the use of fossil fuels, encouraging industrial and consumer recycling, and requiring business to clean up after itself. . . . Faced with declining profits, corporate capital resists innovations that add to the cost of doing business. Lacking an alternative to jobs, many trade unions and their members have opposed ecological protections. . . . Nor have some unions been willing to insist on strict health protections lest plants be shut down and jobs lost. . . .

In light of the mounting evidence of ecological crisis, the idea that economic policy can no longer fail to incorporate fairly sweeping ecological perspectives seems to us to be incontrovertible. However, if this argument is accepted, we can no longer rely on growth to address problems such as technologically induced unemployment and to improve living standards. Yet, in a remarkable example of failure of political imagination and will, uncontrolled growth remains the basis of world and national efforts to resolve long-term economic woes in nearly every major country. The ideological hegemony of growth economics, combined with the powerful threat of globalization, has virtually eliminated from public debate the characteristic industrial-era imperatives of social justice and equality. . . . In addition, what might be called the spread of *virulent nationalism* in nearly all industrialized countries as well as in Eastern Europe militates against international efforts to deal with ecological disaster. . . .

Given ecological and economic crisis and world economic and political restructuring, there is an urgent need for thinking that refuses to remain mired in the impossibilities of the present. For to insist in advance that possibilities are limited to the givens of the social and political world leads to the conclusion that no genuine transformation is possible, which in turn gives rise to the dark conservatism that holds that change is not desirable, and even is evil. . . .

The Need to Reduce Working Hours

There has been no significant reduction in working hours since the implementation of the eight-hour day through collective bargaining and the 1938 enactment of the federal wage and hour law. Since then, we have witnessed a slow increase of working time despite the most profoundly labor-displacing era of technological change since the industrial revolution. People are laboring their lives away, which, perhaps as much as unemployment and poverty, has resulted in many serious family and health problems. In turn, the lengthening of working hours has contributed to unemployment and poverty among those excluded from the labor system.

Therefore, there is an urgent need for a sharp reduction in the workweek from its current forty hours—a reduction of, *initially,* at least ten hours. The thirty-hour week at *no reduction in pay* would create new jobs only if overtime was eliminated for most categories of labor. And, although some people may prefer flexible working arrangements that are more compatible with child-rearing needs or personal preference, the basic workday should, to begin with, be reduced to six hours, both as a health and safety measure and in order to provide more freedom from labor in everyday life. Finally, we envision a progressive reduction of working hours as technological transformation and the elimination of what might be termed make-work in both private and public employment reduces the amount of labor necessary for the production of goods and services. That is, productivity gains would not necessarily, as in the past, be shared between employers and employees in the form of increased income, but would result first in fewer laboring hours.

Obviously, restricting laboring hours raises some important questions. . . .

The experience of the German labor movement is instructive in this regard. In 1985, the Metalworkers Union (IG Metall), which represents auto, steel, and metal fabricating workers, struck for reduced hours. . . . Employers yielded to the demand for a thirty-five-hour workweek, to be implemented in stages over five years. Gradually, other sectors have adopted the shorter workweek. . . . The competitive position of German industries is not suffering because of this innovation, in part because of the tremendous productivity of German workers made possible by cutting-edge technologies that have been widely introduced in production. Moreover, in countries such as Germany where the social wage includes substantial government-administered health benefits and guaranteed income and pensions, labor costs to employers may be lower than in the United States, which does not have these state-sponsored provisions, even when wages are higher. In the United States, employers have shouldered much of the burden of the welfare state, spending as much as 40 percent of wages on fringe benefits. . . .

The privatization of welfare and antediluvian social policies, especially the lack of national health care and a strong pension system, places onerous burdens on enterprise labor costs to provide these social wages. Of course, to require industries to be "competitive" presumably entails a sharp curtailment of these company-paid benefits. . . . Clearly, U.S. workers and their unions gave back many of their previously won gains in the 1980s, but these concessions failed to reverse capital flight. Although movements of capital, especially from north to south, are often meant to reduce labor costs, there are other factors that motivate such shifts. One of them is historical. . . .

Regulating Capital

In some countries, capital may not freely export jobs without consultation with unions and the government. Clearly, reducing working hours without simultaneously addressing the issue of capital flight is unthinkable. In 1988 the U.S. Congress passed modest plant-closing legislation requiring employers only to notify employees and the community of their plans to close a facility. This law could be strengthened to compel collective bargaining with unions and local governments over the conditions of capital flight, including the extent of compensation and effects on the community. To discourage plant closings, employers could be required to pay substantial compensation to displaced workers and to communities, and they could be required to offer transfer rights to their employees. Unions have sought to protect jobs by persuading Congress to pass the so-called domestic content bill according to which a percentage of the components of commodities (autos and garments, for example) sold in the United States would have to be produced by U.S. workers. This provision has been incorporated into the North American Free Trade Agreement (NAFTA) for some items; it could be extended to become a basis of plant-closing legislation.

The most important issue raised by our proposals is international coordination of labor demands. . . . In the face of global competition, it is nothing short of suicidal for labor to remain in competition with itself. Unless these issues are addressed, discussions of the need for shorter hours can never advance beyond the proposal stage.

The question of living standards strikes at the heart of the cultural dimension of this issue. For millions of Americans, working almost all the time is the only way they can maintain their homes and provide for the care and education of their children. Here we offer three suggestions. First, single-family, privately owned homes should not be the most important source of new housing. Publicly financed, affordable multiple-dwelling rental housing would lift an enormous burden from the shoulders of working people. The value of their homes—whether they were cooperatively owned or rented—would no longer depend on market fluctuations that have in recent years severely reduced equity in millions of homes and, perhaps more egregiously, spurred lender-provoked evictions. Second, we need free,

publicly provided child care services like those in many European countries. Since mortgage payments or rent plus child care absorb as much as 50 percent of the income of many households, they bear on laboring practices. Third, the United States could adopt the European system of treating postsecondary education as a public resource therefore a public expense.

At the same time, we would propose that higher education be a right rather than a privilege reserved for a minority of the population, as it is in most of Europe and the countries of the Americas. . . . In most of the world, all education is paid for by the state, but access to education is severely regulated. . . .

Since the 1960s, U.S. colleges and universities have been more accessible to students than they used to be. Some 50 percent of high school students enter some kind of postsecondary education program; about half of them go to community colleges and technical schools. Dropout rates, however, are enormous, and sometimes as high as 70 percent. Plainly, if the revolution in scientific and technical knowledge has occurred, fairly high levels of educational achievement are now a necessity for larger numbers of young people. Just as secondary education became a right at the beginning of the twentieth century, so higher education must become a right at the turn of the twenty-first.

Paradoxically, just at the historical juncture when knowledge work becomes more important, U.S. colleges and universities have entered into a period of downsizing due to budget constraints. Public universities have been especially affected by massive cutbacks. . . .

We are arguing that the only chance to maintain and advance our living standards is by means of a bold, intelligent reassertion of the values of more equality and more high-quality public services as the basis of social policy. The fifteen-year bipartisan experiment in deregulation has failed, miserably, to reach any of its major goals, except that of lining the pockets of a small cabal of business interests. In the wake of deregulation many small businesses have failed, workers have lost their jobs, and services have deteriorated.

Moreover, the private sector has failed to provide moderate-cost housing, day care, and education while the public sector has been ruthlessly gutted in an orgy of cost-cutting measures.

There Is Still Work to Do

Despite labor-saving technologies, there is still much work to do. Our roads, bridges, water systems, schools, and cities need rebuilding and repair. We need a mass-transit system to counteract traffic jams and the deleterious effects of auto and truck emissions. With a growing population, we require more garbage collection, cleaner streets, and refurbished parks, forests, and other public spaces. And, as always, there is an urgent need to reclaim "wilderness" areas that have been subjected to industrial and real-estate development. Surely we require a new, balanced development policy, since the long-awaited era of the post-paper, steel, and fossil fuels society seems still far in the future. We could spend vastly more funds to research, develop, and produce solar, windmill, and water power.

This work is frequently labor-intensive and physically hard, but, because it improves the quality of life, it is worth doing. And because much of it is onerous but necessary, pay should be higher than for many other occupations. Workers should be paid on a principle of what might be called *reverse renumeration,* that is, paying more for jobs that are more unpleasant but enhance "public goods": manual, routine, or dirty tasks such as cleaning the streets and parks and collecting garbage; heavy work and routine mass production tasks; caring for children, older people, and the sick. If we are serious about the arrival of the so-called post-cold war era, paying for these services should pose few additional burdens on ordinary incomes because the military budget still hovers around $300 billion. Even a 50 percent reduction in the military budget and transfer of funds would result in a net increase in jobs, since much current military spending is extremely capital intensive. At the same time, a long-range commitment to expanding public services such as mass transit is expensive.

We should replace the current tax system, which favors the rich, and reintroduce the progressive tax, dropping the fiction that tax incentives are a major impetus to new investment within the U.S. economy. As the experience of the 1980s—when the Reagan administration presided over not only a tax giveaway but also one of the sharpest redistributive tax measures in history—amply demonstrates, putting more money in the pockets of the rich does not guarantee new investment within the borders of the United States or, indeed, better-paying jobs.

If we were committed to abolishing the hierarchical division of intellectual and manual labor, such tasks could be shared through a program of universal public service. A new commitment to universal public education to prepare the multivalenced worker would replace the current focus on specialization. Many if not all tasks in what is conventionally regarded as "menial" work could be shared among a wider portion of the labor force. In this regime of task equalization, every person would be obliged to perform some of the least desirable tasks, regardless of accumulated credentials; these jobs would not be permanently assigned to any class of people.

This is a long-term perspective, but in the wake of the objective possibilities inherent in new technologies for eliminating vast quantities of manual and clerical labor in both "advanced" and developing areas of the world, the question before us is whether the polity is prepared to tolerate *permanent mass unwork* or whether share-work values and programs will begin to bridge the gulf between knowledge-based labor and manual and clerical work.

A Guaranteed Income

Accordingly, if unwork is fated to be no longer the exception to the rule of nearly full employment, we need an entirely new approach to the social wage and, more generally, "welfare" policy. If there is work to be done, everyone should do some of it; additional remuneration would depend on the kinds of work an individual performs. But shorter work days, longer vacations, and earlier retirement imply that most of us should never work anything like "full time" as measured by the standards of the industrializing era. We need a political and social commitment to a national guaranteed income that is equal to the historical level of material culture. That is, everyone would be guaranteed a standard of living that meets basic nutritional, housing, and recreational requirements. Everyone would assume the responsibilities of producing and maintaining public goods, so no able citizen would be freed of the obligation to work. This would place a large burden on the private production sectors to induce people to engage in routine labor, presumably at wage rates higher than the guaranteed income and equal to tasks in the public sector. These rates would constitute a further incentive to invest in labor-saving technologies, which would free people from routine tasks without plunging them into a state of penury.

There would be no welfare system because the distinction between workers and "idlers" would disappear. Services such as health care (including counseling and therapy), education, and social work would expand and be paid for through general tax levies, but, assuming a new perspective on "jobs" and the division of labor, would shift their emphasis from work toward solving problems, exploring possibilities, and finding new ethical and social meaning.

School curricula, for example, could concentrate on broadening students' cultural purview: music, athletics, art, and science would assume a more central place in the curriculum and there would be a renewed emphasis on the aesthetic as well as the vocational aspects of traditional crafts. We suppose this would lead to a revival of what has become known as "leisure studies": psychologists and sociologists would study, prescriptively as well as analytically, what people do with their time, no longer described in precisely the same terms as it was thirty years ago. Concomitantly, space and time themselves become objects both of knowledge and, in the more conventional science fiction sense, of personal and social exploration. Consequently, lifelong learning, travel, avocations, small business, and artisanship

take on new significance as they become possible for all people, not just the middle and upper classes. Some may choose to participate in the technoculture as a critical component of the exercise of the right to *pleasure* as well as work. Others may avoid the technological construction of social and personal meaning.

Only when social policy has been transformed can the conditions that have produced the ecological crisis—consumerism, for example—be redirected. . . .

Clearly, there is an urgent need for a new ethic that addresses the proliferation of waste in our communities. . . .

A New Research Agenda

In the immediate as well as the medium-range future, we need a renewed public commitment to scientific and technological research. Research would not be confined to developing new products or motivated exclusively by considerations such as enhancing the national economy in an era of global competition. Fighting disease, protecting the environment, and finding new ways to construe time and space in both "wilderness" and urban settings would absorb considerable resources. . . .

It is by now evident that among the costs of a relentless pursuit of industrialization was pollution of our water and air. Congress has established a national cleanup fund and imposed some regulation on industrial polluters, but so far has not been willing to reexamine the historical costs of industrial enterprise. More than forty years ago, R. William Kapp took the first steps when he argued that we have failed to calculate the "social costs of private enterprise"[4]: when a coal mine or a metal plant is abandoned, even if the employer is required to "clean up" the surrounding area, people from miles around have already suffered the deleterious effects of air and water pollution. To be sure, public agencies have made our drinking water safer; the Clean Air Act imposed regulations on employers but provided extremely small inspection and enforcement teams.

Even if they were vigorously enforced, these measures are all after the fact. They take for granted the historical regime of industrial production that requires huge quantities of fossil fuels, employs large-scale power plants, and disposes of waste in large dumps that pollute the water bed. . . .

We do not yet fully possess the knowledge required to fundamentally and radically shift the basis of our production regimes: wind, solar, and hydro energy are still in their infancy; we have only scant experience with radically decentralized small-scale production methods, except in agriculture; and alternatives to the life-threatening (as well as life-enhancing) effects of medical treatment require more work (some alternative methods, based on traditional cultures and organicist worldviews, have proven effective against certain diseases, while others remain hypothetical). . . .

In the interest of averting the health crisis as well as the ecological crisis—not for saving labor costs—we need to deploy cybernetic and other labor-saving technologies in conjunction with the development of small-scale, ecologically sound production regimes. Undoubtedly we will find that some of these technologies, especially those that use large quantities of energy, are ecologically problematic; others will certainly prove to be ecologically beneficial.

One of the major issues in our emerging ecological crisis is how to reduce the amount of nonbiodegradable synthetic materials that have replaced cotton, wool, and wood products in packaging, furniture, clothing, cars, houses, and appliances. . . .

Clearly, the inevitability of both the jobless future and the ecological crisis demands a conclusive cultural shift, for we cannot simply legislate a change of this scope. But this shift cannot be achieved without a national and international effort to reduce the degree to which people would be required to give up some components of the "good life" associated with consumer goods. For just as we do not advocate nonwork without adequate income, we see the need to mobilize the scientific and technological revolution to meet ecological and health needs. In the last instance,

this becomes the basic goal of science and technology policy.

Ending Endless Work

To render the workplace rational entails a transformation of what we mean by rationality in production, including our conception of skill and its implied "other," unskill; a transformation of what we mean by mental as opposed to physical labor and judgment of who has the capacity to make decisions under regimes of advanced technologies.

Politics as rational discourse—as opposed to a naked struggle for power—awaits social and economic emancipation. Among the constitutive elements of freedom is *self-managed time*. Our argument in this book is that there are for the first time in human history the material preconditions for the emergence of the individual and, potentially, for a popular politics. The core material precondition is that labor need no longer occupy a central place in our collective lives, nor in our imagination. We do not advocate the emancipation from labor as a purely negative freedom. Its positive content is that, unlike the regime of work without end, it stages the objective possibility of citizenship. . . .

The scope of popular governance would extend from the workplace to the neighborhood. For as Ernest Mandel has argued, there is no possibility of worker self-management, much less the self-management of society, without ample time for decision making.[5] Thus, in order to realize a program of democratization, we must create a new civil society in which freedom consists in the first place (but only in the first place) in the liberation of time from the external constraints imposed by nature and other persons on the individual.

The development of the individual—not economic growth, cost cutting, or profits—must be the fundamental goal of scientific and technological innovation. The crucial obstacle to the achievement of this democratic objective is the persistence of the dogma of work, which increasingly appears, in its religious-ethical and instrumental-rational modalities, as an obvious instrument of domination.

NOTES

1. David Lodge, *Nice Work* (New York, Macmillan, 1986), 85–86.

2. William DiFazio, *Longshoremen* (South Hadley, Mass.: Bergin and Garvey, 1985).

3. "Companies Cutting Funds for Scientific Research," *New York Times,* June 30, 1993.

4. R. William Kapp, *The Social Costs of Private Enterprise* (New York: Schocken, 1951).

5. Ernest Mandel, *Late Capitalism* (London: New Left Books, 1979).

36

The Future of Women and Work: Ending the Twentieth Century

JOAN ACKER

How do global economic and work trends affect women, especially women in the United States? Acker explores both positive and negative views of women's work, and concludes with proposals for altering the negative.

As you read, ask yourself the following questions:

1. In what ways are global capitalistic economic restructuring affecting the working lives of American women?

2. How have the changes affecting working women affected you as a female? As a male?

GLOSSARY

Hegemony Dominance.
Masculinity Qualities characteristic of males.

Introduction

AS THE 20TH CENTURY ends, the ongoing global transformation of the division of labor and the organization of production and paid

work continues. What does this mean for women and their work in the United States? What does it mean for gender/race/class relations? Will new opportunities abound in the new participatory workplaces of the future? Will the gender wage gap decline as women and men become more equally distributed in jobs that no longer require physical strength? Or will the supply of "good" jobs continue to disappear, pushing many women and some men into routine work at very low wages and hastening the demise of the family wage for men? What will happen to African American, Hispanic, and other minority women? Although these questions will be answered conclusively only in the future, at the moment, in the early 1990s, we can see how large-scale changes are affecting women's prospects. We can also speculate about what may happen if nothing interrupts the present trajectory, while recognizing that such trajectories are often interrupted.

In this reading, I first sketch general trends in the economy and working life and then look at

Reprinted from Sociological Perspectives, *Vol. 35, No. 1, pp. 53–68. Copyright ©1992 Pacific Sociological Association. Reprinted by permission of JAI Press, Inc.*

how these trends are manifested in the work lives of women in the United States. I conclude with some proposals for altering these trends.

Global Changes

Social and political commentators, generally accepting the notion that a global transformation is in process, analyze its character in a number of ways. . . . Most experts see these trends as complexly interconnected. While opinions differ about what is changing and how much (Wood 1989), all contend that new things are happening. Among the claims on which some agreement exists are the following.

Capital is becoming increasingly internationalized and production is increasingly oriented toward international markets within which competition is intense (Taylor 1991). In the interests of competition and efficiency, both private and public employers pursue lower labor costs and greater "flexibility" (Standing 1989). The search for lower labor costs leads to a shifting global division of labor that is, at the same time, a shifting gender division of labor (Nash 1983; Standing 1989). Technological innovations facilitate these shifts, while they also contribute to creating new products and new markets. Accompanying these changes are the increasing employment of women in the world wage economy, growth of the service sector, and increasing demand for certain kinds of highly skilled labor within a general pattern of deskilling. Political processes facilitate the reorganization of production, including deregulation of business and industry, erosion of labor regulations, decreasing power of organized labor, reductions in welfare state spending, and privatization of social programs and state-owned production. For workers, both women and men, these changes have mixed consequences; some benefit, while others see their present conditions and future prospects undermined.

THE NEGATIVE VIEW

Looking at these developments, one commentator (Standing 1989) has labelled them the "feminization of labor." Not only is the world's paid labor force increasingly composed of women, but the conditions of work are becoming "feminized" for men as well. Feminization means that a declining proportion of jobs are "good" male jobs that carry the guarantee of lifetime employment with adequate wages and pension guarantees. In highly industrialized countries, technological change results in the deskilling of labor and facilitates the movement of production to areas of cheap labor. Old careers for skilled workers disappear. The family wage sufficient for a single worker to support dependents is fast disappearing; women's paid work is becoming necessary for family survival. Analyzing the United States, Bluestone and Harrison (1988) call this process *"The Great U-Turn."*

Employers achieve the flexibility they need to remain competitive through part-time, contract, and temporary work. On a global level, more and more people are unemployed, surviving in the informal economy. Unemployment, which creates pools of workers eager for jobs at any wage, also increases the flexibility of employers. Protections against arbitrary dismissal and autocratic control are also disappearing. Organized labor in many countries has decreasing power to protect workers, either because of direct attacks on the right to organize or because changes in production are undermining its bases of power in the old blue-collar, male working class. Finally, the welfare state, the bulwark of protection for workers against the risks of the market (Esping-Andersen 1990), is everywhere under attack on the grounds that too much protection undermines the viability of market capitalism. Already-vulnerable groups, such as women and minorities, suffer the most. All these developments are occurring in the core capitalist countries as well as in the periphery. At the same time, the experience in every country is different, shaped by preexisting political, social, and cultural patterns of power.

THE POSITIVE VIEW

Alongside dire predictions about deteriorating conditions for many workers are descriptions of how restructuring has made possible the retention and creation of good jobs for workers as well as for managers and professionals. Restructuring of employment, increased skills and work flexibility, and more worker participation in decision-making are positive consequences of new technology and increased competition. Improved work content and working conditions can result from new technology (Marshall 1987). As old, routine work processes are automated, and as information technology and service industries develop, jobs become more complex and more interesting. New methods of production lead to higher productivity. These changes should lead to higher standards of living, at least for those in the countries that win in global competition.

All of this requires workers with higher-level, more flexible skills. The key to increasing productivity, and to the prosperity that follows, lies in the education of workers and the reorganization of production to allow the most effective use of education (Marshall 1987). Looking to Japan for inspiration, new management thinking emphasizes the importance of teamwork and problem solving. This goes along with reduction of hierarchy and the placement of decision responsibility at lower levels and within work groups. Both Japan and Germany have used these strategies, which have contributed to their success in the process of global transformation.

The positive and negative views of contemporary changes in work and production are not necessarily mutually exclusive. Rising opportunities for professionals, managers, and skilled technicians can emerge side by side with a declining supply of good jobs for the majority of a population. A review of the evidence indicates that this has happened in the United States (Bluestone and Harrison 1988; Appelbaum and Albin 1990) and that these developments have differing impacts along lines of race, class, and gender (Higginbotham 1992).

Consequences for Women in the United States

What does that new world look like for women here in the United States? Most women now have to work to survive, as the alternative of total support from a husband becomes more and more untenable. In 1990, approximately 56 percent of women over the age of 16 were in the paid labor force. Over the preceding 20 years, women's situations at work have improved in minor ways. Sex segregation at the national occupational level has declined somewhat, as has the gender wage gap. Women have broken into many fields formerly monopolized by men and have increased as a proportion of those in most professional and managerial jobs, although they tend to be confined to lower levels and feminine subspecialties. Yet, the majority of women are still in the large, low-wage "women's job" categories where benefits are relatively low and promotion unlikely. The question about the future is whether or not this will change: what kinds of jobs, under what conditions, and at what levels of pay will be available to women? I examine these questions in terms of four characteristics of the present transformations in work: the restructuring of employment, changing skill demands, increasing flexibility, and the reduction of hierarchy along with greater employee participation. . . .

RESTRUCTURING OF EMPLOYMENT

Employment restructuring occurred first in manufacturing production as certain processes in particular industries were moved to areas with cheap labor in the United States as well as in other countries, and as whole industries declined in the United States. The number of blue-collar jobs in female-predominant industries such as garment manufacture and in male-predominant industries such as steel was sharply reduced. This restructuring had a severe impact on women as well as men, and it particularly hit African American men in industrial cities (Wilson 1991), contributing to the emergence of severe ghetto poverty. During the 1980s, however, United States employment increased as service-sector jobs multiplied rapidly. Seventy-nine percent of the 27 million new jobs created between 1973 and 1987 were in services (Appelbaum and Albin 1990). This was the American "miracle"; the creation of millions of new jobs even as the old bases of high employment in manufacturing were eroding.

The service sector consists of finance, real estate, insurance, transportation, public utilities, retail and wholesale trade, personal and business services, amusement, health, education, legal and social services, and government. The United States service sector has two tiers (Appelbaum and Albin 1990). According to Appelbaum and Albin, the knowledge-information tier has high technology, high productivity growth, high wages, and is capital-intensive. The other tier has low productivity growth, low technology, and low wages and is labor-intensive. Women are the majority of workers in both sectors and, in both tiers, their wages are lower than the wages of men in comparable jobs. This sector is also structured by race, with minorities disproportionately found in the lower tier. In addition to the service sector, there are service jobs within manufacturing.[1]

During the 1980s, the service sector appeared to be immune to the troubles affecting manufacturing, and, despite heavy investments in computer technology that often reorganized and accelerated work processes, employment did not decline (Roach 1991). Deregulation during the 1980s, leading to hostile competition, takeovers, buyouts, and questionable lending practices, produced crises in this sector. Now, in 1991, service industries, like those in manufacturing, are also restructuring their employment, with the aim of increasing efficiency and reducing their labor costs (Roach 1991). At the same time, reorganization of management in manufacturing is underway. These changes will significantly affect women because they constitute at least 62 percent of service-sector employment (Sweeney and Nussbaum 1989), and because they have made some gains at the entry and middle levels of management.

Restructuring is occurring in communications, banks and other financial services, airlines, entertainment, and retail sales, and in the managerial levels of other kinds of firms. For example, banking jobs declined by 260,000 positions between 1984 and 1991 (Zonana 1991), with cuts fueled by mergers and other retrenching efforts in response to losses and failures. Overall, service-sector employment lost ground through much of 1991 (Norris 1991). Preliminary Department of Labor statistics show that certain service occupations, as distinguished from service-sector

jobs, declined in 1990–1991. For example, technical, sales, and administrative support positions dropped by 804,000 between March 1990 and March 1991 (U.S. Department of Labor 1991). Women workers accounted for 75 percent of this decline. Longer trends are evident in employment decline in clerical work between 1988 and 1990, in some instances continuing declines that had begun in the early 1980s. The number of bookkeepers, accounting and auditing clerks, telephone operators, bank tellers, general office clerks, secretaries, and typists all declined over these years.[2] Earlier, declines in these occupations were more than compensated for by increases in other administrative support and clerical areas, but this growth seems to have ended in 1991.

Downsizing of management has not yet appeared in aggregate statistics, with the exception of managers in the federal government. However, newspapers and business journals provide case reports and quote top managers on the topic. . . .

Although some of the restructuring of employment in the private service sector is fueled by economic recession, much of it probably represents permanent change in the organization of work and the labor process, made possible by dramatic changes in technology (Baran 1990). Early expectations that computerization would drastically reduce clerical employment were not at first fulfilled. In the insurance industry, for example, discrete functions were first computerized, but this did not eliminate all the other tasks in the processing of claims (Baran 1990). Volume increased and so did employment. Recently, new systems approaches using computerized master files and on-line data entry eliminated many steps in this process (Baran 1990). The elimination of functions creates the conditions for the reduction of the work force. In another example, a national retail chain reduced its customer service staff by almost two-thirds in the mid-1980s by installing an on-line computer system to access customer accounts. The time for processing a complaint declined from 10 minutes to 2 minutes.

The restructuring process is also underway in the public sector, which has long been the exemplary employer of women and minorities. The federal government undertook employment reductions during the 1980s, as the Reagan regime tried to decrease the size of the welfare state. This is now happening in state and local governments as well, including school districts, as budgetary problems become critical. Budget-slashing in public services may be a temporary phenomenon, but, in the near-term, it further reduces opportunities for women and minorities. Only the health sector seems to be immune: here employment continues to grow. However, the outlook could change rapidly if the demands for reform of the health care system are realized. One central proposal is to radically simplify funding and billing. Drastic reductions in clerical and professional staffs in insurance companies and in hospital and clinic billing departments could follow such a reform.

The consequences for women and minorities are obvious. Women hold the majority of service-sector jobs and most of their job growth has been in this area. Minority women found opportunities in the service sector that enabled some to move away from domestic and factory labor. Women's concentration in these jobs protected them against the consequences of restructuring in manufacturing and pointed to continuing rising employment opportunities. As recently as 1987, Kuhn and Bluestone could write "The greatest cause for optimism lies in the sheer number of jobs being created in occupations and industries that are currently majority female" (1987:21). In 1988, the Bureau of Labor Statistics was predicting growth of employment in every service-sector industry. Recent trends make such optimism highly suspect. In addition, as Kuhn and Bluestone (1987) point out, those service-sector jobs are disproportionately at the lower level of work hierarchies, exactly where many of the systems-analysis-generated cuts are coming. Those jobs in the second tier of the service sector that have to do with personal services are more difficult to automate and thus may remain plentiful, but these are the jobs with low pay and few or no benefits. Women of color are disproportionately caught in these dead-end jobs.

Restructuring of manufacturing meant that women's affirmative action path to better jobs through entering male-dominated blue-collar jobs became more and more unrealistic as these jobs disappeared through automation or transfer to other countries. Similarly, movement of women into male-dominated middle-level management and professional jobs may be less and less an option as these jobs also contract. As middle management is cut, there are fewer mobility paths out of the lowest level jobs. Promotion ladders are also truncated as higher-level jobs require more training; employers will look for trained personnel among new graduates rather than promote personnel from the lower levels of their organizations. Reductions in middle-level management will probably also lead to a decrease in entry-level management jobs.

CHANGES IN SKILL DEMANDS

New technology is not always used to replace workers' skills and to deskill workers, as critics of Braverman's (1974) deskilling thesis have pointed out (see, for example, Beechey 1987). Often, it requires an upgrading of skills. Some experts believe that not enough skilled workers are being trained in the United States and predict a shortage of such workers in the near future (Marshall and Brock 1990). Forecasts of a shortage of skilled workers are linked to calls to American business to give up its deskilling, low-wage, short-term profit strategy and to embrace instead a longer-term, high-productivity growth strategy. If these prescriptions are followed, an increase in better jobs, at least for some, might be expected. But would these good jobs be available to women as well as to men?

The answer is equivocal. Skill has gendered meanings. As Cockburn (1985) has demonstrated, skill is tied to masculinity in ways that define women as unskilled. In Cockburn's studies, as computerization and other new technologies were introduced into a variety of work processes, men remained in control of the new machines and boundaries between male and female, skilled and unskilled, were redrawn (Cockburn 1985; see also Hacker 1990).

Women's jobs are often perceived by managers, male colleagues, and women workers themselves as requiring less skill than comparable men's jobs when assessed by job evaluation methods (Acker 1989). Thus, new technology may facilitate the upgrading of skill, but this is not necessarily recognized nor rewarded. For example, the development of computerized systems in office work sometimes results in new and more complex jobs that combine components of routine and skilled clerical work along with some previously professional tasks. Baran (1990) documents this for insurance firms in the United States.

Of course, computerization has resulted in deskilling as well as in reskilling. In some cases, managers' jobs have been deskilled as lower-level workers take over reorganized managerial tasks and as computer monitoring of errors and numbers of key strokes or transactions replaces human supervision. Lower-level jobs have also been deskilled or eliminated. New computer technology can be designed to assist increased autonomy and participation in decision making, or it can be designed to eliminate decision making and to closely control workers (Appelbaum and Albin 1989). In the example of [a] West Coast communications firm . . . both processes were happening. On the one hand, middle-level managers were no longer needed to monitor and plan the work of the large, clerical work force. That task was being done by computers. On the other hand, technical experts and professionals could use computers in their new problem-solving and autonomous work groups.

The introduction of new technology may also increase skill and income differences among women-predominant jobs. Women's jobs are becoming increasingly polarized into routine, low-wage, highly controlled work and non-routine, relatively autonomous, higher-wage jobs (Kuhn and Bluestone 1987; Sweeney and Nussbaum 1989), reflecting the two tiers in the service sector discussed above. For example, two women-predominant occupations that are predicted to expand most rapidly during the 1990s are nursing and nurses' aides. As Glazer (1991) points out, the skill level of nurses is rising as new technology changes the nature of nursing work and as the profession increasingly requires at least a baccalaureate as certification. At the same time, the work of aides is still defined as unskilled and its pay remains near minimum wage levels (Sweeney and Nussbaum 1989).

Many lower-level women's jobs that cannot be automated still require high skill and responsibility, but are devalued in our society and vastly underrewarded. The skills that are necessary for caring work and that must be learned through experience rather than through technical, classroom training are those most devalued (see, for example, Acker 1989). An obvious example is child-care work, another occupational area that is expected to expand during the 1990s, in which pay levels are generally not much above minimum wage. Although comparable-worth successes in the public sector have improved wages for many care workers employed there, these efforts have slowed recently and have hardly touched the private sector. Low levels of union organization contribute to low wages, because without union organization, relatively powerless women have no mechanism for pushing for pay levels that reward these skill requirements.

FLEXIBLE WORK ORGANIZATION

"Flexibility" is one of private-sector management's answers to the problem of maximizing returns in highly competitive markets. For the public-sector manager, flexibility can mean lower labor costs and budgetary savings. Flexibility has a number of meanings (Standing 1989; Wood 1989), and these meanings have gender implications (Walby 1989; Jenson 1989). Flexibility refers to firms, the organization of work, and to workers' skills. In management theory, the flexible firm should be able to respond rapidly to changes in costs, demand, and technology in a rapidly evolving, competitive world. Workers should have a range of skills and job boundaries should be flexible to make such a response possible. Jenson (1989) points out that the new, flexible workers are apt to be men because patterns of sexual segregation are such that men's jobs have been defined as skilled, while women's have been regarded as unskilled or semi-skilled.

Those chosen for new skill training are apt to be those who are already seen as skilled. Cockburn's (1985) studies, discussed above, support this conjecture. There is no reason to think that this will change. The type of flexibility sought through multiple skilling of workers and adaptability to markets is unlikely to benefit women in manufacturing to any great degree, especially women of color and immigrant women. Moreover, as in the banking and insurance examples noted above, although women may develop greater skill and the flexibility to do a variety of tasks, this may not improve their income situation. Gender assumptions allow managers to short-circuit the expected relationship between improved productivity and increased wages.

Another meaning of flexibility is that employers need to be able to increase and decrease the labor force as conditions change. One way to achieve this flexibility is to employ part-time and/or temporary workers who can be easily dismissed and who do not qualify for the same job protections or benefits that labor contracts and legislation grant to regular full-time employees. Women have always comprised the bulk of this "peripheral" or contingent labor force, but the jobs of this type are increasing, particularly with the explosion of the service sector discussed above. According to Sweeney and Nussbaum (1989), 29 percent of the work force was in such jobs in 1988. Part-time workers, counted separately from other contingent workers, constituted 16.5 percent of the work force in 1988, rising to 20 percent by 1991. Women accounted for 65 percent of this part-time labor force (U.S. Department of Labor 1991).

Of course, skilled professionals as well as clerical and service workers are part-time and temporary workers. For many women, part-time work is a good solution to problems of combining paid and unpaid responsibilities. For the successful, high-paid professional, independent contracting may have many advantages; on the other hand, it may constitute an undesirable alternative for professionals who cannot find jobs because of organizational downsizing and restructuring. For the low-paid, routine worker, part-time employment has many disadvantages, even though

it may make it easier to combine family care and paid work. Flexibility for the employer means frequent job changes, part-time, temporary work, and insecurity for the worker. Lack of benefits in many part-time jobs, inadequate unemployment compensation, absence of other income supports, lack of reasonable child care and parental leave increase the insecurity and stress for women and men, but especially for women. Again, women of color are most vulnerable (Eitzen and Zinn 1992).

REDUCTION OF HIERARCHY AND INCREASED PARTICIPATION

Reduction of hierarchy, as I have argued elsewhere (Acker 1990; also see Kanter 1977), may be a precondition for increased gender equality in the workplace. It is also a precondition for a more contributory and productive organization of work that can achieve flexibility and the effective utilization of new skills. Is it possible that feminist and management goals are converging? To reduce hierarchy, some middle-level management jobs must be eliminated or changed, and jobs that were once closely directed and supervised must be reorganized so that workers may take responsibility for their work, feel empowered to make suggestions that will improve productivity, and feel committed to the workplace and organization. In this scenario, a considerable number of jobs disappear, while the quality of others is enhanced. Since men are likely to have more seniority than women, women are apt to lose their jobs first. Those who are left may be better off because group-organized responsibility and decision-making can empower workers and make work more interesting and satisfying. Experience in Sweden supports this view.

Reduction of hierarchy and increased worker responsibility can also be a new form of increased exploitation and control. Expansion of duties and responsibilities can increase the amount of work that members of a group are expected to do. I take my examples from unpublished accounts from Sweden. In a group of home care workers in Sweden, job redesign resulted in group responsibility for scheduling, ordering supplies, and for contacts with the relatives of patients.

These women workers learned the new tasks, including how to use computers to facilitate the work. However, while they shared previous supervisory responsibilities and hierarchy was reduced, no new workers were added to the group. As a consequence, they experienced a self-organized speed-up. There was less time to spend with patients and less time to confer with each other. Control was also transferred from the supervisor to the group. With more work to do, the absence of a group member became more problematic. The workers, now doing their own scheduling, had to find ways of managing such problems. The increased skills and responsibility were gratifying, but worries about the quality of care and sharing of responsibility and control increased tension and dissatisfaction. The public employing organization saved money because supervisory staff were no longer necessary.

In other experiments in group organized work in the care sector, workers were added to groups as tasks increased, allowing the quality of care to be maintained and facilitating ways of sharing the work of absent group members. In these cases, quality of service and worker morale improved, but savings were probably less.

TRENDS IN WAGES

Restructuring is having at least two effects on wages that have significance for the future. The wage gap between women and men is declining and jobs with low wages are increasing. While the wage gap between women and men has declined by a few percentage points since 1979, this may not be a cause for great celebration. Three-tenths of the 5-percentage-point decline between 1979 and 1987 was due to the dropping wages of men (National Committee on Pay Equity 1989). Between 1990 and 1991, the wages of white men fell again, even further diminishing the gender and race gap (DeParle 1991). The decline in white men's wages can be attributed to the disappearance of many high-wage production jobs in the course of the global restructuring of manufacturing. A major question is whether the restructuring of management and the service sector will further erode the incomes of white men and undermine women's wage gains as well.

Some of the growth in service-sector jobs was in professional, managerial, and technical fields in which women had relatively good salaries, although lower than those of men. However, 37 percent of job growth between 1979 and 1987 was in service industries with median incomes below $15,000 per year (Appelbaum and Albin 1990). One study estimates that, in 1984, 25 million adult workers were low-wage earners, defined as working seven or more months at wages of $5.80 per hour or less in 1988 dollars (Institute for Women's Policy Research 1990). Both the number and proportion of these workers increased during the 1980s, especially among women and people of color. In sum, the earnings consequences of economic restructuring are, on the whole, negative for ordinary people, as many others have also observed (Phillips 1990).

Can We Avoid a Bleak Future?

I have drawn a pessimistic picture of the impact of capitalist restructuring, on women workers in the United States. As I finish this reading in November, 1991, the signs of a severe recession abound, leading me to fear that the trends I have pointed to will only worsen. But, even if the economy has an upturn soon, the underlying processes will continue. Women's recent gains in the service sector and management are apt to stagnate or decline as restructuring continues. At the same time, class and race divisions among women are likely to become greater. The old organization of distribution (Acker 1988) from the male family wage to wives and children is clearly breaking down, as men's incomes decline and two incomes become even more necessary for family survival. No new organization of distribution, such as child allowances, that would provide for more family security in an increasingly low-wage work world has yet been achieved. Nor is there any coherent labor-market policy in the United States to prepare workers for new kinds of work or to create new jobs that are good jobs. With new forms of flexible work that have no long-term employment guarantees and few employee benefits, new forms of redistribution are needed (Standing 1989). Old types of unemployment insurance and retirement benefits linked to continuous and long-term employment do not suffice to guarantee a secure daily life and old age. Workers need protection from the dangers of the free-labor market (Esping-Andersen 1990). These protections are, of course, exactly what has been stripped away with the attack on labor unions and on welfare-state provisions.

The ideology supporting and interpreting these changes, as I argued above, looks more like a nineteenth-century hegemonic masculinity than a twenty-first-century vision of gender equality. This ideology and the organization of power it supports present severe difficulties to those who propose public policies for dealing with these new conditions. Nevertheless, this ideology is being challenged and proposals for reform are now emerging.

Some of these proposals are coming from women workers who have been able to organize in their own interests in the public sector where some protections for organization still exist. They have also been able to form coalitions with women in their communities, as the comparable-worth movement attests. With so many more women in the paid labor market, the structural conditions for more organization exist. In a utopian view of the future, women might be the leaders in a movement for changes that will save the United States from becoming another Third World country that keeps its dominance only because it has a big gun for hire.

In this utopia, the reborn hegemonic masculinity gives way as the recognition of the necessity to do something about our domestic problems returns. Military spending is drastically cut and the savings are allocated to rebuilding our physical infrastructure and our social services. Employment in the public and the private sectors increases, replacing jobs lost to restructuring. Unions are still strong in the public sector, and more women and minorities get good jobs there. New laws protecting the rights of workers are passed and union organization grows in the private sector. Affirmative action efforts are

revitalized and comparable-worth reforms of pay structures become commonplace. Rapid change demands flexible workers, and business and public employers collaborate in developing new forms of worker training and education. In many workplaces, the work process is reorganized so that workers have more skill and more responsibility. Controls are reduced or transformed—more people work because it is intrinsically satisfying. New income transfer methods are devised—perhaps a guaranteed annual income is instituted (Standing 1989)—so that people can move from job to job without terror and be assured an income when they stay at home to care for family or when they work part-time. All of this would not necessarily end male domination or achieve equality in the workplace. But, it might create a platform of security from which further changes could come.

NOTES

1. The service sector must be distinguished from service occupations. Although the Bureau of Labor Statistics Category "Service Occupations" is confined to protective, health, and personal services, if service occupations are more broadly defined to include many of the occupations in the service sector, it is clear that many of these jobs are also found within the manufacturing sector. Manufacturers, for example, hire systems analysts and lawyers.

2. Calculated from *Statistical Abstract of the United States,* 110th edition, 1990: 389–391, and U.S. Department of Labor, *Employment and Earnings* 38: 185–190.

REFERENCES

Acker, Joan, 1988. "Class, Gender, and the Relations of Distribution." *Signs* 13: 473–497.

———, 1989. *Doing Comparable Worth.* Philadelphia: Temple University Press.

———, 1990. "Thinking about Wages: The Gendered Wage Gap in Swedish Banks." *Gender and Society* 5: 390–407.

———, 1992. "Gendering Organizational Theory." In *Gendering Organizational Theory,* edited by Albert Mills and Peta Tancred. Newbury Park: Sage.

Appelbaum, Eileen, and Peter Albin, 1990. "Differential Characteristics of Employment Growth in

Service Industries." Pp. 36–53 in *Labor Market Adjustments to Structural Change and Technological Progress,* edited by Eileen Appelbaum and Ronald Schettkat. New York: Praeger.

Baran, Barbara, 1990. "The New Economy: Female Labor and the Office of the Future." Pp. 517–534 in *Women, Class, and the Feminist Imagination,* edited by Karen V. Hansen and Ilene J. Philipson. Philadelphia: Temple University Press.

Beechey, Veronica, 1987. *Unequal Work.* London: Verso.

Bluestone, Barry, and Bennett Harrison, 1988. *The Great U-Turn: Corporate Restructuring and the Polarizing of America.* New York: Basic Books.

Braverman, Harry, 1974. *Labor and Monopoly Capital.* New York: Monthly Review Press.

Cockburn, Cynthia, 1985. *Machinery of Dominance.* London: Pluto.

Connell, R. W., 1988. *Gender and Power.* Stanford: Stanford University Press.

DeParle, Jason, 1991. "Poverty Rate Rose Sharply Last Year as Incomes Slipped." *New York Times,* 27 September 1991.

Eitzen, D. Stanley, and Maxine Baca Zinn, 1992. "Structural Transformation and Systems of Inequality." Pp. 178–182 in *Race, Class, and Gender,* edited by Margaret L. Andersen and Patricia Hill Collins. Belmont, CA: Wadsworth.

Esping-Andersen, Gosta, 1990. *Three Worlds of Welfare Capitalism.* Princeton: Princeton University Press.

Glazer, Nona, 1991. "'Between a Rock and a Hard Place': Women's Professional Organizations in Nursing and Class, Racial, and Ethnic Inequalities." *Gender and Society* 5: 351–372.

Hacker, Sally, 1990. *Doing It the Hard Way.* Boston: Unwin Hyman.

Higginbotham, Elizabeth, 1992. "We Were Never on a Pedestal: Women of Color Continue to Struggle with Poverty, Racism and Sexism." Pp. 183–190 in *Race, Class and Gender,* edited by Margaret L. Andersen and Patricia Hill Collins. Belmont, CA: Wadsworth.

Institute for Women's Policy Studies and National Displaced Homemaker's Network, 1990. *Low Wage Jobs and Workers: Trends and Options for Change.* Washington, DC: National Displaced Homemaker's Network.

Jenson, Jane, 1989. "The Talents of Women, the Skills of Men: Flexible Specialization and Women." Pp. 141–155 in *The Transformation of Work?* edited by Stephen Wood. London: Unwin Hyman.

Kanter, Rosabeth Moss, 1977. *Men and Women of the Corporation.* New York: Basic Books.

Kuhn, Sarah, and Barry Bluestone, 1987. "Economic Restructuring and the Female Labor Market: The Impact of Industrial Change on Women." Pp. 3–32 in *Women, Households, and the Economy,* edited by Lourdes Beneria and Catharine R. Stimpson. New Brunswick: Rutgers University Press.

Marshall, Ray, 1987. *Unheard Voices: Labor and Economic Policy in a Competitive World.* New York: Basic Books.

Marshall, Ray, and William E. Brock, 1990. *America's Choice: High Skills or Low Wages.* Rochester, NY: National Center on Education and the Economy's Commission on the Skills of the American Workforce.

Nash, June, 1983. "The Impact of the Changing International Division of Labor on Different Sectors of the Labor Force." Pp. 3–38 in *Women, Men, and the International Division of Labor,* edited by June Nash and Maria Patricia Fernandez-Kelly. Albany: State University of New York Press.

National Committee on Pay Equity, 1989. *Briefing Paper #1.* Washington, DC: National Committee on Pay Equity.

Norris, Floyd, 1991. "Services: A Boom Area Goes Bust." *The New York Times,* 6 October 1991, 3, 1.

Pateman, Carole, 1988. *The Sexual Contract.* Cambridge: Polity.

Phillips, Kevin, 1990. *The Politics of Rich and Poor.* New York: Random House.

Porter, Michael E., 1990. *The Competitive Advantage of Nations.* New York: The Free Press.

Roach, Steven S., 1991. "Services under Siege: The Restructuring Imperative." *Harvard Business Review* 69: 82–92.

Standing, Guy, 1989. "Global Feminization through Flexible Labor." *World Development* 17: 1077–1095.

Sweeney, John J., and Karen Nussbaum, 1989. *Solutions for the New Work Force.* Cabin John, MD: Seven Locks Press.

U.S. Department of Labor, Bureau of Labor Statistics, 1991. *Employment and Earnings.* Washington, DC: U.S. Government Printing Office.

Walby, Sylvia, 1989. "Flexibility and the Changing Sexual Division of Labor." Pp. 127–140 in *The Transformation of Work?* edited by Stephen Wood. London: Unwin Hyman.

Wilson, William Julius, 1991. "Studying Inner-City Social Dislocations." *American Sociological Review* 56: 1–14.

Zonana, Victor F., 1991. "Banking Industry Implodes." *Los Angeles Times,* 19 August 1991, Section A:1.

37

A Broader Vision:
Social Policy Options in Cross-National Perspective

WILLIAM JULIUS WILSON

This reading outlines a number of seemingly non-radical suggestions for dealing with inner-city unemployment and welfare, yet there appears to be a lack of political support for such. What do you see as the answer to this contradiction? How do Wilson's ideas tie in with those suggested in the second reading of this chapter?

As you read, ask yourself the following questions:

1. What are some of the solutions to the jobs problem in the United States proposed by the author?

2. Why do you agree/disagree with each of the author's proposals?

GLOSSARY

Structural and cultural constraints Limitations on individuals' experiences because of their neighborhood, economic opportunities, and value systems.

... THE UNITED STATES CAN learn from industrial democracies like Japan and Germany. These countries have developed policies designed to increase the number of workers with "higher-order thinking skills," including policies that require young people to meet high performance standards before they can graduate from secondary schools and that hold each school responsible for meeting these national standards. . . .

Students who meet high standards are not only prepared for work, they are ready for technical training and other kinds of postsecondary education. Currently, there are no national standards for secondary students or schools in the United States. . . . A commitment to a system of national performance standards for every public school in the United States would be an impor-

tant first step in addressing the huge gap in educational performance between the schools in advantaged and disadvantaged neighborhoods. . . .

. . . It is important to discuss immediate solutions to the jobs problem in the United States. Because of their level of training and education, the inner-city poor and other disadvantaged workers mainly have access only to jobs that pay the minimum wage or less and are not covered by health insurance. However, recent policies of the federal government could make such jobs more attractive. The United States Congress enacted an expansion of the earned income tax credit (EITC) in 1993. . . . This expansion . . . reflected a recognition that wages for low-paying work have eroded and that other policies to aid the working poor . . . have become weaker.

However, even when the most recent expansion of the EITC is fully in effect in 1996 . . . it will still fall notably short of compensating for the sharp drop in the value of the minimum wage and the marked reductions in AFDC benefits to low-income working families since the early 1970s. . . .

If this benefit is paid on a monthly basis and is combined with universal health care, the condition of workers in the low-wage sector would improve significantly and would approach that of comparable workers in Europe. The passage of universal health care is crucial in removing from the welfare rolls single mothers who are trapped in a public-assistance nightmare by the health care needs of their children. It would also make low-paying jobs more attractive for all low-skilled workers and therefore improve the rate of employment.

The mismatch between residence and the location of jobs is a special problem for some workers in America because, unlike in Europe, the public transportation system is weak and expensive. This presents a special problem for inner-city blacks because they have less access to private automobiles and, unlike Mexicans, do not have a network system that supports organized car pools. Accordingly, they depend heavily on public transportation and therefore have difficulty getting to the suburbs, where jobs are more plentiful and employment growth is greater. Until public transit systems are improved in metropolitan areas, the creation of privately subsidized car-pool and van-pool networks to carry inner-city residents to the areas of employment, particularly suburban areas, would be a relatively inexpensive way to increase work opportunities.

In the inner-city ghettos, the problems of spatial mismatch have been aggravated by the breakdown in the informal job information network. In neighborhoods in which a substantial number of adults are working, people are more likely to learn about job openings or be recommended for jobs by working kin, relatives, friends, and acquaintances. Job referrals from current employees are important in the American labor market. . . . Individuals in jobless ghettos are less likely to gain employment through this process. But the creation of for-profit or not-for-profit job information and placement centers in various parts of the inner city not only could significantly improve awareness of the availability of employment in the metropolitan area but could also serve to refer workers to employers. . . .

As much of the foregoing economic analysis suggests, however, the central problem facing inner-city workers is not improving the flow of information about the availability of jobs, or getting to where the jobs are, or becoming job-ready. The central problem is that the demand for labor has shifted away from low-skilled workers because of structural changes in the economy. . . .

If firms in the private sector cannot use or refuse to hire low-skilled adults who are willing to take minimum-wage or subminimum-wage jobs, then the jobs problem for inner-city workers cannot be adequately addressed without considering a policy of public-sector employment of last resort. Indeed, until current changes in the labor market are reversed or until the skills of the next generation can be upgraded before it enters the labor market, many workers, especially those who are not in the official labor force, will not be able to find jobs unless the government becomes an employer of last resort. . . .

. . . Given the current need for public jobs to enhance the employment opportunities of low-skilled workers, what should be the nature of these jobs and how should they be implemented? Three thoughtful recent proposals for the creation of public jobs deserve serious consideration. One calls for the creation of public-sector infrastructure maintenance jobs, the second for public service jobs for less-skilled workers, and the third, which combines aspects of the first two, for WPA-style jobs of the kind created during the Franklin D. Roosevelt administration. . . .

None of the immediate solutions I am proposing involves retraining workers for higher-paying positions in the highly technological global economy. The need to retain low-skilled workers is generally recognized by policymakers and informed observers in both Europe and the United States. However, the most serious discussions about training for the new economy have focused on young people and their transition from school to work or from school to postsecondary training. The cost of retraining adult workers is considerable, and none of the industrial democracies has advanced convincing proposals indicating how to implement such a program effectively. Moreover, a heavy emphasis on skill development and job retraining is likely to end up mainly benefiting those who already have a good many skills that only need to be upgraded.

Also, none of my immediate solutions offers a remedy for the growing wage inequality in the United States. The long-term solutions I have presented, which include those that prepare the next generation to move into the new jobs created in the global economy, are designed to combat that problem. But my specific recommendations for immediate action would address the employment problems of many low-skilled workers, including those from the inner city. They would confront the current and serious problem of the disappearance of work in the inner-city ghetto. The jobs created would not be high-wage jobs but, with universal health insurance, a child care program, and earned income tax credits attached, they would enable workers and their families to live at least decently and avoid joblessness and the problems associated with it. The United States Congress has already expanded the earned income tax credit. Universal health insurance and some kind of flexible child care program would be costly, but these programs would support *all* Americans. Furthermore, the nation has recognized the need for such social benefits and there remains considerable public support in favor of moving forward on both programs—especially in the area of health insurance. . . .

Programs proposed to increase employment opportunities, such as the creation of WPA-style jobs, should be aimed at broad segments of the U.S. population, not just inner-city workers, in order to provide the needed solid political base of support. In the new, highly integrated global economy, an increasing number of Americans across racial, ethnic, and income groups are experiencing declining real incomes, increasing job displacement, and growing economic insecurity. The unprecedented level of inner-city joblessness represents one important aspect of the broader economic dislocations that cut across racial and ethnic groups in the United States. Accordingly, when promoting economic and social reforms, it hardly seems politically wise to focus mainly on the most disadvantaged groups while ignoring other segments of the population that have also been adversely affected by global economic changes. . . .

The solutions I have outlined were developed with the idea of providing a policy framework that would be suitable for and could be easily adopted by a reform coalition. The long-term solutions, which include the development of a system of national performance standards in public schools, family policies to reinforce the learning system in the schools, a national system of school-to-work transition, and ways to promote city-suburban in-

tegration and cooperation, would be beneficial to and could draw the support of a broad range of groups in America. The short-term solutions, which range from the development of job information and placement centers and subsidized car pools in the ghetto to the creation of WPA-style jobs, are more relevant to low-income Americans, but they are the kinds of opportunity-enhancing programs that Americans of all racial and class backgrounds tend to support. . . .

The long-term solutions that I have advanced would reduce the likelihood that a new generation of jobless workers would be produced from the youngsters now in school and preschool. . . .

My framework for long-term and immediate solutions is based on the notion that the problems of jobless ghettos cannot be separated from those of the rest of the nation. . . . Their most important contribution would be their effect on the children of the ghetto, who would be able to anticipate a future of economic mobility and share the hopes and aspirations that so many of their fellow citizens experience as part of the American way of life.

QUESTIONS FOR DISCUSSION

For further discussion of this topic, see the Wadsworth Sociology Resource Center, "Virtual Society," *http://sociology.wadsworth.com,* under *Sociological Footprints,* by Cargan and Ballantine. You can respond to the discussion questions there or enter your own comments in the online chat forum.

SUGGESTED READINGS AND SOCIOLOGY INTERNET RESOURCES

See the Wadsworth Sociology Resource Center, "Virtual Society," *http://sociology.wadsworth. com,* for additional links, suggestions for further reading, and learning tools related to this chapter.

Either from the "Virtual Society" website or directly from your web browser, you may access InfoTrac College Edition, an online university library that includes over 700 popular and scholarly journals in which you can find articles related to the topics in this chapter.

CHAPTER 10

Politics

POWER AND ITS IMPLICATIONS

IN STUDYING THE INSTITUTION OF POLITICS, sociologists concern themselves with the groups involved in the political process and the conditions that tend to generate political involvement or apathy. These interests in turn lead to several related lines of inquiry: (1) the conditions that lead to or prevent political change, (2) the problems of democracy in the bureaucratic organization, (3) economic power and its effect on the political structure, and (4) the factors in voter participation and ideology. These interests are revealed in the readings in this chapter.

In the first reading, Lipset takes up the first of these lines of inquiry by examining the need for creating and maintaining democracy.

In the second reading, Philip Meyer takes up the the second line of inquiry — the issue of political authority. This reading is about a study that revealed much more obedience in doing an activity than expected despite not enjoying what they were doing. This study is about authority, which is used to control bureaucratic organizations, and it gives us a better understanding of the forces that exert influence, such as the mass media and politicians, through their access to modern-day communication techniques.

The next reading by Clawson, Newstadt, and Scott turns to the third line of inquiry by revealing that PAC money is a major factor in political influence. They illustrate the influence that the corporate elite have via their PAC donations.

With the political campaign now reduced to a personal effort via a media-run candidacy needing lots of money, it is not surprising that the corporate world has attained political power via its political action committee (PAC) contributions.

With growth in diversity and numbers, the American Right has been adding to its political influence. The important factor here, as noted by Jerome Himmelstein in the last line of inquiry and final reading, is to realize that many different elements make up the Right, thus creating widespread ideological-political demands.

In reading this chapter, keep in mind that change is a normative aspect of social institutions and that many factors encourage, delay, or prevent change, such as socialized beliefs, the media, and individuals or corporations with political power via authority and/or funds. Though political change is the main theme of this chapter, this does not mean that the other areas of interest to political sociologists are ignored. Voter participation is declining, and so we must ask what factors lead to the current lowest level of political participation of any democracy. Does voter apathy spring from belief that all is well, that it makes no difference since both parties are more alike than unalike or both ignore the public's wishes, artificial barriers to voter participation, such as registration requirements and elections held on a workday, and/or the recognition of the power of those other factors just mentioned? All this leads us to a final area of concern and a final question: Is it possible to have a democratic system without a strong political party system to articulate a set of principles? Lipset, in the first reading, says no.

38

The Social Requisites of Democracy Revisited: 1993 Presidential Address

SEYMOUR MARTIN LIPSET

Since the end of World War II, there has been a movement toward the creation of new states. In recent years there has also been a movement toward more democratic states. By looking at the conditions for a democracy, Lipset indicates what these states need if they are truly to achieve democracy. Considering these ideas, what do you see as the future of world affairs?

As you read, ask yourself the following questions:

1. *What are some of the requisites needed to maintain democracy?*
2. *Based on the requisites needed to maintain democracy, how is democracy doing? Why?*

GLOSSARY

Efficacy Producing desired results.
Bourgeois Member of the middle class.
Facade Superficial appearance.
Meritocratic Hiring on the basis of ability rather than patronage.
Totalitarian Absolute control by the state.

THE RECENT EXPANSION OF democracy, what Huntington (1991) has called "the third wave," began in the mid-1970s in Southern Europe. Then, in the early and mid-1980s, it spread to Latin America and to Asian countries like Korea, Thailand, and the Philippines, and then in the late 1980s and early 1990s to Eastern Europe, the Soviet Union, and parts of sub-Saharan Africa. Not long ago, the overwhelming majority of the members of the United Nations had authoritarian systems. As of the end of 1993, over half, 107 out of 186 countries, have com-

petitive elections and various guarantees of political and individual rights—that is more than twice the number of two decades earlier in 1970 (Karatnycky 1994:6; *Freedom Review* 1993:3–4, 10). The move toward democracy is not a simple one. . . . Countries that previously have had authoritarian regimes may find it difficult to set up a legitimate democratic system, since their traditions and beliefs may be incompatible with the workings of democracy.

In his classic work *Capitalism, Socialism, and Democracy*, Schumpeter (1950) defined democracy as "that institutional arrangement for arriving at political decisions in which individuals acquire the power to decide by means of a competitive struggle for the people's vote" (p. 250). This definition is quite broad and my discussion here cannot hope to investigate it exhaustively. Instead, I focus here on . . . the factors and processes affecting the prospects for the institutionalization of democracy.

How Does Democracy Arise?

POLITICS IN IMPOVERISHED COUNTRIES

In discussing democracy, I want to clarify my biases and assumptions at the outset. I agree with the basic concerns of the founding fathers of the United States—that government, a powerful state, is to be feared (or suspected, to use the lawyer's term), and that it is necessary to find means to control governments through checks and balances. In our time, as economists have

From the American Sociological Review, *1994, Vol. 59 (February:1–22).*

documented, this has been particularly evident in low-income nations. The "Kuznets curve" (Kuznets 1955; 1963; 1976), although still debated, indicates that when a less developed nation starts to grow and urbanize, income distribution worsens, but then becomes more equitable as the economy industrializes.[1] . . . Before development, the class income structure resembles an elongated pyramid, very fat at the bottom, narrowing or thin toward the middle and top (Lipset 1981:51). Under such conditions, the state is a major, usually *the* most important, source of capital, income, power, and status. This is particularly true in statist systems, but also characterizes many so-called free market economies. For a person or governing body to be willing to give up control because of an election outcome is astonishing behavior, not normal, not on the surface a "rational choice," particularly in new, less stable, less legitimate politics.

Marx frequently noted that intense inequality is associated with scarcity, and therefore that socialism, which he believed would be an egalitarian and democratic system with a politically weak state, could only occur under conditions of abundance (Marx 1958:8–9). To try to move toward socialism under conditions of material scarcity would result in sociological abortions and in repression. The Communists proved him correct. Weffort (1992), a Brazilian scholar of democracy, has argued strongly that, although "the political equality of citizens, . . . is . . . possible in societies marked by a high degree of [economic] inequality," the contradiction between political and economic inequality "opens the field for tensions, institutional distortions, instability, and recurrent violence . . . [and may prevent] the consolidation of democracy" (p. 22). Contemporary social scientists find that greater affluence and higher rates of well-being have been correlated with the presence of democratic institutions (Lipset, Seong, and Torres 1993:156–58; see also Diamond 1992a). Beyond the impact of national

wealth and economic stratification, contemporary social scientists also agree with Tocqueville's analysis, that social equality, perceived as equality of status and respect for individuals regardless of economic condition, is highly conducive for democracy (Tocqueville 1976: vol. 2, 162–216). . . . Weffort (1992) emphasized, "such a 'minimal' social condition is absent from many new democracies, . . . [which can] help to explain these countries' typical democratic instability" (p. 18).

THE ECONOMY AND THE POLITY

In the nineteenth century, many political theorists noted the relationship between a market economy and democracy (Lipset 1992: 2). As Glassman (1991) has documented, "Marxists, classical capitalist economists, even monarchists accepted the link between industrial capitalism and parliamentary democracy" (p. 65). Such an economy, including a substantial independent peasantry, produces a middle class that can stand up against the state and provide the resources for independent groups. . . . Schumpeter (1950) held that, "modern democracy is a product of the capitalist process" (p. 297). Moore (1966), noting his agreement with the Marxists, concluded, "No bourgeois, no democracy" (p. 418).

Berger (1992), from the conservative side, noted . . . "[t]here have been numerous cases of *non*democratic market economies" (p. 9). That is, capitalism has been a necessary, but not sufficient condition (Diamond 1993a). As reported earlier (Diamond, Linz, and Lipset 1988:xxi), those democracies "most advanced in their capitalist development (size of market sector of the economy, autonomy of their entrepreneurial class) are also those that have been most exposed to pressures for democracy."

Waisman (1992:140–55), seeking to explain why some capitalist societies, particularly in Latin America, have not been democratic, has suggested that . . . a free market needs democracy and vice versa.

But while the movement toward a market economy and the growth of an independent

---

[1]These generalizations do not apply to the East Asian NICS, South Korea, Taiwan, and Singapore.

middle-class have weakened state power and enlarged human rights and the rule of law, it has been the working class, particularly in the West, that has demanded the expansion of suffrage and the rights of parties. As John Stephens (1993) noted, "Capitalist development is associated with the rise of democracy in part because it is associated with the transformation of the class structure strengthening the working class" (p. 438).

Corruption, a major problem of governance, is inherent in systems built on poverty (Klitgaard 1991:86–98). The state must allocate resources it controls, such as jobs, contracts, and investment capital. When the state is poor, it emphasizes particularistic, personalistic criteria. The elimination of personal "networking" on resources controlled or influenced by the state is obviously impossible. Formulating laws and norms to reduce the impact of personal networks, rules that require the application of impersonal meritocratic standards, is desirable; but doing so has taken a long time to institutionalize in the now-wealthy countries, and has usually gone against the traditions and needs of people in less affluent ones. Hence, as Jefferson, Madison, and others argued in the late eighteenth century, the less the state has to do the better; the fewer economic resources the state can directly control, the greater the possibilities for a free polity.

Therefore, a competitive market economy can be justified sociologically and politically as the best way to reduce the impact of nepotistic networks. The wider the scope of market forces, the less room there will be for rent-seeking by elites with privileged access to state power and resources. Beyond limiting the power of the state, however, standards of propriety should be increased in new and poor regimes, and explicit objective standards should be applied in allocating aid, loans, and other sources of capital from outside the state. Doing this, of course, would be facilitated by an efficient civil service selected by meritocratic standards. It took many decades for civil service reforms to take hold in Britain, the United States, and various European countries (Johnston 1991:53–56). To change the norms and rules in contemporary impoverished countries will not be achieved easily. . . .

THE CENTRALITY OF POLITICAL CULTURE

Democracy requires a supportive culture, the acceptance by the citizenry and political elites of principles underlying freedom of speech, media, assembly, religion, of the rights of opposition parties, of the rule of law, of human rights, and the like. . . . Such norms do not evolve overnight. Attempts to move from authoritarianism to democracy have failed after most upheavals from the French Revolution in 1789 to the February Revolution in Russia in 1917, from those in most new nations in Latin America in the nineteenth century to those in Africa and Asia after World War II. Linz (1988) and Huntington (1991) noted that the two previous waves of democratization were followed by "reverse waves" which witnessed the revival of authoritarianism. "Only four of the seventeen countries that adopted democratic institutions between 1915 and 1931 maintained them throughout the 1920s and 1930s . . . . [O]ne-third of the 32 working democracies in the world in 1958 had become authoritarian by the mid-1970s" (Huntington 1991:17–21).

These experiences do not bode well for the current efforts in the former Communist states of Eastern Europe or in Latin America and Africa. And the most recent report by Freedom House concludes: "As 1993 draws to a close, freedom around the world is in retreat while violence, repression, and state control are on the increase. The trend marks the first increase in five years . . ." (Karatnycky 1994:4). A "reverse wave" in the making is most apparent in sub-Saharan Africa, where "9 countries showed improvement while 18 registered a decline" (p. 6). And in Russia, a proto-fascist movement led all other parties, albeit with 24 percent of the vote, in the December 1993 elections, while the Communists and their allies secured over 15 percent.

Almost everywhere that the institutionalization of democracy has occurred, the process has been a gradual one in which opposition and individual rights have emerged in the give and take of politics, (Sklar 1987:714). . . .

As a result, democratic systems developed gradually, at first with suffrage, limited by and linked to property and/or literacy. Elites yielded slowly in admitting the masses to the franchise and in tolerating and institutionalizing opposition rights. . . . As Dahl (1971:36–37) has emphasized, parties such as the Liberals and Conservatives in nineteenth-century Europe, formed for the purpose of securing a parliamentary majority rather than to win the support of a mass electorate, were not pressed to engage in populist demagoguery.

Comparative politics suggest that the more the sources of power, status, and wealth are concentrated in the state, the harder it is to institutionalize democracy. Under such conditions the political struggle tends to approach a zero-sum game in which the defeated lose all. The greater the importance of the central state as a source of prestige and advantage, the less likely it is that those in power—or the forces of opposition—will accept rules of the game that institutionalize party conflict and could result in the turnover of those in office. Hence, once again it may be noted, the chances for democracy are greatest where, as in the early United States and to a lesser degree in other Western nations, the interaction between politics and economy is limited and segmented. In Northern Europe, democratization let the monarchy and the aristocracy retain their elite status, even though their powers were curtailed. In the United States, the central state was not a major source of privilege for the first half-century or more, and those at the center thus could yield office easily.

Democracy has never developed anywhere by plan, except when it was imposed by a democratic conqueror, as in post-World War II Germany and Japan. From the United States to Northern Europe, freedom, suffrage, and the rule of law grew in a piecemeal, not in a planned, fashion. To legitimate themselves, governmental parties, even though they did not like it, ultimately had to recognize the right of oppositions

to exist and compete freely. Almost all the heads of young democracies, from John Adams and Thomas Jefferson to Indira Gandhi, attempted to suppress their opponents. As noted before, most new democracies are soon overthrown, as in France prior to 1871, in various parts of Europe after 1848, in Eastern, Central, and Southern Europe after World War I, and repeatedly in Latin America and Africa. Democratic successes have reflected the varying strengths of minority political groups and lucky constellations, as much or more than commitments by new office holders to the democratic process.

Cross-national historical evaluations of the correlates of democracy have found that cultural factors appear even more important than economic ones (Lipset et al. 1993:168–70). . . . Dahl (1970:6), Kennan (1977:41–43), and Lewis (1993:93–94) have emphasized that the first group of countries that became democratic in the nineteenth century (about 20 or so) were Northwest European or settled by Northwest Europeans. "The evidence has yet to be produced that it is the natural form of rule for peoples outside these narrow perimeters" (Kennan 1977:41–43).[2] Lewis (1993), an authority on the Middle East, has reiterated Kennan's point: "No such [democratic] system has originated in any other cultural tradition; it remains to be seen whether such a system transplanted and adapted in another culture can long survive" (pp. 93–94).

More particularly, recent statistical analyses of the aggregate correlates of political regimes have indicated that having once been a British colony is the variable most highly correlated with democracy (Lipset et al. 1993:168). As Weiner (1987) has pointed out, beyond the experiences in the Americas and Australasia in the nineteenth century, "every country with a population of at least 1 million (and almost all the smaller countries as well) that has emerged from colonial rule and has had a continuous democratic experience is a former British colony" (p. 20). The factors

[2]That evidence, of course, has emerged in recent years in South and East Asia, Latin America, and various countries descended from Southern Europe.

underlying this relationship are not simple (Smith 1978). In the British/non-British comparison, many former British colonies, such as those in North America before the revolution or India and Nigeria in more recent times, had elections, parties, and the rule of law before they became independent. In contrast, the Spanish, Portuguese, French, Dutch, and Belgian colonies, and former Soviet-controlled countries did not allow for the gradual incorporation of "out groups" into the polity. Hence democratization was much more gradual and successful in the ex-British colonies than elsewhere; their pre-independence experiences were important as a kind of socialization process and helped to ease the transition to freedom.

RELIGIOUS TRADITION

Religious tradition has been a major differentiating factor in transformations to democracy (Huntington 1993:25–29). Historically, there have been negative relationships between democracy and Catholicism, Orthodox Christianity, Islam, and Confucianism; conversely Protestantism and democracy have been positively interlinked. These differences have been explained by (1) the much greater emphasis on individualism in Protestantism and (2) the traditionally close links between religion and the state in the other four religions. Tocqueville (1975) and Bryce (1901) emphasized that democracy is furthered by a separation of religious and political beliefs, so that political stands are not required to meet absolute standards set down by the church. . . .

Protestants, particularly the non–state-related sects, have been less authoritarian, more congregational, participatory, and individualistic. Catholic countries, however, have contributed significantly to the third wave of democratization during the 1970s and 1980s, reflecting "the major changes in the doctrine, appeal, and social and political commitments of the Catholic Church that occurred . . . in the 1960s and 1970s" (Huntington 1991:281, 77–85). The changes that have occurred are primarily a result of the delegitimation of so-called ultra-rightist or clerical fascism in Catholic thought and poli-

tics, an outgrowth of the defeat of fascism in Europe, and considerable economic growth in many major Catholic lands in post-war decades, countries such as Italy, Spain, Quebec, Brazil, and Chile.

Conversely, Moslem (particularly Arab) states have not taken part in the third wave of democratization. Almost all remain authoritarian. Growth of democracy in the near future in most of these countries is doubtful because "notions of political freedom are not held in common . . .; they are alien to Islam" (Vatikiotis 1988:118). As Wright (1992) has stated, Islam "offers not only a set of spiritual beliefs, but a set of rules by which to govern society" (p. 133). . . .

Kazancigil (1991) has offered parallel explanations of the weakness of democracy in Islam with those for Orthodox Christian lands as flowing from their failures "to dissociate the religious from the political spheres" (p. 345). In Eastern Europe, particularly Russia, the Orthodox Church has closely linked the two. As Guroff and Guroff (1993) emphasized: "The Church has always been an organ of the Russian state, both under the Tsar and under the Soviet Union. . . . Neither in Tsarist Russia, nor in the Soviet Union has the Orthodox Church played an active role in the protection of human rights or religious tolerance" (pp. 10–11).

Noting that in Confucian China "no church or cultural organization . . . existed independently of the state" (p. 25), and that "Islam has emphasized the identity between the religious and political communities," Eisenstadt (1968) stressed the resultant "important similarity between the Chinese and Islamic societies" (p. 27). Huntington (1993) reported that "no scholarly disagreement exists regarding the proposition that traditional Confucianism was either undemocratic or antidemocratic" (p. 15; see also Whyte 1992:60). Lucian Pye (1968; see also Pye with Pye 1985) has pointed to the similarities between Confucian and Communist beliefs about "authority's rights to arrogance . . . both have been equally absolute . . . upholding the monopolies of officialdom. . . . It is significant that . . . both Confucianism and Maoism in ideological content, have explicitly stressed the problems of authority and order" (Pye 1968:16). Though somewhat less pessimistic, He Baogang's (1992) evaluation of cultural factors in mainland China concluded that "evidence reveals that the antidemocratic culture is currently stronger than the factors related to a democratic one" (p. 134). Only Japan, the most diluted Confucian country, "had sustained experience with democratic government prior to 1990, . . . [although its] democracy was the product of an American presence" (Huntington 1991:15). The others—Korea, Vietnam, Singapore, and Taiwan—were autocratic. . . . The situation, of course, has changed in recent years in response to rapid economic growth, reflecting the ways in which economic changes can impact on the political system undermining autocracy.

But India, a Hindu country that became democratic prior to industrialization, is different:

> The most salient feature of Indian civilization, from the point of view of our discussion, is that it is probably the only complete, highly differentiated civilization which throughout history has maintained its cultural identity without being tied to a given political framework. . . . [T]o a much greater degree than in many other historical imperial civilizations politics were conceived in secular forms. . . . Because of the relative dissociation between the cultural and the political order, the process of modernization could get underway in India without being hampered by too specific a traditional-cultural orientation toward the political sphere. (Eisenstadt 1968:32)

These generalizations about culture do not auger well for the future of the third wave of democracy in the former Communist countries. The Catholic Church played a substantial role in Poland's move away from Soviet Communism. But as noted previously, historically deeply religious Catholic areas have not been among the most amenable to democratic ideas. Poland is now troubled by conflicts flowing from increasing Church efforts to affect politics in Eastern Europe even as it relaxes its policies in Western Europe and most of the Americas. Orthodox Christianity

is hegemonic in Russia and Belarus. The Ukraine is dominated by both the Catholic and Orthodox Churches. And fascists and Communists are strong in Russia and the Ukraine. Moslems are a significant group in the Central Asian parts of the former Soviet Union, the majority in some—these areas are among the consistently least democratic of the successor Soviet states. Led by the Orthodox Serbians, but helped by Catholic Croats and Bosnian Moslems, the former Yugoslavia is being torn apart along ethnic and religious lines with no peaceful, much less democratic, end in sight. We are fooling ourselves if we ignore the continuing dysfunctional effects of a number of cultural values and the institutions linked to them.

But belief systems change; and the rise of capitalism, a large middle class, an organized working class, increased wealth, and education are associated with secularism and the institutions of civil society which help create autonomy for the state and facilitate other preconditions for democracy. In recent years, nowhere has this been more apparent than in the economically successful Confucian states of East Asia—states once thought of as nearly hopeless candidates for both development and democracy. Tu (1993) noted their totally "unprecedented dynamism in democratization and marketization. Singapore, South Korea, and Taiwan all successfully conducted national elections in 1992, clearly indicating that democracy in Confucian societies is not only possible but also practical" (p. viii). Nathan and Shi (1993), reporting on "the first scientifically valid national sample survey done in China on political behavior and attitudes," stated: "When compared to residents of some of the most stable, long-established democracies in the world, the Chinese population scored lower on the variables we looked at, but not so low as to justify the conclusion that democracy is out of reach" (p. 116). Surveys which have been done in Russia offer similar positive conclusions (Gibson and Duch 1993), but the December, 1993 election in which racist nationalists and pro-Communists did well indicate much more is needed. Democracy is not taking root in much

of the former Soviet Union, the less industrialized Moslem states, nor many nations in Africa.
The end is not in sight for many of the efforts at
new democracies; the requisite cultural changes
are clearly not established enough to justify the
conclusion that the "third wave" will not be reversed. According to the Freedom House survey,
during 1993 there were "42 countries register-
ing a decline in their level of freedom [political
rights and civil liberties] and 19 recording gains"
(Karatnycky 1994:5).[3]

Institutionalization

New democracies must be institutionalized, con-
solidated, and become legitimate. They face many
problems. . . .

LEGITIMACY

Political stability in democratic systems cannot
rely on force. The alternative to force is legitimacy, an accepted systemic "title to rule. . . ."

Weber (1946), the fountainhead of legitimacy
theory, named three ways by which an authority
may gain legitimacy. These may be summarized:

(1) *Traditional*—through "always" having possessed the authority, the best example being the
title held in monarchical societies.

(2) *Rational-legal*—when authority is obeyed
because of a popular acceptance of the appropriateness of the system of rules under which they
have won and held office. In the United States,
the Constitution is the basis of all authority.

(3) *Charismatic*—when authority rests upon
faith in a leader who is believed to be endowed
with great personal worth, either from God, as in
the case of a religious prophet or simply from the
display of extraordinary talents. The "cult of personality" surrounding many leaders is an illustration of this (pp. 78–79).

Legitimacy is best gained by prolonged effectiveness, effectiveness being the actual performance of the government and the extent

to which it satisfies the basic needs of most
of the population and key power groups (such
as the military and economic leaders) (Lipset
[1960] 1981:64–70; Lipset 1979:16–23; Linz
1978:67–74; Linz 1988:79–85; Diamond et al.
1990:9–16). This generalization, however, is of
no help to new systems for which the best immediate institutional advice is to separate the
source and the agent of authority.

The importance of this separation cannot be
underestimated. The agent of authority may be
strongly opposed by the electorate and may
be changed by the will of the voters, but the
essence of the rules, the symbol of authority,
must remain respected and unchallenged. Hence,
citizens obey the laws and rules, even while dis-
liking those who enforce them. . . .

Rational-legal legitimacy is weak in most new
democratic systems, since the law had previously
operated in the interests of a foreign exploiter or
domestic dictator. Efforts to construct rational-
legal legitimacy necessarily involve extending the
rule of the law and the prestige of the courts,
which should be as independent from the rest of
the polity as possible. As Ackerman (1992:60–62)
and Weingast (1993) note, in new democracies,
these requirements imply the need to draw up a
"liberal" constitution *as soon as possible*. The constitution can provide a basis for legitimacy, for
limitations on state power, and for political and
economic rights. . . .

The postwar democratic regimes of the formerly fascist states, created, like the Weimar Republic, under the auspices of the conquerors,
clearly had no legitimacy at their outset. But they
had the advantage of the subsequent postwar
"economic miracles" which produced jobs and
a steadily rising standard of living. These new
regimes have been economically viable for over
four decades. The stability of these democratic
systems is also linked to the discrediting of
anti-democratic right-wing tendencies—these
forces were identified with fascism and military
defeat.

To reiterate, if democratic governments which
lack traditional legitimacy are to survive, they
must be effective, or as in the example of some

---

[3]In the Freedom House survey, a country may move up
or down with respect to measures of freedom without changing its status as a democratic or authoritarian system.

new Latin American and post-communist democracies, may have acquired a kind of negative legitimacy—an inoculation against authoritarianism because of the viciousness of the previous dictatorial regimes. Newly independent countries that are post-revolutionary, post-coup, or post-authoritarian regimes are inherently low in legitimacy. Thus most of the democracies established in Europe after World War I as a result of the overthrow of the Austro-Hungarian, German, and Czarist Russian empires did not last. . . .

All other things being equal, an assumption rarely achieved, nontraditional authoritarian regimes are more brittle than democratic ones. By definition, they are less legitimate; they rely on force rather than belief to retain power. Hence, it may be assumed that as systems they are prone to be disliked and rejected by major segments of the population. . . .

The record, as in the case of the Soviet Union, seems to contradict this, since that regime remained in power for three-quarters of a century. However, a brittle, unpopular system need not collapse. Repressive police authority, a powerful army, and a willingness by rulers to use brute force may maintain a regime's power almost indefinitely. The breakdown of such a system may require a major catalytic event, a defeat in war, a drastic economic decline, or a break in the unity of the government elite. In the Soviet Union, a variety of economic and social data available before Gorbachev came to power indicated enormous weaknesses—declines in productivity and increases in mortality—that suggested serious malfunctions in the system; the size and scope of its secret police attested to low legitimacy. . . .

In contrast to autocracies, democratic systems rely on and seek to activate popular support and constantly compete for such backing. Government ineffectiveness need not spill into other parts of the society and economy. Opposition actually serves as a communication mechanism, focusing attention on societal and governmental problems. Freedom of opposition encourages a free flow of information about the economy as well as about the polity. . . .

Non-traditional authoritarian regimes seek to gain legitimacy through cults of personality (e.g., Napoleon, Toussant, Diaz, Mussolini, Hitler). New autocrats lack the means to establish legal-rational legitimacy through the rule of law. Communist governments, whose Marxist ideology explicitly denied the importance of "great men" in history and stressed the role of materialist forces and "the people," were forced to resort to charismatic legitimacy. Their efforts produced the cults of Lenin, Stalin, Mao, Tito, Castro, Ho, Kim, and others. . . .

But charismatic legitimacy is inherently unstable. As mentioned earlier, a political system operates best when the source of authority is clearly separated from the agent of authority. If the ruler and his or her policies are seen as oppressive or exploitive, the regime and its rules will also be rejected. People will not feel obligated to conform or to be honest; force alone cannot convey a "title to rule."

Executive and Electoral Systems . . .

EXECUTIVE SYSTEMS

In considering the relation of government structure to legitimacy it has been suggested that republics with powerful presidents will, all other things being equal, be more unstable than parliamentary ones in which powerless royalty or elected heads of state try to act out the role of a constitutional monarch. In the former, where the executive is chief of state, symbolic authority and effective power are combined in one person, while in the latter they are divided. With a single top office, it is difficult for the public to separate feelings about the regime from those held toward the policy makers. The difficulties in institutionalizing democracy in the many Latin American presidential regimes over the last century and a half may reflect this problem. The United States presents a special case, in which, despite combining the symbolic authority and power into the Presidency, the Constitution has been so hallowed by ideology and prolonged effectiveness for over 200 years, that it, rather than those who occupy the offices it specifies, has become the accepted ultimate source of authority. . . .

Evaluation of the relative worth of presidential and parliamentary systems must also consider the nature of each type. In presidential regimes, the power to enact legislation, pass budgets and appropriations, and make high level appointments are divided among the president and (usually two) legislative Houses; parliamentary regimes are unitary regimes, in which the prime minister and cabinet can have their way legislatively. A prime minister with a parliamentary majority, as usually occurs in most Commonwealth nations and a number of countries in Europe, is much more powerful and less constrained than a constitutional president who can only propose while Congress disposes (Lijphart 1984:4–20). The weak, divided-authority system has worked in the United States, although it has produced much frustration and alienation at times. But, as noted, the system has repeatedly broken down in Latin America, although one could argue that this is explained not by the constitutional arrangements, but by cultural legacies and lower levels of productivity. Many parliamentary systems have failed to produce stable governments because they lack operating legislative majorities. . . . There is no consensus among political scientists as to which system, presidential or parliamentary, is superior, since it is possible to point to many failures for both types.

ELECTORAL SYSTEMS

The procedures for choosing and changing administrations also affect legitimacy (Lipset 1979:293–306). Elections that offer the voters an effective way to change the government and vote the incumbents out will provide more stability; electoral decisions will be more readily accepted in those systems in which electoral rules, distribution of forces, or varying party strengths make change more difficult.

Electoral systems that emphasize single-member districts, such as those in the United States and in much of the Commonwealth, press the electorate

to choose between two major parties. The voters know that if they turn against the government party, they can replace it with the opposition. The parties in such systems are heterogeneous coalitions, and while many voters frequently opt for the "lesser evil," since the opposition usually promises to reverse course, incumbents can be punished for unpopular policies or for happening to preside over depressing events.

In systems with proportional representation, the electorate may not be able to determine the composition of the government. In this type, a representation is assigned to parties which corresponds to their proportions of the vote. . . . Where no party has a majority, alliances may be formed out of diverse forces. A party in a government coalition may gain votes, but may then be excluded from the new cabinet formed after the election. Small, opportunistic, or special interest parties may hold the balance of power and determine the shape and policies of post-election coalitions. The tendency toward instability and lack of choice in proportional systems can be reduced by setting up a minimum vote for representation, such as the five percent cut-off that exists in Germany and Russia. In any case, electoral systems, whether based on single-member districts or proportional representation, cannot guarantee particular types of partisan results (Lipset 1979:293–306; Gladdish 1993).

Civil Society and Political Parties

CIVIL SOCIETY AS A POLITICAL BASE

More important than electoral rules in encouraging a stable system is a strong civil society—the presence of myriad "mediating institutions," including "groups, media, and networks" (Diamond 1993b:4), that operate independently between individuals and the state. These constitute "subunits, capable of opposing and countervailing the state" (Gellner 1991:500). . . .

Citizen groups must become the bases of—the sources of support for—the institutionalized political parties which are a necessary condition for—part of the very definition of—a modern democracy. . . .

A fully operative civil society is likely to also be a participant one. Organizations stimulate interests and activity in the larger polity; they can be consulted by political institutions about projects that affect them and their members, and they can transfer this information to the citizenry. Civil organizations reduce resistance to unanticipated changes because they prevent the isolation of political institutions from the polity and can smooth over, or at least recognize, interest differences early on. . . .

Totalitarian systems, however, do not have effective civil societies. Instead, they either seek to eliminate groups mediating between the individual and the state or to control these groups so there is no competition. And while by so doing they may undermine the possibility for *organized* opposition, they also reduce group effectiveness generally, and reduce the education of individuals for innovative activities (i.e., Tocqueville's "civil partnerships" [1976, vol. 2:124]). . . .

The countries of Eastern Europe and the former Soviet Union, however, are faced with the consequences of the absence of modern civil society, a lack that makes it difficult to institutionalize democratic polities. These countries have not had the opportunity to form the civil groups necessary to coalesce into stable political parties, except through churches in some nations, such as Poland, and assorted small autonomous illegal networks (Sadowski 1993:171–80). Instead, they have had to create parties "from scratch." Ideologically splintered groups must oppose the former Communists, who have been well organized for many years and have constructed their own coalitions. "Instead of consolidation, there is fragmentation: 67 parties fought Poland's most recent general election, 74 Romania's" (*Economist* 1993a:4). As a result, the former Communists (now "socialists") have either been voted in as the majority party in parliament, as in Lithuania, or have become the largest party heading up a coalition cabinet, as in Poland. In January 1992, the Communist-backed candidate for pres-

ident in Bulgaria garnered 43 percent of the vote (Malia 1992:73). These situations are, of course, exacerbated by the fact that replacing command economies by market processes is difficult, and frequently conditions worsen before they begin to improve.

Recent surveys indicate other continuing effects of 45 to 75 years of Communist rule. An overwhelming majority (about 70 percent) of the population in nearly all of the countries in Eastern Europe agree that "the state should provide a place of work, as well as a national health service, housing, education, and other services" (*Economist* 1993a:5). . . .

POLITICAL PARTIES AS MEDIATORS

Political parties themselves must be viewed as the most important mediating institutions between the citizenry and the state (Lipset 1993). And a crucial condition for a stable democracy is that major parties exist that have an almost permanent significant base of support. That support must be able to survive clear-cut policy failures by the parties. If this commitment does not exist, parties may be totally wiped out, thus eliminating effective opposition. The Republicans in the United States, for example, though declining sharply in electoral support, remained a major opposition party in the early 1930s, despite the fact that the Great Depression started under their rule and reached severe economic depths in unemployment, bankruptcy, and stock market instability never seen before.

If, as in new democracies, parties do not command such allegiance, they can be easily eliminated. The Hamiltonian Federalist party, which competed in the early years of the American Republic with the Jeffersonian Democratic-Republicans, declined sharply after losing the Presidency in 1800 and soon died out (Lipset 1979:40–41; Dauer 1953). . . . It may be argued then, that having at least two parties with an uncritically loyal mass base comes close to being a necessary condition for a stable democracy. Democracy requires strong parties that can offer alternative policies and criticize each other. Historically, the cross-cutting cleavages of impoverished India linked to allegiances of caste, linguistic, and religious groupings have contributed to the institutionalization of democracy by producing "strong commitment to parties" on the part of a large majority (Das Gupta 1989:95; Diamond 1989b:19). More recently, volatility and decay in the party system has been associated with a decline in the quality and stability of democracy in India (Kohli 1992).

SOURCES OF POLITICAL PARTY SUPPORT

. . . In *Party Systems and Voter Alignments* (1967), we analyzed modern political divisions in Europe as outgrowths of two revolutions, the National Revolution and the Industrial Revolution. These transformations created social cleavages that became linked to party divisions and voting behavior. The first was political, and resulted in *center-periphery* conflicts between the national state and culture and assorted subordinate ones, such as ethnic, linguistic, or religious groups often located in the peripheries, the outlying regions. This political revolution also led to *state-church* conflicts — struggles between the state, which sought to dominate, and the church, which tried to maintain its historic corporate rights. The Industrial Revolution was economic and gave rise to *land-industry* conflicts between the landed elite and the growing bourgeois class. This was followed by the *capitalist-worker* conflicts — the struggles on which Marx focused.

These four sources of conflict, *center-periphery, state-church, land-industry,* and *capitalist-worker,* have continued to some extent in the contemporary world, and have provided a framework for the party systems of the democratic polities, particularly in Europe. Class became the most salient source of conflict and voting, particularly after the extension of the suffrage to all adult males (Lipset and Rokkan 1967). Both Tocqueville (1976:vol. 2, 89–93), in the early nineteenth century and Bryce (1901:335), at the end of it, noted that at the bottom of the American political party conflict lay the struggle between

aristocratic and democratic interests and sentiments. . . . Given all the transformations in Western society over the first half of the twentieth century, it is noteworthy how little the formal party systems changed. Essentially the conflicts had become institutionalized — the Western party systems of the 1990s resemble those of pre-World War II. . . .

Beginning in the mid-1960s, the Western world appears to have entered a new political phase. It is characterized by the rise of so-called "post-materialistic issues, a clean environment, use of nuclear power, a better culture, equal status for women and minorities, the quality of education, international relations, greater democratization, and a more permissive morality, particularly as affecting familial and sexual issues" (Lipset 1981:503–21). These have been perceived by some social analysts as the social consequences of an emerging third "revolution," the Post-Industrial Revolution, which is introducing new bases of social and political conflict. Inglehart (1990) and others have pointed to new cross-cutting lines of conflict — an *industrial-ecology* conflict — between the adherents of the industrial society's emphasis on production (who also hold conservative positions on social issues) and those who espouse the post-industrial emphasis on the quality-of-life and liberal social views when dealing with ecology, feminism, and nuclear energy. Quality-of-life concerns are difficult to formulate as party issues, but groups such as the Green parties and the New Left or New Politics — all educated middle class groups — have sought to foster them. . . .

The one traditional basis of party differentiation that seems clearly to be emerging in Russia is the center-periphery conflict, the first one that developed in Western society. The second, church-state (or church-secular), is also taking shape to varying degrees. Land-industry (or rural-urban) tension is somewhat apparent. Ironically, the capitalist-worker conflict is as yet the weakest, perhaps because a capitalist class and an independently organized working-class do not yet exist. Unless stable parties can be formed, competitive democratic politics is not likely to last in many of the new Eastern European and Central Asian polities. . . .

The Rule of Law and Economic Order

Finally, order and predictability are important for the economy, polity, and society. The Canadian Fathers of Confederation, who drew up the newly unified country's first constitution in 1867, described the Constitution's objective as "peace, order, and good government" (Lipset 1990b:xiii). Basically, they were talking about the need for the "rule of law," for establishing rules of "due process," and an independent judiciary. Where power is arbitrary, personal, and unpredictable, the citizenry will not know how to behave; it will fear that any action could produce an unforeseen risk. Essentially, the rule of law means: (1) that people and institutions will be treated equally by the institutions administering the law—the courts, the police, and the civil service; and (2) that people and institutions can predict with reasonable certainty the consequences of their actions, at least as far as the state is concerned. . . .

In discussing "the social requisites of democracy," I have repeatedly stressed the relationship between the level of economic development and the presence of democratic government. As noted, a host of empirical studies has continued to find significant correlations between socioeconomic variables (such as GNP, educational attainments, level of health care) on the one hand, and political outcomes (such as free polities and human rights) on the other (Lipset et al. 1993; Diamond 1992a; Inkeles 1991; Bollen and Jackman 1985a; Bollen and Jackman 1985b; Bollen 1979; 1980; Flora 1973; Flanigan and Fogelman 1971; Olsen 1968; Neubauer 1967; Cutright 1963). . . .

Clearly, socioeconomic correlations are merely associational, and do not necessarily indicate cause. Other variables, such as the force of historical incidents in domestic politics, cultural factors, events in neighboring countries, diffusion effects from elsewhere, leadership, and movement behavior can also affect the nature of the polity.

Thus, the outcome of the Spanish Civil War, determined in part by other European states, placed Spain in an authoritarian mold, much as the allocation of Eastern Europe to the Soviet Union after World War II determined the political future of that area and that Western nations would seek to prevent the electoral victories of Communist-aligned forces. Currently, international agencies and foreign governments are more likely to endorse pluralistic regimes. . . .

Conclusion

Democracy is an international cause. A host of democratic governments and parties, as well as various non-governmental organizations (NGOs) dedicated to human rights, are working and providing funds to create and sustain democratic forces in newly liberalized governments and to press autocratic ones to change (*Economist* 1993c:46). Various international agencies and units, like the European Community, NATO, the World Bank, and the International Monetary Fund (IMF), are requiring a democratic system as a condition for membership or aid. A diffusion, a contagion, or demonstration effect seems operative, as many have noted, one that encourages democracies to press for change and authoritarian rulers to give in. It is becoming both uncouth and unprofitable to avoid free elections, particularly in Latin America, East Asia, Eastern Europe, and to some extent in Africa (Ake 1991:33). Yet the proclamation of elections does not ensure their integrity. The outside world can help, but the basis for institutionalized opposition, for interest and value articulation, must come from within.

Results of research suggest that we be cautious about the long-term stability of democracy in many of the newer systems given their low level of legitimacy. As the Brazilian scholar Francisco Weffort (1992) has reminded us, "In the 1980s, the age of new democracies, the processes of political democratization occurred at the same moment in which those countries suffered the experience of a profound and prolonged economic crisis that resulted in social exclusion and massive poverty. . . . Some of those countries are building a political democracy on top of a minefield of social apartheid . . ." (p. 20). Such conditions could easily lead to breakdowns of democracy as have already occurred in Algeria, Haiti, Nigeria, and Peru, and to the deterioration of democratic functioning in countries like Brazil, Egypt, Kenya, the Philippines, and the former Yugoslavia, and some of the trans-Ural republics or "facade democracies," as well as the revival of anti-democratic movements on the right and left in Russia and in other formerly Communist states.

What new democracies need, above all, to attain legitimacy is efficacy—particularly in the economic arena, but also in the polity. . . .

REFERENCES

Ake, Claude. 1991. "Rethinking African Democracy." *Journal of Democracy* 2(1):32–47.

Ackerman, Bruce. 1992. *The Future of Liberal Revolution*. New Haven, CT: Yale University.

Amalrik, André. 1970. *Will the Soviet Union Survive Until 1984?* New York: Harper and Row.

Berger, Peter. 1986. *The Capitalist Revolution*. New York: Basic Books.

———. 1992. "The Uncertain Triumph of Democratic Capitalism." *Journal of Democracy* 3(3):7–17.

Bryce, James. 1901. *Study in History and Jurisprudence*. New York: Oxford University.

Dahl, Robert. 1970. *After the Revolution: Authority in a Good Society*. New Haven, CT: Yale University.

———. 1971. *Polyarchy: Participation and Opposition*. New Haven, CT: Yale University.

Das Gupta, Jyotirindra. 1989. "India: Democratic Becoming and Combined Development." Pp. 53–104 in *Democracy in Developing Countries: Asia,* edited by L. Diamond, J. Linz, and S. M. Lipset. Boulder, CO: Lynne Rienner.

Dauer, Manning. 1953. *The Adams Federalists*. Baltimore, MD: Johns Hopkins.

Diamond, Larry. 1989b. "Introduction: Persistence, Erosion, Breakdown and Renewal." Pp. 1–52 in *Democracy in Developing Countries: Asia,* edited by L. Diamond, J. Linz, and S. M. Lipset. Boulder, CO: Lynne Rienner.

———. 1993a. "Economic Liberalization and Democracy." The Hoover Institution, Stanford University, Stanford, CA. Unpublished manuscript.

Diamond, Larry, Juan Linz, and Seymour Martin Lipset, eds. 1988. *Democracy in Developing Countries: Africa*. Boulder, CO: Lynne Rienner.

Economist. 1993a. Survey on Eastern Europe. March 13:1–22.

Economist. 1993b. "Russia Into the Swamp." May 22:59–60.

Economist. 1993c. "Aid for Africa: If You're Good." May 29:46.

Eisenstadt, Shmuel N. 1968. "The Protestant Ethic Theses in the Framework of Sociological Theory and Weber's Work." Pp. 3–45 in *The Protestant Ethic and Modernization: A Comparative View,* edited by S. N. Eisenstadt. New York: Basic Books.

Feshbach, Murray. 1978. "Population and Manpower Trends in the U.S.S.R." Paper presented at the conference on the Soviet Union Today, sponsored by the Kennan Institute for Advanced Russian Studies, Woodrow Wilson International Center for Scholars, Apr., Washington, DC.

———. 1982. "Issues in Soviet Health Problems." Pp. 203–27 in *Soviet Economy in the 1980s: Problems and Prospects, Part 2.* U.S. Congress, Joint Economic Committee. 97th Cong., 2d sess., 31 Dec. Washington, DC: Government Printing Office.

———. 1983. "Soviet Population, Labor Force and Health." Pp. 91–138 in *The Political Economy of the Soviet Union.* U.S. Congress, Joint Hearings of the House Committee on Foreign Affairs and Joint Economic Committee. 98th Cong., 1st sess., 26 July and 29 Sept. Washington, DC: Government Printing Office.

Gibson, James L. and Raymond M. Duch. 1993. "Emerging Democratic Values in Soviet Political Culture." Pp. 69–94 in *Public Opinion and Regime Change,* edited by A. A. Miller, W. M. Reisinger, and V. Hesli, Boulder, CO: Westview.

Gladdish, Ken. 1993. "The Primacy of the Particular." *Journal of Democracy* 4(1):53–65.

Glassman, Ronald. 1991. *China in Transition: Communism, Capitalism and Democracy.* Westport, CT: Praeger.

Guroff, Gregory and A. Guroff. 1993. "The Paradox of Russian National Identity." (Russian Littoral Project, Working Paper No. 16). College Park and Baltimore, MD: University of Maryland-College Park and The Johns Hopkins University SAIS.

He Baogang. 1992. "Democratization: Antidemocratic and Democratic Elements in the Political Culture of China." *Australian Journal of Political Science* 27:120–36.

Huntington, Samuel. 1968. *Political Order in Changing Societies*. New Haven, CT: Yale University.

———. 1991. *The Third Wave: Democratization in the Late Twentieth Century.* Norman, OK: University of Oklahoma.

———. 1993. "The Clash of Civilizations." *Foreign Affairs* 72(3):22–49.

Inglehart, Ronald. 1990. *Culture Shift in Advanced Industrial Society.* Princeton, NJ: Princeton University.

Karatnycky, Adrian. 1994. "Freedom in Retreat." *Freedom Review* 25(1):4–9.

Kazancigil, Ali. 1991. "Democracy in Muslim Lands: Turkey in Comparative Perspective." *International Social Science Journal* 43:343–60.

Kennan, George. 1977. *Clouds of Danger: Current Realities of American Foreign Policy.* Boston, MA: Little, Brown.

Klitgaard, Robert. 1991. "Strategies for Reform." *Journal of Democracy* 2(4):86–100.

Kohli, Atul. 1992. "Indian Democracy: Stress and Resilience." *Journal of Democracy* 3(1):52–64.

Kuznets, Simon. 1955. "Economic Growth and Income Inequality." *American Economic Review* 45:1–28.

———. 1963. "Quantitative Aspects of the Economic Growth of Nations: VIII, The Distribution of Income by Size." *Economic Development and Cultural Change* 11:1–80.

———. 1976. *Modern Economic Growth: Rate, Structure and Spread.* New Haven, CT: Yale University.

Lewis, Bernard. 1993. "Islam and Liberal Democracy." *Atlantic Monthly.* 271(2):89–98.

Lijphart, Arend. 1977. *Democracy in Plural Societies: A Comparative Exploration.* New Haven, CT: Yale University.

———. 1984. *Democracies, Patterns of Majoritarian and Consensus Government in Twenty-One Countries.* New Haven, CT: Yale University.

Linz, Juan J. 1988. "Legitimacy of Democracy and the Socioeconomic System." Pp. 65–97 in *Comparing Pluralist Democracies: Strains on Legitimacy,* edited by M. Dogan. Boulder, CO: Westview.

Lipset, Seymour Martin [1960]1981. *Political Man: The Social Bases of Politics.* Expanded ed. Baltimore, MD: Johns Hopkins.

———. 1990b. *Continental Divide: The Values and Institutions of the United States and Canada.* New York: Routledge.

———. 1992. "Conditions of the Democratic Order and Social Change: A Comparative Discussion." Pp. 1–14 in *Studies in Human Society: Democracy and Modernity,* edited by S. N. Eisenstadt. New York: E. J. Brill.

———. 1993. "Reflections on Capitalism, Socialism and Democracy." *Journal of Democracy* 4(2):43–53.

Lipset, Seymour Martin and Stein Rokkan. 1967. "Cleavage Structures, Party Systems and Voter Alignments." Pp. 1–64 in *Party Systems and Voter Alignments,* edited by S. M. Lipset and S. Rokkan. New York: Free Press.

Lipset, Seymour Martin, Kyoung-Ryung Seong and John Charles Torres. 1993. "A Comparative Analysis of the Social Requisites of Democracy." *International Social Science Journal* 45:155–75.

Malia, Martin. 1992. "Leninist Endgame." *Daedalus* 121(2):57–75.

Marx, Karl. 1958. *Capital.* Vol. 1. Moscow, Russia: Foreign Languages Publishing House.

Moore, Barrington. 1966. *Social Origins of Dictatorship and Democracy: Lord and Peasant in the Making of the Modern World.* Boston, MA: Beacon.

Nathan, Andrew J. and Tao Shi. 1993. "Cultural Requisites for Democracy in China: Findings from a Survey." *Daedalus* 122:95–124.

Pye, Lucian W. 1968. *The Spirit of Chinese Politics.* Cambridge, MA: Massachusetts Institute of Technology.

Sadowski, Christine M. 1993. "Autonomous Groups as Agents of Democratic Change in Communist and Post-Communist Eastern Europe." Pp. 163–95 in *Political Culture and Developing Countries,* edited by L. Diamond. Boulder, CO: Lynne Rienner.

Schumpeter, Joseph. 1950. *Capitalism, Socialism, and Democracy.* 3rd ed. New York: Harper and Row.

Smith, Tony. 1978. "A Comparative Study of French and British Decolonization." *Comparative Studies in Society and History* 20(1):70–102.

Stephens, John D. 1993. "Capitalist Development and Democracy: Empirical Research on the Social Origins of Democracy." Pp. 409–47 in *The Idea of Democracy,* edited by D. Copp, J. Hampton, and J. Roemer. Cambridge, England: Cambridge University.

Tocqueville, Alexis de. 1976. *Democracy in America.* Vols. 1 and 2. New York: Knopf.

Todd, Emanuel. 1979. *The Final Fall: Essays on the Decomposition of the Soviet Sphere.* New York: Karz.

Tu, Wei-ming. 1993. "Introduction: Cultural Perspectives." *Daedalus* 122:vii–xxii.

Waisman, Carlos. 1992. "Capitalism, the Market and Economy." Pp. 140–55 in *Reexamining Democracy,* edited by G. Marks and L. Diamond. Newbury Park, CA: Sage.

Weber, Max. 1946. *From Max Weber: Essays in Sociology.* Edited and translated by H. H. Gerth and C. W. Mills. New York: Oxford University.

Weffort, Francisco C. 1992. "New Democracies, Which Democracies?" (Working Paper #198). The Woodrow Wilson Center, Latin American Program, Washington, DC.

Weiner, Myron. 1987. "Empirical Democratic Theory." Pp. 3–34 in *Competitive Elections in Developing Countries,* edited by M. Weiner and E. Ozbudun. Durham, NC: Duke University.

Weingast, Barry. 1993. "The Political Foundations of Democracy and the Rule of Law." The Hoover Institution, Stanford, CA: Unpublished manuscript.

Wright, Robin. 1992. "Islam and Democracy." *Foreign Affairs* 71(3):131–45.

39

If Hitler Asked You to Electrocute a Stranger, Would You? Probably.

PHILIP MEYER

Many have wondered how a former corporal (Hitler) could manage to influence so many people and get them to commit atrocities such as those that occurred in the concentration camps. This reading reveals the roles of authority and charisma in those decisions.

As you read, ask yourself the following questions:

1. *What were two of the findings of the Milgram study that were surprising?*
2. *In the position of the testees, how would you have reacted? Why?*

GLOSSARY

Pathological Disordered in behavior.

Macabre Gruesome or horrible.

Sadistic Deriving pleasure from inflicting pain.

IN THE BEGINNING, STANLEY Milgram was worried about the Nazi problem. He doesn't worry much about the Nazis anymore. He worries about you and me, and, perhaps, himself a little bit too.

Stanley Milgram is a social psychologist, and when he began his career at Yale University in 1960 he had a plan to prove, scientifically, that Germans are different. The Germans-are-different hypothesis has been used by historians, such as William L. Shirer, to explain the systematic destruction of the Jews by the Third Reich. One madman could decide to destroy the Jews and even create a master plan for getting it done. But to implement it on the scale that Hitler did meant that thousands of other people had to go along with the scheme and help to do the work. The Shirer thesis, which Milgram set out to test, is that Germans have a basic character flaw which explains the whole thing, and this flaw is a readiness to obey authority without question, no matter what outrageous acts the authority commands.

The appealing thing about this theory is that it makes those of us who are not Germans feel better about the whole business. Obviously, you and I are not Hitler, and it seems equally obvious that we would never do Hitler's dirty work for him. But now, because of Stanley Milgram, we are compelled to wonder. Milgram developed a laboratory experiment which provided a systematic way to measure obedience. His plan was to try it out in New Haven on Americans and then go to Germany and try it out on Germans. He was strongly motivated by scientific curiosity, but there was also some moral content in his decision to pursue this line of research, which was, in turn, colored by his own Jewish background. If he could show that Germans are more obedient than Americans, he could then vary the conditions of the experiment and try to find out just what it is that makes some people more obedient than others. With this understanding, the world might, conceivably, be just a little bit better.

But he never took his experiment to Germany. He never took it any farther than Bridgeport. The first finding, also the most unexpected and disturbing finding, was that we Americans are an obedient people: not blindly obedient, and not blissfully obedient, just obedient. "I found so much obedience," says Milgram softly, a little sadly, "I hardly saw the need for taking the experiment to Germany."

There is something of the theater director in Milgram, and his technique, which he learned from one of the old masters in experimental psychology, Solomon Asch, is to stage a play with every line rehearsed, every prop carefully selected, and everybody an actor except one person. That one person is the subject of the experiment. The subject, of course, does not know he is in a play. He thinks he is in real life. The value of this technique is that the experimenter, as though he were God, can change a prop here, vary a line there, and see how the subject responds. Milgram eventually had to change a lot of the script just to get people to stop obeying. They were obeying so much, the experiment wasn't working—it was like trying to measure oven temperature with a freezer thermometer.

The experiment worked like this: If you were an innocent subject in Milgram's melodrama, you read an ad in the newspaper or received one in the mail asking for volunteers for an educational experiment. The job would take about an hour and pay $4.50. So you make an appointment and go to an old Romanesque stone structure on High Street with the imposing name of The Yale Interaction Laboratory. It looks something like a broadcasting studio. Inside, you meet a young, crew-cut man in a laboratory coat who says he is Jack Williams, the experimenter. There is another citizen, fiftyish, Irish face, an accountant, a little overweight, and very mild and harmless-looking. This other citizen seems nervous and plays with his hat while the two of you sit in chairs side by side and are told that the $4.50 checks are yours no matter what happens. Then you listen to Jack Williams explain the experiment.

It is about learning, says Jack Williams in a quiet, knowledgeable way. Science does not know much about the conditions under which people learn and this experiment is to find out about negative reinforcement. Negative reinforcement is getting punished when you do something wrong, as opposed to positive reinforcement which is getting rewarded when you do something right. The negative reinforcement in this case is electric shock. You notice a book on the table titled, *The Teaching-Learning Process,* and you assume that this has something to do with the experiment.

Then Jack Williams takes two pieces of paper, puts them in a hat, and shakes them up. One piece of paper is supposed to say, "Teacher" and the other, "Learner." Draw one and you will see which you will be. The mild-looking accountant draws one, holds it close to his vest like a poker player, looks at it, and says, "Learner." You look at yours. It says, "Teacher." You do not know that the drawing is rigged, and both slips say "Teacher." The experimenter beckons to the mild-mannered "learner."

"Want to step right in here and have a seat, please?" he says. "You can leave your coat on the back of that chair . . . roll up your right sleeve, please. Now what I want to do is strap down your arms to avoid excessive movement on your part during the experiment. This electrode is connected to the shock generator in the next room.

"And this electrode paste," he says, squeezing some stuff out of a plastic bottle and putting it on the man's arm, "is to provide a good contact and to avoid a blister or burn. Are there any questions now before we go into the next room?"

You don't have any, but the strapped-in "learner" does.

"I do think I should say this," says the learner. "About two years ago, I was at the veterans' hospital . . . they detected a heart condition. Nothing serious, but as long as I'm having these shocks, how strong are they—how dangerous are they?"

Williams, the experimenter, shakes his head casually. "Oh, no," he says. "Although they may be painful, they're not dangerous. Anything else?"

Nothing else. And so you play the game. The game is for you to read a series of word pairs: for example, blue-girl, nice-day, fat-neck. When you finish the list, you read just the first word in each pair and then a multiple-choice list of four other words, including the second word of the pair. The learner, from his remote, strapped-in position, pushes one of four switches to indicate which of the four answers he thinks is the right one. If he gets it right, nothing happens and you go on to the next one. If he gets it wrong, you push a switch that buzzes and gives him an electric shock. And then you go to the next word. You start with 15 volts and increase the number of volts by 15 for each wrong answer. The control board goes from 15 volts on one end to 450 volts on the other. So that you know what you are doing, you get a test shock yourself, at 45 volts. It hurts. To further keep you aware of what you are doing to that man in there, the board has verbal descriptions of the shock levels, ranging from "Slight Shock" at the left-hand side, through "Intense Shock" in the middle, to "Danger: Severe Shock" toward the far right. Finally, at the very end, under 435- and 450-volt switches, there are three ambiguous X's. If, at any point, you hesitate, Mr. Williams calmly tells you to go on. If you still hesitate, he tells you again.

Except for some terrifying details, which will be explained in a moment, this is the experiment. The object is to find the shock level at which you disobey the experimenter and refuse to pull the switch.

When Stanley Milgram first wrote this script, he took it to fourteen Yale psychology majors and asked them what they thought would happen. He put it this way: Out of one hundred persons in the teacher's predicament, how would their break-off points be distributed along the 15-to-450 volt scale? They thought a few would break off very early; most would quit someplace in the middle, and a few would go all the way to the end. The highest estimate of the number out of one hundred who would go all the way to the end was three. Milgram then informally polled some of his fellow scholars in the psychology department. They agreed that a very few would go to the end. Milgram thought so too.

"I'll tell you quite frankly," he says, "before I began this experiment, before any shock generator was built, I thought that most people would break off at 'Strong Shock' or 'Very Strong Shock.' You would get only a very, very small proportion of people going out to the end of the shock generator, and they would constitute a pathological fringe."

In his pilot experiments, Milgram used Yale students as subjects. Each of them pushed the shock switches one by one, all the way to the end of the board.

So he rewrote the script to include some protests from the learner. At first, they were mild, gentlemanly, Yalie protests, but "it didn't seem to have as much effect as I thought it would or should," Milgram recalls. "So we had more violent protestations on the part of the person getting the shock. All of the time, of course, what we were trying to do was not to create a macabre situation, but simply to generate disobedience. And that was one of the first findings. This was not only a technical deficiency of the experiment, that we didn't get disobedience. It really was the finding: that obedience would be much greater than we had assumed it would be and disobedience would be much more difficult than we had assumed."

As it turned out, the situation did become rather macabre. The only meaningful way to generate disobedience was to have the victim protest with great anguish, noise, and vehemence. The protests were tape-recorded so that all the teachers ordinarily would hear the same sounds and nuances, and they started with a grunt at 75 volts, proceeded through a "Hey, that really hurts," at 125 volts, got desperate with, "I can't stand the pain, don't do that," at 180 volts, reached complaints of heart trouble at 195, an agonized scream at 285, a refusal to answer at 315, and only heartrending, ominous silence after that.

Still, 65 percent of the subjects, twenty- to fifty-year-old American males, everyday, ordinary people, like you and me, obediently kept pushing those levers in the belief that they were shocking the mild-mannered learner, whose name was Mr. Wallace, and who was chosen for the role because of his innocent appearance, all the way up to 450 volts.

Milgram was now getting enough disobedience so that he had something he could measure. The next step was to vary the circumstances to see what would encourage or discourage obedience. There seemed very little left in the way of discouragement. The victim was already screaming at the top of his lungs and feigning a heart attack. So whatever new impediment to obedience reached the brain of the subject had to travel by some route other than the ear. Milligan thought of one.

He put the learner in the same room with the teacher. He stopped strapping the learner's hand down. He rewrote the script so that at 150 volts the learner took his hand off the shock plate and declared that he wanted out of the experiment. He rewrote the script some more so that the experimenter then told the teacher to grasp the learner's hand and physically force it down on the plate to give Mr. Wallace his unwanted electric shock.

"I had the feeling that very few people would go on at that point, if any," Milgram says. "I

thought that would be the limit of obedience that you find in the laboratory."

It wasn't.

Although seven years have now gone by, Milgram still remembers the first person to walk into the laboratory in the newly rewritten script. He was a construction worker, a very short man. "He was so small," says Milgram, "that when he sat on the chair in front of the shock generator, his feet didn't reach the floor. When the experimenter told him to push the victim's hand down and give the shock, he turned to the experimenter, and he turned to the victim, his elbow went up, he fell down on the hand of the victim, his feet kind of tugged to one side, and he said, 'Like this, boss?' ZZUMPH!"

The experiment was played out to its bitter end. Milgram tried it with forty different subjects. And 30 percent of them obeyed the experimenter and kept on obeying.

"The protests of the victim were strong and vehement, he was screaming his guts out, he refused to participate, and you had to physically struggle with him in order to get his hand down on the shock generator," Milgram remembers. But twelve out of forty did it.

Milgram took his experiment out of New Haven. Not to Germany, just twenty miles down the road to Bridgeport. Maybe, he reasoned, the people obeyed because of the prestigious setting of Yale University. If they couldn't trust a center of learning that had been there for two centuries, whom could they trust? So he moved the experiment to an untrustworthy setting.

The new setting was a suite of three rooms in a run-down office building in Bridgeport. The only identification was a sign with a fictitious name: "Research Associates of Bridgeport." Questions about professional connections got only vague answers about "research for industry."

Obedience was less in Bridgeport. Forty-eight percent of the subjects stayed for the maximum shock, compared to 65 percent at Yale. But this was enough to prove that far more than Yale's prestige was behind the obedient behavior.

For more than seven years now, Stanley Milgram had been trying to figure out what makes ordinary American citizens so obedient. The most obvious answer—that people are mean, nasty, brutish, and sadistic—won't do. The subjects who gave the shocks to Mr. Wallace to the end of the board did not enjoy it. They groaned, protested, fidgeted, argued, and in some cases, were seized by fits of nervous, agitated giggling.

"They even try to get out of it," says Milgram, "but they are somehow engaged in something from which they cannot liberate themselves. They are locked into a structure, and they do not have the skills or inner resources to disengage themselves."

Milgram, because he mistakenly had assumed that he would have trouble getting people to obey the orders to shock Mr. Wallace, went to a lot of trouble to create a realistic situation.

There was crew-cut Jack Williams and his grey laboratory coat. Not white, which might denote a medical technician, but ambiguously authoritative grey. Then there was the book on the table, and the other appurtenances of the laboratory which emitted the silent message that things were being performed here in the name of science, and were therefore great and good.

But the nicest touch of all was the shock generator. When Milgram started out, he had only a $300 grant from the Higgins Fund of Yale University. Later he got more ample support from the National Science Foundation, but in the beginning he had to create this authentic-looking machine with very scarce resources except for his own imagination. So he went to New York and roamed around the electronic shops until he found some little black switches at Lafayette Radio for a dollar apiece. He bought thirty of them. The generator was a metal box, about the size of a small footlocker, and he drilled the thirty holes for the thirty switches himself in a Yale machine shop. But the fine detail was left to professional industrial engravers. So he ended up with a splendid-looking control panel dominated by the row of switches, each labeled with its voltage, and each having its own red light that flashed on when the switch was pulled. Other things happened when a switch was pushed. Besides the ZZUMPH-ing noise, a blue light la-

beled "voltage energizer" went on, and a needle on a dial labeled "voltage" flicked from left to right. Relays inside the box clicked. Finally, in the upper left-hand corner of the control panel was this inscription, engraved in precise block letters:

SHOCK GENERATOR TYPE ZLB
DYSON INSTRUMENT COMPANY
WALTHAM, MASS.
OUTPUT: 15 VOLTS–450 VOLTS

One day a man from the Lehigh Valley Electronics Company of Pennsylvania was passing through the laboratory, and he stopped to admire the shock generator.

"This is a very fine shock generator," he said. "But who is this Dyson Instrument Company?" Milgram felt proud at that, since Dyson Instrument Company existed only in the recesses of his imagination.

When you consider the seeming authenticity of the situation, you can appreciate the agony some of the subjects went through. It was pure conflict. As Milgram explains to his students, "When a parent says, 'Don't strike old ladies,' you are learning two things: the content, and, also, to obey authority. This experiment creates conflicts between the two elements."

Subjects in the experiment were not asked to give the 450-volt shock more than three times. By that time, it seemed evident that they would go on indefinitely. "No one," says Milgram, "who got within five shocks of the end ever broke off. By that point, he had resolved the conflict."

Why do so many people resolve the conflict in favor of obedience?

Milgram's theory assumes that people behave in two different operating modes as different as ice and water. He does not rely on Freud or sex or toilet-training hang-ups for this theory. All he says is that ordinarily we operate in a state of autonomy, which means we pretty much have and assert control over what we do. But in certain circumstances, we operate under what Milgram calls a state of agency (after agent, n. . . . one who acts for or in the place of another by authority from him; a substitute; a deputy.—*Webster's Collegiate Dictionary*). A state of agency, to Milgram, is nothing more than a frame of mind.

"There's nothing bad about it, there's nothing good about it," he says. "It's a natural circumstance of living with other people. . . . I think of a state of agency as a real transformation of a person; if a person has different properties when he's in that state, just as water can turn to ice under certain conditions of temperature, a person can move to the state of mind that I call agency . . . the critical thing is that you see yourself as the instrument of the execution of another person's wishes. You do not see yourself as acting on your own. And there's a real transformation, a real change of properties of the person."

To achieve this change, you have to be in a situation where there seems to be a ruling authority whose commands are relevant to some legitimate purpose; the authority's power is not unlimited.

But situations can be and have been structured to make people do unusual things, and not just in Milgram's laboratory. The reason, says Milgram, is that no action, in and of itself, contains meaning.

"The meaning always depends on your definition of the situation. Take an action like killing another person. It sounds bad.

"But then we say the other person was about to destroy a hundred children, and the only way to stop him was to kill him. Well, that sounds good.

"Or, you take destroying your own life. It sounds very bad. Yet, in the Second World War, thousands of persons thought it was a good thing to destroy your own life. It was set in the proper context. You sipped some saki from a whistling cup, recited a few haiku. You said, 'May my death be as clean and as quick as the shattering of crystal.' And it almost seemed like a good, noble thing to do, to crash your kamikaze plane into an aircraft carrier. But the main thing was, the definition of what a kamikaze pilot was doing had been determined by the relevant authority. Now,

once you are in a state of agency, you allow the authority to determine, to define what the situation is. The meaning of your actions is altered."

So, for most subjects in Milgram's laboratory experiments, the act of giving Mr. Wallace his painful shock was necessary, even though unpleasant, and besides they were doing it on behalf of somebody else and it was for science. There was still strain and conflict, of course. Most people resolved it by grimly sticking to their task and obeying. But some broke out. Milgram tried varying the conditions of the experiment to see what would help break people out of their state of agency.

"The results, as seen and felt in the laboratory," he has written, "are disturbing. They raise the possibility that human nature, or more specifically the kind of character produced in American democratic society, cannot be counted on to insulate its citizens from brutality and inhumane treatment at the direction of malevolent authority. A substantial proportion of people do what they are told to do, irrespective of the content of the act and without limitations of conscience, so long as they perceive that the command comes from a legitimate authority. If in this study, an anonymous experimenter can successfully command adults to subdue a fifty-year-old man and force on him painful electric shocks against his protest, one can only wonder what government, with its vastly greater authority and prestige, can command of its subjects."

This is a nice statement, but it falls short of summing up the full meaning of Milgram's work. It leaves some questions still unanswered.

The first question is this: Should we really be surprised and alarmed that people obey? Wouldn't it be even more alarming if they all refused to obey? Without obedience to a relevant ruling authority there could not be a civil society. And without a civil society, as Thomas Hobbes pointed out in the seventeenth century, we would live in a condition of war, "of every man against every other man," and life would be "solitary, poor, nasty, brutish, and short."

In the middle of one of Stanley Milgram's lectures at CUNY recently, some mini-skirted undergraduates started whispering and giggling in the back of the room. He told them to cut it out. Since he was the relevant authority in that time and place, they obeyed, and most people in the room were glad that they obeyed.

This was not, of course, a conflict situation. Nothing in the coeds' social upbringing made it a matter of conscience for them to whisper and giggle. But a case can be made that in a conflict situation it is all the more important to obey. Take the case of war, for example. Would we really want a situation in which every participant in a war, direct, or indirect—from front-line soldiers to the people who sell coffee and cigarettes to employees at the Concertina barbed-wire factory in Kansas—stops and consults his conscience before each action? It is asking for an awful lot of mental strain and anguish from an awful lot of people. The value of having civil order is that one can do his duty, or whatever interests him, or whatever seems to benefit him at the moment, and leave the agonizing to others. When Francis Gary Powers was being tried by a Soviet military tribunal after his U-2 spy plane was shot down, the presiding judge asked if he had thought about the possibility that his flight might have provoked a war. Powers replied with Hobbesian clarity: "The people who sent me should think of these things. My job was to carry out orders. I do not think it was my responsibility to make such decisions."

It was not his responsibility. And it is quite possible that if everyone felt responsible for each of the ultimate consequences of his own tiny contributions to complex chains of events, then society simply would not work. Milgram, fully conscious of the moral and social implications of his research, believes that people should feel responsible for their actions. If someone else had invented the experiment, and if he had been the naive subject, he feels certain that he would have been among the disobedient minority.

"There is no very good solution to this," he admits, thoughtfully. "To simply and categorically say that you won't obey authority may resolve your personal conflict, but it creates more problems for society which may be more serious in the long run. But I have no doubt that to disobey is the proper thing to do in this [the laboratory] situation. It is the only reasonable value judgment to make."

The conflict between the need to obey the relevant ruling authority and the need to follow your conscience becomes sharpest if you insist on living by an ethical system based on a rigid code—a code that seeks to answer all questions in advance of their being raised. Code ethics cannot solve the obedience problem. Stanley Milgram seems to be a situation ethicist, and situation ethics does offer a way out: When you feel conflict, you examine the situation and then make a choice among the competing evils. You may act with a presumption in favor of obedience, but reserve the possibility that you will disobey whenever obedience demands a flagrant and outrageous affront to conscience. This, by the way, is the philosophical position of many who resist the draft. In World War II, they would have fought. Vietnam is a different, an outrageously different, situation.

Life can be difficult for the situation ethicist, because he does not see the world in straight lines, while the social system too often assumes such a God-given, squared-off structure. If your moral code includes an injunction against all war, you may be deferred as a conscientious objector. If you merely oppose this particular war, you may not be deferred.

Stanley Milgram has his problems, too. He believes that in the laboratory situation he would not have shocked Mr. Wallace. His professional critics reply that in his real-life situation he has done the equivalent. He has placed innocent and naive subjects under great emotional strain and pressure in selfish obedience to his quest for knowledge. When you raise this issue with Milgram, he has an answer ready. There is, he explains patiently, a critical difference between his naive subjects and the man in the electric chair. The man in the electric chair (in the mind of the naive subject) is helpless, strapped in. But the naive subject is free to go at any time.

Immediately after he offers this distinction, Milgram anticipates the objection.

"It's quite true," he says, "that this is almost a philosophic position, because we have learned that some people are psychologically incapable of disengaging themselves. But that doesn't relieve them of the moral responsibility."

The parallel is exquisite. "The tension problem was unexpected," says Milgram in his defense. But he went on anyway. The naive subjects didn't expect the screaming protests from the strapped-in learner. But that went on.

"I had to make a judgment," says Milgram. "I had to ask myself, was this harming the person or not? My judgment is that it was not. Even in the extreme cases, I wouldn't say that permanent damage results."

Sound familiar? "The shocks may be painful," the experimenter kept saying, "but they're not dangerous."

After the series of experiments was completed, Milgram sent a report of the results to his subjects and a questionnaire, asking whether they were glad or sorry to have been in the experiment. Eighty-three and seven-tenths percent said they were glad and only 1.3 percent were sorry; 15 percent were neither sorry nor glad. However, Milgram could not be sure at the time of the experiment that only 1.3 percent would be sorry.

Kurt Vonnegut Jr. put one paragraph in the preface to *Mother Night,* in 1966, which pretty much says it for the people with their fingers on the shock-generator switches, for you and me, and maybe even for Milgram. "If I'd been born in Germany," Vonnegut says, "I suppose I would have *been* a Nazi, bopping Jews and gypsies and Poles around, leaving boots sticking out of snowbanks, warming myself with my sweetly virtuous insides. So it goes."

Just so. One thing that happened to Milgram back in New Haven during the days of the experiment was that he kept running into people he'd watched from behind the one-way glass. It gave him a funny feeling, seeing those people going about their everyday business in New Haven and knowing what they would do to Mr. Wallace if ordered to. Now that his research results are in and you've thought about it, you can get this funny feeling too. You don't need one-way glass. A glance in your own mirror may serve just as well.

40
Money Changes Everything

DAN CLAWSON, ALAN NEWSTADT, AND DENISE SCOTT

The United States claims to be a representative democracy. The implication of this reading is that our political representation is far narrower than trumpeted. What can be done to bring about our democratic claims? Do you think it will happen?

As you read, ask yourself the following questions:

1. *What effects have PACs had on campaign financing?*
2. *How would you change the financing of political campaigns?*

GLOSSARY

Hegemony Leadership dominance.

IN THE PAST TWENTY years political action committees, or PACs, have transformed campaign finance. . . .

Most analyses of campaign finance focus on the candidates who receive the money, not on the people and political action committees that give it. PACs are entities that collect money from many contributors, pool it, and then make donations to candidates. Donors may give to a PAC because they are in basic agreement with its aims, but once they have donated they lose direct control over their money, trusting the PAC to decide which candidates should receive contributions. . . .

Why Does the Air Stink?

Everybody wants clean air. Who could oppose it? "I spent seven years of my life trying to stop the Clean Air Act," explained the PAC director for a major corporation that is a heavy-duty polluter.

Nonetheless, he was perfectly willing to use his corporation's PAC to contribute to members of Congress who voted for the act:

> How a person votes on the final piece of legislation often is not representative of what they have done. Somebody will do a lot of things during the process. How many guys voted against the Clean Air Act? But during the process some of them were very sympathetic to some of our concerns.

In the world of Congress and political action committees things are not always what they seem. Members of Congress want to vote for clean air, but they also want to receive campaign contributions from corporate PACs and pass a law that business accepts as "reasonable." The compromise solution to this dilemma is to gut the bill by crafting dozens of loopholes inserted in private meetings or in subcommittee hearings that don't receive much (if any) attention in the press. Then the public vote on the final bill can be nearly unanimous: members of Congress can assure their constituents that they voted for the final bill and their corporate PAC contributors that they helped weaken the bill in private. We can use the Clean Air Act of 1990 to introduce and explain this process.

The public strongly supports clean air and is unimpressed when corporate officials and apologists trot out their normal arguments: "corporations are already doing all they reasonably can to improve environmental quality"; "we need to balance the costs against the benefits"; "people will lose their jobs if we make controls any stricter." The original Clean Air Act was passed

In Money Talks: Corporate PACs and Political Influence. *Basic Books, 1992.*

in 1970, revised in 1977, and not revised again until 1990. Although the initial goal of its supporters was to have us breathing clean air by 1975, the deadline for compliance has been repeatedly extended—and the 1990 legislation provides a new set of deadlines to be reached sometime far in the future.

Because corporations control the production process unless the government specifically intervenes, any delay in government action leaves corporations free to do as they choose. Not only have laws been slow to come, but corporations have fought to delay or subvert implementation. The 1970 law ordered the Environmental Protection Agency (EPA) to regulate the hundreds of poisonous chemicals that are emitted by corporations, but as William Greider notes, "in twenty years of stalling, dodging, and fighting off court orders, the EPA has managed to issue regulatory standards for a total of seven toxics."

Corporations have done exceptionally well politically, given the problem they face: the interests of business often are diametrically opposed to those of the public. Clean air laws and amendments have been few and far between, enforcement is ineffective, and the penalties for infractions are minimal. . . .

This corporate struggle for the right to pollute takes place on many fronts. One front is public relations: the Chemical Manufacturers Association took out a two-page Earth Day ad in the *Washington Post* to demonstrate its concern for the environment; coincidentally many of the corporate signers are also on the EPA's list of high-risk producers. Another front is research: expert studies delay action while more information is gathered. The federally funded National Acid Precipitation Assessment Program (NAPAP) took ten years and $600 million to figure out whether acid rain was a problem. Both business and the Reagan administration argued that no action should be taken until the study was completed. The study was discredited when its summary of findings minimized the impact of acid rain—even though this did not accurately represent the expert research in the report. But the key site of struggle has been Congress, where

for years corporations have succeeded in defeating environmental legislation. In 1987 utility companies were offered a compromise bill on acid rain, but they "were very adamant that they had beat the thing since 1981 and they could always beat it," according to Representative Edward Madigan (R-Ill.). Throughout the 1980s the utilities defeated all efforts at change. . . .

The stage was set for a revision of the Clean Air Act when George Bush was elected as "the environmental president" and George Mitchell, a strong supporter of environmentalism, became the Senate majority leader. But what sort of clean air bill would it be? "What we wanted," said Richard Ayres, head of the environmentalists' Clean Air Coalition, "is a health-based standard—one-in-1-million cancer risk." Such a standard would require corporations to clean up their plants until the cancer risk from their operations was reduced to one in a million. "The Senate bill still has the requirement," Ayres said, "but there are forty pages of extensions and exceptions and qualifications and loopholes that largely render the health standard a nullity." Greider reports, for example, that "according to the EPA, there are now twenty-six coke ovens that pose a cancer risk greater than 1 in 1000 and six where the risk is greater than 1 in 100. Yet the new clean-air bill will give the steel industry another thirty years to deal with the problem."

This change from what the bill was supposed to do to what it did do came about through what corporate executives like to call the "access" process. The main aim of most corporate political action committee contributions is to help corporate executives attain "access" to key members of Congress and their staffs. Corporate executives (and corporate PAC money) work to persuade the member of Congress to accept a carefully predesigned loophole that sounds innocent but effectively undercuts the stated intention of the bill. Representative Dingell (D-Mich.), chair of the House Committee on Energy and Commerce, is a strong industry supporter; one of the people we interviewed called him "the point man for the Business Roundtable on clean air." Representative Waxman (D-Calif.), chair of the Subcommittee on Health and the Environment, is an environmentalist. Observers of the Clean Air Act legislative process expected a confrontation and contested votes on the floor of Congress.

The problem for corporations was that, as one Republican staff aide said, "If any bill has the blessing of Waxman and the environmental groups, unless it is totally in outer space, who's going to vote against it?" But corporations successfully minimized public votes. Somehow Waxman was persuaded to make behind-the-scenes compromises with Dingell so members didn't have to publicly side with business against the environment during an election year. Often the access process leads to loopholes that protect a single corporation, but for "clean" air most special deals targeted entire industries, not specific companies. The initial bill, for example, required cars to be able to use strictly specified cleaner fuels. But the auto industry wanted the rules loosened, and Congress eventually modified the bill by incorporating a variant of a formula suggested by the head of General Motors' fuels and lubricants department.

Nor did corporations stop fighting after they gutted the bill through amendments. Business pressed the EPA for favorable regulations to implement the law: "The cost of this legislation could vary dramatically, depending on how EPA interprets it," said William D. Fay, vice president of the National Coal Association, who headed the hilariously misnamed Clean Air Working Group, an industry coalition that fought to weaken the legislation. An EPA aide working on acid rain regulations reported, "We're having a hard time getting our work done because of the number of phone calls we're getting" from corporations and their lawyers.

Corporations trying to convince federal regulators to adopt the "right" regulations don't rely exclusively on the cogency of their arguments. They often exert pressure on a member of Congress to intervene for them at the EPA or other agency. Senators and representatives regularly intervene on behalf of constituents and contributors by doing everything from straightening out a social security problem to asking a regulatory

agency to explain why it is pressuring a company. This process—like campaign finance—usually follows accepted etiquette. In addressing a regulatory agency the senator does not say, "Lay off my campaign contributors, or I'll cut your budget." One standard phrasing for letters asks regulators to resolve the problem "as quickly as possible within applicable rules and regulations." No matter how mild and careful the inquiry, the agency receiving the request is certain to give it extra attention; only after careful consideration will they refuse to make any accommodation.

The power disparity between business and environmentalists is enormous during the legislative process but even larger thereafter. When the Clean Air Act passed, corporations and industry groups offered positions, typically with large pay increases, to congressional staff members who wrote the law. The former congressional staff members who work for corporations know how to evade the law and can persuasively claim to EPA that they know what Congress intended. Environmental organizations pay substantially less than Congress and can't afford large staffs. They are rarely able to become involved in the details of the administrative process or influence implementation and enforcement.

Having pushed Congress for a law, and the Environmental Protection Agency for regulations, allowing as much pollution as possible, business then went to the Quayle Council for rules allowing even more pollution. Vice President J. Danforth Quayle's Council, technically the Council on Competitiveness, was created by President Bush specifically to help reduce regulations on business. Quayle told the *Boston Globe* "that his council has an 'open door' to business groups and that he has a bias against regulations." The Council reviews, and can override, all federal regulations, including those by the EPA setting the limits at which a chemical is subject to regulation. The council also recommended that corporations be allowed to increase their polluting emissions if a state did not object within seven days of the proposed increase. Corporations thus have multiple opportunities to win. If they lose in Congress, they can win at the regulatory agency; if they lose there, they can try again at the Quayle Council. If they lose there,. they can try to reduce the money available to enforce regulations, tie up the issue in the courts, or accept a minimal fine.

The operation of the Quayle Council probably would have received little publicity, but reporters discovered that the executive director of the Council, Allan Hubbard, had a clear conflict of interest. Hubbard chaired the biweekly White House meetings on the Clean Air Act. He owns half of World Wide Chemical, received an average of more than a million dollars a year in profits from it while directing the Council, and continues to attend quarterly stockholder meetings. According to the *Boston Globe*, "Records on file with the Indianapolis Air Pollution Control Board show that World Wide Chemical emitted 17,000 to 19,000 pounds of chemicals into the air last year." The company "does not have the permit required to release the emissions," "is putting out nearly four times the allowable emissions without a permit, and could be subject to a $2,500-a-day penalty," according to David Jordan, director of the Indianapolis Air Pollution Board. . . .

The real issue is the system of business-government relations, and especially of campaign finance, that offers business so many opportunities to craft loopholes, undermine regulations, and subvert enforcement. Still worse, many of these actions take place outside of public scrutiny.

THE CANDIDATES' PERSPECTIVE

. . . Money has always been a critically important factor in campaigns, but the shift to expensive technology has made it the dominant factor. Today money is the key to victory and substitutes for everything else—instead of door-to-door canvassers, a good television spot; instead of a committee of respected long-time party workers who know the local area, a paid political consultant and media expert. To be a viable political candidate, one must possess—or be able to raise—huge sums. Nor is this a one-time requirement; each reelection campaign requires new infusions of cash.

The quest for money is never ending. Challengers must have money to be viable contenders; incumbents can seldom predict when they might face a tight race. In 1988 the average winning candidate for the House of Representatives spent $388,000; for the Senate, $3,745,000. Although the Congress, especially the Senate, has many millionaires, few candidates have fortunes large enough to finance repeated campaigns out of their own pockets. It would take the entire congressional salary for 3.1 years for a member of the House, or 29.9 years for a senator, to pay for a single reelection campaign. Most members are therefore in no position to say, "Asking people for money is just too big a hassle. Forget it. I'll pay for it myself." They must raise the money from others, and the pressure to do so never lets up. To pay for an average winning campaign, representatives need to raise $3,700 and senators $12,000 during *every week* of their term of office.

Increasingly incumbents use money to win elections before voters get involved. Senator Rudy Boschwitz (R-Minn.) spent $6 million getting reelected in 1984 and had raised $1.5 million of it by the beginning of the year, effectively discouraging the most promising Democratic challengers. . . .

Fundraising isn't popular with the public, but candidates keep emphasizing it because it works: the champion money raiser wins almost regardless of the merits. *Almost* is an important qualifier here, as Boschwitz would be the first to attest: in his 1990 race he outspent his opponent by about five to one and lost nonetheless. . . .

It is not only that senators leave committee hearings for the more crucial task of calling people to beg for money. They also chase all over the country because reelection is more dependent on meetings with rich people two thousand miles from home than it is on meetings with their own constituents.

. . . Do members of Congress incur any obligations in seeking and accepting these campaign contributions? Bob Dole, Republican leader in the Senate and George Bush's main rival for the 1988 Republican presidential nomination, was quoted by the *Wall Street Journal* as saying, "When the Political Action Committees give money, they expect something in return other than good government." One unusually outspoken business donor, Charles Keating, made the same point: "One question among the many raised in recent weeks had to do with whether my financial support in any way influenced several political figures to take up my cause. I want to say in the most forceful way I can, I certainly hope so." . . .

THE CURRENT LAW

The law, however, regulates fundraising and limits the amount that any one individual or organization may (legally) contribute. According to current law:

1. A *candidate* may donate an unlimited amount of personal funds to his or her *own* campaign. The Supreme Court has ruled this is protected as free speech.
2. Individuals may not contribute more than $1,000 per candidate per election, nor more than $25,000 in total in a given two-year election cycle.
3. Political action committees may contribute up to $5,000 per candidate per election. Since most candidates face primaries, an individual may contribute $2,000 and a PAC $10,000 to the candidate during a two-year election cycle. PACs may give to an unlimited number of candidates and hence may give an unlimited amount of money.
4. Individuals may contribute up to $5,000 per year to a political action committee.
5. Candidates must disclose the full amount they have received, the donor and identifying information for any individual contribution of $200 or more, the name of the PAC and donation amount for any PAC contribution however small, and all disbursements. PACs must disclose any donation they make to a candidate, no matter how small. They must also disclose the total amount received by the PAC and the names and positions of all contributors who give the PAC more than $200 in a year.
6. Sponsoring organizations, including corporations and unions, may pay all the expenses of creating and operating a PAC. Thus a corporation may pay the cost of the rent, tele-

phones, postage, supplies, and air travel for all PAC activities; the salaries of full-time corporate employees who work exclusively on the PAC; and the salaries of all managers who listen to a presentation about the PAC. However, the PAC money itself—the money used to contribute to candidates—must come from voluntary donations by individual contributors. The corporation may not legally take a portion of its profits and put it directly into the PAC.

7. Corporations may establish and control the PAC and solicit stockholders and/or managerial employees for contributions to the PAC. It is technically possible for corporations to solicit hourly (or nonmanagerial) employees and for unions to solicit managers, but these practices are so much more tightly regulated and restricted that in practice cross-solicitation is rare.
8. The Federal Election Commission (FEC) is to monitor candidates and contributors and enforce the rules.

These are the key rules governing fundraising, but the history of campaign finance is that as time goes on, loopholes develop. . . . What is generally regarded as the most important current loophole is that there are no reporting requirements or limits for contributions given to political parties as opposed to candidates. Such money is ostensibly to be used to promote party building and get-out-the-vote drives; in 1988 literally hundreds of individuals gave $100,000 or more in unreported "soft money" donations. Many of these loopholes are neither accidents nor oversights. Three Democrats and three Republicans serve as federal election commissioners, and commissioners are notorious party loyalists. Because it requires a majority to investigate a suspected violation, the FEC not only fails to punish violations, it fails to investigate them.

CORPORATE PACS

The Federal Election Commission categorizes PACs as corporate, labor, trade-health-membership, and nonconnected. Nonconnected PACs are unaffiliated with any other organization: they are formed exclusively for the purpose of raising and contributing money. Most subsist by direct-mail fundraising targeted at people with a commitment to a single issue (abortion or the environment) or philosophical position (liberalism or conservatism). Other PACs are affiliated with an already existing organization, and that organization — whether a corporation, union, trade, or membership association — pays the expenses associated with operating the PAC and decides what will happen to the money the PAC collects.

Candidates increasingly rely on PACs because they can easily solicit a large number of PACs, each of which is relatively likely to make a major contribution. "From 1976–88, PAC donations rose from 22 per cent to 40 per cent of House campaign receipts, and from 15 per cent to 22 per cent of Senate receipts." Almost half of all House members (205 of the 435) "received at least 50 per cent of their campaign contributions from PACs." The reliance of PACs is greater in the House than in the Senate: PACs give more to Senate candidates, but Senate races are more expensive than House races, so a larger fraction of total Senate-race receipts comes from individual contributions.

Although other sorts of PACs deserve study, we believe the most important part of this story concerns corporate PACs, the subject of this book. We focus on corporate PACs for three interrelated reasons. First, they are the largest concentrated source of campaign money and the fastest growing. In 1988 corporate PACs contributed more than $50 million, all trade-membership-health PACs combined less than $40 million, labor PACs less than $35 million, and nonconnected PACs less than $20 million. Moreover, these figures understate the importance of corporate decisions about money because industry trade associations are controlled by corporations and follow their lead. In addition, corporate executives have high incomes and make many individual contributions; a handful of labor leaders may attempt to do the same on a reduced scale, but rank-and-file workers are unlikely to do so. Second, corporations have disproportionate power in U.S. society, magnifying the importance of the money they contribute. Fi-

nally, corporate PACs have enormous untapped fundraising potential. They are in a position to coerce their donors in a way no other kind of PAC can and, if the need arose, could dramatically increase the amount of money they raise. . . .

Corporate PACs follow two very different strategies, pragmatic and ideological. . . .

Pragmatic donations are given specifically to advance the short-run interests of the donor, primarily to enable the corporation to gain a chance to meet with the member and argue its case. Because the aim of these donations is to gain "access" to powerful members of Congress, the money is given without regard to whether or not the member needs it and with little consideration of the member's political stance on large issues. The corporation's only concern is that the member will be willing and able to help them out — and virtually all members, regardless of party, are willing to cooperate in this access process. Perhaps the most memorable characterization of this strategy was by Jay Gould, nineteenth-century robber baron and owner of the Erie Railroad: "In a Republican district I was a Republican; in a Democratic district, a Democrat; in a doubtful district I was doubtful; but I was always for Erie."

Ideological donations, on the other hand, are made to influence the political composition of the Congress. From this perspective, contributions should meet two conditions: (1) they should be directed to politically congenial "pro-free enterprise" candidates who face opponents unsympathetic to business (in practice, these are always conservatives); and (2) they should be targeted at competitive races where money can potentially influence the election outcome. The member's willingness to do the company favors doesn't matter, and even a conservative "free enterprise" philosophy wouldn't be sufficient: if the two opponents' views were the same, then the election couldn't influence the ideological composition of Congress. Most incumbents are reelected: in some years as many as 98 percent of all House members running are reelected. Precisely because incumbents will probably be reelected even without PAC support, ideological corporations usually give to nonincumbents, either challengers or candidates for open seats.

Virtually all corporations use some combinations of pragmatic and ideological strategies. The simplest method of classifying PACs is by the proportion of money they give to incumbents: the higher this proportion, the more pragmatic the corporation. . . . In 1988 about a third (36 percent) of the largest corporate PACs gave more than 90 percent of their money to incumbents, and another third (34 percent) gave 80 to 90 percent to incumbents. Although roughly a third gave less than 80 percent to incumbents, only eight corporate PACs gave less than 50 percent of their money to incumbents (that is, more than 50 percent to nonincumbents). . . . The pragmatic emphasis of recent years is a change from 1980, when a large number of corporations followed an ideological approach.

Our Research

. . . Our quantitative analyses concentrate on Democrats and Republicans in general-election contests for congressional seats. We focus on the 309 corporate PACs that made the largest contributions in the period from 1975 to 1988. As might be expected, these are almost exclusively very large corporations: on average in 1984 they had $6.7 billion in sales and 48,000 employees. . . . Moreover, not all firms with large PACs have huge sales, so our sample includes about twenty-five "small" firms with 1984 revenues of less than $500 million. . . .

On average, in 1988 these PACs gave 52.7 percent of their money to Republicans and 47.3 percent to Democrats. They gave 83.6 percent of their money to incumbents, 10.2 percent to candidates for open seats, and 6.2 percent to challengers.

The PAC officials we interviewed were selected from this set of the 309 largest corporate PACs and were representative of the larger sample in terms of both economic and political characteristics. . . .

A third source of original data supplements our quantitative analyses of the 309 largest corporate PACs and our 38 in-depth interviews. In November and December of 1986 we mailed surveys to a random sample of ninety-four directors of large corporate PACs, achieving a response rate of 58 percent. For the most part, we use this to place our interview comments in context: if a PAC director tells us a story of being pressured by a candidate, how typical is this? How many other PAC directors report similar experiences? Finally, our original data also are supplemented by books, articles, and newspaper accounts about campaign finance. . . .

Overview and Background

. . . We argue that corporate PACs differ from other PACs in two ways: (1) as employees, managers can be—and are—coerced to contribute; and (2) corporate PACs are not democratically controlled by their contributors (even in theory). . . .

We argue that PAC contributions are best understood as gifts, not bribes. They create a generalized sense of obligation and an expectation that "if I scratch your back you scratch mine.". . .

A corporation uses the member of Congress's sense of indebtedness for past contributions to help it gain access to the member. In committee hearings and private meetings the corporation then persuades the member to make "minor" changes in a bill, which exempt a particular company or industry from some specific provision.

Even some corporations are troubled by this "access" approach, and . . . consider the alternative: donations to close races intended to change the ideological composition of the Congress. In the late 1980s and early 1990s only a small number of corporations used this as their primary strategy, but most corporations make some such donations. In the 1980 election a large group of corporations pursued an ideological strategy. We argue this was one of the reasons for the conservative successes of that period . . .

Do competing firms or industries oppose each other in Washington, such that one business's political donations oppose and cancel out those of the next corporation or industry? More generally, how much power does business have in U.S. society, and how does its political power relate to its economic activity? . . .

The PAC directors we interviewed are not very worried about reform: they don't expect meaningful changes in campaign funding laws, and they assume that if "reforms" are enacted, they will be easily evaded. . . .

Three interrelated points, . . . First, power is exercised in many loose and subtle ways, not simply through the visible use of force and threats. Power may in fact be most effective, and most limiting, when it structures the conditions for action—even though in these circumstances it may be hard to recognize. Thus PAC contributions can and do exercise enormous influence through creating a sense of obligation, even if there is no explicit agreement to perform a specific service in return for a donation. Second, business is different from, and more powerful than, other groups in the society. As a result, corporations and their PACs are frequently treated differently than others would be. Other groups could not match business power simply by raising equivalent amounts of PAC money. Third, this does not mean that business always wins, or that it wins automatically. If it did, corporate PACs would be unnecessary. Business must engage in a constant struggle to maintain its dominance. This is a class struggle just as surely as are strikes and mass mobilizations, even though it is rarely thought of in these terms.

WHAT IS POWER?

Our analysis is based on an understanding of power that differs from that usually articulated by both business and politicians. The corporate PAC directors we interviewed insisted that they have no power. . . .

The executives who expressed these views used the word *power* in roughly the same sense that it is usually used within political science, which is also the way the term was defined by Max Weber, the classical sociological theorist. Power, according to this common conception, is the ability to make someone do something against his or her will. If that is what power means, then corporations rarely have power in relation to members of Congress. As one corporate senior vice president said to us, "You certainly aren't going to be able to buy anybody for $500 or $1,000 or $10,000. It's a joke." In this regard we agree with the corporate officials we interviewed: a PAC is not in a position to say to a member of Congress, "Either you vote for this bill, or we will defeat your bid for reelection." Rarely do they even say, "Vote for this bill, or you won't get any money from us." . . . Therefore, if power is the ability to make someone do something against his or her will, then PAC donations rarely give corporations power over members of Congress.

This definition of power as the ability to make someone do something against his or her will is what Steven Lukes calls a *one-dimensional view of power. A two-dimensional view* recognizes the existence of nondecisions: a potential issue never gets articulated or, if articulated by someone somewhere, never receives serious consideration. . . . A two-dimensional view of power makes the same point: in some situations no one notices power is being exercised—because there is no overt conflict.

Even this model of power is too restrictive, however, because it still focuses on discrete decisions and nondecisions. . . . Such models do not recognize "the idea that the most fundamental use of power in society is its use in structuring the basic manner in which social agents interact with one another." . . . Similarly, the mere presence of a powerful social agent alters social space for others and causes them to orient to the powerful agent. One of the executives we interviewed took it for granted that "if we go see the congressman who represents [a city where the company has a major plant], where 10,000 of our employees are also his constituents, we don't need a PAC to go see him." The corporation is so important in that area that the member has to orient himself or herself in relation to the corporation and its concerns. In a different sense, the mere act of accepting a campaign contribution changes the way a member relates to a PAC, creating a sense of obligation and need to reciprocate. The PAC contribution has altered the member's social space, his or her awareness of the company and wish to help it, even if no explicit commitments have been made.

BUSINESS IS DIFFERENT

Power therefore is not just the ability to force people to do something against their will; it is most effective (and least recognized) when it shapes the field of action. Moreover, business's vast resources, influence on the economy, and general legitimacy place it on a different footing from other so-called special interests. Business donors are often treated differently from other campaign contributors. When a member of Congress accepts a $1,000 donation from a corporate PAC, goes to a committee hearing, and proposes "minor" changes in a bill's wording, those changes are often accepted without discussion or examination. The changes "clarify" the language of the bill, perhaps legalizing higher levels of pollution for a specific pollutant or exempting the company from some tax. The media do not report on this change, and no one speaks against it. . . .

Even groups with great social legitimacy encounter more opposition and controversy than business faces for proposals that are virtually without public support. Contrast the largely unopposed commitment of more than $500 billion for the bailout of savings and loan associations with the sharp debate, close votes, and defeats for the rights of men and women to take *unpaid* parental leaves. Although the classic phrase for something noncontroversial that everyone must support is to call it a "motherhood" issue, and it would cost little to guarantee every woman the right to an unpaid parental leave, nonetheless this measure generated intense scrutiny and controversy, ultimately going down to defeat. Few people are prepared to publicly defend pollution or tax evasion, but business is routinely able to win pollution exemptions and tax loopholes. Although cumulatively these provisions may trouble people, individually most are allowed to pass without scrutiny. *No* analysis of corporate political activity makes sense unless it begins with a recognition that the PAC is a vital element of corporate power, but it does not operate by itself. The PAC donation is always backed by the wider range of business power and influence.

Corporations are different from other special-interest groups not only because business has far more resources, but also because of this acceptance and legitimacy. When people feel that "the system" is screwing them, they tend to blame politicians, the government, the media — but rarely business. Although much of the public is outraged at the way money influences elections and public policy, the issue is almost always posed in terms of what politicians do or don't do. This pervasive double standard largely exempts business from criticism. . . .

Many people who are outraged that members of Congress recently raised their pay to $125,100 are apparently unconcerned about corporate executives' pay. One study calculated that CEOs at the largest U.S. companies are paid an average of $2.8 million a year, 150 times more than the average U.S. worker and 22 times as much as members of Congress. More anger is directed at Congress for delaying new environmental laws than at the companies who fight every step of the way to stall and subvert the legislation. When members of Congress do favors for large campaign contributors, the anger is directed at the senators who went along, not at the business owner who paid the money (and usually initiated the pressure). The focus is on the member's receipt of thousands of dollars, not on the business's receipt of millions (or hundreds of millions) in tax breaks or special treatment. It is widely held that "politics is dirty," but companies' getting away with murder — quite literally — generates little public comment and condemnation. This disparity is evidence of business's success in shaping public perceptions. Lee Atwater, George Bush's campaign manager for the 1988 presidential election, saw this as a key to Republican success:

> In the 1980 campaign, we were able to make the establishment, insofar as it is bad, the government. In other words, big government was the enemy, not big business. If the people think the problem is that taxes are too high, and the government interferes too much, then we are doing our job. But, if they get to the point where they say that the real problem is that rich people aren't paying taxes . . . then the Democrats are going to be in good shape.

. . . We argue corporations are so different, and so dominant that they exercise a special kind of power, what Antonio Gramsci called *hegemony*.

Hegemony can be regarded as the ultimate example of a field of power that structures what people and groups do. It is sometimes referred to as a world view — a way of thinking about the world that influences every action and makes it difficult to even consider alternatives. But in Gramsci's analysis it is much more than this; it is a culture and set of institutions that structure life patterns and coerce a particular way of life. . . .

Hegemony is most successful and most powerful if it is unrecognized. . . . In some sense gender relations in the 1950s embodied a hegemony even more powerful than that of race relations. Betty Friedan titled the first chapter of *The Feminine Mystique* "The Problem That Has No Name" because women literally did not have a name for and did not recognize the existence of their oppression. Women as well as men denied the existence of inequality or oppression and denied the systematic exercise of power to maintain unequal relations.

We argue that today business has enormous power and exercises effective hegemony, even though (perhaps because) this is largely undiscussed and unrecognized. *Politically* business power today is similar to white treatment of blacks in 1959: business may sincerely deny its power, but many of the groups it exercises power over recognize it, feel dominated, resent this, and fight the power as best they can. *Economically* business power is more similar to gender relations in 1959: virtually no one sees this power as problematic. If the issue is brought to people's attention, many still don't see a problem: "Well, so what? How else could it be? Maybe we don't like it, but that's just the way things are." . . .

Hegemony is never absolute. . . . A hegemonic power is usually opposed by a counter-hegemony. . . .

THE LIMITS TO BUSINESS POWER

We have argued that power is more than winning an open conflict, and business is different from other groups because of its pervasive influence on our society — the way it shapes the social space for all other actors. These two arguments, however, are joined with a third: a recognition of, in fact an insistence on, the limits to business

power. We stress the power of business, but business does not feel powerful. . . .

Executives believe that corporations are constantly under attack, primarily because government simply doesn't understand that business is crucial to everything society does but can easily be crippled by well-intentioned but unrealistic government policies. A widespread view among the people we interviewed is that "far and away the vast majority of things that we do are literally to protect ourselves from public policy that is poorly crafted and nonresponsive to the needs and realities and circumstances of our company." These misguided policies, they feel, can come from many sources — labor unions, environmentalists, the pressure of unrealistic public-interest groups, the government's constant need for money, or the weight of its oppressive bureaucracy. Simply maintaining equilibrium requires a pervasive effort: if attention slips for even a minute, an onerous regulation will be imposed or a precious resource taken away. . . . But evidently the corporation agrees . . . since it devotes significant resources to political action of many kinds, including the awareness and involvement of top officials. Chief executive officers and members of the board of directors repeatedly express similar views.

Both of these views — the business view of vulnerability and our insistence on their power — are correct. . . .

Perhaps once upon a time business could simply make its wishes known and receive what it wanted; today corporations must form PACs, lobby actively, make their case to the public, run advocacy ads, and engage in a multitude of behaviors that they wish were unnecessary. From the outside we are impressed with the high success rates over a wide range of issues and with the lack of a credible challenge to the general authority of business. From the inside they are impressed with the serious consequences of occasional losses and with the continuing effort needed to maintain their privileged position.

Business power does not rest *only* on PAC donations, but the PAC is a crucial aspect of business power. A football analogy can be made: business's vast resources and its influence on the economy may be equivalent to a powerful offensive line that is able to clear out the opposition and create a huge opening, but someone then has to take the ball and run through that opening. The PAC and the government relations operation are, in this analogy, like a football running back. When they carry the ball they have to move quickly, dodge attempts to tackle them, and if necessary fight off an opponent and keep going. The analogy breaks down, however, because it implies a contest between two evenly matched opponents. Most of the time the situation approximates a contest between an NFL team and high school opponents. The opponents just don't have the same muscle. Often they are simply intimidated or have learned through past experience the best thing to do is get out of the way. Occasionally, however, the outclassed opponents will have so much courage and determination that they will be at least able to score, if not to win.

41
The Vicissitudes and Variety of the American Right

JEROME L. HIMMELSTEIN

This reading carries the theme of political influence a step in another direction. It is Himmelstein's thesis that the political Right consists of varied elements, and thus has been growing in numbers and influence. Consider the various groups of the political Right, the "truisms" of their beliefs, and whether there will be conflict among them as each one seeks more influence.

As you read, ask yourself the following questions:

1. Give one example of each of the three different elements of the American Right.

2. How have the changes brought about by the American Right affected you? Why?

GLOSSARY

Denominational Divided into sects.
Sectarianism A particular faith.

Some General Comments
on an Unruly Topic . . .

THE AMERICAN RIGHT DEFIES any overall, unified explanation. Accordingly, I shall not attempt one. Instead I shall content myself with identifying several elements of the Right, describing what has happened with each since the early Reagan years, and raising what I think are the important questions to ask about their impact in the coming years. . . .

To begin, here are some general points worth keeping in mind while studying the Right. First, *there is no such thing as the Right*—one unified movement or set of social processes. There are many elements to the Right, connected to each other to be sure, but each having its own history, leaders, organizations, and agenda. . . .

Paper for presentation at the 1995 meetings of the American Sociological Association, August 22, Washington, D.C. Reprinted by permission of the author.

Second, *everything looks different over time:* Political reality is slippery and ever changing. George Orwell once warned that "whoever is winning at the moment will always seem to be invincible" even if they are not and warned of the fallacy of believing that "because this was happening, nothing else could happen." That's good advice to keep in mind, whether the Right is ascendant or in relative eclipse.[1] . . .

. . . the Right at the end of the Reagan years was "entrenched," its "forward momentum . . . was largely exhausted and hence the prospects of the Right making further major gains seemed slight."[2]

Third, one reason that political reality is slippery is that *small changes can have big results.* The current ascendant position of the Right is due almost entirely to the 1994 elections. Had Republicans not won control of both Houses of Congress, the Right would still be a political presence, but clearly the Contract with America would not be the national agenda. Those election victories, however, were in themselves a small change that had immense consequences. They involved only a small shift in popular vote: According to exit polls published in the *New York Times,* Republicans got 50 percent of the vote in congressional races, up a mere three points from the average for the previous seven elections. This increase, moreover, did not even represent a simple shift in 3 percent of the voters from one party to another. It involved as well a changing pattern of voter turnout—a slight increase among more affluent voters and a slight decrease among lower income and minority voters. Furthermore, polls and interviews both before and after the election suggest that voter choices had little ideological meaning: Few voters had heard of the Contract with America, the Republican's main electoral selling point, much less approved of it unambiguously.[3]

How the Various Elements of the Right Have Fared . . .

THE CONSERVATIVE MOVEMENT

The conservative intellectual and political movement began in the 1950s as a delayed reaction to the New Deal. The founding of *National Review* in 1955 may be seen as a seminal event in this movement's history. Supported by waves of right-wing activism, corporate money, and close ties to the Republican Party, it developed as a network of organizations and journals dedicated to developing conservative ideas, recruiting new members on college campuses and elsewhere, and promoting conservatives for political office. By the late 1970s it had begun to develop a dense organizational infrastructure, to which the name "New Right" was often applied. By the mid-1980s it had become less a movement and more a "counter-establishment," "counter-intelligentsia," or "counter-counterculture" as various observers called it, though what "establishment" it was countering is by now unclear.[4]

This organizational infrastructure provides careers and connections for new generations of would-be conservative leaders and intellectuals. At the same time, it ensures that on virtually any issue there is an organization capable of presenting a ready-made right-wing position in a way that carries authority. This is one of the few constants in the world of the Right and constitutes its ideological backbone. When political opportunity beckons, a whole network of organizations is ready to respond. . . .

THE REPUBLICAN PARTY

By the early Reagan years the Republican Party had developed several advantages over the Democrats. Rebuilding after Watergate, the GOP had built up strong national party organizations, which outraised their opponents' national organizations several times over. This financial advantage translated into advantages in political technology and expertise, the ability to track elections closely, identify swing constituencies, develop effective propaganda, and target resources where they would have the most impact.

The Republicans also have been more unified ideologically—a more consistently conservative party than the Democrats are a liberal party. This perhaps explains another GOP advantage—raising money from "small" donors. It seems clear that the GOP does a better job of attracting the money of the ideologically conservative middle class than the Democrats do the money of the liberal middle class. The former has transferred their financial commitments from independent conservative groups to the party; the latter have not.

During the final Reagan years the Republican Party closed what might be called the "public opinion gap." After decades of lagging behind the Democrats in number of self-identified supporters, the GOP drew almost even, though neither party came close to enjoying majority support among the public. Similarly, the Republicans became identified for the first time by a plurality of Americans as the party best able to secure economic prosperity, a title held consistently by the Democrats up through the early 1980s.

Even with these advantages, the Republican Party noticeably failed to make itself the majority party in the 1980s. Despite holding the presidency for three terms, it made no consistent gains elsewhere. On almost all levels of elective office, the GOP was no better off at the end of the Reagan years than at the beginning. Not only did the party fail to dent the power of Democratic incumbency, it failed to make consistent headway in picking up open seats. This was most striking in the South, where an early shift of white voters—first to George Wallace and then to GOP presidential candidates—was not consistently followed by a shift in voting down the line.

One reason may be ideological. When all is said and done, most Americans expect the state to play an active role in economic life, which is to say, in improving their life chances. U.S. elections, however, are not clearly ideological. In the end, voters choose between candidates, not parties or ideologies.

Incumbents tend to win elections because they raise enough money and do enough favors for their constituents to discourage strong opponents. You can't beat someone with no one, even if someone happens to be unpopular. Where incumbents fail to discourage strong opposing candidates, however, they are often vulnerable.[5]

From this perspective, the key to whether the 1994 elections represent the future of electoral politics or a mere glitch is how effectively the Re-

publicans can recruit strong candidates, capable of raising lots of money and running credible campaigns. Republicans have an uneven history of doing this for two reasons. First, conservatively inclined potential candidates have often sought careers outside politics, in business for instance. Second, in many parts of the country, politically ambitious persons of no strong ideological conviction simply have found the Democratic Party to be a better bet. How both these factors have changed, if they have, may determine the ability of the GOP to recruit good candidates, and this in turn may determine whether they keep and expand their Congressional Majorities, providing an ever more hospitable political vehicle for the Right. This may be especially telling in the South, where political opportunity for Republicans seems finally to have percolated down to congressional, state, and local levels.[6]

THE RELIGIOUS RIGHT

The Right has built itself on successive bursts of grassroots activism since the 1950s, each with distinctive social roots. In the 1980s, the politicization of evangelical Christians provided the conservative movement an especially powerful wave of activism. Although over-hyped in the wake of the 1980 elections, this New Religious Right had become a palpable political reality by mid-decade. Evangelical Christians, especially religiously observant ones, shifted their political identification and their votes to the Republican Party during the Reagan years and have stayed there. Just as important, they provided the activists who peopled a growing array of issue-oriented movements, including those opposed to the Equal Rights Amendment and abortion.

Yet by the late 1980s, the New Religious Right seemed to have hit a wall. Pat Robertson's candidacy for the Republican presidential nomination, flopped badly despite having raised large amounts of money. Robertson did not transcend a distinctly sectarian appeal, doing well among charismatic Christians, but not among other religious conservatives. At the same time, the televangelists, from whose ranks several of the leaders of the New Religious Right had come, found their ministries in fiscal crisis or faced personal scandals. Most important, nearly all the original organizations of the movement were either formally defunct or mere shadows of their former selves: the Moral Majority, Religious Roundtable, Christian Voice, National Christian Action Council, Freedom Council, and the American Coalition for Traditional Values.[7]

Since 1987, however, the organizations of the New Religious Right have been born again. Most notable is the Christian Coalition, but the Family Research Council and the American Freedom Coalition have joined a few older groups, such as Concerned Women of America to push the religious right's position. The resurgence of the religious right makes sense given the persistence of the conditions that gave birth to it in the first place: (1) the declining importance of denominational distinctions in the religious world and the growing polarization of the theological conservative and theological liberal/secular; (2) the continuing growth of conservative churches and the growing density of their religious organizations; and (3) the persistence of a variety of "moral," "social," or "cultural" issues that provide focal points for the general conservative evangelical critique of American society.

One key to the future of the religious right is whether it can overcome the narrow sectarianism of some of the original New Religious Right groups while maintaining support of core activists.

Older groups were often narrowly rooted in Fundamentalist or Charismatic constituencies; the Moral Majority, for example, recruited most of its state leaders from the Baptist Bible Fellowship, resulting in a rhetoric that was often intolerant and dogmatic. The newer organizations seek a broader constituency and thus have supplemented sectarian religious rhetoric with a more pluralist secular language. Thus school prayer is now justified as a matter of student rights as well as of putting religious values back into school; battles over the content of school textbooks are presented as a matter of parental rights and control over curriculum as well as of countering secular humanism; school voucher plans are framed as a matter of school choice, etc. In each case the rhetoric of freedom and rights muscles in next to the rhetoric of morality and religion. Ralph Reed, executive director of the Christian Coalition, makes broad ecumenical appeals across religious lines, while Pat Robertson, president of the same organization, spins arcane conspiracy theories that draw heavily on classic antisemitism. The question is, can it work? Can the religious right appeal to the true believer and to a broad public?[8]

CORPORATE CONSERVATISM

Big business played a major role in the rise of the Right in the late 1970s and early 1980s. Faced with declining profits and growing international competition, the growth of regulatory agencies with broad authority, and growing public mistrust of business, corporations organized on a broad basis to defeat labor law reform and a consumer protection agency and later helped pass the Reagan tax cuts. In the process, effective classwide lobbying organizations like the Business Roundtable came to the fore. Capitalist money flowed to a growing network of right-wing public policy organizations, changing the shape of the think-tank universe. Most interestingly, corporate political action committees began funneling a substantial chunk of their contributions away from congressional incumbents to conservative Republican challengers and contenders for open seats.

By 1984, however, this corporate conservative mobilization had ebbed. Large corporate interests had achieved much of their agreed upon agenda, and they disagreed about the emerging economic issues of the mid-1980s, such as trade legislation or budget deficits. In addition, after the 1982 recession, there seemed to be little realistic opportunity for Republicans to take total control of Congress. Emblematic of the shift, corporate PAC money moved back to almost solely supporting incumbents, with Democrats getting about half the money. Big business had moved away from a conservative strategy of pushing the political universe to the right in favor of a more pragmatic strategy of maintaining access to whomever is in political power.[9]

The 1994 elections and their aftermath witnessed a move back again from corporate pragmatism to corporate conservatism. As Republicans developed an increasingly credible campaign to take control of both houses of Congress, corporate PAC money moved to Republican challengers and open-seat candidates in the summer and fall of 1994. Since the elections, business representatives working through the Thursday Group have met systematically with congressional Republicans to help push through the Contract with America, especially those provisions for business tax cuts, reductions in regulation, and limits on the scope of civil lawsuits. In addition, corporate interests, always a strong lobbying presence, seemed to have played an even stronger role on Capitol Hill. In several cases, Republican-controlled congressional committees have farmed out the job of drafting legislation directly to law firms representing business interest.[10]

The key issue here is how the dialectic of corporate conservatism and corporate pragmatism will play out in the future. The politics of big business has always been a mix of the two: Corporations have often tended to pursue a pragmatic strategy, working with whomever is in political power, except when faced with crises or exceptional political opportunities. This has won them disdain from conservative Republican leaders: House majority leader Armey, always favoring "ag" words, refers to many corporate leaders as "prags." With Republicans controlling both houses of Congress and after 1996 perhaps the White House as well, corporate pragmatism and conservatism may work in the same direction. Corporations may have no reason to stick with the Democratic Party—a huge problem and a great opportunity for Democrats, who have always competed ferociously for business support.[11]

The Radical Right

The less radical right has always been more or less closely shadowed by the more radical right. The militias and the harder core white suprema-

cist, survivalist, and Christian Identity groups from which they have emerged in the last three years are thus nothing new. What has always distinguished the more radical right has been one or more of the following: (1) broad, detailed conspiracy theories of history, (2) overt racial or ethnic bigotry, (3) the advocacy and practice of violence.[12]

The crucial point about the relationship between Right and the Radical Right is that they have always had one. . . . There have always been ideological affinities and overlapping social networks. As a result, the Right has always had to draw lines—to construct boundaries—where none naturally exist; and these boundaries have always been permeable. Fantastic conspiracies about communists, international bankers, and illuminati (a staple of the right for generations), are not wholly separable from a more nuanced critique of secular culture and liberal elites. Overt racial and ethnic bigotry too easily blend with talk about nationalism and American values. The strident condemnation of persons and groups said to be doing immoral things and destroying American society can easily shade over into advocating violence against them.

Because of their affinities, the silly conspiracies, overt bigotry, and violence of the more radical Right have always threatened to discredit the less radical Right. This is one problem the contemporary militias pose. Born of the most extreme racist, survivalist, and Christian Identity groups, they have spread out well beyond these groups and moderated just enough to provide a broad bridge between them and the less radical Right. The ties have become more numerous and hence lines are harder to draw.[13]

In addition, the militias present a second problem, a distinctly ideological one, for the less radical Right. One of the most striking things about the militias is that their attack on the state focuses on its repressive functions—the FBI, the ATF, the CIA, even the military. All of these are seen as part of a gigantic conspiracy to deprive Americans of their liberty. This makes them a less-than-perfect ally for the less radical right who have championed the forces of law and order and who have aimed their antistatism at the social welfare functions of the state. . . .

Put another way, the growing antilaw enforcement binge on the Right clashes with the Right's usual and powerful appeal to law and order. This is the distinctive problem the militias pose and why conservative journals like *National Review* are yet again puzzling over how and where to "draw the line" between Right and Radical Right.[14]

Conclusion

This paper has not sought either a global theory of the American Right nor a grand set of predictions about its future. Indeed, it has suggested that given the complexity of the Right and the tendency for small changes to have large consequences in American politics, neither is possible. Instead, it has sought to provide a way of organizing our thinking about the American Right by distinguishing several of its elements and identifying central issues with regard to each.

NOTES

1. Sonia Orwell and Ian Angus, ed., *The Collected Essays, Journalism, and Letters of George Orwell* (New York: Harcourt, Brace, Jovanovich, 1968), Volume 4, pp. 174, 324.

2. Jerome L. Himmelstein, *To The Right: The Transformation of American Conservatism* (Berkeley: University of California, 1990), p. 199, and "If They Did So Well, Why Do They Feel So Bad? The Right After the Reagan Years," Paper presented at the Ninth Presidential Conference: Ronald Reagan, the Fortieth President, Hofstra University, April 22–24, 1992.

3. "Portrait of the Electorate: Who Voted for Whom in the House," *New York Times*, November 13, 1994, p. 24; "Low-Income Voters' Turnout Fell in 1994, Census Reports," *New York Times*, June 11, 1995, p. 28; Richard L. Berke, "Victories Were Captured by G.O.P. Candidates, Not the Party's Platform," *New York Times*, November 10, 1994, p. B1; Richard L. Berke, "Poll Finds Public Doubts Key Parts of G.O.P.'s Agenda," *New York Times*, February 28, 1995, p. A1; Sara Rimer, "Other Party to Contract Isn't Impressed," *New York Times*, April 12, 1995, p. A1.

4. For a general overview of the development of the conservative movement, see Himmelstein, *To The Right*, pp. 13–94. For recent writing on the conservative policy elite, see James Atlas, "The Counter

Counterculture," *New York Times Magazine*, February 12, 1995, pp. 30–38; "The Right-Wing Media Machine," special issue of *Extra!*, March/April, 1995; "Burgeoning Conservative Think Tanks," special report, National Committee for Responsive Philanthropy, Spring 1991.

5. For a general discussion of the changing balance of power between the two major parties, see Himmelstein, *To The Right*, pp. 165–197, 206–211. For observations on the role of candidate quality in the 1994 elections, see David E. Rosenbaum, "G.O.P. Unleashes Its New Weapon: Winning Candidates," *New York Times*, November 13, 1994, p. E1. For data on small donors, see Stephen Engleberg, "Small Donors Filling G.O.P. War Chests," *New York Times*, April 29, 1995, p. 8.

6. On the 1994 elections in the South, see Peter Applebome, "The Rising G.O.P. Tide Overwhelms the Democratic Levees in the South," *New York Times*, November 11, 1994, p. A27; Ronald Smothers, "G.O.P. Gains in South Spread to Local Level," *New York Times*, April 11, 1995, p. A16; R. W. Apple Jr., "Louisiana, a Democratic Rock, Feels the Surge of a G.O.P. Tide," *New York Times*, April 29, 1995, p. 1.

7. For a general discussion of the religious right, see Himmelstein, *To The Right*, pp. 97–128, 203–204.

8. For the rebirth and possible transformation of the religious right, see Matthew C. Moen, *The Transformation of the Christian Right* (Tuscaloosa: University of Alabama, 1992). For a discussion of the conflicting ecumenical and sectarian faces of the religious right, see Frank Rich, "Bait and Switch II," *New York Times*, April 6, 1995, p. A31.

9. For a detailed analysis of corporate conservatism, see Himmelstein, *To The Right*, pp. 129–164, 204–206.

10. Stephen Engleberg, "Business Leaves the Lobby and Sits at Congress's Table," *New York Times*, March 31, 1995, p. A1; Engleberg, "100 Days of Dreams Come True for Lobbyists in Congress," *New York Times*, April 14, 1995; Thomas Ferguson, "G.O.P. $$$ Talked; Did Voters Listen?," *The Nation*, December 26, 1994, pp. 792–798.

11. For a more detailed discussion of corporate pragmatism and corporate conservatism, Jerome L. Himmelstein, "Two Faces of Business Power," *Business and the Contemporary World* 4 (4), 1992, 88–94 and Himmelstein, *Doing Good and Looking Good: Corporate Philanthropy and Business Power* (Indianapolis: Indiana University, forthcoming).

12. For an early effort to distinguish less and more radical Rights, see Arnold Forster and Benjamin R. Epstein, *Danger on the Right* (New York: Random House, 1964). For recent work on the militias and the Patriot Movement, see Chip Berlet and Matthew N. Lyons, "Militia Nation," *The Progressive*, June, 1995, pp. 22–25; Loretta J. Ross, "Saying It with a Gun,"

The Progressive, June, 1995, pp. 26–27; Michael Kelly, "The Road to Paranoia," *The New Yorker,* June 19, 1995, pp. 60–75; Marc Cooper, "Montana's Mother of All Militias," *The Nation,* May 22, 1995, pp. 714–722; Philip Weiss, "Off the Grid," *New York Times Magazine,* pp. 24–33+.

13. For linkages between the militias and mainstream politics, see James Ridgeway, "The Posse Goes to Washington: How the Militias and Far Right Got a Foothold on Capitol Hill," *Village Voice,* May 23, 1995, pp. 17–19.

14. "Lost Right?," *National Review,* July 10, 1995, pp. 14, 16.

QUESTIONS FOR DISCUSSION

For further discussion of this topic, see the Wadsworth Sociology Resource Center, "Virtual Society," *http://sociology.wadsworth.com,* under *Sociological Foot-prints,* by Cargan and Ballantine. You can respond to the discussion questions there or enter your own comments in the online chat forum.

SUGGESTED READINGS AND SOCIOLOGY INTERNET RESOURCES

See the Wadsworth Sociology Resource Center, "Virtual Society," *http://sociology.wadsworth. com,* for additional links, suggestions for further reading, and learning tools related to this chapter.

Either from the "Virtual Society" website or directly from your web browser, you may access InfoTrac College Edition, an online university library that includes over 700 popular and scholarly journals in which you can find articles related to the topics in this chapter.

Part IV

SOME PROCESSES
OF SOCIAL LIFE

A N EARLY PIONEER IN sociology, Auguste Comte, divided sociology into two major parts: statics and dynamics. Statics is the study of order, whereas dynamics is the study of social progress. Although the terms utilized by Comte have changed, this basic division of sociology remains in use today as the study of social structure and functions and social change. This section deals with some of the processes or dynamics in society.

In the preceding sections, we noted that individuals are transformed into social beings through group interactions in a specific cultural context. We saw that socialization not only transforms people into social beings but also makes them viable members of society by imbuing them with the culture of the society. The major institutions are an essential part of the process. But society does not consist just of structure and institutions; it also involves the process of intermingling and the results of that interaction. Results that bring about change can either aid or hinder individuals or society in meeting their needs.

The intermingling of the world's peoples has produced group relations that sometimes lead to stereotyping, prejudice, and discrimination based on cultural or physical differences between people. The readings in Chapter 11 on minority relations discuss what is happening to some specific minorities who are attempting to change their experiences and situations.

To function smoothly, society requires normative behavior of its members. The socialization process generally results in individuals who follow the norms of society. However, different socialization processes will result in different interpretations of the norms and produce what we call *deviance*. Deviants may be any people or groups whom society has labeled as different. In Chapter 12, we read about myths concerning deviants and about the effects of labeling.

Chapter 13 deals with population and the urban scene, two topics that are related in many ways. Population growth has led to the development of urban areas that fulfill necessary functions for the population. But there is another side to population growth. Unchecked growth can mean enormous consumption of energy, which creates numerous problems. Trends in migration patterns reflect economic and social problems as people seek better opportunities. These problems are explored in this chapter.

The theme of Chapter 14 is change. All societies undergo change, but change can be so rapid that a society may find itself with few norms to guide behavior. This situation can result in such collective behaviors as fads, panic, and even revolution. Another effect of rapid change may be the inability of a society to keep its beliefs and value systems in line with changes in its material system. Groups of people dissatisfied with the societal system join together to form social movements directed toward accomplishing desired changes. Some people fear change and its consequences for their lives, and some change may be disruptive to routines. However, change is not all bad, or dysfunctional. In fact, the possibilities open to us through societal change can be both exciting and fulfilling.

CHAPTER 11

Minority Relations

"I MET THE STRANGER AND THEM IS US!"*

"WE" AND "THEY," IN-GROUPS AND OUT-GROUPS—these are categories people use to order their complex world. How easy it is to judge by generalization and stereotype rather than fact, and to maintain our ethnocentric attitude favoring the "we," the in-group, *our* group. If categorization serves as the primary means of ordering our lives, is that bad? Potentially, yes. In the process of categorizing, we accept some people and ideas, and rule out others—often by using arbitrary criteria that discriminate against individuals who belong to certain groups.

*Rudyard Kipling.

The problem is that this consciousness of difference has created policies from pluralism and separation to slavery and extermination; some of these policies, in turn, have led to group self-consciousness and to reactions ranging from social protest to insurrection. Minority groups are often identified by distinct physical or cultural characteristics, and as a result they are singled out for differential and unequal treatment.

To understand why prejudice and discrimination occur, social scientists provide explanations on three levels: individual, group, and societal. Individuals exhibit prejudice through their beliefs and actions. Social psychologists studying prejudiced individuals have even developed a set of characteristics to describe the most prejudiced personality type. Using a minority group as a scapegoat for venting frustration and aggression is one expression of individual prejudice.

Prejudicial attitudes can be seen in many forms and against many different people. Prejudice often surfaces at the group level in discriminatory acts, such as restricting access to housing or jobs, unequal education, or other exclusionary measures. Usually, those who practice prejudice and discrimination need no evidence to support the biased acts, but they will use any available evidence to support their actions, even if it is unsupported or false. Often people discriminate on the basis of stereotypes that have become part of the common lore. Stereotypes are general beliefs, unchecked and unproven, directed against a group of people in society.

The United States, land of opportunity, has proven to be anything but that for segments of the population. Founded on principles of freedom and justice, certain groups, including Africans and Asians, were left out of the formula. Africans came as slaves and have a long road to travel before the remnants of their suppression—prejudice, discrimination, racism—are overcome. Asians often supplied needed labor for building the railroad, but when economic times were hard they were no longer welcomed by the dominant group. Similar patterns of race relations between dominant and subordinate groups occur in countries around the world. Consider the current strife in German cities as disaffected youth attack foreign workers.

The first reading lays the foundation for understanding minority relations; Robert Merton's classic piece on prejudice and discrimination shows the relationship between these two key concepts, setting the stage for the other readings in this section.

The second reading, by Lewis Killian, traces the civil rights movement from the 1960s to the present, giving a pessimistic outlook for the future. Although hopes for change and a better life were high in the 1960s, economic stagnation reduced prospects for improvement of minority groups on the road to equality. Perhaps improved economic conditions will bring change.

The topic of minority relations is relevant throughout the world. Eduardo Bonilla-Silva and Mary Hovsepian discuss the problems that occur when workers from developing countries seek opportunities in western Europe and the United States and the rise of hostilities that greet these workers and their families, especially when economic conditions in the host countries are poor.

What has been referred to as "the black underclass"[1] is due mainly to historical discrimination and economic trends. These problems have causes that lie in the economic and social structure of society; they are difficult to change, and in fact are increasing in many urban areas. Many frustrated, disaffected youth who see no opportunities and face daily hostile environments find belonging, acceptance, and protection in gang membership. Hispanics are the fastest-growing minority group in the United States, and experience prejudice and discrimination because of differences in culture and language. Moore traces the history of these groups and their current isolation and stigmatization in society, factors that will not make change easy. Moore and Pinderhughes outline economic and social problems faced by Hispanics living in the United States today.

Although we may not all come in direct contact with some minority groups, we are all affected by changing gender and age relationships. Economic problems have been eased for most of the elderly population in the United States, but social isolation and stigmatization can still be problematic. Longino points to the increasing issues related to this group as it grows in size, but also to ways of coping with problems that arise.

If racism, prejudice, stereotyping, and discrimination are to be reduced, then individuals and groups must recognize the effects of prejudice and discrimination embedded in the societal structure on their own beliefs and actions. Only in this manner can they overcome their own ethnocentrism and view others from a cultural relativist's perspective. As you consider these readings, think of ways that this can be done in our society. Consider why some minorities have had more success than others. Also note other examples you have observed of groups segregating themselves for self-protection, of false stereotyping, and of economic conflicts between groups that lead to discrimination.

NOTE

1. William Julius Wilson, The black underclass, *Wilson Quarterly* (Spring 1984).

42
Discrimination and the American Creed

ROBERT K. MERTON

Two basic concepts necessary to gain an under-standing of minority relations are prejudice *(an attitude) and* discrimination *(an action). These concepts describe the attitudes and actions of individuals and groups toward one another. We may think of these as separate concepts, but Robert K. Merton, an important structural functional theorist, points out how they are interrelated.*

As you read this classic piece, consider the following:

1. What is "the American creed"?

2. Using the four types discussed by Merton, how might prejudice and discrimination be reduced?

GLOSSARY

Prejudice Attitude; rigid and irrational generalization about a category of people.

Discrimination Behavior; treating categories of people differently, usually unequally.

American creed Set of values and precepts imbedded in American culture to which Americans are expected to conform.

SET FORTH IN THE Declaration of Independence, the preamble of the Constitution, and the Bill of Rights, the American creed has since often been misstated. This part of the cultural heritage does *not* include the patently false assertion that all human beings are created equal in capacity or endowment. It does *not* imply that an Einstein and a moron are equal in intellectual capacity or that Joe Louis and a small, frail Columbia

professor (or a Mississippian Congressman) are equally endowed with brawny arms harboring muscles as strong as iron bands. It does *not* proclaim universal equality of innate intellectual or physical endowment.

Instead, the creed asserts the indefeasible principle of the human right to full equity—the right of equitable access to justice, freedom, and opportunity, irrespective of race or religion or ethnic origin. It proclaims further the universalist doctrine of the dignity of the individual, irrespective of the groups of which he is a part. It is a creed announcing full moral equities for all, not an absurd myth affirming the equality of intellectual and physical capacity of all people everywhere. And it goes on to say that although individuals differ in innate endowment, they do so as individuals, not by virtue of their group memberships.

Viewed sociologically, the creed is a set of values and precepts embedded in American culture to which Americans are expected to conform. It is a complex of affirmations, rooted in the historical past and ceremonially celebrated in the present, partly enacted in the laws of the land and partly not. Like all creeds, it is a profession of faith, a part of cultural tradition sanctified by the larger traditions of which it is a part.

It would be a mistaken sociological assertion, however, to suggest that the creed is a fixed and static cultural constant, unmodified in the course of time, just as it would be an error to imply that as an integral part of the culture, it evenly blan-

From "Discrimination and the American Creed" by Robert K. Merton, in Discrimination and the National Welfare, *Robert M. MacIver, ed. Copyright 1949 by the Institute for Religious and Social Studies. Reprinted by permission of HarperCollins Publishers, Inc.*

A Typology of Ethnic Prejudice and Discrimination

	Attitude Dimension:* Prejudice and Non-Prejudice	Behavior Dimension:* Discrimination and Non-Discrimination
Type I: Unprejudiced Non-Discriminator	+	+
Type II: Unprejudiced Discriminator	+	−
Type III: Prejudiced Non-Discriminator	−	+
Type IV: Prejudiced Discriminator	−	−

*Where (+) = conformity to the creed and (−) = deviation from the creed.

kets all subcultures of the national society. It is indeed dynamic, subject to change and in turn promoting change in other spheres of culture and society. It is, moreover, unevenly distributed throughout the society, being institutionalized as an integral part of local culture in some regions of the society and rejected in others. . . .

With respect to actual practices: conduct may or may not conform to the creed. And further, this being the salient consideration: *conduct may or may not conform with individuals' own belief concerning the moral claims of all people to equal opportunity.*

Stated in formal sociological terms, this asserts that attitudes and overt behavior vary independently. *Prejudicial attitudes need not coincide with discriminatory behavior.* The implications of this statement can be drawn out in terms of a logical syntax whereby the variables are diversely combined, as can be seen in the typology at the top of this page.

By exploring the interrelations between prejudice and discrimination, we can identify four major types in terms of their attitudes toward the creed and their behavior with respect to it. Each type is found in every region and social class, although in varying numbers. By examining each type, we shall be better prepared to understand their interdependence and the appropriate types of action for curbing ethnic discrimination. The folk-labels for each type are intended to aid in their prompt recognition.

Type I: The Unprejudiced Non-Discriminator or All-Weather Liberal

These are the racial and ethnic liberals who adhere to the creed in both belief and practice. They are neither prejudiced nor given to discrimination. Their orientation toward the creed is fixed and stable. Whatever the environing situation, they are likely to abide by their beliefs: hence, the *all-weather* liberal.

These make up the strategic group that *can* act as the spearhead for the progressive extension of the creed into effective practice. They represent the solid foundation both for the measure of ethnic equities that now exist and for the future enlargement of these equities. Integrated with the creed in both belief and practice, they would seem most motivated to influence others toward the same democratic outlook. They represent a reservoir of culturally legitimatized goodwill that can be channeled into an active program for extending belief in the creed and conformity with it in practice.

Most important, as we shall see presently, the all-weather liberals comprise the group that can so reward others for conforming with the creed as to transform deviants into conformers. They alone can provide the positive social environment for the other types who will no longer find it expedient or rewarding to retain their prejudices or discriminatory practices.

Although ethnic liberals are a *potential* force for the successive extension of the American creed, they do not fully realize this potentiality in actual fact, for a variety of reasons. Among the limitations on effective action are several fallacies to which the ethnic liberal seems peculiarly subject. First among these is the *fallacy of group soliloquies*. Ethnic liberals are busily engaged in talking to themselves. Repeatedly, the same groups of like-minded liberals seek each other out, hold periodic meetings in which they engage in mutual exhortation, and thus lend social and psychological support to one another. But however much these unwittingly self-selected audiences may reinforce the creed among themselves, they do not thus appreciably diffuse the creed in belief or practice to groups that depart from it in one respect or the other.

More, these group soliloquies in which there is typically wholehearted agreement among fellow-liberals tend to promote another fallacy limiting effective action. This is the *fallacy of unanimity*. Continued association with like-minded individuals tends to produce the illusion that a large measure of consensus has been achieved in the community at large. The unanimity regarding essential cultural axioms that obtains in these small groups provokes an overestimation of the strength of the movement and of its effective inroads upon the larger population, which does not necessarily share these creedal axioms. Many also mistake participation in the groups of like-minded individuals for effective action. Discussion accordingly takes the place of action. The reinforcement of the creed for oneself is mistaken for the extension of the creed among those outside the limited circle of ethnic liberals. . . .

plicitly acquiescing in expressions of ethnic prejudice by others or in the practice of discrimination by others. This is the expediency of the timid: the liberal who hesitates to speak up against discrimination for fear he might lose esteem or be otherwise penalized by his prejudiced associates. Or his expediency may take the form of grasping at advantages in social and economic competition deriving solely from the ethnic status of competitors. Thus the expediency of the self-assertive: the employer, himself not an anti-Semite or Negrophobe, who refuses to hire Jewish or Negro workers because "it might hurt business"; the trade union leader who expediently advocates racial discrimination in order not to lose the support of powerful Negrophobes in his union.

In varying degrees, fair-weather liberals suffer from guilt and shame for departing from their own effective beliefs in the American creed. Each deviation through which they derive a limited reward from passively acquiescing in or actively supporting discrimination contributes cumulatively to this fund of guilt. They are, therefore, peculiarly vulnerable to the efforts of the all-weather liberals who would help them bring conduct into accord with beliefs, thus removing this source of guilt. They are the most amenable to cure, because basically they want to be cured. Theirs is a split conscience that motivates them to cooperate actively with people who will help remove the source of internal conflict. They thus represent the strategic group promising the largest returns for the least effort. Persistent reaffirmation of the creed will only intensify their conflict but a long regimen in a favorable social climate can be expected to transform fair-weather liberals into all-weather liberals.

Type II: The Unprejudiced Discriminator or Fair-Weather Liberal

The fair-weather liberal is the man of expediency who, despite his own freedom from prejudice, supports discriminatory practices when it is the easier or more profitable course. Expediency may take the form of holding his silence and thus im-

Type III: The Prejudiced Non-Discriminator or Fair-Weather Illiberal

The fair-weather illiberal is the reluctant conformist to the creed, the man of prejudice who does not believe in the creed but conforms to it

in practice through fear of sanctions that might otherwise be visited upon him. You know him well: the prejudiced employer who discriminates against racial or ethnic groups until a Fair Employment Practice Commission, able and willing to enforce the law, puts the fear of punishment into him; the trade union leader, himself deeply prejudiced, who does away with Jim Crow in his union because the rank-and-file demands that it be done away with; the businessman who forgoes his own prejudices when he finds a profitable market among the very people he hates, fears, or despises; the timid bigot who will not express his prejudices when he is in the presence of powerful men who vigorously and effectively affirm their belief in the American creed.

It should be clear that the fair-weather illiberal is the precise counterpart of the fair-weather liberal. Both are men of expediency, to be sure, but expediency dictates different courses of behavior in the two cases. The timid bigot conforms to the creed only when there is danger or loss in deviations, just as the timid liberal deviates from the creed only when there is danger or loss in conforming. *Superficial similarity in behavior of the two in the same situation should not be permitted to cloak a basic difference in the meaning of this outwardly similar behavior,* a difference that is as important for social policy as it is for social science. Whereas the timid bigot is under strain when he conforms to the creed, the timid liberal is under strain when he deviates. For ethnic prejudice has deep roots in the character structure of the fair-weather bigot, and this will find overt expression unless there are powerful countervailing forces—institutional, legal, and interpersonal. He does not accept the moral legitimacy of the creed; he conforms because he must, and will cease to conform when the pressure is removed. The fair-weather liberal, on the other hand, is effectively committed to the creed and does not require strong institutional pressure to conform; continuing interpersonal relations with all-weather liberals may be sufficient.

This is one critical point at which the traditional formulation of the problem of ethnic discrimination as a departure from the creed can lead to serious errors of theory and practice. *Overt behavioral deviation (or conformity) may signify importantly different situations, depending upon the underlying motivations.* Knowing simply that ethnic discrimination is rife in a community does not therefore point to appropriate lines of social policy. It is necessary to know also the distribution of ethnic prejudices and basic motivations for these prejudices as well. Communities with the same amount of overt discrimination may represent vastly different types of problems, dependent on whether the population is comprised of a large nucleus of fair-weather liberals ready to abandon their discriminatory practices under slight interpersonal pressure or a large nucleus of fair-weather illiberals who will abandon discrimination only if major changes in the local institutional setting can be effected. Any statement of the problem as a gulf between creedal ideals and prevailing practice is thus seen to be overly simplified in the precise sense of masking this decisive difference between the type of discrimination exhibited by the fair-weather liberal and by the fair-weather illiberal. . . .

Type IV: The Prejudiced Discriminator or the All-Weather Illiberal

This type, too, is not unknown to you. He is the confirmed illiberal, the bigot pure and unashamed, the man of prejudice consistent in his departures from the American creed. In some measure, he is found everywhere in the land, though in varying numbers. He derives large social and psychological gains from his conviction that "any white man (including the village idiot) is 'better' than any nigger (including George Washington Carver)." He considers differential treatment of Negro and white not as "discrimination," in the sense of unfair treatment, but as "discriminating," in the sense of showing acute discernment. For him, it is as clear that one "ought" to accord a Negro and a white different treatment in a wide diversity of situations as it is clear to the population at large that one "ought" to accord a child and an adult different treatment in many situations.

This illustrates anew my reason for questioning the applicability of the usual formula of the American dilemma as a gap between lofty creed and low conduct. For the confirmed illiberal, ethnic discrimination does *not* represent a discrepancy between *his* ideals and *his* behavior. His ideals proclaim the right, even the duty, of discrimination. Accordingly, his behavior does not entail a sense of social deviation, with the resultant strains that this would involve. The ethnic illiberal is as much a conformist as the ethnic liberal. He is merely conforming to a different cultural and institutional pattern that is centered, not on the creed, but on a doctrine of essential inequality of status ascribed to those of diverse ethnic and racial origins. To overlook this is to overlook the well-known *fact* that our national culture is divided into a number of local subcultures that are not consistent among themselves in all respects. And again, to fail to take this fact of different subcultures into account is to open the door for all manner of errors of social policy in attempting to control the problems of racial and ethnic discrimination.

This view of the all-weather illiberal has one immediate implication with wide bearing upon social policies and sociological theory oriented toward the problem of discrimination. The extreme importance of the social surroundings of the confirmed illiberal at once becomes apparent. For as these surroundings vary, so, in some measure, does the problem of the consistent illiberal. The illiberal, living in those cultural regions where the American creed is widely repudiated and is no effective part of the subculture, has his private ethnic attitudes and practices supported by the local mores, the local institutions, and the local power structure. The illiberal in cultural areas dominated by a large measure of adherence to the American creed is in a social environment where he is isolated and receives small social support for his beliefs and practices. In both instances, the *individual* is an illiberal, to be sure, but he represents two significantly different *sociological types*. In the first instance, he is a *social conformist*, with strong moral and institutional reinforcement, whereas in the second, he is a *social deviant*, lacking strong social corroboration. In the one case, his discrimination involves him in further integration with his network of social relations; in the other, it threatens to cut him off from sustaining interpersonal ties. In the first cultural context, personal change in his ethnic behavior involves alienating himself from people significant to him; in the second context, this change of personal outlook may mean fuller incorporation in groups significant to him. In the first situation, modification of his ethnic views requires him to take the path of greatest resistance whereas in the second, it may mean the path of least resistance. From all this, we may surmise that any social policy aimed at changing the behavior and perhaps the attitudes of the all-weather illiberal will have to take into systematic account the cultural and social structure of the area in which he lives.

43

Race Relations and the Nineties: Where Are the Dreams of the Sixties?

LEWIS M. KILLIAN

The prospects for racial equality, a principle supported by most Americans, seem bleak according to Lewis Killian, a long-time scholar of race relations. As we reach the end of the 1990s, bringing about the economic reforms needed to reduce class and racial inequalities looks dim. Killian traces the civil rights movements that have led to the current stagnation.

Keep in mind the following questions as you read:

1. Is there hope for change in the race relations situation?

2. What changes are likely in the early twenty-first century?

GLOSSARY

Civil rights movement Movement to gain social, political, and economic rights for all Americans.

Black nationalism and black power Blacks joining together as a powerful force to overthrow stigmatization and inequality.

Nonviolence A strategy advocated by Martin Luther King, Jr., and others to bring about change without the use of violence.

Racism Negative attitudes and behaviors that lead to subordination and prejudice against minorities. Often used to mean prejudice and discrimination.

ONE OF THE MOST inspiring events of the 1960s occurred on August 28, 1963, when Martin Luther King, Jr., stood on the steps of the Lincoln Memorial and declared, "I have a dream my four little children will one day live in a nation where they will not be judged by the color of their skin but by the content of their character. I

have a dream today!" (1986:219). This was an era of brave, optimistic dreams. Those dreams began to take shape ten years before King's memorable speech, as the school desegregation decision of 1954 gave rise to brave hopes in the hearts of segregated, downtrodden blacks. The concept "a revolution of rising expectations" well described their situation. Victories over white southern resistance in Montgomery, Tallahassee, Little Rock, and New Orleans provided black Americans and their white allies with a sense of empowerment. The sit-ins of the early sixties, still nonviolent, still interracial, showed that the Movement could not be suppressed. As King said in his address, "Nineteen sixty-three is not an end, but a beginning. And those who hope that the Negro needed to blow off steam and will now be content will have a rude awakening if the nation returns to business as usual" (1986:219). Yet as we stand on the brink of the nineties the dreams have dimmed and the nation has indeed returned to business as usual. In his last presidential address to SCLC, in 1972, King urged again, "Let us be dissatisfied until men and women, however black they may be, will be judged on the basis of the content of their character and not on the basis of the color of their skin" (1986:251). How sorely pained he would be were he to witness the state of ethnic relations today!

A Succession of Dreams

The Civil Rights vision formulated by King and his lieutenants was the first of a series of dreams.

© *The University of North Carolina Press.* Social Forces, *September 1990, 69(1):1–13. Reprinted by permission of* Social Forces *and the author.*

It was symbolized by its famous slogan, "black and white together."

Before the tragic end of his career King did place greater and greater emphasis on economic equality, particularly as he saw segregation diminishing while black unemployment and poverty persisted. He called for full employment, a guaranteed annual income, redistribution of wealth, and skepticism toward the capitalistic economy. "A true revolution of values," he declared, "will soon look uneasily on the glaring contrast of poverty and wealth" (1986:241).

During the height of the Civil Rights Movement the courage of the workers and the vicious violence of the white southern resistance engendered a national orgy of guilt and fear that provided the catalyst for the passage of the Civil Rights Act of 1964 and the Voting Rights Act the next year. The basic economic changes required if laws mandating desegregation and equal opportunity were to have more than a minimal effect had not come about, however. Moreover, the urban insurrections, brought to the forefront of the news by the Watts riot of 1965, awakened the nation to the fact that blacks were still far from content. King's dream of a revolution fueled by love and fought with nonviolence faded in the smoke of ghetto fires. The competing dream of Black Power dominated the last half of the decade.

The Black Power Dream

The vision of black power as the way out of inequality and stigmatization had deep roots in black history in the United States. Even as King was emerging as a national black hero the nationalistic message preached so eloquently by Malcolm X resonated in the consciousness of hundreds of impatient, angry black people who saw no chance of entering white middle-class society. As far back as the middle of the nineteenth century Black Nationalism had been a strong ideological undercurrent in the United States, surfacing, as William J. Wilson has argued, when intense frustration and disillusionment follow a span of heightened expectations (1973:50).

As limited as its human and financial resources may have been, the Black Power Movement was, with the aid of a titillated white press, able to drown out the voices of the leaders of the Civil Rights Movement. "Power," not "love"; "defensive violence" and "any means necessary," not nonviolence; and "soul" or "blood," not "black and white together," were the cries resounding in the ghettos and repeated on the nation's television screens.

Even more so than do most social movements, the Black Power Movement failed to achieve its stated objectives. Blacks did not win even veto power in the politics of the nation, the states, or the cities. The sort of power they gained in predominantly black cities and congressional districts resulted from demographic changes, not from concessions to the demands of the movement. Neither the extravagant dream of a black republic in the South nor the moderate one of viable all-black municipalities, such as Soul City, North Carolina, was realized. Real advances in self-chosen separatism, going beyond the historically black churches and fraternal organizations, primarily took the form of black studies departments, black cultural centers, and a few all-black dormitories in predominantly white universities.

Despite its near failure, the significance of the short-lived black power movement has been greatly underestimated. First of all, its very demise dramatized the lengths that the white power structure, at all levels, would go to suppress blacks who did not remain meek and mild. Of even greater importance was the change in the terms of the ongoing debate between whites and blacks and within each community about the future of blacks in U.S. society. King's shining grail of integration had lost its luster, at least for the time being. Assimilation had long been the dominant theme among both black Americans and white liberals, but now both its feasibility and desirability were being questioned. Various forms of pluralism gained legitimacy. At best, assimilation was a dream to be deferred until after a period of benign race consciousness.

The theme of black consciousness underscored the pervasive persistence of ethnic diversity in the society. It was accompanied by a novel concept, that of ethnic group rights. The idea of civil rights, individual rights based on citizenship, was now supplemented by the idea of rights based on membership in an ethnic group with a collective claim to being or having been an oppressed minority.

This seed fell on fertile soil, for other ethnic groups, not only Latinos and Native Americans, but also what Michael Novak called the "unmeltable white ethnics," began to advance their claims (1972). A system of competitive pluralism coupled with what Barbara Lal has called "compulsory ethnicity" arose along with "the institutionalization of ethnic identification as a basis for the assertion of collective claims concerning the distribution of scarce resources" (1983:167). In this connection we should note that since 1980 citizens filling out the schedule of the decennial census have been called on to specify the ancestral group with which they identify—application of the rule of descent has received bureaucratic sanction at the federal level. Yet those are probably the most inaccurate data to be found in the census volumes. Careful research has shown about one-third of respondents are likely to change their ethnic responses from year to year. And, ironically, Stanley Lieberson suggests that a new ethnic group is now growing in the United States—unhyphenated whites. He identifies them for statistical purposes as that one sixth of the population who, in 1980, identified themselves simply as "American" or refused to report any ancestry (1982).

One of the last demands addressed to United States society in the spirit of black power was the call for reparations. The Black Manifesto read by James Forman on the steps of New York's Riverside Church on May 4, 1969, was not an angry, quixotic whim of Forman and the few associates who accompanied him. It was a document drawn up by the National Black Economic Development Conference at a meeting set up by the Interreligious Foundation for Community Organization. The latter was created by most of the mainline Protestant denominations in the nation.

In the Manifesto Forman and others charged that the white Christian churches and the Jewish synagogues were part and parcel of the capitalistic system which had exploited the resources, the minds, the bodies, and the labor of blacks for centuries. The NBEDC was demanding $500 million in reparations. The melodramatic rhetoric of the Manifesto proclaimed that this came to "$15 per nigger," but the demand was not for the distribution of such a pittance to 30 million black individuals. Instead it called for the establishment of a southern land bank to enable displaced black farmers to establish farm cooperatives; black-controlled publishing houses and audio-visual networks; skills training centers; and other such collective enterprises. Whether such projects would have succeeded is beside the point. What is important is that the demand for reparations called for compensation to a group in the name of ethnic group rights; it was not a plea for the funding of "black capitalism."

The Dreams Fade

The principal outcome of the Manifesto was an outpouring of resolutions by churches. As Arnold Schucter observed, the "great orgy of American guilt" seemed to have subsided by that time, as had the urban insurrections (1970:28). Cointel-Pro was decimating the ranks of the Black Power Movement, and agents provocateurs were giving it a terrorist image. Even whites who had finally begun sympathizing with the goals of the Civil Rights Movement were asking, "Haven't we done enough for the blacks?"

Already the trend toward white acceptance of the principle of racial equality, particularly as applied to education and equal job opportunity, was discernible in public opinion polls. It was widely agreed that "white racism" was a terrible evil—but what did this mean?

The term "racism," usually meaning "white racism," became a catchword after the Kerner Commission declared in the summary of its Report to the President on the causes of civil disorders, "White racism is essentially responsible for the explosive mixture which has been accumulating in our cities since the end of World War II" (1968:203). But who are the white racists—particularly in the eyes of the majority of whites who now claim to accept the principle of racial equality? It is not they themselves but those Klansmen and American Nazis and Skinheads. They themselves are innocent, for they have accepted the victories of the Civil Rights Movement. They don't object to sharing public accommodations with blacks and they will let their children go to school with them as long as there aren't too many. They believe that blacks should have equal job opportunities and if a lot of them remain poor it must be because they don't take advantage of the changes open to them. Schucter was all too accurate when he wrote in 1970, "We are faced with a society in which racism has become institutionalized even though the majority of Americans vehemently protest their innocence" (1970:28).

The Fruits of the Dreams

By the beginning of the 1970s both the Civil Rights Movement and the Black Power Movement were comparatively dormant. As pointed out earlier the Black Power Movement had consequences of greater significance than is generally recognized. These consequences are seen primarily in the world view of blacks in the U.S., symbolized by the fact that the vast majority now call themselves "Black," not "Negro," and some even prefer "African American." Concretely, black power is seen only in the political realm and then only dimly. There is an important but still very small black congressional caucus, but there has been no black senator since the defeat of Edward Brooke in 1972. Numerous blacks have been elected to city, county, and state offices, but not until 25 years after the passage of the Voting Rights Act did a state elect a black governor. Black political power remains dependent upon a high degree of black residential concentration and the drawing of electoral boundaries to reflect that concentration.

The Civil Rights Movement, despite its apparent triumph with the passage of the Civil Rights Act and the Voting Rights Act, still won only intermediate objectives. *De jure* segregation was struck down. *De facto* segregation in public places was greatly reduced, but the illusion of equality

created in the forum was not reflected at the hearth; American homes, neighborhoods, and private clubs remained highly segregated. King's dream of a society where people would not be judged on the basis of their color remained woefully unfulfilled. Even Latinos, a newer minority in many areas, find it easier to escape from the barrio than do blacks from the ghetto.

But what were the objectives unattained by either movement? Let us look again for a moment at the response of the white churches to the Black Manifesto. They placed new emphasis on preaching and teaching against "racism" and on welcoming blacks into the pews of white churches; they raised money—not a great deal—to put into the ghettos to aid the poor, the disenfranchised, and the uneducated. But the lesson of the failed dreams of the 1960s is that it is not sensitivity training, nor token integration, nor welfare that is needed to eradicate the destructive consequences of ethnic discrimination. It is drastic economic reform and that revolution of values that Martin Luther King said would "look uneasily on the glaring contrast of poverty and wealth." What the dreams did not produce was a society where black and white children would not only sit beside each other in school but also achieve equal gains in learning; a society where blacks would not only have equal rights to jobs but also have jobs; a society where poverty not only would ignore ethnic boundaries but also would actually diminish. How much closer are we to that sort of society than we were in 1970?

A Glass Half Empty

Many times after the publication of *The Impossible Revolution?* I heard myself characterized as a chronic pessimist who would always see a glass as half empty, never as half full (Killian 1968). The analogy itself is flawed, of course—in life good or bad is never stable but is always rising or falling.

Today when I look about me, particularly in the South, and see whites and blacks eating, shopping, studying, working, and playing in each other's presence in places where once they were cruelly segregated, I think, "How great and wonderful the progress since 1954!" But when I look at the little clumps of black people sitting, talking, huddling together even in supposedly integrated settings, I wonder if we have not progressed only to that condition which Cayton and Drake called "the equality of anonymity" (1945:102). When I drive through a still segregated and often very poor black residential area, and when I look at the economic indicators, I am even more pessimistic. Indeed, I am convinced that the glass is surely half empty, for the level of black well-being is falling.

Reports and Reports

Testifying before the Kerner Commission in 1968 Kenneth B. Clark said, "I read that report . . . of the 1919 riot in Chicago, and it is as if I were reading the report of the investigating committee on the Harlem riot of '35, the report of the investigating committee on the Harlem riot of '43, the report of the McCone Commission on the Watts riot. I must again in candor say to you members of this committee it is a kind of Alice in Wonderland — with the same moving picture shown over and over again, the same analysis, the same recommendations, and the same inaction" (1977:ix).

Now we have the latest of the massive, comprehensive studies of how blacks are faring, *Blacks in American Society,* put together by a team of distinguished social scientists (Jaynes et al. 1989). It comes 70 years after the Chicago research, 45 years after *An American Dilemma,* and 20 years after the Kerner report. This volume starts out with refreshing honesty. While acknowledging many improvements, the authors declare, "We also describe the continuance of conditions of poverty, segregation, discrimination, and social fragmentation of the most serious proportions" (1989:ix).

The analysis of trends since the great migration of blacks out of southern agriculture beginning in 1939 leads to the conclusion that the place of blacks in the American economy has been, and remains, that of a reserve army of labor. They have enjoyed some progress during periods of prosperity and high employment, usually war-induced. "But," says the study, "after

initial reports of rising relative black economic status, black gains have stagnated on many measures of economic position since the early 1970s" (1989:274). Two examples of this stagnation are given. Poverty rates for blacks increased from 29.7% in 1974 to 31% in 1985. Blacks' real per capita income in 1984 was one-third higher than in 1968 but still stood in the same relationship to white income as in 1971 — 57%. Yet it is important to note that poverty had increased among whites also, from 7.3% in 1974 to 11% in 1985.

There is no need to repeat the much cited evidence of the accentuated differences in status among blacks, with some segments gaining drastically relative to whites and others losing ground. The major source of inequality within the black community, the authors note, is the increased fraction of black men with no earnings at all. The major reasons for black economic inequality are (1) the concentration of black workers, particularly men, in low-paying jobs and (2) the relatively high proportion of unemployed blacks, many of them not even in the labor force. In fact, while between 1973 and 1986 black men with jobs continued to approach whites in position on the occupational ladder and in hourly wage rates, the gains were offset by employment losses. The optimistic reports about employment rates released almost every month from Washington rarely note that black unemployment still continues at a rate twice that of whites and that the rates are based on persons in the labor force, not including the bitter, discouraged dropouts.

Jaynes and his coauthors reject the explanation that it is transfer payments — the much-maligned "welfare" — causing people to drop out of the labor force. They offer instead a structural explanation: "The shifting industrial base of the U.S. economy from blue-collar manufacturing to service industries, the slowdown in economic growth, and the consequent decline in real wages could be expected to produce a period of economic and social distress. For displaced and educationally or spatially misplaced workers, the rise in unemployment and increased competition for moderate-to-high-paying jobs might well lead to a rise in the number of discouraged workers" (1989:310).

The most ominous of the statistics drawn together by this committee pertain to poverty rates. We have seen a dramatic decline from the unbelievably high rates, in 1939, of 93% and 65% for black and white people, respectively. By 1974 the rates were 30% for blacks and 9% for whites, but by 1986 rates for both groups were higher, 31% and 11%. Even more alarming is the prevalence of poverty among black children—44% in 1985, compared with 16% among white children.

The pessimistic conclusion of the chapter "Blacks in the Economy" reads, "The economic fortunes of blacks are strongly tied (more so than those of whites) to a strong economy and vigorously enforced policies against discrimination. Without these conditions, the black middle class may persist, but it is doubtful it can grow or thrive. And the position of lower status blacks cannot be expected to improve" (1989:324).

To add my own pessimistic coda, a "strong economy" must achieve more than merely providing low-paying jobs in the service sector to replace those lost in the industrial sector through automation or export. Yet this seems to be what many secure people accept as a measure of solving the problem of unemployment. Moreover, with the insecurity felt by many whites it cannot be expected that they will willingly share with blacks the risk of falling into poverty.

Hence the crescendo of rhetoric decrying growing white racism and calling for more affirmative action programs is simplistic, avoiding the main problems facing both blacks and the society. In addition to focusing on the economic nature of these problems, we must also consider the changes in the nature of what is now called "racism."

How Much and What Kind of "Racism"?

Although there is no doubt that "racism" subsumes a multitude of sins, the term itself is very imprecise. Scholars defining it usually list a number of varieties. Since the 1960s it seems to have replaced the older concepts "prejudice" and "discrimination" to denote those negative attitudes and behaviors that result in the subordination and oppression of some groups which are socially defined as "races."

Focusing on the attitudinal components of racism, Schuman, Steeh, and Bobo found a paradox in the attitudes of white Americans in public opinion surveys from 1942 through 1983 (1985). On the one hand, they found strong positive trends toward acceptance of the principle of racial equality and the rejection of absolute segregation. On the other hand, questions concerning governmental implementation of these abstract principles got relatively low levels of support, and there are few signs showing that such support has increased over time. In 1989 the authors of *Blacks in American Society,* who found no reason to disagree with this observation, added their own finding that measures of black alienation from white society suggest an increase from the late 1960s to the 1980s (1989:131).

Many theoretical explanations have been advanced for the paradox disclosed by Schuman and his associates. One theory focuses on the level of abstract principles, seeing agreement with them as evidence of a strong progressive trend. It underplays the contradictory aspect of the findings as well as the absence of proportional structural changes in society, such as the persistence of massive residential segregation.

A sharply contrasting explanation holds that underlying "racist" attitudes have not changed. Agreement with abstract principles of racial equality constitutes only lip service conforming to a new cultural norm rendering crude, overt expression of racial prejudice less than respectable. Racial prejudice is now expressed symbolically. Opposition to school busing, open housing laws, and affirmative action, as well as failure to vote for a black candidate for public office, is to be explained primarily in terms of symbolic, covert racism. The more complex explanation of competing values such as objections to governmental intrusion, individualism, and genuine concern about what happens to one's children is rejected out of hand.

In *Racial Formation in the United States* Michael Omi and Howard Winant similarly give little credence to attitudinal expressions of support for abstract principles unless they are paralleled by support for implementation (1986). Unlike other sociologists they offer a theory of how the persistence of covert racism has affected racial politics in what they define as still being a "racial state."

They concede that a great transformation in ideas about race took place in the United States during the 1950s and 1960s. This had two major consequences. One was new, self-conscious racial identities which persisted even after the movements through which they were forged disintegrated. The second they call the "rearticulation of racial ideology" in reaction to the partial victories of the Civil Rights Movement and, I would add, of the Women's Liberation Movement. The conservative, right-wing trend in U.S. politics rests on racism, they suggest. "As the right sees it," they say, "racial problems today center on the new forms of racial injustice which originated in the great transformation. This new injustice confers group rights on racial minority groups, thus granting a new form of privilege — that of preferential treatment" (1986:114). Further developing this theme Omi and Winant assert, "In this scenario, the victims of racial discrimination have dramatically shifted from racial minorities to whites, particularly white males" (1986:114). They make a persuasive case that even though they were alluded to by code words, racial issues were central to support for President Reagan in his two elections and for President George Bush. Who can question that the Republican Party's "southern strategy" has included a strong component of this rearticulated racial ideology, one appealing not only to voters such as those who elected David Duke to the Louisiana legislature, but also to numerous white voters outside the South?

This pessimistic view of the United States as basically a racial state in which racism changes its face but does not disappear is frightening to anyone who hopes for movement toward greater equality in the 1990s. An even more ominous view of a majority of the electorate is offered by Edna Bonacich, who attaches more importance to class as a factor than do Omi and Winant. She asserts, "The United States is an immensely un-

equal society in terms of distribution of material wealth, and consequently in the distribution of all the benefits and privileges that accrue to wealth. . . . This inequality is vast irrespective of race." Granting that people of color suffer disproportionately, she goes on to say, "I believe that racial inequality is inextricably tied to overall inequality and to an ideology that endorses vast inequality as justified and desirable." She concludes, "And even if some kind of racial parity at the level of averages could be achieved, the amount of suffering at the bottom would remain undiminished, hence unconscionable" (1989:80).

Bonacich cites dramatic statistics demonstrating the vastness of inequality and its frightening growth. In 1987, for example, 6.7 million American workers living on the minimum wage had incomes of $9,968 a year, while Lee Iacocca was paid over $20 million, or $9,615 an hour. In 1986 there were 26 billionaires in the country; in 1987, 49. "The Culture of Inequality" which Michael Lewis identified in 1978 is more entrenched than ever (1978).

In the 1960s James Baldwin asked, "Who wants to be integrated into a burning house?" The house is still burning, being slowly consumed by the heat of greed and fear. Speculators gamble with the nation's wealth but pass the bill to the government when the dice roll against them. The CEOs of corporations have learned to live comfortably with affirmative action at the middle levels of the occupational scale but are equally comfortable with reductions in the total size of their work force. Often unnoted in optimistic studies of affirmative action is that increases in minority shares of employment are usually accompanied by contraction in the number of all persons employed. The size of the piece of the pie is not as critical in these times as is the shrinking of the fraction of the pie left for the have-nots in a class-polarized society.

The ideology of inequality Bonacich addresses is sustained also by the insecurities of people who have left the work force and are living on fixed incomes, either from interest and dividends or from those transfer payments now known as "entitlements." They do not see as their enemies the 0.5% of the families who in 1983 held 35% of the net wealth of the nation. Instead they fear the faceless people at the bottom of the heap. Have they not been told in campaign after campaign that it is the demands of the poor for welfare, social services, and higher wages that might cause higher taxes and increased prices? Polarization does not start near the top of the income distribution but near the bottom. One of Jonathan Rieder's subjects in *Canarsie* put it exactly, "We never join the have-a-littles with the have-nots to fight the haves. We make sure the have-a-littles fight the have-nots" (1985:119).

Now, ironically, as the crisis of capitalism in America intensifies, the attention of American voters is distracted by the failures of socialist polities and economies abroad, as if that made their own plight less perilous and their own future more secure. The prospects for a racial rejection of the culture of inequality, with its concomitant acceptance of racial inequality, seem dim. To me, some pessimistic warnings from the past seem more appropriate today than do optimistic predictions for the 1990s.

Voices from the Past

When I look at the retreat of the federal government from vigorous enforcement of civil rights laws, I am reminded of the warning of Frederick Douglass, issued as he witnessed a similar retreat. He wrote, "No man can be truly free whose liberty is dependent upon the thought, feeling, and actions of others, and who has himself no means in his own hands for guarding, protecting, defending, and maintaining that liberty" (1962:539).

During the Civil Rights and Black Power movements black Americans were catalysts in producing a national orgy of guilt, but they did not attain the sort of power Douglass described. After laws promising equal opportunity were passed, blacks lacked the political clout to get succeeding congresses to pass laws to implement these promises. They were forced to depend, instead, on sympathetic bureaucrats in the executive branch and a narrow majority in a relatively friendly Supreme Court to promote implemen-

tation in the absence of majority popular support. Now we see the administrative and judicial support fading because of the growing strength of a political party that does not depend on minority voters for victory and often appears downright hostile to their interests. As inadequate as were the responses of the liberal Kennedy, Johnson, and Carter administrations, they were magnificent when compared to those of elected officials who use "liberal" as a code word signifying softness on crime, welfare fraud, pauperism, reverse discrimination, and the spread of communism in the Third World.

Yet the black middle class still prospers relative to its past condition as the gap between the haves and have-nots grows in the black community, just as in the white. Here I am reminded of Stokely Carmichael's quip in the 1960s, "To most whites Black Power seems to mean that the Mau Mau are coming to the suburbs at night" (1966:5). Today we might say that, to most whites, actually accepting blacks as residents of their neighborhoods seems to mean that drug-ridden welfare recipients from the ghetto will be on their doorstep tomorrow. Julius Lester wrote at about the same time, "The black middle class is aware of its precarious position between the ghetto blacks and white society; and its members know that because they are black, they are dispensable" (1968:34). Even the qualified black person who seems to have achieved equality is regarded as the "exceptional" black and even then is often suspected of reaching that level because of affirmative action. Until the plight of the underclass is alleviated its shadow will continue to blight the lives and fortunes of those blacks who have partially escaped the bonds of past discrimination.

Lester said that the black middle class knew that it was dispensable. Sidney Willhelm asked, "Who needs the Negro?" (1970). Although asked in 1970, his question is still horrendously relevant today. Writing before the export of semi-skilled jobs to Third World countries became another threat to workers in the United States, he warned about automation: "The Negro becomes a victim of neglect as he becomes useless to an emerging economy of automation. With the onset of automation the Negro moves out of his historical state of oppression into one of uselessness. Increasingly, he is not so much economically exploited as he is irrelevant" (1970:162).

Although he and Willhelm have been highly critical of each other's work, William J. Wilson pointed to the same problem in *The Declining Significance of Race*. He observed, "Representing the very center of the New American economy, corporate industries are characterized by vertically integrated production processes and technologically progressive systems of production and distribution. The growth of production depends more on technical progress and increases in physical capital per worker than on the growth of employment." He added, "In short, an increasing number of corporate sector workers have become redundant because the demand for labor is decreased in the short run by the gap between productivity and the demand for goods" (1978:96–97). Perhaps we must ask today, "Who needs people, except as consumers?" Willhelm characterized our situation as one in which "the new standard of living entails both production and distribution of goods without, however, involving either a producer or distributor through large-scale employment" (1978:203). Hence this oft-disparaged but frighteningly accurate prophet among sociologists advanced a truly radical proposition: "It will be incumbent for a society relying upon automation and dedicated to the well being of human beings to accept a new economic gauge, namely: *services are to be rendered and goods produced, distributed and consumed in keeping with a designated standard of living*" (1978:203). This is the sort of change in perspective of which Wilson said, in 1987, "It will require a radicalism that neither Democrat nor Republican parties have as yet been realistic enough to propose" (1987:139). Instead what we continue to see is platforms that imply that if profits are kept high and taxes low so that investment is encouraged, plenty of jobs will be created. Then, if blacks will get an education and develop the right attitudes toward work and the family, they can enjoy that portion of the prosperity that trickles down to them. This, unfortunately, is the dream of many white voters today.

It is not, however, an accurate vision of things to come but a rose-tinted stereotype of an industrial era which is gone forever and was never good for minority workers.

Dreams or Nightmares for the 1990s?

During those years after 1940 during which I studied, taught, and lived race relations in the South, I had my own dream. It was that my fellow white southerners, most of whom I knew as good, kind people, would have peeled from their eyes the veil which kept them from knowing what they were doing to black people. Someday they would see, I hoped, how segregation and discrimination, no matter how paternalistic, left cruel injuries which would handicap both current and future generations.

During the decade after 1954 I thought I was beginning to see that veil thinning under the assault of the Civil Rights Movement. I did witness heartening changes, but then I saw new complacency, with white America asking, "How much more are we supposed to do for them?" Now I see a new veil blinding people whom I still believe to be good-spirited. They are blind to institutional discrimination and to the poverty increasing in our nation even more rapidly than 20 years ago. Ironically, the behavior of most white people, particularly in the South, has changed more than have their attitudes. Now they mix with their black fellow citizens, yet blacks still remain largely invisible to them. They admit selected, acceptable blacks to their company as individuals but ignore the tragedy of the masses who yearly become more separated and alienated from what appears to those on top as an affluent society. Poverty, black and white, is concealed in a way different from when the rich and the poor lived closer to each other. It is known to many Americans only by flitting images in the mass media. In the meantime defense of what security and prosperity one does enjoy, rather than concern for the social problems threatening the nation, anchors successful political appeals with the dominant theme, "no new taxes."

In 1961 James E. Conant wrote in *Slums and Suburbs*, "We are allowing social dynamite to accumulate in our large cities" (p. 2). In the 1960s there were explosions of that dynamite, but its potential for destruction was far from exhausted. Now, in 1990, more dynamite is accumulating and in more cities.

Yet at this very time there appears to be a new basis for optimism. Many Americans are celebrating the end of the Cold War and looking forward to a "peace dividend." The case for deferring spending on domestic programs because of the demands for military defense loses its cogency. Journalists and novelists ask, "Who will be the enemy now that the Soviet Union is no longer the evil empire?"

There has been another cold war, however—a war of heartless neglect of the burgeoning needs of the truly disadvantaged in our own affluent society. A bright new dream would feature the end of this cold war and the beginning of a new war on poverty. We can expect increasingly urgent demands for a concerted attack on underemployment, undereducation, crime which preys on the poor, and the hopelessness that causes young people to drop out not only from school but also from the labor force. But these problems cannot be adequately addressed with the meager surplus left after the requirements for deficit reduction and new foreign aid are met. New taxes and a more equitable distribution of wealth will be required. But what if the response of the "haves" and the "have-littles" to this summons for self-sacrifice is a new wave of blaming the victim? If this is the case, the new enemy will be our own underclass.

REFERENCES

Bonacich, Edna, 1989. "Inequality in America: The Failure of the American System for People of Color." *Sociological Spectrum* 9:77–101.

Carmichael, Stokely, 1966. "What We Want." *New York Review of Books*, September 22.

Cayton, Horace W., and St. Clair Drake, 1945. *Black Metropolis*. Harcourt Brace.

Clark, Kenneth, 1977. P. ix in the Preface to *Commission Politics*, by Michael Lipsky and David J. Olson. *Transaction*.

Conant, James B., 1961. *Slums and Suburbs*. McGraw-Hill.

Douglass, Frederick, 1962. *Life and Times of Frederick Douglass*. Collier.

Jaynes, Gerald, and Robin M. Williams, 1989. *Blacks and American Society*. National Academy Press.

Killian, Lewis M., 1968. *The Impossible Revolution? Black Power and the American Dream*. Random House.

King, Martin Luther, Jr., (1963), 1986. "I Have a Dream." In *A Testament of Hope: The Essential Writings of Martin Luther King, Jr.*, edited by James M. Washington. Harper & Row.

———, (1967). 1986. "A Time to Break Silence." Loc. cit.

———, (1972). 1986. "Where Do We Go from Here?" Loc. cit.

Lal, Barbara Lallis, 1983. "Perspectives on Ethnicity: Old Wines in New Bottles." *Ethnic and Racial Studies* 6:154–173.

Lewis, Michael, 1978. *The Culture of Inequality*. University of Massachusetts Press.

Lieberson, Stanley, 1982. "A New Ethnic Group in the United States." Pp. 259–267 in *Majority and Minority*, edited by Norman R. Yetman. 4th ed. Allyn & Bacon.

Myrdal, Gunnar, 1944. *An American Dilemma*. Harper.

Novak, Michael, 1972. *The Rise of the Unmeltable Ethnics*. Macmillan.

Omi, Michael, and Howard Winant, 1986. *Racial Formation in the United States from the 1960's to the 1980's*. Routledge & Kegan Paul.

Report of the National Advisory Commission on Civil Disorders, 1968. Bantam.

Rieder, Jonathan, 1985. *Canarsie: The Jews and Italians of Brooklyn Against Liberalism*. Harvard University Press.

Schucter, Arnold, 1970. *Reparations: The Black Manifesto and Its Challenge to White America*. J. B. Lippincott.

Schuman, Howard, Charlotte Steeh, and Lawrence Bobo, 1985. *Racial Attitudes in America*. Harvard University Press.

Willhelm, Sydney, 1970. *Who Needs the Negro?* Schenkman.

Wilson, William J., 1973. *Power, Racism, and Privilege*. Macmillan.

———, 1980. *The Declining Significance of Race*. University of Chicago Press.

———, 1987. *The Truly Disadvantaged*. University of Chicago Press.

44

"This Is a White Country": The Racial Ideology of the Western Nations of the World-System

EDUARDO BONILLA-SILVA AND MARY HOVSEPIAN

Racial inequalities and racism are found in many countries. Bonilla-Silva and Hovsepian provide evidence that the world-system of capitalist economies, including rich and poor nations, has resulted in increased racial tensions. As citizens of poorer countries seek job opportunities in the Western world, for instance, racist backlash increases.

Note the following as you read this selection:

1. What is the result of the globalization of race relations?

2. What can be done to reduce racial backlash in the Western world?

GLOSSARY

Racism The belief that one category of people is superior or inferior to another.
GNP A nation's gross national product.
Neonazis Modern-day white supremacist groups.

International Context: The Internationalization of the Economy and the Globalization of Race Relations

THE NATIONAL CAPITALIST ECONOMIES of the world have formed a "world-system" for over 600 years (Braudel 1979; Wallerstein 1974; Hopkins and Wallerstein 1996). The extension of that system into Africa, the Americas, and Asia in the 16th century involved the racialization of the peoples of the entire world (Balibar and Wallerstein 1991; Rodney 1981). In order to dominate the "new world," European nations developed a structure of knowledge-meaning that created the notion of the "West" (Hopkins and Wallerstein 1996). This intellectual construction facilitated the expansion of the world-system by *racializing* the inhabitants of peripheral and core nations (Rodney 1981). The concept of the West crystallized a set of binary oppositions that defined the peoples of Western and of non-Western nations: human/subhuman, developed/underdeveloped, civilized/barbarian, rational/instinctive, Christian/heathen, superior/inferior, and clean/unclean (Markus 1994). By defining non-Western nations in this fashion, core nations were able to conquer, exploit, and massacre Indian, African, and Asiatic peoples without much guilt and to use their natural resources to advance their own social, economic, and political interests—including the development of democratic regimes with extensive citizenship rights for all (white) citizens (Berkhoffer 1979; Gunder-Frank 1978; Hopkins and Wallerstein 1996; Rodney 1981).

This Western discourse was not—and is not—just a set of ideas revolving in the heads of Europeans. This discourse was an essential component in the structuration of various kinds of social relations of domination and subordination between "Western" and non-Western peoples, between whites and nonwhites in the world-system (Balibar and Wallerstein 1991; Bonilla-Silva 1997; Spoonley 1988). Racism (racial ideology), as I have suggested elsewhere, is not a free-floating ideology. Racism is always anchored in real practices and it reinforces social relations among

Unpublished manuscript (excerpts pp. 7–12, 24–25). Used by permission of the author.

racialized subjects in a social order, that is, it supports a racialized social structure (Bonilla-Silva 1997). Thus, for example, the racial ideology of Canada, Australia, and the United States is the direct product of their own racial situations. Even Western countries that did not have historical racial minorities, such as the Netherlands, France, or England, established racial structures in their colonies which have shaped the way in which they have dealt with "colonial immigrants" and other immigrants of color. For instance, although France had by 1930 the highest level of foreigners of any country in the world with seven percent, Arabs, who were neither the largest immigrant group nor the last arrivals, were the object of the most severe antipathies and found themselves at the bottom of the occupational structure (Stora 1996).

Since the mid-1960s, the capitalist world-system has experienced a systemic transformation or, properly speaking, a crisis, that has produced a dramatic restructuring, the famous "globalization" that we hear about almost every day (Amin 1992). The central features of this transformation are the "decline in the importance of territorially based mass production, the globalization of finance and technology, and the increased specialization and diversity of markets" (Kaldor 1996). Each of these elements is a result of the serious world-systemic crisis of accumulation in the late 1960s and early 1970s that produced drastic shifts in the loci of production (from center to peripheries), investment (from productive to financial), and the countries spending a significant portion of their GNP on military expenditures (by incorporating peripheral and semi-peripheral nation-states as central actors in the military race). Although advocates of capitalism interpret these various changes as progressive and speak of a "global village," this new stage in the world-system should be characterized as "the empire of chaos" (Amin 1992).

The chaos produced by the restructuring of the world-system has had local (plant relocations), national (downsizing of the labor force of large multinational companies in the core and "shock therapies" in the periphery), and international repercussions (NAFTA, new world-level economic and political arrangements, etc.). The dislocations caused by these changes and labor recruitment policies by some core nation-states have led to monumental migrations of people from the Third World into core states and the deterioration of the status of workers in the Western world (Cohen 1997).[1] Although a substantial part of this migration is legal and even sponsored by the core states, increasingly since the 1970s the migration has been illegal (Wallerstein 1996).

This new international order has led to the globalization of race and race relations and the intensification and diversification of the numbers of racial Others in the Western world. Although race has fractured countries such as the United States, Australia, New Zealand, and Canada since their inception, it was until recently a marginal social category in most Western nations. Today, as a direct result of the international movement of peoples, all Western nations have interiorized the Other, colonial and otherwise (Miles 1993; Winant 1994). In European nations such as Luxembourg, Belgium, Austria, the Netherlands, France, and Germany, the geographical distance between the "uncivilized" and the "civilized" has been "bridged" through what Balibar calls the "interiorization of the exterior" (Balibar 1991).[2] Accordingly, today immigrants and minorities of color in Europe constitute anywhere between 1.4 percent of the population, as in Italy, to 27.5 percent, as in Luxembourg (Castles and Miller 1993: 80).

Although many analysts conceive these immigrants as basically workers who have been racialized as an "underclass" (Castells 1979; Castles and Miller 1993; Cohen 1987; Loomis 1990; Spoonley 1996), I contend that the racialized character of their experience is deeper and in line with 500 years of Western history (Potts 1990; Jayasuriya 1996). For example, in England, although European (white) workers were viewed as easily assimilable, a clear stigma was attached to Caribbean workers whose absorption into the social body was deemed "very difficult" (Royal Commission 1949, as cited by Layton-Henry

1994: 284). In France, even before the development of the fascist National Front Party, French workers had racist views and feelings toward "black" (Algerians and Caribbean) workers and were among the first to oppose immigration (Grillo 1991). Finally, since immigration is not a new phenomenon in these countries, and in many, a substantial proportion of the immigrants are white (two-thirds of those in Europe and most of those in England), the "immigrants" that matter are those defined as "black," "non-Western," "unchristian." Accordingly, for example, although Belgium has over half a million French and Italian "foreigners," it targets its 250,000 Arabs and the "blacks" as the objects of scapegoating. It is also significant that studies of the various "immigrants" show that darker immigrants (Caribbeans, Arabs, and Southern Europeans) are viewed and treated much worse than White immigrants (Castles and Kosack 1984: 443–446). In England, although immigration restrictions were imposed on all groups, political leaders have said that immigration from "Canada, Australia, and New Zealand formed no part of the [immigration] problem" (Saggar 1992: 105) and that earlier migrations of Irish, in contrast to those of Jews and blacks (19th century until 1960), did not produce major reactions from the body politic (Solomos 1989).

Despite the different legal status of these people of color in Western nations (guest workers, asylum seekers, "aliens," or "citizens"), they have a number of similarities. First, in economic terms, all experience a racialized class status characterized by segmented labor market experiences—even segmentation in middle class occupations, overrepresentation in manual and "underclass" locations in the class structure, and significantly higher levels of unemployment (Berrier 1985; Castles and Miller 1993; Loomis 1993). They also experience very little occupational mobility even among second-generation "immigrants." Second, they tend to live in ethnic quarters or ghettos and are more likely to rent rather than to own their houses (Loomis 1990). This is partly due to discrimination in the housing markets (Loomis 1990; Massey and Denton 1993; Suárez-

Orozco 1994). Finally, all people of color in Europe (Turks, Arabs, Native peoples, blacks from the Caribbean and Africa, etc.), whether immigrant or not, experience what Suárez-Orozco has termed as "expressive exploitation" or the psychological aspects of depreciation—derogatory attitudes, stereotyping and related behavior, and racially motivated violence. In short, people of color in the historically white countries of the West experience a status and are treated as second-class citizens (Layton-Henry 1990), a status that resembles that of the historical racial minorities in Western nations such as the United States, Canada, Australia, and New Zealand. . . .

Conclusions

In the postmodern world no one is racist except for Nazis and neonazis and members of white supremacist groups.[3] Yet racial minorities and immigrants of color are experiencing a racial backlash all over the Western world. That backlash is evident in attacks on affirmative action-type policies, the growth in racial violence, the increase in electoral support for populist racial parties, and the move to the right by mainstream parties on racially-perceived matters such as immigration. Faced with economic insecurity, restructuring, transnationalism, and new political alignments, whites in the Western world are struggling—ideologically and practically—to maintain what they regard as their "rights" to cultural, social, political, economic, and psychological advantages as white, "civilized," and "Christian" citizens over racial minorities, immigrants, or any representative of the Other.

The apparent contradiction between a racial backlash and a Western world that pretends to be cosmopolitan, multicultural, and raceless (Guibernau 1996) is explained by the fact that contemporary racial ideology combines abstract and technocratic liberalism with ethnonational and culturalist elements. Laissez faire racism, which ideologically equalizes the races ("We are all equal!") although in fact they remain unequal, provides the ammunition for whites to feel *moral indignation, anger, resentment,* and even *hate* toward minorities and the programs viewed as providing "preferential" treatment to them. Therefore, this new racial ideology allows whites in the West to defend their racial privilege without appearing to be "racist" (Bonilla-Silva, Forman, and Padín 1998; Feagin and Vera 1995; Wetherell and Potter 1992). Contemporary racial struggle is waged with a new racial language and new racial ideas. Instead of the biologically based racism of the past, the new racial ideology allows even racists such as Enoch Powell to express racial resentment—evident in statements such as the one below—in a way that is acceptable to most whites in the West.

> The spectacle which I cannot help seeing . . . is that of Britain which has lost, quite suddenly, in the space of less than a generation, all consciousness and conviction of being a nation: the web which binds it to its past has been torn asunder, and what has made the spectacle the more impressive has been the indifference, not to say levity, with which the change has been greeted (Enoch Powell's statement to *The Guardian* in 1981, cited in Saggar 1992: 176).

This new racism is not a hangover from the past, an articulation of the New Right, a simple case of scapegoating, or something affecting only workers. I suggest that the new racism is world-systemic and affects all Western nations although its specific articulations vary by locality. The reason why racism in all Western nations has a similar macro-racial discourse is because these nations share a history of racial imperialism and the notion of the West, have real but differing racial structures (Bonilla-Silva 1997), and have a significant presence of the Other, either through immigration, as in most European nations, or through their history of constitution as nations, as in New World nations. . . .

Although anti-racist organizations have surfaced in all Western nations (e.g., *Lichterketten* in Germany, SOS Racism in Britain, etc.), they have not been able to mount a counteroffensive rooted in the recognition of the materiality of racialized discourse and behavior. In too many places anti-racist campaigns have been highly ritualistic (can-

dlelights, commercials, etc.) or very narrowly defined (against certain fascist groups) with little concern for developing a broader political agenda. Unless anti-racist organizations understand the centrality and meaning of race and the new racism, they will not be able to develop a progressive agenda around a reconceptualized notion of citizenship that includes both the idea of equality of rights and the equality of status (Ansell 1997; Jayasuriya 1996; Tlati 1996). Failure to do so, regardless of talks about racial reconciliation (as President Clinton has proposed in the United States), liberal views on "cosmopolitanism" (as many Europeans suggest), or programs based on an abstract liberal universalism, will maintain people of color in the West as denizens.

REFERENCES

Amin, Samir. 1992. *Empire of Chaos*. New York: Monthly Review Press.

Balibar, Etienne, and Immanuel Wallerstein. 1991. *Race, Nation, Class: Ambiguous Identities*. London and New York: Verso.

Berkhoffer, Robert F., Jr. 1979. *The White Man's Indian*. New York: Vintage Books.

Berrier, Robert J. 1985. "The French Textile Industry: A Segmented Labor Market." In *Guests Come To Stay: The Effects of European Labor Migration on Sending and Receiving Countries,* edited by Rosemarie Rogers, pp. 51–68. Boulder and London: Westview Press.

Bonilla-Silva, Eduardo. 1997. "Rethinking Racism: Toward a Structural Interpretation." *American Sociological Review,* Vol. 62, No. 3, pp. 465–480.

Braudel, Fernand. 1979. *The Perspective of the World: Civilization and Capitalism, 15th-18th Century,* Vol. III. New York: Harper & Row.

Castles, Stephen. 1994. "Democracy and Multicultural Citizenship. Australian Debates and their Relevance for Western Europe." In *From Aliens to Citizens: Redefining the Status of Immigrants in Europe,* edited by Rainer Baubock, pp. 3–27. Germany: Avebury.

Castles, Stephen, and Mark J. Miller. 1993. *The Age of Migration: International Population Movements in the Modern World*. Hong Kong: MacMillan.

Cohen, Robin. 1997. *Global Diasporas: An Introduction*. Seattle: University of Washington Press.

Hopkins, Terrence K., and Immanuel Wallerstein. 1996. "The World System: Is There a Crisis?" In *The Age of Transition: Trajectory of the World-System, 1945–2025,* pp. 1–12. Lechhardt, Australia: Pluto Press.

Gunder-Frank, Andre. 1978. *World Accumulation: 1492–1789.* New York: Monthly Review Press.

Jayasuriya, Laksiri. 1996. "Immigration and Settlement in Australia: An Overview and Critique of Multiculturalism." In *Immigration and Integration in Post-Industrial Societies: Theoretical Analysis and Policy-Related Research,* edited by Naomi Cameron, pp. 206–226. London: St. Martin's Press.

Kaldor, Mary. 1996. "Cosmopolitanism Versus Nationalism: The New Divide." In *Europe's New Nationalism: States and Minorities in Conflict,* edited by Richard Caplan and John Feffer, pp. 42–58. New York: Oxford University Press.

Loomis, Terrence. 1990. *Pacific Migrant Labour, Class and Racism in New Zealand: Fresh Off the Boat.* Aldershot, England: Avebury.

Markus, Andrew. 1994. *Australian Race Relations, 1788–1993.* St. Leonards, Australia: Allen & Unwin.

Potts, Lydia. 1990. *The World Labor Market: A History of Immigration.* London: Zed Books.

Rodney, Walter. 1981 [1972]. *How Europe Underdeveloped Africa.* Washington, D.C.: Howard University Press.

Saggar, Shamit. 1991. *Race and Politics in Britain.* London: Harvester/Wheatsheaf.

Schoenbaum, David, and Elizabeth Pond. 1996. *The German Question and Other German Questions.* New York: St. Martin's Press.

Solomos, John. 1989. *Race and Racism in Contemporary Britain.* London: MacMillan.

Spoonley, Paul. 1988. *Racism and Ethnicity.* Auckland, New Zealand: Oxford University Press.

Stora, Benjamin. 1996. "Locate, Isolate, Place under Surveillance: Algerian Migration to France in the 1930s." In *Franco-Arab Encounters: Studies in Memory of David C. Gordon,* edited by L. Carl Brown and Matthew S. Gordon, pp. 373–391. Beirut, Lebanon: American University of Beirut.

Tlati, Soraya. 1996. "French Nationalism and the Issue of North African Immigration." In *Franco-Arab Encounters: Studies in Memory of David C. Gordon,* edited by L. Carl Brown and Matthew S. Gordon, pp. 392–414. Beirut, Lebanon: American University of Beirut.

Wallerstein, Immanuel. 1996. "The Global Picture, 1945–90." In *The Age of Transition: Trajectory of the World-System, 1945–2025,* edited by Terence K. Hopkins and Immanuel Wallerstein, pp. 209–225. Lechhardt, Australia: Pluto Press.

Winant, Howard. 1994. *Racial Conditions: Politics, Theory, Comparisons.* Minneapolis, Minnesota: University of Minnesota Press.

Zainu'ddin, Ailsa. 1968. *A Short History of Indonesia.* Melbourne, Australia: Cassell.

NOTES

1. Although monumental transfers of *labor power* and of some *workers* (Africans) have occurred in the world-system since the 16th century, this is the first time that monumental transfers of *workers* from the periphery to the center are occurring (Potts 1990).

2. It is important to point out that immigration did not cause racism in these Western nations although "it did give post-War racism a new focus and created new targets for racial victimization and 'scapegoating'" (MacLaughlin 1993: 24). Older racial traditions allowed bringing in immigrants of color and assigning them to subordinate slots in these societies (Miles 1987: 165).

3. Jesse Daniels shows in her book *White Lies* (1996) that this so-called extremist discourse is connected to mainstream racial discourse on a variety of issues and that it helps "liberal" white folks to maintain white supremacy without having to be personally invested in fighting racial others.

45

In the Barrios:
Latinos and the Underclass Debate

JOAN MOORE AND RAQUEL PINDERHUGHES

In an important work by sociologist William Julius Wilson called The Truly Disadvantaged, *the term* underclass *was introduced to mean persistent poverty due largely to economic restructuring. The use of this term has been debated by scholars. Some see it blaming the victim, the poor, for their condition because of their values and behaviors. Others see it as a debate over who is responsible for the poor — the individuals themselves or society? Behavioral pathology or economic structure?*

In this discussion by Moore and Pinderhughes, the concept of underclass is considered as it applies to Latinos.

As you read, consider the following:

1. Does the term underclass *apply to the Latino population?*

2. What makes the Latino population unique as a minority group in the United States?

GLOSSARY

Underclass Meaning is debated, but it often refers to the poorest of the poor in the United States.
Polarization of the labor market High- and low-level jobs but few in the middle.
Rustbelt Area of the country (Midwest) where jobs are being lost.
Sunbelt Area of the country (mostly south) where jobs are increasing.
Informal economic activities Outside government control, small-scale.

IN THE PUBLICATION *THE Truly Disadvantaged,* William Julius Wilson's seminal work on persistent, concentrated poverty in Chicago's

black neighborhoods, Wilson used the term "underclass" to refer to the new face of poverty, and traced its origins to economic restructuring. He emphasized the impact of persistent, concentrated poverty not only on individuals but on communities.

. . . The term "Hispanic" is used particularly by state bureaucracies to refer to individuals who reside in the United States who were born in, or trace their ancestry back to, one of twenty-three Spanish-speaking nations. Many of these individuals prefer to use the term "Latino." . . .

No matter what the details, when one examines the history of the term underclass among sociologists, it is clear that Wilson's 1987 work seriously jolted the somewhat chaotic and unfocused study of poverty in the United States. He described sharply increasing rates of what he called "pathology" in Chicago's black ghettos. By this, Wilson referred specifically to female headship, declining marriage rates, illegitimate births, welfare dependency, school dropouts, and youth crime. The changes in the communities he examined were so dramatic that he considered them something quite new.

Two of the causes of this new poverty were particularly important, and his work shifted the terms of the debate in two respects. First, Wilson argued effectively that dramatic increases in joblessness and long-term poverty in the inner city were a result of major economic shifts — economic restructuring. "Restructuring" referred to changes in the global economy that led to deindustrialization, loss and relocation of jobs, and a

Joan Moore and Raquel Pinderhughes, eds., In the Barrios: Latinos and the Underclass Debate. *New York: Russell Sage Foundation, 1993. Excerpts from pp. xi to xxxix.*

decline in the number of middle-level jobs—a polarization of the labor market. Second, he further fueled the debate about the causes and consequences of persistent poverty by introducing two neighborhood-level factors into the discussion. He argued that the outmigration of middle- and working-class people from the urban ghetto contributed to the concentration of poverty. These "concentration effects" meant that ghetto neighborhoods showed sharply increased proportions of very poor people. This, in turn, meant that residents in neighborhoods of concentrated poverty were isolated from "mainstream" institutions and role models. As a result, Wilson postulates, the likelihood of their engaging in "underclass behavior" was increased. Thus the social life of poor communities deteriorated because poverty intensified. . . .

The Latino Population— Some Background

American minorities have been incorporated into the general social fabric in a variety of ways. Just as Chicago's black ghettos reflect a history of slavery, Jim Crow legislation, and struggles for civil and economic rights, so the nation's Latino barrios reflect a history of conquest, immigration, and a struggle to maintain cultural identity.

In 1990 there were some 22 million Latinos residing in the United States, approximately 9 percent of the total population. Of these, 61 percent were Mexican in origin, 12 percent Puerto Rican, and 5 percent Cuban. These three groups were the largest, yet 13 percent of Latinos were of Central and South American origin and another 9 percent were classified as "other Hispanics."[1] Latinos were among the fastest-growing segments of the American population, increasing by 7.6 million, or 53 percent, between 1980 and 1990. There are predictions that Latinos will outnumber blacks by the twenty-first century. If Latino immigration and fertility continue at their current rate, there will be over 54 million Latinos in the United States by the year 2020.

This is an old population: as early as the sixteenth century, Spanish explorers settled what is now the American Southwest. In 1848, Spanish and Mexican settlers who lived in that region became United States citizens as a result of the Mexican-American War. Although the aftermath of conquest left a small elite population, the precarious position of the masses combined with the peculiarities of southwestern economic development to lay the foundation for poverty in the current period (see Barrera 1979; Moore and Pachon 1985).

In addition to those Mexicans who were incorporated into the United States after the Treaty of Guadalupe Hidalgo, Mexicans have continually crossed the border into the United States, where they have been used as a source of cheap labor by U.S. employers. The volume of immigration from Mexico has been highly dependent on fluctuations in certain segments of the U.S. economy. This dependence became glaringly obvious earlier in this century. During the Great Depression of the 1930s state and local governments "repatriated" hundreds of thousands of unemployed Mexicans, and just a few years later World War II labor shortages reversed the process as Mexican contract-laborers (*braceros*) were eagerly sought. A little later, in the 1950s, massive deportations recurred when "operation Wetback" repatriated hundreds of thousands of Mexicans. Once again, in the 1980s, hundreds of thousands crossed the border to work in the United States, despite increasingly restrictive legislation.

High levels of immigration and high fertility mean that the Mexican-origin population is quite young—on the average, 9.5 years younger than the non-Latino population—and the typical household is large, with 3.8 persons, as compared with 2.6 persons in non-Latino households (U.S. Bureau of the Census 1991b). Heavy immigration, problems in schooling, and industrial changes in the Southwest combine to constrain advancement. The occupational structure remains relatively steady, and though there is a growing middle class, there is also a growing number of very poor people. . . .

Over the past three decades the economic status of Puerto Ricans dropped precipitously. By 1990, 38 percent of all Puerto Rican families were below the poverty line. A growing proportion of these families were concentrated in poor urban neighborhoods located in declining industrial centers in the Northeast and Midwest, which experienced massive economic restructuring and diminished employment opportunities for those with less education and weaker skills. The rising poverty rate has also been linked to a dramatic increase in female-headed households. Recent studies show that the majority of recent migrants were not previously employed on the island. Many were single women who migrated with their young children (Falcon and Gurak 1991). Currently, Puerto Ricans are the most economically disadvantaged group of all Latinos. As a group they are poorer than African Americans.

Unlike other Latino migrants, who entered the United States as subordinate workers and were viewed as sources of cheap labor, the first large waves of Cuban refugees were educated middle- and upper-class professionals. Arriving in large numbers after Castro's 1959 revolution, Cubans were welcomed by the federal government as bona fide political refugees fleeing communism and were assisted in ways that significantly contributed to their economic well-being. Cubans had access to job-training programs and placement services, housing subsidies, English-language programs, and small-business loans. Federal and state assistance contributed to the growth of a vigorous enclave economy (with Cubans owning many of the businesses and hiring fellow Cubans) and also to the emergence of Miami as a center for Latin American trade. Cubans have the highest family income of all Latino groups. Nevertheless, in 1990, 16.9 percent of the Cuban population lived below the poverty line.

In recent years large numbers of Salvadorans and Guatemalans have come to the United States in search of refuge from political repression. But unlike Cubans, few have been recognized by the U.S. government as bona fide refugees. Their settlement and position in the labor market have been influenced by their undocumented (illegal) status. Dominicans have also come in large numbers to East Coast cities, many also arriving as undocumented workers. Working for the lowest wages and minimum job security, undocumented workers are among the poorest in the nation.

Despite their long history and large numbers, Latinos have been an "invisible minority" in the United States. Until recently, few social scientists and policy analysts concerned with understanding stratification and social problems in the United States have noticed them. Because they were almost exclusively concerned with relations between blacks and whites, social scientists were primarily concerned with generating demographic information on the nation's black and white populations, providing almost no information on other groups.[2] Consequently, it has been difficult, sometimes impossible, to obtain accurate data about Latinos.

Latinos began to be considered an important minority group when census figures showed a huge increase in the population. By 1980 there were significant Latino communities in almost every metropolitan area in the nation. As a group, Latinos have low education, low family incomes, and are more clustered in low-paid, less-skilled occupations. Most Latinos live in cities, and poverty has become an increasing problem. On the whole, Latinos are more likely to live in poverty than the general U.S. population: poverty is widespread for all Latino subgroups except Cubans. They were affected by structural factors that influenced the socioeconomic status of all U.S. workers. In 1990, 28 percent were poor as compared with 13 percent of all Americans and 32 percent of African Americans (U.S. Bureau of the Census 1991b). Puerto Ricans were particularly likely to be poor. . . .

The Importance of Economic Restructuring

The meaning of economic restructuring has shaped the debate about the urban underclass. . . .

First, there is the "Rustbelt in the Sunbelt" phenomenon. Some researchers have argued that deindustrialization has been limited to the Rustbelt, and that the causal chain adduced by Wilson therefore does not apply outside that region. But the fact is that many Sunbelt cities developed manufacturing industries, particularly during and after World War II. Thus Rustbelt-style economic restructuring—deindustrialization, in particular—has also affected them deeply. In the late 1970s and early 1980s cities like Los Angeles experienced a major wave of plant closings that put a fair number of Latinos out of work (Morales 1985; Soja, Morales, and Wolff 1983).

Second, there has been significant reindustrialization and many new jobs in many of these cities, a trend that is easily overlooked. Most of the expanding low-wage service and manufacturing industries, like electronics and garment manufacturing, employ Latinos (McCarthy and Valdez 1986; Muller and Espenshade 1986), and some depend almost completely on immigrant labor working at minimum wage (Fernandez-Kelly and Sassen 1991). In short, neither the Rustbelt nor the Sunbelt has seen uniform economic restructuring.

Third, Latinos are affected by the "global cities" phenomenon, particularly evident in New York and Chicago. This term refers to a particular mix of new jobs and populations and an expansion of both high- and low-paid service jobs (see Sassen-Koob 1984). When large multinational corporations centralize their service functions, upper-level service jobs expand. The growing corporate elite want more restaurants, more entertainment, more clothing, and more care for their homes and children, but these new consumer services usually pay low wages and offer only temporary and part-time work. The new service workers in turn generate their own demand for low-cost goods and services. Many of them are Latino immigrants and they create what Sassen calls a "Third World city . . . located in dense groupings spread all over the city": this new "city" also provides new jobs (1989, p. 70).

Los Angeles . . . has experienced many of these patterns.[3] The loss of manufacturing jobs has been far less visible than in New York or

Chicago, for although traditional manufacturing declined, until the 1990s high-tech manufacturing did not. Moreover, Los Angeles' international financial and trade functions flourished (Soja 1987). The real difference between Los Angeles on the one hand and New York and Chicago on the other was that more poor people in Los Angeles seemed to be working.[4] In all three cities internationalization had similar consequences for the *structure* of jobs for the poor. More of the immigrants pouring into Los Angeles were finding jobs, while the poor residents of New York and Chicago were not.

Fourth, even though the deindustrialization framework remains of overarching importance in understanding variations in the urban context of Latino poverty, we must also understand that economic restructuring shows many different faces. It is different in economically specialized cities. Houston, for example, has been called "the oil capital of the world," and most of the devastating economic shifts in that city were due to "crisis and reorganization in the world oil-gas industry" (Hill and Feagin 1987, p. 174). Miami is another special case. The economic changes that have swept Miami have little to do with deindustrialization, or with Europe or the Pacific Rim, and much to do with the overpowering influence of its Cuban population, its important "enclave economy," and its "Latino Rim" functions (see Portes and Stepick 1993).

Finally, economic change has a different effect in peripheral areas. Both Albuquerque and Tucson are regional centers in an economically peripheral area. Historically, these two cities served the ranches, farms, and mines of their desert hinterlands. Since World War II, both became military centers, with substantial high-tech defense industrialization.[5] Both cities are accustomed to having a large, poor Latino population, whose poverty is rarely viewed as a crisis. In Tucson, for example, unemployment for Mexican Americans has been low, and there is stable year-round income. But both cities remain marginal to the national economy, and this means that the fate of their poor depends more on local factors.

Laredo has many features in common with other cities along the Texas border, with its substantial military installations, and agricultural and tourist functions. All of these cities have been affected by general swings in the American and Texan economy. These border communities have long been the poorest in the nation, and their largely Mexican American populations have suffered even more from recent economic downturns. They are peripheral to the U.S. economy, but the important point is that their economic well-being is intimately tied to the Mexican economy. They were devastated by the collapse of the peso in the 1980s. They are also more involved than most American cities in international trade in illicit goods, and poverty in Laredo has been deeply affected by smuggling. Though Texas has a long history of discrimination against Mexican Americans, race is not an issue within Laredo itself, where most of the population — elite as well as poor — is of Mexican descent. . . .

The Informal and Illicit Economies

The growth of an informal economy is part and parcel of late twentieth-century economic restructuring. Particularly in global cities, a variety of "informal" economic activities proliferates — activities that are small-scale, informally organized, and largely outside government regulations (cf. Portes, Castells, and Benton 1989). Some low-wage reindustrialization, for example, makes use of new arrangements in well-established industries (like home work in the garment industry, as seamstresses take their work home with them). Small-scale individual activities such as street vending and "handyman") house repairs and alterations affect communities in peripheral as well as global cities. . . . These money-generating activities are easily ignored by researchers who rely exclusively on aggregate data sources: they never make their way into the statistics on labor-market participation, because they are "off the books." But they play a significant role in the everyday life of many African American neighborhoods as well as in the barrios.

And, finally, there are illicit activities — most notoriously, a burgeoning drug market. There is not much doubt that the new poverty in the United States has often been accompanied by a resurgence of illicit economic activities (see Fagan, forthcoming, for details on five cities). It is important to note that most of the Latino communities . . . have been able to contain or encapsulate such activities so that they do not dominate neighborhood life. But in most of them there is also little doubt that illicit economic activities form an "expanded industry." They rarely provide more than a pittance for the average worker: but for a very small fraction of barrio households they are part of the battery of survival strategies.

Researchers often neglect this aspect of the underclass debate because it is regarded as stigmatizing. However, some . . . make it clear that the neglect of significant income-generating activities curtails our understanding of the full range of survival strategies in poor communities. At the worst (as in Laredo) it means that we ignore a significant aspect of community life, including its ramifications in producing yet more overpolicing of the barrios. Even more important, many of these communities have been able to encapsulate illicit economic activities so that they are less disruptive. This capacity warrants further analysis.

Immigration

Immigration — both international and from Puerto Rico — is of major significance for poor Latino communities in almost every city in every region of the country. Further, there is every reason to believe that immigration will continue to be important.[6]

First, it has important economic consequences. Immigration is a central feature of the economic life of global cities: for example, Los Angeles has been called the "capital of the Third World" because of its huge Latino and Asian immigration (Rieff 1991). In our sample, those cities most bound to world trends (New York, Los Angeles,

Chicago, Houston, and Miami) experienced massive Latino immigration in the 1980s. In the Los Angeles, Houston, and Miami communities . . . immigration is a major factor in the labor market, and the residents of the "second settlement" Puerto Rican communities described in New York and Chicago operate within a context of both racial and ethnic change and of increased Latino immigration. The restructured economy provides marginal jobs for immigrant workers, and wage scales seem to drop for native-born Latinos in areas where immigration is high.[7] This is a more complicated scenario than the simple loss of jobs accompanying Rustbelt deindustrialization. Immigrants are ineligible for most government benefits, are usually highly motivated, and are driven to take even the poorest-paying jobs. They are also more vulnerable to labor-market swings.

These may be construed as rather negative consequences, but in addition, immigrants have been a constructive force in many cities. For example, these authors point to the economic vitality of immigrant-serving businesses. Socially and culturally, there are references . . . to the revival of language and of traditional social controls, the strengthening of networks, and the emergence of new community institutions. Recent research in Chicago (van Haitsma 1991) focuses on the "hard work" ethos of many Mexican immigrants and the extensive resource base provided by kinship networks, a pattern that is echoed and amplified. . . . Most of Tucson's Chicano poor—not just immigrants—are involved in such helping networks.

Though immigrants have been less important in the peripheral cities of Albuquerque, Laredo, and Tucson, each of these cities is special in some way. Albuquerque has attracted few Mexican immigrants, but it draws on a historical Latino labor pool—English-speaking rural *Manitos*—who are as economically exploitable as are Spanish-speaking immigrants from Mexico. Until recently Tucson was also largely bypassed by most Mexican immigrants. Instead, there is an old, relatively self-contained set of cross-border net-

works, with well-established pathways of family movement and mutual aid. Similar networks also exist in Laredo. Laredo's location on the border means that many of its workers are commuters— people who work in Laredo but live in Mexico.

In recent years, immigration has not been very significant in most African American communities, and as a consequence it is underemphasized in the underclass debate. It is also often interpreted as wholly negative. This is partly because the positive effects can be understood only by researchers who study immigrant communities themselves, partly because in some places large numbers of immigrants have strained public resources, and partly because immigrants have occasionally become a source of tension among poor minority populations. Though the specific contouring of immigration effects varies from place to place, in each city . . . immigration is a highly significant dimension of Latino poverty, both at the citywide level and also in the neighborhoods. It is an issue of overriding importance for the understanding of Latino poverty, and thus for the understanding of American urban poverty in general. . . .

The concentration of poverty comes about not only because of market forces or the departure of the middle classes for better housing; in Houston, Rodriguez shows that restructuring in real estate had the effect of concentrating poverty. Concentrated poverty can also result from government planning. Chicago's decision decades ago to build a concentration of high-rise housing projects right next to one another is a clear case in point. Another is in New York's largely Latino South Bronx, where the city's ten-year-plan created neighborhoods in which the least enterprising of the poor are concentrated, and in which a set of undesirable "Not-In-My-Back-Yard" institutions, such as drug-treatment clinics and permanent shelters for the homeless, were located. These neighborhoods are likely to remain as pockets of unrelieved property for many generations to come (Vergara 1991). It was not industrial decline and the exodus of stable working people that created these pockets: the cities of

Chicago and New York chose to segregate their problem populations in permanent buildings in those neighborhoods. . . .

In addition, studies demonstrate that it is not just poverty that gets concentrated. Most immigrants are poor, and most settle in poor communities, thus further concentrating poverty. But, as Rodriguez shows, immigrant communities may be economically, culturally, and socially vital. Social isolation early in the immigration process, he argues, can strengthen group cohesion and lead to community development, rather than to deterioration. Los Angeles portrays institution-building among immigrants in poor communities, and institutional "resilience" characterizes many of the communities . . . especially New York and Chicago. Vélez-Ibáñez's analysis of poverty in Tucson points to the overwhelming importance of "funds of knowledge" shared in interdependent household clusters. Although a priori it makes sociological sense that concentrated poverty should destroy communities, these studies offer evidence that a different pattern emerges under certain circumstances. To use Grenier and Stepick's term, "social capital" also becomes concentrated.

In short, the concentration of poverty need not plunge a neighborhood into disarray. . . . This line of reasoning raises other issues. If it isn't just demographic shifts that weaken neighborhoods, then what is it? These questions strike at the heart of the underclass debate. The old, rancorous controversy about the usefulness of the "culture of poverty" concept questioned whether the poor adhered to a special set of self-defeating values, and if so, whether those values were powerful enough to make poverty self-perpetuating. That argument faded as research focused more effectively on the situational and structural sources of poverty. We do not intend to revive this controversy. It is all too easy to attribute the differences between Latino and black poverty to "the culture." This line can be invidious, pitting one poor population against another in its insinuation that Latino poverty is somehow "better" than black poverty. (Ironically, this would reverse another outdated contention—i.e., that Latinos are poor *because* of their culture.) . . .

Other Aspects of Urban Space . . .

Where a poor neighborhood is located makes a difference.

First, some are targets for "gentrification." This is traditionally viewed as a market process by which old neighborhoods are revitalized and unfortunate poor people displaced. But there is a different perspective. Sassen (1989) argues that gentrification is best understood in the context of restructuring, globalization, and politics. It doesn't happen everywhere . . . gentrification, along with downtown revitalization and expansion, affects Latino neighborhoods in Chicago, Albuquerque, New York, and west side Los Angeles. In Houston, a variant of "gentrification" is documented. Apartment owners who were eager to rent to Latino immigrants when a recession raised their vacancy rates were equally eager to "upgrade" their tenants when the economy recovered and the demand for housing rose once again. Latinos were "gentrified" out of the buildings.

Second, Latinos are an expanding population in many cities, and they rub up against other populations. Most of the allusions to living space center on ethnic frictions accompanying the expansion of Latino areas of residence. Ethnic succession is explicit in Albuquerque and in Chicago. . . . It is implicit in East Los Angeles, with the Mexicanization of Chicano communities, and in Houston, with the immigration of Central Americans to Mexican American neighborhoods and the manipulated succession of Anglos and Latinos. In Albuquerque and East Los Angeles, Latinos are "filling-in" areas of the city, in a late phase of ethnic succession. Ethnic succession is *not* an issue in Laredo because the city's population is primarily of Mexican origin. It is crucial in Miami, where new groups if immigrants are establishing themselves within the Latino community: newer immigrants tend to move into areas vacated by earlier Cuban arrivals, who leave for the suburbs. In Brooklyn, a different kind of urban ecological function is filled by the Puerto Rican barrio—that of an ethnic buffer between African American and Anglo communities. Los Angeles' Westlake area is most strongly affected by its location near downtown: it is intensely involved in both gentrification and problems of ethnic succession. Here the Central Americans displaced a prior population, and, in turn, their nascent communities are pressured by an expanding Koreatown to the west and by gentrification from the north and from downtown.

These details are important in themselves, but they also have implications for existing theories of how cities grow and how ethnic groups become segregated (and segregation is closely allied to poverty). Most such theories take the late nineteenth-century industrial city as a point of departure—a city with a strong central business district and clearly demarcated suburbs. In these models, immigrants initially settle in deteriorating neighborhoods near downtown. Meanwhile, earlier generations of immigrants, their predecessors in those neighborhoods, leapfrog out to "areas of second settlement," often on the edge of the city. . . .

Thus it is no surprise that the "traditional" Rustbelt pattern of ethnic location and ethnic succession fails to appear in most cities discussed in this volume. New Latino immigrants are as likely to settle initially in communities on the edge of town (near the new jobs) as they are to move near downtown; or their initial settlement may be steered by housing entrepreneurs, as in Houston. The new ecology of jobs, housing, and shopping malls has made even the old Rustbelt cities like Chicago less clearly focused on a central downtown business district.

Housing for the Latino poor is equally distinctive. Poor communities in which one-third to one-half of the homes are owner-occupied would seem on the face of it to provide a different ambience from public housing—like the infamous phalanx of projects on Chicago's South Side that form part of Wilson's focus. . . .

Finally, space is especially important when we consider Mexican American communities on the

border. Mexican Americans in most border communities have important relationships with kin living across the border in Mexico, and this is certainly the case in Tucson and Laredo. But space is also important in economic matters. Shopping, working, and recreation are conditioned by the proximity of alternative opportunities on both sides of the border. And in Laredo the opportunities for illicit economic transactions also depend on location. The Laredo barrios in which illicit activities are most concentrated are located right on the Rio Grande River, where cross-border transactions are easier.

In sum, when we consider poor minority neighborhoods, we are drawn into a variety of issues that go well beyond the question of how poverty gets concentrated because middle-class families move out. We must look at the role of urban policy in addition to the role of the market. We must look at the factors that promote and sustain segregation. We must look at how housing is allocated, and where neighborhoods are located within cities. And, finally, we must look at how the location of a neighborhood facilitates its residents' activity in licit and illicit market activities.

REFERENCES

AFL-CIO Industrial Union Department 1986. *The Polarization of America*. Washington, DC: AFL-CIO Industrial Union Department.

Auletta, Ken, 1982. *The Underclass*. New York: Random House.

Barrera, Mario, 1979. *Race and Class in the Southwest*. Notre Dame, IN: University of Notre Dame Press.

Bluestone, Barry, and Bennett Harrison, 1982. *The Deindustrialization of America*. New York: Basic Books.

Chenault, Lawrence Royce, 1938. *The Puerto Rican Migrant in New York*. New York: Columbia University Press.

Clark, Margaret, 1959. *Health in the Mexican American Culture*. Berkeley: University of California Press.

Crawford, Fred, 1961. *The Forgotten Egg*. San Antonio, TX: Good Samaritan Center.

Edmundson, Munro S., 1957. *Los Manitos: A Study of Institutional Values.* New Orleans: Tulane University, Middle American Research Institute.

Ellwood, David T., 1988. *Poor Support: Poverty in the American Family.* New York: Basic Books.

Falcon, Luis, and Douglas Gurak, 1991. "Features of the Hispanic Underclass: Puerto Ricans and Dominicans in New York." Unpublished manuscript.

Fernandez-Kelly, Patricia, and Saskia Sassen, 1991. "A Collaborative Study of Hispanic Women in the Garment and Electronics Industries: Executive Summary." New York: New York University, Center for Latin American and Caribbean Studies.

Galarza, Ernesto, 1965. Merchants of Labor. San Jose, CA: The Rosicrucian Press, Ltd.

Goldschmidt, Walter, 1947. *As You Sow.* New York: Harcourt, Brace.

Gosnell, Patricia Aran, 1949. *Puerto Ricans in New York City.* New York: New York University Press.

Handlin, Oscar, 1959. *The Newcomers: Negroes and Puerto Ricans.* Cambridge, MA: Harvard University Press.

Hill, Richard Child, and Joe R. Feagin, 1987. "Detroit and Houston: Two Cities in Global Perspective." In Michael Peter Smith and Joe R. Feagin, eds. In *The Capitalist City,* pp. 155–177. New York: Basil Blackwell.

Kluckhohn, Florence, and Fred Strodtbeck, 1961. *Variations in Value Orientations.* Evanston, IL: Row, Peterson.

Leonard, Olen, and Charles Loomis, 1938. *Culture of a Contemporary Rural Community: El Cerito, NM.* Washington, DC: U.S. Department of Agriculture.

Levy, Frank, 1977. "How Big Is the Underclass?" Working Paper 0090-1. Washington, DC: Urban Institute.

Maldonado-Denis, Manuel, 1972. *Puerto Rico: A Sociohistoric Interpretation.* New York: Random House.

Massey, Douglas, and Mitchell Eggers, 1990. "The Ecology of Inequality: Minorities and the Concentration of Poverty." *American Journal of Sociology* 95: 1153–1188.

Matza, David, 1966. "The Disreputable Poor." In Reinhardt Bendix and Seymour Martin Lipset, eds. *Class, Status and Power,* pp. 289–302. New York: Free Press.

McCarthy, Kevin, and R. B. Valdez, 1986. *Current and Future Effects of Mexican Immigration in California.* Santa Monica, CA: Rand Corporation.

McWilliams, Carey, 1949. *North From Mexico.* New York: J. B. Lippincott.

Menefee, Seldon, and Orin Cassmore, 1940. *The Pecan Shellers of San Antonio.* Washington: WPA, Division of Research.

Mills, C. Wright, Clarence Senior, and Rose K. Goldsen, 1950. *The Puerto Rican Journey.* New York: Harper.

Montiel, Miguel, 1970. "The Social Science Myth of the Mexican American Family." *El Grito* 3:56–63.

Moore, Joan, 1989. "Is There a Hispanic Underclass?" *Social Science Quarterly* 70:265–283.

Moore, Joan, and Harry Pachon, 1985. *Hispanics in the United States.* Englewood Cliffs, NJ: Prentice Hall.

Morales, Julio, 1986. *Puerto Rican Poverty and Migration: We Just Had to Try Elsewhere.* New York: Praeger.

Morales, Rebecca, 1985. "Transitional Labor: Undocumented Workers in the Los Angeles Automobile Industry." *International Migration Review* 17:570–96.

Morris, Michael, 1989. "From the Culture of Poverty to the Underclass: An Analysis of a Shift in Public Language." *The American Sociologist* 20: 123–133.

Muller, Thomas, and Thomas J. Espenshade, 1986. *The Fourth Wave.* Washington, DC: Urban Institute Press.

Murray, Charles, 1984. *Losing Ground.* New York: Basic Books.

Padilla, Elena, 1958. *Up From Puerto Rico.* New York: Columbia University Press.

Perry, David, and Alfred Watkins, 1977. *The Rise of the Sunbelt Cities.* Beverly Hills, CA: Sage.

Portes, Alejandro, Manuel Castells, and Lauren A. Benton, 1989. *The Informal Economy.* Baltimore: Johns Hopkins University Press.

Portes, Alejandro, and Alex Stepick, 1993. *City on the Edge: The Transformation of Miami.* Berkeley: University of California Press.

Rand, Christopher, 1958. *The Puerto Ricans.* New York: Oxford University Press.

Ricketts, Erol, and Isabel V. Sawhill, 1988. "Defining and Measuring the Underclass." *Journal of Policy Analysis and Management* 7:316–325.

Reiff, David, 1991. *Los Angeles: Capital of the Third World.* New York: Simon and Schuster.

Rodriguez, Clara, 1989. *Puerto Ricans: Born in the U.S.A.* Boston: Unwin Hyman.

Romano-V, Octavio I, 1968. "The Anthropology and Sociology of the Mexican Americans." *El Grito* 2:13–26.

Russell, George, 1977. "The American Underclass." *Time Magazine* 110 (August 28):14–27.

Sanchez, George, 1940. *Forgotten People: A Study of New Mexicans*. Albuquerque: University of New Mexico Press.

Sassen, Saskia, 1989. "New Trends in the Sociospatial Organization of the New York City Economy." In Robert Beauregard, ed. *Economic Restructuring and Political Response*. Newberry Park, CA.

Sassen-Koob, Saskia, 1984. "The New Labor Demand in Global Cities." In Michael Smith, ed. *Cities in Transformation*. Beverly Hills, CA: Sage.

Saunders, Lyle, 1954. *Cultural Differences and Medical Care*. New York: Russell Sage Foundation.

Senior, Clarence Ollson, 1965. *Our Citizens from the Caribbean*. New York: McGraw Hill.

Soja, Edward, 1987. "Economic Restructuring and the Internationalization of the Los Angeles Region." In Michael Peter Smith and Joe R. Feagin, eds. *The Capitalist City*, pp. 178–198. New York: Basil Blackwell.

Stevens Arroyo, Antonio M., 1974. *The Political Philosophy of Pedro Abizu Campos: Its Theory and Practice*. Ibero American Language and Area Center. New York: New York University Press.

Sullivan, Mercer L., 1989a. *Getting Paid: Youth Crime and Work in the Inner City*. Ithaca: Cornell University Press.

Taylor, Paul, 1928. *Mexican Labor in the U.S.: Imperial Valley*. Berkeley: University of California Publications in Economics.

———, 1930. *Mexican Labor in the U.S.: Dimit County, Winter Garden District, South Texas*. Berkeley: University of California Publications in Economics.

———, 1934. *An American-Mexican Frontier*. Chapel Hill, NC: University of North Carolina Press.

U.S. Bureau of the Census, 1991b. *The Hispanic Population in the United States: March 1991*. Current Population Reports, Series P-20, No. 455. Washington, DC: U.S. Government Printing Office.

Vaca, Nick, 1970. "The Mexican American in the Social Sciences." *El Grito* 3:17–52.

Vergara, Camilo Jose, 1991. "Lessons Learned: Lessons Forgotten: Rebuilding New York City's Poor Communities." *The Livable City* 15:3–9.

Wagenheim, Kal, 1975. *A Survey of Puerto Ricans on the U.S. Mainland*. New York: Praeger.

Wakefield, Dan, 1959. *Island in the City*. New York: Corinth Books.

Wilson, William Julius, 1987. *The Truly Disadvantaged: The Inner City, the Underclass, and Public Policy*. Chicago: The University of Chicago Press.

———, 1990. "Social Theory and Public Agenda Research: The Challenge of Studying Inner-city Social Dislocations." Paper presented at Annual Meeting of the American Sociological Association.

NOTES

1. Tabulations from the 1990 census provided by the Population Division of the U.S. Bureau of the Census in June 1992.

2. A perfect case in point is the Panel Study on Income Dynamics (PSID), probably the most important social science data set for analyzing poverty and social mobility over time. Since its inception, the PSID oversampled blacks, but it was not until 1990 that an effort was made to sample Hispanics.

3. The headquarters of 24 percent of the world's largest multinational corporations were located in Chicago, New York, and Los Angeles (Smith and Feagin 1987).

4. A census study shows that in poor Los Angeles neighborhoods both population and median household income increased between 1970 and 1980, whereas in Chicago and New York they declined precipitously (Weicher 1990). "Poor neighborhoods" refers to "groups of continuous low-income census tracts with 20,000 or more residents in the aggregate" in which 20 percent or more of the population were beneath the poverty line in both 1970 and 1980 (pp. 69–70). In Los Angeles, population in such neighborhoods increased by 13.5 percent, and median household income by 4.2 percent. But in Chicago population decreased by 25.9 percent and by 31.7 percent in New York. Median household income decreased by 26.8 percent in Chicago and by 22.8 percent in New York.

5. In Albuquerque, Kirtland Air Force Base has been a major employer, along with the Los Alamos and Sandia research complexes. In Tucson, Davis-Monthan Air Force Base is also very large, and Hughes Aircraft dominates a thriving defense industry (Luckingham 1982).

6. The 1986 Immigration and Control Act apparently reduced undocumented immigration primarily through its legalization provisions, which permitted some 2.7 million undocumented workers to regularize their status (Fix 1991). Though this may have been responsible for a decline in the number of undocumented immigrants apprehended in the late 1980s, a resurgence of apprehensions in 1990 indicates that the pressure for immigration remains high.

7. In New York, some of the high labor-force dropout rates among Puerto Ricans may be accounted for by competition with Dominican workers.

46
Myths of an Aging America

CHARLES F. LONGINO, JR.

Because the elderly receive differential treatment in many aspects of their lives, they are sometimes considered a minority group. Longino outlines the problems that present themselves as this group increases in size.

Consider the following questions as you read:

1. What are population trends affecting the elderly as a group?

2. What is needed to improve the lives of this group?

GLOSSARY

The demographic imperative The inevitability that as populations grow, feeding, health care, and other necessities will become strained.

Population bomb Exploding world population, making care and feeding difficult.

IF DEMOGRAPHERS EVER SAT around campfires telling scary stories, their favorite would be called "The Demographic Imperative." It's a dark and terrible story about the transformation of a carefree, youthful American into a nation of ancient, sickly wretches with no one to care for them. Like the best scary stories, it is firmly based in fact. But like many horror stories, it's not so scary when viewed in the proper light.

The story begins at the dawn of the twentieth century, when most Americans were rural dwellers under age 25. Only 3.1 million Americans—just 4 percent of the total population—were aged 65 and older. The life expectancy of a newborn white child was only about 50 years then, and it was less than 35 for blacks and others.

Then a miracle happened. In just three decades, the average life expectancy for whites shot up to about 60, and the life expectancy for others approached 50. The number of elderly people more than doubled to 6.7 million, about 5 percent of the total population. By this time, the largest share of Americans were urban dwellers. With the Great Depression in full swing, they were hungry for a better life. In response to the growing needs of the elderly, the Social Security Act passed in 1935.

Now the story moves forward another 30 years, to 1960. While America entered a new cultural phase that worshipped youth, its elderly population more than doubled again. There were 16.7 million Americans aged 65 and older in 1960, about 9 percent of the total population. Life expectancy reached about 70 for whites and over 60 for others.

As the baby-boom generation continued to swell hospital nurseries, demographers began worrying that the miracle of longer life might turn into a uncontrollable beast. They began talking about a "population bomb" and anticipating the hardship that growing numbers of elderly Americans would place upon society. More than two-thirds of Americans lived in urban areas in 1960. In response to heightened concern for the plight of the elderly, the Medicare Amendment to the Social Security Act passed in 1965.

In the last 30 or so years, the elderly population has nearly doubled again. Now 31 million people, or 12 percent of the total population, are aged 65 and older. The American landscape has

Reprinted from American Demographics, *August 1994, Vol. 16, No. 8, pp. 36–42. Reprinted by permisson.*

Doubling Time

The number of elderly Americans has doubled every 30 years.

(U.S. population aged 65 and older in millions, 1900-2025)

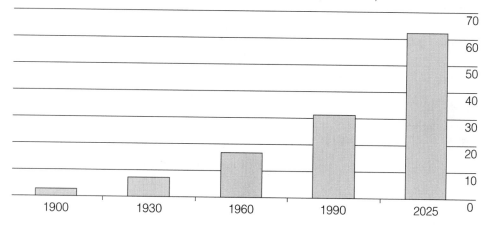

Source: Decennial censuses 1900-90, projection for 2025 is from the Census Bureau's Current Population Reports: "Population Projections of the United States, by Age, Sex, Race and Hispanic Origin: 1992 to 2050," Series P-25-1092

changed dramatically since 1900: the total population has more than tripled, and urban areas are now home to more than 75 percent of Americans.

In another 35 years, the elderly population should double again. The Census Bureau anticipates that 62 million people, or almost one in five Americans, will be aged 65 and older by 2025. And by 2045, the elderly population will reach 77 million, more than the total population of the U.S. in 1900. And now comes the scary part of the demographic imperative.

An Overwhelming Burden?

At this point in the story, demographers lower their voices and begin talking about the changing characteristics of the elderly population. As they speak, shivers run down the spines of spellbound listeners. The elderly population is not only growing rapidly, it's also getting older.

The oldest old, aged 85 and older, are increasing at a faster rate than the total elderly population. In 1990, fewer than one in ten elderly persons was aged 85 or older. By 2045, the oldest old will be one in five. Increasing longevity and the steady movement of baby boomers into this oldest age group will drive this trend.

The demographic imperative reminds us that disability increases as people age. Only 9 percent of people aged 65 to 69 need help with any personal care, such as eating, bathing, going to the bathroom, or dressing; 45 percent of people aged 85 and older need help. And because women live longer than men, many of the oldest old are widows. There are 1.5 women for every man aged 65 and older. Among the oldest old, there are 2.6 women for every man.

Even though women live longer, they are more vulnerable than men to chronic conditions. And their low incomes make their medical care a public rather than a private issue. When these grim statistics are projected into the future, many researchers conclude that Americans are growing older and becoming more sickly. They worry that Americans will be less capable of caring for themselves, both physically and economically.

Finding able-bodied people to provide that care could be a problem, because the supply of

Depending on You

Between 1990 and 2025, elderly people
will account for all of America's
increase in dependency.

(dependency ratio* for children younger than 18 and the
elderly aged 65 and older, 1990 and 2025)

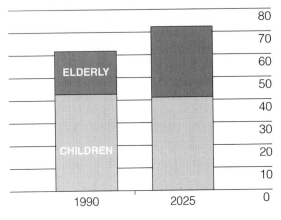

*The dependency ratio is the number of children
younger than 18 and the number of adults aged 65 and
older compared with 100 adults of working age.

Source: 1990 census and the Census Bureau's Current
Population Reports: "Population Projections of the United
States, by Age, Sex, Race, and Hispanic Origin: 1992 to
2050" Series P-25-1092

caregivers is not keeping pace with the growth in the older population. The number of elderly persons for every 100 adults of working age (aged 18 to 64) is called the old-age dependency ratio. In 1990, there were 20 elderly persons for every 100 working-aged adults. But the burden of caring for the elderly may become much heavier as the baby-boom generation enters its retirement years. When the youngest boomers approach retirement age in 2025, there will be 32 elderly persons for every 100 people of working age.

The rising cost of caring for the elderly deepens the darkness of the demographic imperative. Without serious health-care reform, the U.S. could spend 20 percent of its Gross Domestic Product on health care by the year 2000. Currently, more than one-third of total healthcare expenditures are spent on the 12 percent of the population

aged 65 and older. And public funds are the major source of payment for the elderly population.

The deepening gloom of the demographic imperative is global in proportions. Rising health-care costs were absorbed in the growing national economy of the 1950s and 1960s. But national economic trends became starkly negative during the late 1970s and 1980s. The industrialized world is caught up in an extended period of recession, economic restructuring, and stagnation. The results are lower tax revenues and greater demands on the national treasury.

The demographic imperative concludes that tomorrow's elderly will be whiplashed by converging trends. Their numbers and proportions are growing inexorably as the wealth of the nation deteriorates. Our already serious long-term-care problem promises to grow to crisis proportions. In the horrifying final vision of the demographic imperative, the United States becomes a 21st-century Calcutta, with futuristic Mother Theresas ministering to the dying elderly on the streets of Cincinnati.

The Wild Card

Many researchers have become accustomed to the story of the demographic imperative. In fact, they're so convinced of it that anyone who would question its dire predictions is in danger of being dismissed as a crackpot. Almost everyone foresees that the demand for long-term care will increase, but there is room for disagreement about how much it will increase—and how to address the problem. The story of the demographic imperative and its many depressing details indeed deserves to be challenged.

The biggest problem with the demographic imperative is that it ignores generational effects. Yet generational change will transform the older population, the caregiving population, and society as a whole. Researchers have become fixated on certain age groups. They characterize the population aged 85 and older as a high-risk group and underestimate differences in coming generations. Yet it is likely that several factors will work to reduce disability among the elderly, in-

Dependency Now

In 1990, 20 states had fewer than 20 elderly for every 100 working-age adults.
(ratio of population aged 65 and older to population aged 18 to 64, by state, 1990)

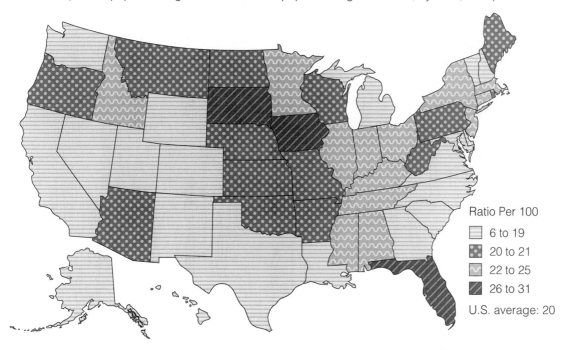

Ratio Per 100

- 6 to 19
- 20 to 21
- 22 to 25
- 26 to 31

U.S. average: 20

Source: 1990 census

cluding improved health, new forms of service delivery, and improved technology.

Will baby boomers, who popularized healthy lifestyles, be healthier in old age? Since 1953, per capita tobacco consumption has declined by 40 percent in the United States. Butter consumption is down by one-third, whole milk and cream are down one-fourth, and saturated animal fats for cooking are down 40 percent. Consumption of vegetable oils and fish has increased. These behaviors may cut the rates of chronic diseases in old age, potentially reducing the use of health services and lengthening the average person's productive life.

These positive trends may already be paying healthy dividends. The prevalence of chronic disability among the elderly declined 4 percent between 1984 and 1989. Furthermore, the declines were greatest among those aged 85 and older. These improvements may be due to in-

creasing levels of education and income, as a new generation moves into the eldest age group.

Of course, modest declines in disability will not be enough to offset the increases in disability caused by the growth of the elderly population. But even though absolute numbers of the chronically disabled will certainly continue to rise, the expansion may not be as great as the catastrophe some demographic doomsayers predict.

Another progressive and rapid change among the elderly is their increasing desire to live independently of children and other relatives. The independent elderly are less likely to use caregivers and more likely to use paid help or products and services to get what they need. In 1990, about 9 million Americans aged 65 and older lived alone. By 2010, that number is expected to approach 13 million, according to *American Demographics* projections. Three out of four of

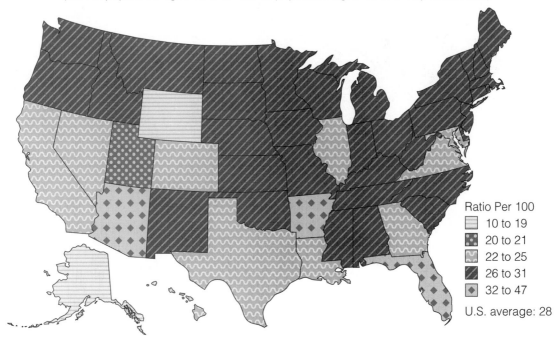

Dependency in 2020

When baby boomers retire, most states will have at least 28 elderly for every
100 working-age adults.
(ratio of population aged 65 and older to population aged 18 to 64, by state, 2020)

Ratio Per 100
10 to 19
20 to 21
22 to 25
26 to 31
32 to 47

U.S. average: 28

Source: Census Bureau, Current Population Reports, P25-1111

those householders will be women. But older men living alone—especially those aged 75 and older—are a rapidly growing segment.

Today, elderly people in their 60s and 70s are well-endowed with family resources—because they are the parents of the baby boom. The supply of family caregivers will decline as the baby-boom generation retires, because this group has fewer children. On the other hand, the burden of caregiving will be slightly offset by future declines in childbearing. In 1990, there were 42 children under age 18 for every 100 adults aged 18 to 64. By 2025, that number will slip to 41.

Assistive Devices

The depletion of caregivers is being accelerated by the continued movement of women out of the home and into the labor force. Yet gerontol-

ogists have not detected a shift from family care-givers to paid caregivers; instead, the use of assistive devices and housing modifications is rising sharply, while the long-term use of personal assistants alone is declining significantly.

Greater residential independence, combined with the development and use of personal-assistance technologies, seem to be part of a modern elderly person's longterm adaptive process. Unless there is some kind of interdependence that preserves self-respect and self-determination, dependency on family members or others will be a far less attractive alternative than technically supported self-care for those who can afford it. As the high-risk population grows—and the horror story tells us that it surely will—a marketplace for assistive technology will grow with it.

Sensory technology, such as eyeglasses and hearing aids, has been around for a long time.

Down with Disability

The number of older Americans increased almost 15 percent in the 1980s but the share of elderly with limited disabilities actually declined.

(percent change in number of adults aged 65 and older with daily living disability by type of activity limitations, 1982-89)

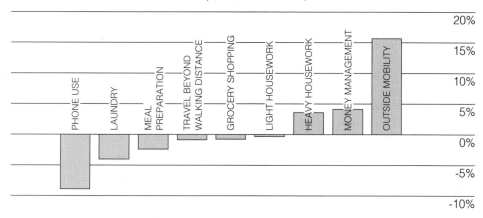

Source: Kenneth G. Manton, et al., and "Changes in the Use of Personal Assistance and Special Equipment from 1982 to 1989: Results from the 1982 and 1989 NLTCS," *The Gerontologist,* Vol. 33, No. 2, 1993

Now, electric wheelchairs, sensory aids for telephone equipment, voice-activated devices for adjusting lighting and temperature at home, and biomedical devices to strengthen or replace legs, arms, toes, and fingers are becoming commonplace. This category of technology will be a growth industry in the next century as the baby boom reaches advanced age.

In the 21st century, older Americans may be better prepared to live independently. Incremental increases in Social Security will protect older householders from inflation and recessions. Aging baby-boom women will also have resources their mothers never had. Because they delayed marriage and experienced high divorce rates, many have lived independently for years. They are used to keeping their homes, managing their money, and tending to emergencies by themselves. Virtually all baby-boom women have worked outside the home for at least part of their lives, and many will have their own pension incomes.

Americans are notorious for addressing problems only when they become national crises. The Social Security adjustments in the early 1980s averted such a crisis; the health-care reform legislation of the early 1990s is another example. When the long-term care crisis mounts in the next century, the U.S. public will come to the rescue by advancing policies to reform caregiving. Even in the next decade, we will see policy discussion that will set the stage for those dramatic events. The ideas that will become a part of the national debate in 2010 are forming today.

Businesses can also help to defuse the demographic imperative. Private firms that produce and distribute services and goods to enhance independence longer into old age will find growing markets, and the competition they generate will make these products available at a decreasing cost. This will primarily benefit the smaller generation following the boomers. And because the baby boom came later in many Third World countries than in the United States, overseas markets will provide expanded opportunities for assistive devices long after the American baby-boom generation is gone.

Those who wring their hands in despair for the calamity that will eventually befall the baby boom in old age should relax a bit. Despite their competitive struggle for education, jobs, and housing, boomers have always had political clout. When they turned 18, they got the vote. Boomers stopped the Vietnam War, relaunched the feminist movement, celebrated the first Earth Day, and raised the drinking age before their kids became teenagers. In 2010, the baby boom will demand changes in long-term-care policy. They will want better support in their old age, and they will have it.

When Chicken Little said that the sky was falling, he lost credibility. The same fate may await those who crow about the demographic imperative without questioning its assumptions. The imperative is frequently used as a strategy to keep public support high for special interests that serve the elderly. These interests will certainly have a place in America's future, but they may not bear as much of a burden as the demographic imperative now assigns them. The future may not be quite as scary as we think.

QUESTIONS FOR DISCUSSION

For further discussion of this topic, see the Wadsworth Sociology Resource Center, "Virtual Society," *http://sociology.wadsworth.com*, under *Sociological Footprints*, by Cargan and Ballantine. You can respond to the discussion questions there or enter your own comments in the online chat forum.

SUGGESTED READINGS AND
SOCIOLOGY INTERNET RESOURCES

See the Wadsworth Sociology Resource Center, "Virtual Society," *http://sociology.wadsworth. com*, for additional links, suggestions for further reading, and learning tools related to this chapter.

Either from the "Virtual Society" website or directly from your web browser, you may access InfoTrac College Edition, an online university library that includes over 700 popular and scholarly journals in which you can find articles related to the topics in this chapter.

CHAPTER 12

Deviance

WHAT DOES IT MEAN?

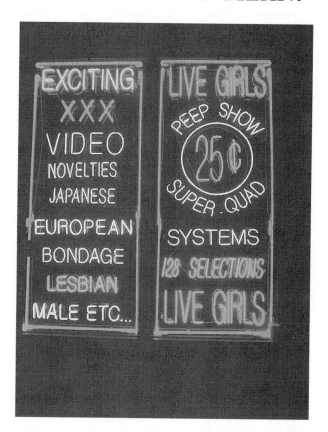

DEVIANCE IS ONE OF THE MOST myth-laden of social issues. The trouble with myths is that they often contain a grain of truth, which can make their distortions difficult to demonstrate. Most often, myths are simplistic explanations based on the individual's own set of biases.

One common myth is based on the idea that crime is genetic; that is, that there is a relationship between the committing of crime and one's race or genes, or that criminals reflect a distinctive physical type. In fact, crime, like all human behavior, is learned — there is no monopoly on crime by any race, ethnic group, or physical type. In the first reading of this chapter, Merton points out that, contrary to genetic propensities for crime, societies themselves can stimulate crime. Societies create and define certain goals as desirable, and individuals want to attain these goals. However, they may not have the means to do so. Merton discusses ways — legitimate and illegitimate — that people use to reach desired goals in society.

Another myth concerns the kind of crimes being committed. The mass media concentrate on crimes that elicit reader/listener interest — crime against persons. In fact, FBI reports show that most crime is against property. Related to this myth is the equally misleading myth that crime rates are increasing. It is true that crime statistics appear to show an increase, but they do not tell the whole story. Some of the increase simply reflects improved methods of data collection.

It is this idea that has led to the proposal that some behaviors labeled as crimes should be decriminalized. That is, the "crime" label should be removed from crimes that have no victims, such as prostitution, pot smoking, and pornography. On the other hand, some areas of deviant behavior have many victims but have not been labeled as social problems, such as white-collar crime. White-collar crime is indeed criminal behavior, because it fits the definition of a crime — the legal description of a socially harmful act or an act for which the law provides a penalty. In the second reading of this chapter, Mokhiber notes the extensiveness of this little-known criminal behavior.

A final myth involves the idea that punishment and incarceration reduce crime. When we consider the high incidence of repeated criminal behavior despite severe punishment, we can conclude that there is little or no apparent relationship between incarceration, rehabilitation, and the reduction of crime. In addition, the cost of keeping people in prison is staggering — about $10 billion in known costs per year. The American Bar Association would like to add restitution and community service to penalties as a means of helping with the costs and rehabilitation.

In any case, the public apparently feels that it is important to wreak its revenge against those that dare to violate society's norms. Besides, the public feels safer and probably is safer from a given criminal while he or she is in prison. Politicians — recognizing the popularity of the public's "get tough" stance — are encouraged to promote tougher sentencing programs. But, as Jeffrey Reiman notes in the third reading, such "get tough" programs do little to actually lessen crime since they do not deal with crime prevention. In fact, Reiman shows that crime control is a failure and that apparently nothing is being learned from this failure. It is this failure of prison sentences at rehabilitation and as a deterrent to crime that has led to the suggestion that what is needed is a moral crusade like the War on Drugs. Do you believe that a crusade of this type would reduce crime, or would it just sound good? How would you deal with this problem?

In the final reading in this chapter, D. L. Rosenhan deals with a different aspect of myths — the official definitions of deviance. He notes how labeling persons as deviant causes them to be seen as deviant even if they are not. More important, once labeled as such, they will never again be considered "normal" despite normal behavior. It would seem, then, that the label defines the action rather than the action speaking for itself. When you consider this reading, think about other forms of "deviance" that may be "artifacts of a label."

As you consider these readings, you may want to ask yourself why corporate crime is not considered a social problem, while prostitution, pot smoking, and pornography are. Is this another aspect of the corporate political power noted in Chapter 10? You might also ask yourself whether other areas of behavior should be labeled as social problems.

47

Social Structure and Anomie

ROBERT K. MERTON

Deviance is the breaking of social norms, activity that Merton argues is common in society. This is because individuals have goals—what is desired in society—but not always the legitimate means to achieve the goals. Therefore, they may resort to illegitimate means of achieving goals.

As you read the article ask yourself the following:

1. What is the relationship Merton sets up between goals and means, and what are his "modes of adaptation"?

2. What are examples of the "strain toward anomie" in your community or society?

GLOSSARY

Social structure Relatively stable pattern of social behavior.

Anomie Societal condition in which norms break down and there is little moral guidance for individuals.

Patterns of Cultural Goals and Institutional Norms

AMONG THE SEVERAL ELEMENTS of social and cultural structures, two are of immediate importance. These are analytically separable although they merge in concrete situations. The first consists of culturally defined goals, purposes, and interests, held out as legitimate objectives for all or for diversely located members of the society. The goals are more or less integrated—the degree is a question of empirical fact—and roughly ordered in some hierarchy of value. Involving various degrees of sentiment and significance, the

prevailing goals comprise a frame of aspirational reference. They are the things "worth striving for." They are a basic, though not the exclusive, component of what [cultural anthropologist Ralph] Linton has called "designs for group living." And though some, not all, of these cultural goals are directly related to the biological drives of man, they are not determined by them.

A second element of the cultural structure defines, regulates and controls the acceptable modes of reaching out for these goals. Every social group invariably couples its cultural objectives with regulations, rooted in the mores or institutions, of allowable procedures for moving toward these objectives. These regulatory norms are not necessarily identical with technical or efficiency norms. Many procedures which from the standpoint of particular individuals would be most efficient in securing desired values—the exercise of force, fraud, power—are ruled out of the institutional area of permitted conduct. At times, the disallowed procedures include some which would be efficient for the group itself—*e.g.*, historic taboos on vivisection, on medical experimentation, on the sociological analysis of "sacred" norms—since the criterion of acceptability is not technical efficiency but value-laden sentiments (supported by most members of the group or by those able to promote these sentiments through the composite use of power and propaganda). In all instances, the choices of expedients for striving toward cultural goals is limited by institutionalized norms. . . .

Contemporary American culture appears to approximate the polar type in which great em-

American Sociological Review, *Vol. 3, October 1938, pp. 672–682.*

phasis upon certain success-goals occurs without equivalent emphasis upon institutional means. It would of course be fanciful to assert that accumulated wealth stands alone as a symbol of success just as it would be fanciful to deny that Americans assign it a place high in their scale of values. In some large measure, money has been consecrated as a value in itself, over and above its expenditure for articles of consumption or its use for the enhancement of power. "Money" is peculiarly well adapted to become a symbol of prestige. As [German sociologist Georg] Simmel emphasized, money is highly abstract and impersonal. However acquired, fraudulently or institutionally, it can be used to purchase the same goods and services. The anonymity of an urban society, in conjunction with these peculiarities of money, permits wealth, the sources of which may be unknown to the community in which the plutocrat lives or, if known, to become purified in the course of time, to serve as a symbol of high status. Moreover, in the American Dream there is no final stopping point. The measure of "monetary success" is conveniently indefinite and relative. At each income level, as H. F. Clark found, Americans want just about twenty-five per cent more (but of course this "just a bit more" continues to operate once it is obtained). In this flux of shifting standards, there is no stable resting point, or rather, it is the point which manages always to be "just ahead." An observer of a community in which annual salaries in six figures are not uncommon, reports the anguished words of one victim of the American Dream: "In this town, I'm snubbed socially because I only get a thousand a week. That hurts." . . .

Thus the culture enjoins the acceptance of three cultural axioms: First, all should strive for the same lofty goals since these are open to all; second, present seeming failure is but a way-station to ultimate success; and third, genuine failure consists only in the lessening or withdrawal of ambition. . . .

In sociological paraphrase, these axioms represent, first, the deflection of criticism of the social structure onto one's self among those

so situated in the society that they do not have full and equal access to opportunity; second, the preservation of a structure of social power by having individuals in the lower social strata identify themselves, not with their compeers, but with those at the top (whom they will ultimately join); and third, providing pressures for conformity with the cultural dictates of unslackened ambition by the threat of less than full membership in the society for those who fail to conform.

It is in these terms and through these processes that contemporary American culture continues to be characterized by a heavy emphasis on wealth as a basic symbol of success, without a corresponding emphasis upon the legitimate avenues on which to march toward this goal. How do individuals living in this cultural context respond? And how do our observations bear upon the doctrine that deviant behavior typically derives from biological impulses breaking through the restraints imposed by culture? What, in short, are the consequences for the behavior of people variously situated in a social structure of a culture in which the emphasis on dominant success-goals has become increasingly separated from an equivalent emphasis on institutionalized procedures for seeking these goals?

Types of Individual Adaptation

Turning from these culture patterns, we now examine types of adaptation by individuals within the culture-bearing society. Though our focus is still the cultural and social genesis of varying rates and types of deviant behavior, our perspective shifts from the plane of patterns of cultural values to the plane of types of adaptation to these values among those occupying different positions in the social structure.

We here consider five types of adaptation, as these are schematically set out in the following table, where (+) signifies "acceptance," (−) signifies "rejection," and (±) signifies "rejection of prevailing values and substitution of new values."

A Typology of Modes of Individual Adaptation

Modes of Adaptation	Culture Goals	Institutionalized Means
I. Conformity	+	+
II. Innovation	+	−
III. Ritualism	−	+
IV. Retreatism	−	−
V. Rebellion	±	±

Examination of how the social structure operates to exert pressure upon individuals for one or another of these alternative modes of behavior must be prefaced by the observation that people may shift from one alternative to another as they engage in different spheres of social activities. These categories refer to role behavior in specific types of situations, not to personality. They are types of more or less enduring response, not types of personality organization. To consider these types of adaptation in several spheres of conduct would introduce a complexity unmanageable within the confines of this chapter. For this reason, we shall be primarily concerned with economic activity in the broad sense of "the production, exchange, distribution, and consumption of goods and services" in our competitive society, where wealth has taken on a highly symbolic cast.

I. CONFORMITY

To the extent that a society is stable, adaptation type I—conformity to both cultural goals and institutionalized means—is the most common and widely diffused. Were this not so, the stability and continuity of the society could not be maintained. The mesh of expectancies constituting every social order is sustained by the model behavior of its members representing conformity to the established, though perhaps secularly changing, culture patterns. It is, in fact, only because behavior is typically oriented toward the basic values of the society that we may speak of a human aggregate as comprising a society. Unless there is a deposit of values shared by interacting individuals, there exist social relations, if the disorderly interactions may be so called, but no society. . . .

II. INNOVATION

Great cultural emphasis upon the success-goal invites this mode of adaptation through the use of institutionally proscribed but often effective means of attaining at least the simulacrum of success—wealth and power. This response occurs when the individual has assimilated the cultural emphasis upon the goal without equally internalizing the institutional norms governing ways and means for its attainment.

From the standpoint of psychology, great emotional investment in an objective may be expected to produce a readiness to take risks, and this attitude may be adopted by people in all social strata. From the standpoint of sociology, the question arises, which features of our social structure predispose toward this type of adaptation, thus producing greater frequencies of deviant behavior in one social stratum than in another?

On the top economic levels, the pressure toward innovation not infrequently erases the distinction between business-like strivings this side of the mores and sharp practices beyond the mores. As [economist Thorstein] Veblen observed, "It is not easy in any given case—indeed it is at times impossible until the courts have spoken—to say whether it is an instance of praiseworthy salesmanship or a penitentiary offense." The history of the great American fortunes is threaded with strains toward institutionally dubious innovation as is attested by the many tributes to the robber barons. The reluctant admiration often expressed privately, and not seldom publicly, of these "shrewd, smart, and successful" men is a product of a cultural structure in which the sacrosanct goal virtually consecrates the means. . . .

Living in the age in which the American robber barons flourished, [Ambrose] Bierce could not easily fail to observe what became later known as "white-collar crime." Nevertheless, he was aware that not all of these large and dramatic departures from institutional norms in the top economic strata are known, and possibly fewer deviations among the lesser middle classes come to light. [Edwin H.] Sutherland has repeatedly documented the prevalence of "white-collar criminality" among business men. He notes, further, that many of these crimes were not prosecuted because they were not detected or, if detected, because of "the status of the business man, the trend away from punishment, and the relatively unorganized resentment of the public against white-collar criminals." A study of some 1,700 prevalently middle-class individuals found that "off the record crimes" were common among wholly "respectable" members of society. Ninety-nine per cent of those questioned confessed to having committed one or more of 49 offenses under the penal law of the State of New York, each of these offenses being sufficiently serious to draw a maximum sentence of not less than one year. The mean number of offenses in adult years—this excludes all offenses committed before the age of sixteen—was 18 for men and 11 for women. Fully 64% of the men and 29% of the women acknowledged their guilt on one or more counts of felony which, under the laws of New York is ground for depriving them of all rights of citizenship. One keynote of these findings is expressed by a minister, referring to false statements he made about a commodity he sold, "I tried truth first, but it's not always successful." On the basis of these results, the authors modestly conclude that "the number of acts legally constituting crimes are far in excess of those officially reported. Unlawful behavior, far from being an abnormal social or psychological manifestation, is in truth a very common phenomenon."

But whatever the differential rates of deviant behavior in the several social strata, and we know from many sources that the official crime statistics uniformly showing higher rates in the lower strata are far from complete or reliable, it appears from our analysis that the greatest pressures toward deviation are exerted upon the lower strata. Cases in point permit us to detect the sociological mechanisms involved in producing these pressures. Several researchers have shown that specialized areas of vice and crime constitute a "normal" response to a situation where the cultural emphasis upon pecuniary success has been absorbed, but where there is little access to conventional and legitimate means for becoming successful. The occupational opportunities of people in these areas are largely confined to manual labor and the lesser white-collar jobs. Given the American stigmatization of manual labor *which has been found to hold rather uniformly in all social classes,* and the absence of realistic opportunities for advancement beyond this level, the result is a marked tendency toward deviant behavior. The status of unskilled labor and the consequent low income cannot readily compete *in terms of established standards of worth* with the promises of power and high income from organized vice, rackets, and crime.

For our purposes, these situations exhibit two salient features. First, incentives for success are provided by the established values of the culture *and* second, the avenues available for moving toward this goal are largely limited by the class structure to those of deviant behavior. It is the *combination* of the cultural emphasis and the social structure which produces intense pressure for deviation. Recourse to legitimate channels for "getting in the money" is limited by a class structure which is not fully open at each level to men of good capacity. Despite our persisting open-class-ideology, advance toward the success-goal is relatively rare and notably difficult for those armed with little formal education and few economic resources. The dominant pressure leads toward the gradual attenuation of legitimate, but by and large ineffectual, strivings and the increasing use of illegitimate, but more or less effective, expedients.

Of those located in the lower reaches of the social structure, the culture makes incompatible demands. On the one hand, they are asked to orient their conduct toward the prospect of large wealth—"Every man a king," said Marden and Carnegie and Long—and on the other, they are largely denied effective opportunities to do so institutionally. The consequence of this structural inconsistency is a high rate of deviant behavior. The equilibrium between culturally designated ends and means becomes highly unstable with progressive emphasis on attaining the prestige-laden ends by any means whatsoever. Within this context, Al Capone represents the triumph of amoral intelligence over morally prescribed "failure," when the channels of vertical mobility are closed or narrowed *in a society which places a high premium on economic affluence and social ascent for all its members.*

This last qualification is of central importance. It implies that other aspects of the social structure, besides the extreme emphasis on pecuniary success, must be considered if we are to understand the social sources of deviant behavior. A high frequency of deviant behavior is not generated merely by lack of opportunity or by this exaggerated pecuniary emphasis. A comparatively rigidified class structure, a caste order, may limit opportunities far beyond the point which obtains in American society today. It is only when a system of cultural values extols, virtually above all else, certain *common* success-goals *for the population at large* while the social structure rigorously restricts or completely closes access to approved modes of reaching these goals *for a considerable part of the same population,* that deviant behavior ensues on a large scale. Otherwise said, our egalitarian ideology denies by implication the existence of non-competing individuals and groups in the pursuit of pecuniary success. Instead, the same body of success-symbols is held to apply for all. Goals are held to transcend class lines, not to be bounded by them, yet the actual social organization is such that there exist class differentials in accessibility of the goals. In this setting, a cardinal American virtue, "ambition," promotes a cardinal American vice, "deviant behavior."

This theoretical analysis may help explain the varying correlations between crime and poverty. "Poverty" is not an isolated variable which operates in precisely the same fashion wherever found; it is only one in a complex of identifiably interdependent social and cultural variables. Poverty as such and consequent limitation of oppor-

tunity are not enough to produce a conspicuously high rate of criminal behavior. Even the notorious "poverty in the midst of plenty" will not necessarily lead to this result. But when poverty and associated disadvantages in competing for the culture values approved for *all* members of the society are linked with a cultural emphasis on pecuniary success as a dominant goal, high rates of criminal behavior are the normal outcome. Thus, crude (and not necessarily reliable) crime statistics suggest that poverty is less highly correlated with crime in southeastern Europe than in the United States. The economic life-chances of the poor in these European areas would seem to be even less promising than in this country, so that neither poverty nor its association with limited opportunity is sufficient to account for the varying correlations. However, when we consider the full configuration — poverty, limited opportunity, and the assignment of cultural goals — there appears some basis for explaining the higher correlation between poverty and crime in our society than in others where rigidified class structure is coupled with *differential class symbols of success.* . . .

In societies such as our own, then, the great cultural emphasis on pecuniary success for all and a social structure which unduly limits practical recourse to approved means for many set up a tension toward innovative practices which depart from institutional norms. But this form of adaptation presupposes that individuals have been imperfectly socialized so that they abandon institutional means while retaining the success-aspiration. Among those who have fully internalized the institutional values, however, a comparable situation is more likely to lead to an alternative response in which the goal is abandoned but conformity to the mores persists. This type of response calls for further examination.

III. RITUALISM

The ritualistic type of adaptation can be readily identified. It involves the abandoning or scaling down of the lofty cultural goals of great pecuniary success and rapid social mobility to the point where one's aspirations can be satisfied.

But though one rejects the cultural obligation to attempt "to get ahead in the world," though one draws in one's horizons, one continues to abide almost compulsively by institutional norms.

It is something of a terminological quibble to ask whether this represents genuinely deviant behavior. Since the adaptation is, in effect, an internal decision and since the overt behavior is institutionally permitted, though not culturally preferred, it is not generally considered to represent a social problem. Intimates of individuals making this adaptation may pass judgment in terms of prevailing cultural emphases and may "feel sorry for them," they may, in the individual case, feel that "old Jonesy is certainly in a rut." Whether this is described as deviant behavior or no, it clearly represents a departure from the cultural model in which men are obliged to strive actively, preferably through institutionalized procedures, to move onward and upward in the social hierarchy.

We should expect this type of adaptation to be fairly frequent in a society which makes one's social status largely dependent upon one's achievements. For, as has so often been observed, this ceaseless competitive struggle produces acute status anxiety. One device for allaying these anxieties is to lower one's level of aspiration — permanently. Fear produces inaction, or more accurately, routinized action.

The syndrome of the social ritualist is both familiar and instructive. His implicit life-philosophy finds expression in a series of cultural clichés: "I'm not sticking *my* neck out," "I'm playing safe," "I'm satisfied with what I've got," "Don't aim high and you won't be disappointed." The theme threaded through these attitudes is that high ambitions invite frustration and danger whereas lower aspirations produce satisfaction and security. It is a response to a situation which appears threatening and excites distrust. It is the attitude implicit among workers who carefully regulate their output to a constant quota in an industrial organization where they have occasion to fear that they will "be noticed" by managerial personnel and "something will happen" if their output rises and falls. It is the perspective of the frightened employee, the zealously conformist bureaucrat in the teller's cage of the private banking enterprise or in the front office of the public works enterprise. It is, in short, the mode of adaptation of individually seeking *a private* escape from the dangers and frustrations which seem to them inherent in the competition for major cultural goals by abandoning these goals and clinging all the more closely to the safe routines and the institutional norms.

If we should expect *lower-class* Americans to exhibit Adaptation II — "innovation" — to the frustrations enjoined by the prevailing emphasis on large cultural goals and the fact of small social opportunities, we should expect *lower-middle-class* Americans to be heavily represented among those making Adaptation III, "ritualism." For it is in the lower middle class that parents typically exert continuous pressure upon children to abide by the moral mandates of the society, and where the social climb upward is less likely to meet with success than among the upper middle class. The strong disciplining for conformity with mores reduces the likelihood of Adaptation II and promotes the likelihood of Adaptation III. The severe training leads many to carry a heavy burden of anxiety. The socialization patterns of the lower middle class thus promote the very character structure most predisposed toward ritualism, and it is in this stratum, accordingly, that the adaptive pattern III should most often occur. . . .

IV. RETREATISM

. . . In this category fall some of the adaptive activities of psychotics, autists, pariahs, outcasts, vagrants, vagabonds, tramps, chronic drunkards, and drug addicts. They have relinquished culturally prescribed goals and their behavior does not accord with institutional norms. This is not to say that in some cases the source of their mode of adaptation is not the very social structure which they have in effect repudiated nor that their very existence within an area does not constitute a problem for members of the society.

From the standpoint of its sources in the social structure, this mode of adaptation is most likely to occur when *both* the culture goals and the institutional practices have been thoroughly assimilated by the individual and imbued with affect and high value, but accessible institutional avenues are not productive of success. There results a twofold conflict: the interiorized moral obligation for adopting institutional means conflicts with pressures to resort to illicit means (which may attain the goal) and the individual is shut off from means which are both legitimate and effective. The competitive order is maintained but the frustrated and handicapped individual who cannot cope with this order drops out. Defeatism, quietism, and resignation are manifested in escape mechanisms which ultimately lead him to "escape" from the requirements of the society. It is thus an expedient which arises from continued failure to near the goal by legitimate measures and from an inability to use the illegitimate route because of internalized prohibitions, *this process occurring while the supreme value of the success-goal has not yet been renounced*. The conflict is resolved by abandoning *both* precipitating elements, the goals and the means. The escape is complete, the conflict is eliminated and the individual is asocialized. . . .

This fourth mode of adaptation, then, is that of the socially disinherited who if they have none of the rewards held out by society also have few of the frustrations attendant upon continuing to seek these rewards. It is, moreover, a privatized rather than a collective mode of adaptation. Although people exhibiting this deviant behavior may gravitate toward centers where they come into contact with other deviants and although they may come to share in the subculture of these deviant groups, their adaptations are largely private and isolated rather than unified under the aegis of a new cultural code. The type of collective adaptation remains to be considered.

V. REBELLION

This adaptation leads men outside the environing social structure to envisage and seek to bring into being a new, that is to say, a greatly modified social structure. It presupposes alienation from reigning goals and standards. These come to be regarded as purely arbitrary. And the arbitrary is precisely that which can neither exact allegiance nor possess legitimacy, for it might as well be otherwise. In our society, organized movements for rebellion apparently aim to introduce a social structure in which the cultural standards of success would be sharply modified and provision would be made for a closer correspondence between merit, effort, and reward. . . .

When the institutional system is regarded as the barrier to the satisfaction of legitimized goals, the stage is set for rebellion as an adaptive response. To pass into organized political action, allegiance must not only be withdrawn from the prevailing social structure but must be transferred to new groups possessed of a new myth. The dual function of the myth is to locate the source of large-scale frustrations in the social structure and to portray an alternative structure which would not, presumably, give rise to frustration of the deserving. It is a charter for action. . . .

The Strain Toward Anomie

The social structure we have examined produces a strain toward anomie and deviant behavior. The pressure of such a social order is upon outdoing one's competitors. So long as the sentiments supporting this competitive system are distributed throughout the entire range of activities and are not confined to the final result of "success," the choice of means will remain largely within the ambit of institutional control. When, however, the cultural emphasis shifts from the satisfactions deriving from competition itself to almost exclusive concern with the outcome, the resultant stress makes for the breakdown of the regulatory structure. With this attenuation of institutional controls, there occurs an approximation to the situation erroneously held by the utilitarian philosophers to be typical of society, a situation in which calculations of personal advantage and fear of punishment are the only regulating agencies.

This strain toward anomie does not operate evenly throughout the society. Some effort has been made in the present analysis to suggest the strata most vulnerable to the pressures for deviant behavior and to set forth some of the mechanisms operating to produce those pressures. For purposes of simplifying the problem, monetary success was taken as the major cultural goal, although there are, of course, alternative goals in the repository of common values. The realms of intellectual and artistic achievement, for example, provide alternative career patterns which may not entail large pecuniary rewards. To the

extent that the cultural structure attaches prestige to these alternatives and the social structure permits access to them, the system is somewhat stabilized. Potential deviants may still conform in terms of these auxiliary sets of values.

But the central tendencies toward anomie remain, and it is to these that the analytical scheme here set forth calls particular attention.

48

Crime in the Suites

RUSSELL MOKHIBER

Most criminology theories also leave little room for explaining the white-collar criminal. Perhaps because it is, as Mokhiber notes, a sociopolitical artifact, we have chosen not to recognize the impact of white-collar crime despite its widespread cost.

As you read, ask yourself the following questions:

1. The author claims that crime is a label. Why?

2. What items would you add to the crime label? Why? Subtract from the crime label. Why?

GLOSSARY

Suites Refers to corporate offices.

White collar crime Crime committed by corporate executives/professionals in their work.

ON JUNE 6, SEVENTY federal agents raided the Rocky Flats nuclear weapons facility in Colorado. The decision to invade the bomb plant came on the heels of a lengthy investigation described in FBI agent Jon S. Lipsky's 116-page affidavit, which convinced a federal judge to unleash the agents. In his report, Lipsky accused Rockwell International and the U.S. Department of En-

ergy (DOE) of "knowingly and falsely" stating that the plutonium-processing plant complied with this country's environmental laws. In doing so, the contractor and its government client concealed "serious contamination" at the site. Lipsky charged that Rockwell and DOE secretly dumped hazardous waste into public drinking water and surreptitiously operated an incinerator they said had been shut down.

While scandal at nuclear weapons plants seems almost a regular news feature of late, the capacity demonstrated by the Justice Department in Colorado to deploy an environmental police force—replete with FBI agents, investigators, prosecutors, wiretaps and aerial surveillance—is in fact an unusual thing. The government rarely flexes its legal muscle to prosecute major environmental crimes or, for that matter, corporate crimes generally. For every Rocky Flats, there are dozens of corporate environmental crimes that go undetected, unprosecuted, and unpunished.

"Crime is a sociopolitical artifact, not a natural phenomenon," writes legal scholar Herbert

From Greenpeace, *Vol. 14, No. 5, September/October 1989, pp. 14–17. Reprinted by permission.*

Corporate Crime-Busting: Some Legal and Social Remedies

■ Congress should pass an executive responsibility statute making it a criminal offense for a corporate supervisor willfully or recklessly to fail to oversee an assigned activity that results in criminal conduct.

■ Corporate managers should be required to report to federal authorities a product or process that may cause death or serious injury. This would ensure that R&D departments keep worker health and safety in mind.

■ Congress should require publicly held corporations to report their litigation records—indictments, convictions, sentences, fines, and product-liability lawsuits—to the FBI. This corporate crime database could then be used by communities and prosecutors to inform their fights against criminally inclined companies.

■ At the local level, corporate crimewatch committees should be formed to keep an eye on the activities of neighboring corporations and to keep police and prosecutors on their toes. Victims of corporate crime, such as those who have been injured by the Dalkon Shield, Agent Orange, and asbestos, have formed organizations to lobby for just compensation, strong laws and, where applicable, effective prosecution and strict sentences.

■ Creative penalties should be devised, such as court-ordered adverse publicity. As a condition of probation, for example, a judge could order a company to take out network television advertisements telling viewers about its long criminal record.

■ More than anything else, corporate criminals should do time. They should be jailed alongside the mugger and drug dealer, not in the posh "Club Feds" usually reserved for white-collar crooks.

Packer in *The Limits of the Criminal Law.* "We can have as much or as little crime as we please, depending on what we choose to count as criminal." In this country, we have chosen to have very little corporate crime. Most corporate wrongs against humans and the environment are not considered criminal in the traditional sense—that is, activity that is prohibited by the state and prosecuted to conviction. While corporations like Rockwell International can be criminally prosecuted for serious violations of environmental laws, they usually face less demanding and less visible civil procedures.

On the face of it, this leniency is grossly out of proportion to the effects of the corporate crime wave. Every year, roughly 28,000 deaths and 130,000 serious injuries are caused by dangerous products. At least 100,000 workers die from exposure to deadly chemicals and other safety hazards. Workplace carcinogens are estimated to cause between 23 and 38 percent of all cancer deaths. More than 45,000 Americans die in automobile crashes every year. Many of those deaths either are caused by defects or are easily preventable by a simple redesign.

The financial cost to society is staggering. The National Association of Attorneys General reports that fraud costs the nation's businesses and individuals upwards of $100 billion each year. The Senate Judiciary Committee has estimated that faulty goods, monopolistic practices and other such violations annually cost consumers $174 to $231 billion. Added to this is the $10 to $20 billion a year the Justice Department says taxpayers lose when corporations violate federal regulations. As a rule of thumb, the Bureau of National Affairs estimates that the dollar cost of corporate crime in the United States is more than 10 times greater than the combined total from larcenies, robberies, burglaries, and auto thefts committed by individuals.

The full extent of the corporate crime wave is hidden. Although the federal government tracks street crime month by month, city by city through the FBI's Uniform Crime Reports, it does not track corporate crime. So the government can tell the public whether burglary is up or down in Los Angeles for any given month, but it cannot say the same about insider trading, midnight dumping, consumer defrauding, or illegal polluting.

Still, we do know that corporate crime is pervasive. A 1979 Justice Department study, "Illegal Corporate Behavior," found that 582 corporations surveyed racked up a total of 1554 law violations in just two years. A 1980 *Fortune* magazine survey revealed that 11 percent of 1,043 large companies had been convicted on criminal charges or consent decrees for five offenses: bribery, criminal fraud, illegal political contributions, tax evasion and criminal antitrust. A *U.S. News & World Report* study of the 500 largest corporations found that "115 have been convicted in the last decade of at least one major crime or have paid civil penalties for serious misbehavior" in excess of $50,000. And in 1985, George Washington University Professor Amitai Etzioni found that roughly two-thirds of America's 500 largest companies were involved to some extent in illegal behavior over the preceding 10 years.

By the mid-1930s, evidence was mounting that exposure to asbestos was a threat to human health. In 1982, the Manville Corporation (previously Johns Manville), the nation's largest manufacturer of asbestos, filed for bankruptcy to shelter its assets from 16,500 personal injury lawsuits. In the intervening 50 years, the corporation actively suppressed asbestos studies and hid information from its employees on the dangers of working with asbestos. They even cut workers off from their own health records. "As long as [the employee] is not disabled," rationalized the company's medical director in 1963, "it is felt that he should not be told of his condition so that he can live and work in peace, and the company can benefit from his many years of experience."

Over the next 30 years, 240,000 people — 8000 per year, almost one every hour on average — will die from asbestos-related cancer. The company will pay some $2.5 billion to its victims, a hefty civil penalty. But no asbestos executive has ever been prosecuted for reckless homicide.

Likewise, it was not a "crime," in the traditional sense of the word, for Union Carbide's Bhopal, India, subsidiary to operate a pesticide manufacturing plant so incompetently that in 1984, clouds of deadly methyl isocyanate gas escaped, killing 2000 to 5000 persons and injuring 200,000.

And it is not a "crime" for the tobacco companies knowingly to market a highly addictive drug that kills more than 365,000 Americans a year, 1000 every day. This toll is higher than the number of Americans killed annually by AIDS, heroin, crack, alcohol, car accidents, fire, and murder combined.

And it is not a "crime" to market known cancer-causing pesticides such as Alar. Nor is it a "crime" to dump toxins into the air and water. General Motors (GM), among others, has been campaigning actively against public health for decades. In 1949, the company was convicted of conspiracy to destroy the nation's mass transit systems by buying up and then dismantling electrical transit systems in urban areas around the country.

The environmental consequences of this crime are still felt today. Los Angeles, which in the 1930s boasted an efficient system of electrified public transit that served 56 cities, saw the system destroyed and replaced with diesel buses and a freeway network for GM's cars. The city now has one of the worst air pollution problems in the country, and the Bush administration has proposed exempting it from some provisions of the Clean Air Act.

"What is good for General Motors is not necessarily good for the country," former San Francisco Mayor Joe Alioto told senators in a hearing about the destruction of the electric transit system in the Bay Area. "In the field of transportation, what has been good for General Motors has, in fact, been very, very bad for the country."

With the enormous resources available to them, companies like General Motors can ensure that the laws protecting us from them remain weak. During the last decade, for example, General Motors has successfully opposed amendments that would strengthen federal clean air and federal fuel-efficiency standards. GM has spent more than $1.8 billion lobbying Congress against clean air amendments since 1981, the year the Clean Air Act came due for reauthorization. In addition, GM's political action committee made more than $750,000 in campaign contributions, much of it to legislators who sit on committees with jurisdiction over clean air issues.

Lack of accountability is deeply embedded in the concept of the corporation. Shareholders' liability is limited to the amount of money they invest. Managers' liability is limited to what they choose to know about the operations of the company. And the corporation's liability is limited by Congress (the Price-Anderson Act, for example, caps the liability of nuclear power companies in the aftermath of a nuclear disaster), by insurance and by laws allowing corporations to duck liability by altering their corporate structure (the Manville bankruptcy dodge, for example).

In addition, since the turn of the century, most laws governing corporate behavior give regulators the option of avoiding criminal charges and proceeding with less burdensome and less noticeable civil enforcement. In this way, corporations avoid either admitting or denying that they violated the law and are let off with slap-on-the-wrist fines and consent decrees. For environmental, labor, securities, energy, and food and drug violations, the civil injunction is today the primary method of enforcing the law against big business.

Fines, dismissed by criminologists as "license fees to violate the law," are the customary civil penalty for corporate wrong-doing. "One jail sentence is worth 100 consent decrees," said one federal judge. "Fines are meaningless because the defendant in the end is always reimbursed by the proceeds of his wrongdoing or by his company."

Under civil enforcement, the executives of criminal corporations are freed from the stigma of prosecution and possible jail sentences. "The violations of these laws are crimes," wrote Edwin Sutherland in his 1949 classic, *White-Collar Crime,* "but they are treated as though they are not crimes, with the effect and probably the intention of eliminating the stigma of crime."

Sanctions for egregious corporate crimes rarely match the gravity of the offense, nor do they compare well with the punishment meted out for common street crimes. Not one corporate executive went to jail for marketing thalidomide,

a drug that caused severe birth defects in 8000 babies during the 1960s, but Wallace Richard Stewart of Kentucky was sentenced in July 1983 to 10 years in prison for stealing a pizza. Not one Hooker Chemical manager went to jail, nor was Hooker charged with a criminal offense after the company exposed its workers and Love Canal neighbors to toxics, but under a Texas habitual offenders statute, William Rummel was sentenced to life in prison for stealing a total of $229.11 over a period of nine years. And General Motors was fined a mere $5000 for its mass transit conspiracy, which set back the country's environmental standards for decades.

"No amount of money paid out of corporate assets can address the wrongful acts of the individuals responsible within the organization," says Kenneth Oden, a District Attorney in Austin who has prosecuted a number of occupational homicide cases. "Sometimes the boss needs to be placed in handcuffs and taken to jail." While incarceration of street criminals may have a limited deterrent effect, jail time for corporate executives has a markedly different impact. "I would starve before I would do it again," said one General Electric official, convicted and jailed in a price-fixing scandal.

In February 1983, a worker at Film Recovery Systems' silver extraction plant became nauseated while working in a room with open vats of hydrogen cyanide. He staggered outside the plant, collapsed, and died. The medical examiner reported that he died of "acute cyanide toxicity." A month later the state attorney for Cook County, Chicago, charged three executives of Film Recovery Systems with homicide.

Prosecutors argued that plant employees were forced to work in the equivalent of a huge gas chamber, that the company hired mostly illegal aliens who spoke little English, that the company had scraped skull and crossbones warnings off the side of the cyanide drums, and that ventilation was so inadequate that a thick yellow haze hung inside the plant.

After a two-month trial, each of the three executives was found guilty of murder and reckless conduct, fined $10,000 and sentenced to 25 years in prison for murder and 364 days for reckless conduct. Two operating corporations were found guilty of reckless conduct and involuntary manslaughter and fined $11,000 each.

The Film Recovery Systems case represents the first time a corporate executive has been found guilty of murder in an occupational death case, and public sentiment seems to be calling for more such legal actions. Earlier this year in Torrance, California, the city attorney, citing the fear of a "disaster of Bhopal-like proportions," filed an unusual lawsuit against Mobil Oil. He sought to have Mobil's giant Torrance refinery declared a public nuisance, thus giving the city the authority to regulate it. The lawsuit cites the plant's appalling safety record—127 accidents at the refinery since December 1979, including the fiery deaths of three persons, among them a passing motorist, in an explosion and fire at the tank farm.

The district attorney (DA) for Los Angeles County requires prosecutors to investigate the circumstances of every occupational death or serious injury on the job. In the past four years, the DA has investigated more than 100 such cases and has brought criminal charges in more than two dozen cases. And in Austin, Milwaukee, and New York City, activist prosecutors are hitting employers with homicide charges for death on the job.

In early 1989, the Commonwealth of Massachusetts announced the creation of a statewide Environmental Crimes Task Force that will use prosecutors, scientists, investigators and police officers to target high-priority threats to public health and natural resources. The 34-member strike force will specialize in major cases involving threats to drinking water supplies, harm to wetlands, illegal dumping and toxic discharges into sewage systems.

"This should send a clear message to everyone across the state: If you pollute, we're going to catch you and you'll pay the price," said Massachusetts Environmental Affairs Secretary John DeVillars. "Poisoning someone's water supply or illegally dumping material isn't a victimless crime. It's a costly crime that has a major impact on individuals whose health may suffer. It damages our quality of life."

At the federal level, the Justice Department's Environmental Crimes Section, which was created in 1983, has recorded 520 indictments and more than 400 convictions, bringing in $22 million in fines and more than 240 years of actual jail time. Earlier this year, Ashland Oil was found guilty of violating federal environmental laws in connection with the collapse of an Ashland storage tank that spilled more than 500,000 gallons of oil into the Monongahela River outside of Pittsburgh on January 2, 1988.

The developments described above point to a new willingness on the part of the public and the judicial system to see corporate crime punished fairly. Until now, the law has taught that if you are strong, rich, and corporate, you can inflict the most egregious wrongs on society and continue business as usual. There is no reason why this cannot change. In a just society, the criminal law should also teach that those who poison the air, water and land, injure and kill others, or inflict cancer and birth defects are criminals and should be justly punished.

49

Crime Control in America: Nothing Succeeds Like Failure

JEFFREY REIMAN

The author of this reading implies that crime control is a purposeful failure since it is known what would be more successful. The purposefulness is tied in with Durkheim's theory of the necessity of crime. What do you think of the means for dealing with crime? Why do you think they are not instigated?

As you read, ask yourself the following questions:

1. The author claims that our assaults on the crime problem are a failure. Why?

2. What changes would you make in the criminal justice system? Defend your changes.

GLOSSARY

FBI Crime Index Federal Bureau of Investigation report on criminal offenses "against the person" and "against property."

UCR Uniform Crime Reports, an FBI measure of crime.

Designed to Fail

SOMETHING IN THE AMERICAN grain keeps us from admitting defeat both to ourselves and others. Perhaps it is the heady air of the long-closed frontier trapped in our lungs; whatever it is, it keeps us from confessing to anything more serious than the temporary elusiveness of victory. Americans never confess to having lost a war, although we do admit there are a few we did not win. . . .

Needless to say, we've been putting more police on the streets and more criminals behind bars for the past 20 years, the very period during which crime and drug use have expanded dramatically, according to our newest leaders. But we do not need a presidential announcement to learn that our assaults on the crime problem are a failure. Everyone knows that for all our efforts, intelligence, and money, serious crime remains an enormous social problem. Although rates for some crimes do occasionally decrease from one year to the next, our crime rates remain very high compared with other advanced nations, and they are usually going *up*.

In 1960, the average citizen had less than a 1-in-50 chance of being a victim of one of the crimes on the FBI Index (murder, forcible rape, robbery, aggravated assault, burglary, larceny, or auto theft). In 1970, that chance grew to 1-in-25. In 1986, the FBI reported nearly 5,500 Index crimes per 100,000 citizens, a further increase in the likelihood of victimization to a 1-in-18 chance. And by 1991, this had reached 5,898 per 100,000 citizens—a better than 1-in-17 chance. The FBI reported slight declines in 1992 and 1993, with the rate for 1993 at 5,483 per 100,000, roughly where it was in 1986. Most of the decline was accounted for by a drop in property crimes. Even with these declines, the FBI says in its most recent report: "Every American now has a realistic chance of murder victimization in view of the random nature the crime has assumed."[5] The FBI, however, counts only the crimes that are *reported*. Using "victimization studies" (asking randomly selected citizens whether or not they had been a victim of a crime reported to the police), experts estimate that *unreported* crime is at least double

Reiman, Jeffrey. And the Poor Get Prison: Economic Bias in American Criminal Justice. *Allyn & Bacon, 1976. Reprinted by permission of the publisher.*

the number of reported crimes, and in some cases as much as six times as great. For example, the FBI reported 1,029,580 crimes of violence in 1977, 1,488,140 in 1986, and 1,932,270 in 1992. Victimization surveys, however, reported 5,902,000 crimes of violence in 1977, 5,515,000 in 1986, and 6,621,140 in 1992.[6] . . .

Whichever way you take it—a smaller amount of crime rapidly increasing or a larger amount remaining stable or even occasionally dipping slightly—criminal justice policy is failing to make our lives substantially safer.

How are we to comprehend this failure? It appears that our government is failing to fulfill the most fundamental task of governance: keeping our streets and homes safe, assuring us of what the Founding Fathers called "domestic tranquillity," providing us with the minimal requirement of civilized society. It appears that our new centurions with all their modern equipment and know-how are no more able than the old Roman centurions to hold the line against the forces of barbarism and chaos.

One way to understand this failure is to look at the *excuses* that are offered for it. This we will do—but mainly to show that they do not hold up!

One commonly heard excuse is that we can't reduce crime because our laws and our courts are too lenient. *Translation:* We are failing to reduce crime because we don't have the heart to do what has to be done.

Another excuse points to some feature of modern life, such as urbanization or population growth . . . *Translation:* We are failing to reduce crime because it is impossible to reduce crime.

Some try to excuse our failure by claiming that we simply do not know how to reduce crime. *Translation:* Even though we are doing our best, we are failing to reduce crime because our knowledge of the causes of crime is still too primitive to make our best good enough.

These excuses simply do not pass muster. . . .

Failure is, after all, in the eye of the beholder. The last runner across the finish line has failed in the race only if he or she wanted to win. If the runner wanted to lose, the "failure" is, in fact, a success. Here, I think lies the key to understanding our criminal justice system.

If we look at the system as "wanting" to reduce crime, it is an abysmal failure—and we cannot understand it. If we look at it as *not* wanting to reduce crime, it's a howling success—and all we need to understand is why the goal of the criminal justice system is to fail to reduce crime. If we can understand this, then the system's "failure," as well as its obstinate refusal to implement the policies that could remedy that "failure," becomes perfectly understandable.

In other words, I propose that we can make more sense out of criminal justice policy by assuming that its goal is to maintain crime than by assuming that its goal is to reduce crime! . . .

KNOWN SOURCES OF CRIME

Those youngsters who figure so prominently in arrest statistics are not drawn equally from all economic strata. Although there is much reported and even more unreported crime among middle-class youngsters, the street crime attributed to this age group that makes our city streets a perpetual war zone is largely the work of poor ghetto youth. This is the group at the lowest end of the economic spectrum. This is a group among whom unemployment approaches 50 percent, with underemployment (the rate of persons either jobless or with part-time, low-wage jobs) still higher. This is a group with no realistic chance (for any but a rare individual) to enter college or amass sufficient capital (legally) to start a business or to get into the union-protected, high-wage, skilled job markets. We know that poverty is a *source* of crime, even if we do not know how it *causes* crime—and yet we do virtually nothing to improve the life chances of the vast majority of the inner-city poor. They are as poor as ever and are facing cuts in welfare and other services.

The gap between rich and poor worsened during the eighties. Says *The Economist*, "For all the talk of the fragmentation of America, there is only one division that is dangerously getting worse, and that is the gap between rich and poor."[44] In 1980, the poorest fifth of the nation's families received 5.2 percent of the aggregate income, and the richest fifth received 41.5 percent. By 1992, the share of the poorest fifth had declined to 4.4 percent, while that of the richest fifth had risen to 44.6 percent. During this same period, the share of the top 5 percent rose from 15.3 to 17.6 percent.[45] The Census Bureau reports that the number of poor Americans rose for the third year in a row in 1992, to 36.9 million.[46]

That poverty is a source of crime is not refuted by the large and growing amount of white-collar crime . . . In fact, poverty contributes to crime by creating need, while — at the other end of the spectrum — wealth can contribute to crime by unleashing greed. Some criminologists have argued that economic inequality itself worsens crimes of the poor and of the well off by increasing the opportunities for the well off and increasing the humiliation of the poor.[47] And inequality has worsened in recent years. . . .

Almost a decade later, an article in *Business Week,* looking back at the 1980s, confirms the charge in retrospect: "At the uppermost end of the income scale, tax cuts made aftertax income surge even higher than pretax income. And at the low end of the distribution scale, cuts in income transfers hurt the poor." The article notes also "the extraordinarily high level of child poverty in America today. One in five children under the age of 15 lives in poverty, and a staggering 50% of all black children under the age of six live in poverty."[50]

Moreover, as unemployment has gone up and down over the past decades, unemployment at the bottom of society remains strikingly worse than the national average. For example, over the past 25 years black unemployment has remained slightly more than twice the rate of white unemployment. In 1967, when 3.4 percent of white workers were unemployed, 7.4 percent of black workers were jobless. By 1993, 6 percent of white workers were unemployed and 12.9 percent of blacks were. Among those in the crime-prone ages of 16 to 19, 16.2 percent of white youngsters and 38.9 percent (more than one of every three) black youngsters were jobless.[51]

Writes Todd Clear, professor of criminal justice at Rutgers University, "Let's start investing in things that really reduce crime: good schools, jobs and a future for young parents and their children."[52] Why don't we?

There is more. We know that prison produces more criminals than it cures. We know that more than 70 percent of the inmates in the nation's prisons or jails are not there for the first time. We know that prison inmates are denied autonomy and privacy and subjected to indignities, mortifications, and acts of violence as regular features of their confinement — all of which is heightened by overcrowding. State and federal prisons were at 123.3 percent of design capacity in 1990.[53] The predictable result, as delineated by Robert Johnson and Hans Toch in *The Pains of Imprisonment,* "is that the prison's survivors become tougher, more pugnacious, and less able to feel for themselves and others, while its nonsurvivors become weaker, more susceptible, and less able to control their lives."[54] Prisoners are thus bereft of both training and capacity to handle daily problems in competent and socially constructive ways, inside or outside of prison. Once on the outside, burdened with the stigma of a prison record and rarely trained in a marketable skill, they find few opportunities for noncriminal employment open to them.

Should we then really pretend that we do not *know* why they turn to crime? Can we honestly act as if we do not know that our prison system (combined with our failure to ensure a meaningful postrelease noncriminal alternative for the ex-con) is a *source* of crime? Recidivism does not happen because ex-cons miss their alma mater. In fact, if prisons are built to deter people from crime, one would expect that ex-prisoners would be the most deterred because the deprivations of prison are more real to them than to the rest of us. Recidivism is thus a doubly poignant testimony to the job that prison does in preparing its graduates for crime — and yet we do little to change the nature of prisons or to provide real services to ex-convicts. . . .

[T]he war on crime is a failure and an avoidable one: The American criminal justice system —

by which I mean the entire process from law-making to law enforcing—has done little or nothing to reduce the enormous amount of crime that characterizes our society and threatens our citizens. Over the last several decades, crime has generally risen, although in recent years it has occasionally declined. No doubt demographic changes, most significantly the growth followed by the decrease in the number of youngsters in the crime prone years, have played a role in this. This in itself suggests that criminal justice policy and practice cannot be credited with the recent occasional declines. At the same time, however, neither can it be thought on this basis that public policy cannot reduce the crime we have. . . . [T]here are a number of things we have good reason to believe would succeed in reducing crime—effective gun control, decriminalization of illicit drugs, and, of course, amelioration of poverty—that we refuse to do.

NOTES

5. *UCR-1992,* p. 58; *UCR-1993,* p. 5; the quote is at p. 287.

6. *UCR-1986,* p. 41; *UCR-1991,* pp. 5, 10; *UCR-1992,* p. 58; *Sourcebook-1987,* p. 240; *Sourcebook-1993,* p. 246, Table no. 3.1.

44. "UnAmerican Thoughts," *Economist* (October 26, 1991), p. 23.

45. *StatAbst-1994,* p. 470, Table no. 716.

46. Guy Gugliotta, "Number of Poor Americans Rises for 3rd Year," *Washington Post,* October 5, 1993, p. A6.

47. See, for example, John Braithwaite, "Poverty, Power, and White-Collar Crime," in Kip Schlegel and David Weisburd, eds., *White Collar Crime Reconsidered* (Boston: Northeastern University Press, 1992), pp. 78–107.

50. Karen Pennar, "The Rich Are Richer—And America May Be the Poorer," *Business Week* (November 18, 1991), pp. 85, 88.

51. *StatAbst-1992,* p. 399, Table no. 635; *StatAbst-1994,* p. 416, Table no. 646. See also, *"Racial Gulf:* Blacks' Hopes, Raised by '68 Kerner Report, Are Mainly Unfulfilled," *Wall Street Journal,* February 26, 1988, pp. 1, 9; *StatAbst-1992,* p. 80, Table no. 109; "Today's Native Sons," *Time,* December 1, 1986, pp. 26–29; and *StatAbst-1988,* p. 75, Table no. 113.

52. Todd R. Clear, "'Tougher' Is Dumber," *New York Times,* December 4, 1993, p. 21.

53. BJS, *Correctional Populations in the United States, 1990,* p. 46. Nearly 22 percent of state facilities are under a state or federal consent decree to limit population (ibid., p. 47).

54. Robert Johnson and Hans Toch, "Introduction," in Johnson and Toch, eds., *The Pains of Imprisonment* (Beverly Hills, Calif.: Sage, 1982), pp. 19–20.

50

On Being Sane in Insane Places

D. L. ROSENHAN

Rosenhan reveals a push toward deviance not noted by most theorists—the labeling of behavior as not normal. The question being raised in this reading is "What is normality as opposed to abnormality?"

As you read, ask yourself the following questions:

1. *What were the main factors affecting the treatment of the patients?*
2. *What other institutions are affected by labeling? Why do you say this?*

GLOSSARY

Pseudopatient A pretend patient.

Type 1 error A false-negative error; for example, diagnosing a sick person as healthy.

Type 2 error A false-positive error; for example, diagnosing a healthy person as sick.

Depersonalization Removing a sense of the individual from treatment.

Prima facie evidence Presumption of fact.

Veridical Truthful.

IF SANITY AND INSANITY exist, how shall we know them?

The question is neither capricious nor itself insane. However much we may be personally convinced that we can tell the normal from the abnormal, the evidence is simply not compelling. It is commonplace, for example, to read about murder trials wherein eminent psychiatrists for the defense are contradicted by equally eminent psychiatrists for the prosecution on the matter of the defendant's sanity. More generally, there are a great deal of conflicting data on the reliability, utility, and meaning of such terms as "sanity," "insanity," "mental illness," and "schizophre-

nia."[1] Finally, as early as 1934, Benedict suggested that normality and abnormality are not universal.[2] What is viewed as normal in one culture may be seen as quite aberrant in another. Thus, notions of normality and abnormality may not be quite as accurate as people believe they are.

To raise questions regarding normality and abnormality is in no way to question the fact that some behaviors are deviant or odd. Murder is deviant. So, too, are hallucinations. Nor does raising such questions deny the existence of the personal anguish that is often associated with "mental illness." Anxiety and depression exist. Psychological suffering exists. But normality and abnormality, sanity and insanity, and the diagnoses that flow from them may be less substantive than many believe them to be.

At its heart, the question of whether the sane can be distinguished from the insane (and whether degrees of insanity can be distinguished from each other) is a simple matter: do the salient characteristics that lead to diagnoses reside in the patients themselves or in the environments and contexts in which observers find them? From Bleuler, through Kretchmer, through the formulators of the recently revised *Diagnostic and Statistical Manual* of the American Psychiatric Association, the belief has been strong that patients present symptoms, that those symptoms can be categorized, and, implicitly, that the sane are distinguishable from the insane. More recently, however, this belief has been questioned. Based in part on theoretical and anthropological considerations, but also on philosophical, legal, and therapeutic ones, the view has grown that psy-

Reprinted by permission from Science, *Vol. 179, January 19, 1973, pp. 250–258. Copyright © 1973 by the American Association for the Advancement of Science.*

chological categorization of mental illness is use-less at best and downright harmful, misleading, and pejorative at worst. Psychiatric diagnoses, in this view, are in the minds of the observers and are not valid summaries of characteristics displayed by the observed.[3, 4, 5]

Gains can be made in deciding which of these is more nearly accurate by getting normal people (that is, people who do not have, and have never suffered, symptoms of serious psychiatric disorders) admitted to psychiatric hospitals and then determining whether they were discovered to be sane and, if so, how. If the sanity of such pseudopatients were always detected, there would be prima facie evidence that a sane individual can be distinguished from the insane context in which he is found. Normality (and presumably abnormality) is distinct enough that it can be recognized wherever it occurs, for it is carried within the person. If on the other hand, the sanity of the pseudopatients were never discovered, serious difficulties would arise for those who support traditional modes of psychiatric diagnosis. Given that the hospital staff was not incompetent, that the pseudopatient had been behaving as sanely as he had been outside of the hospital, and that it had never been previously suggested that he belonged in a psychiatric hospital, such an unlikely outcome would support the view that psychiatric diagnosis betrays little about the patient but much about the environment in which an observer finds him.

This reading describes such an experiment. Eight sane people gained secret admission to 12 different hospitals.[6] Their diagnostic experiences constitute the data of the first part of this article; the remainder is devoted to a description of their experiences in psychiatric institutions. Too few psychiatrists and psychologists, even those who have worked in such hospitals, know what the experience is like. They rarely talk about it with former patients, perhaps because they distrust information coming from the previously insane. Those who have worked in psychiatric hospitals are likely to have adapted so thoroughly to the settings that they are insensitive to the impact of that experience. And while there have been occasional reports of researchers who submitted themselves to psychiatric hospitalization,[7] these researchers have commonly remained in the hospitals for short periods of time, often with the knowledge of the hospital staff. It is difficult to know the extent to which they were treated like patients or like research colleagues. Nevertheless, their reports about the inside of the psychiatric hospital have been valuable. This reading extends those efforts.

Pseudopatients and Their Settings

The eight pseudopatients were a varied group. One was a psychology graduate student in his twenties. The remaining seven were older and "established." Among them were three psychologists, a pediatrician, a psychiatrist, a painter, and a housewife. Three pseudopatients were women, five were men. All of them employed pseudonyms, lest their alleged diagnoses embarrass them later. Those who were in mental health professions alleged another occupation in order to avoid the special attentions that might be accorded by staff, as a matter of courtesy or caution, to ailing colleagues.[8] With the exception of myself (I was the first pseudopatient and my presence was known to the hospital administrator and chief psychologist and, so far as I can tell, to them alone), the presence of pseudopatients and the nature of the research program was not known to the hospital staff.[9]

The settings were similarly varied. In order to generalize the findings, admission into a variety of hospitals was sought. The 12 hospitals in the sample were located in five different states on the East and West coasts. Some were old and shabby, some were quite new. Some were research-oriented, others not. Some had good staff-patient ratios, others were quite understaffed. Only one was a strictly private hospital. All of the others were supported by state or federal funds or, in one instance, by university funds.

After calling the hospital for an appointment, the pseudopatient arrived at the admissions office complaining that he had been hearing voices. Asked what the voices said, he replied that they were often unclear, but as far as he could tell they said "empty," "hollow," and "thud." The voices were unfamiliar and were of the same sex as the pseudopatient. The choice of these symptoms was occasioned by their apparent similarity to existential symptoms. Such symptoms are alleged to arise from painful concerns about the perceived meaninglessness of one's life. It is as if the hallucinating person were saying, "My life is empty and hollow." The choice of these symptoms was also determined by the *absence* of a single report of existential psychoses in the literature.

Beyond alleging the symptoms and falsifying name, vocation, and employment, no further alterations of person, history, or circumstances were made. The significant events of the pseudopatient's life history were presented as they had actually occurred. Relationships with parents and siblings, with spouse and children, with people at work and in school, consistent with the aforementioned exceptions, were described as they were or had been. Frustrations and upsets were described along with joys and satisfactions. These facts are important to remember. If anything, they strongly biased the subsequent results in favor of detecting sanity, since none of their histories or current behaviors were seriously pathological in any way.

Immediately upon admission to the psychiatric ward, the pseudopatient ceased simulating *any* symptoms of abnormality. In some cases, there was a brief period of mild nervousness and anxiety, since none of the pseudopatients really believed that they would be admitted so easily. Indeed, their shared fear was that they would be immediately exposed as frauds and greatly embarrassed. Moreover, many of them had never visited a psychiatric ward; even those who had, nevertheless, had some genuine fears about what might happen to them. Their nervousness, then, was quite appropriate to the novelty of the hospital setting, and it abated rapidly.

Apart from that short-lived nervousness, the pseudopatient behaved on the ward as he "normally" behaved. The pseudopatient spoke to patients and staff as he might ordinarily. Because there is uncommonly little to do on a psychiatric ward, he attempted to engage others in conversation. When asked by staff how he was feeling, he indicated that he was fine, that he no longer experienced symptoms. He responded to instructions from attendants, to calls for medication (which was not swallowed), and to dining-hall instructions. Beyond such activities as were available to him on the admissions ward, he spent his time writing down his observations about the ward, its patients, and the staff. Initially these notes were written "secretly," but as it soon became clear that no one much cared, they were subsequently written on standard tablets of paper in such public places as the dayroom. No secret was made of these activities.

The pseudopatient, very much as a true psychiatric patient, entered a hospital with no foreknowledge of when he would be discharged. Each was told that he would have to get out by his own devices, essentially by convincing the staff that he was sane. The psychological stresses associated with hospitalization were considerable, and all but one of the pseudopatients desired to be discharged immediately after being admitted. They were, therefore, motivated not only to behave sanely, but to be paragons of cooperation. That their behavior was in no way disruptive is confirmed by nursing reports, which have been obtained on most of the patients. These reports uniformly indicate that the patients were "friendly," "cooperative," and "exhibited no abnormal indications."

The Normal Are Not Detectably Sane

Despite their public "show" of sanity, the pseudopatients were never detected. Admitted, except in one case, with a diagnosis of schizophrenia,[10] each was discharged with a diagnosis of schizophrenia "in remission." The label "in remission" should in no way be dismissed as a formality, for at no time during any hospitalization had any question been raised about any pseudopatient's simulation. Nor are there any indications in the hospital records that the pseudopatient's status was suspect. Rather, the evidence is strong that, once labeled schizophrenic, the pseudopatient was stuck with that label. If the pseudopatient was to be discharged, he must naturally be "in remission"; but he was not sane, nor, in the institution's view, had he ever been sane.

The uniform failure to recognize sanity cannot be attributed to the quality of the hospitals, for, although there were considerable variations among them, several are considered excellent. Nor can it be alleged that there was simply not enough time to observe the pseudopatients. Length of hospitalization ranged from 7 to 52 days, with an average of 19 days. The pseudopatients were not, in fact, carefully observed, but this failure clearly speaks more to traditions within psychiatric hospitals than to lack of opportunity.

Finally, it cannot be said that the failure to recognize the pseudopatients' sanity was due to the fact that they were not behaving sanely. While there was clearly some tension present in all of them, their daily visitors could detect no serious behavioral consequences—nor, indeed, could other patients. It was quite common for the patients to "detect" the pseudopatients' sanity. During the first three hospitalizations, when accurate counts were kept, 35 of a total of 118 patients on the admissions ward voiced their suspicions, some vigorously. "You're not crazy. You're a journalist, or a professor [referring to the continual notetaking]. You're checking up on the hospital." While most of the patients were reassured by the pseudopatient's insistence that he had been sick before he came in but was fine now, some continued to believe that the pseudopatient was sane throughout his hospitalization.[11] The fact that the patients often recognized normality when staff did not raises important questions.

Failure to detect sanity during the course of hospitalization may be due to the fact that physicians operate with a strong bias toward what statisticians call the type 2 error. This is to say that physicians are more inclined to call a healthy per-

son sick (a false positive, type 2) than a sick person healthy (a false negative, type 1). The reasons for this are not hard to find: it is clearly more dangerous to misdiagnose illness than health. Better to err on the side of caution, to suspect illness even among the healthy.

But what holds for medicine does not hold equally well for psychiatry. Medical illnesses, while unfortunate, are not commonly pejorative. Psychiatric diagnoses, on the contrary, carry with them personal, legal, and social stigmas.[12] It was therefore important to see whether the tendency toward diagnosing the sane insane could be reversed. The following experiment was arranged at a research and teaching hospital whose staff had heard these findings but doubted that such an error could occur in their hospital. The staff was informed that at some time during the following 3 months, one or more pseudopatients would attempt to be admitted into the psychiatric hospital. Each staff member was asked to rate each patient who presented himself at admission or on the ward according to the likelihood that the patient was a pseudopatient. A 10-point scale was used, with a 1 and 2 reflecting high confidence that the patient was a pseudopatient.

Judgments were obtained on 193 patients who were admitted for psychiatric treatment. All staff who had had sustained contact with or primary responsibility for the patient—attendants, nurses, psychiatrists, physicians, and psychologists—were asked to make judgments. Forty-one patients were alleged, with high confidence, to be pseudopatients by at least one member of the staff. Twenty-three were considered suspect by at least one psychiatrist. Nineteen were suspected by one psychiatrist *and* one other staff member. Actually, no genuine pseudopatient (at least from my group) presented himself during this period.

The experiment is instructive. It indicates that the tendency to designate sane people as insane can be reversed when the stakes (in this case, prestige and diagnostic acumen) are high. But what can be said of the 19 people who were suspected of being "sane" by one psychiatrist and another staff member? Were these people truly "sane," or was it rather the case that in the course of avoiding the type 2 error the staff tended to make more errors of the first sort—calling the crazy "sane"? There is no way of knowing. But one thing is certain: any diagnostic process that lends itself so readily to massive errors of this sort cannot be a very reliable one.

The Stickiness of Psychodiagnostic Labels

Beyond the tendency to call the healthy sick—a tendency that accounts better for diagnostic behavior on admission than it does for such behavior after a lengthy period of exposure—the data speak to the massive role of labeling in psychiatric assessment. Having once been labeled schizophrenic, there is nothing the pseudopatient can do to overcome the tag. The tag profoundly colors others' perceptions of him and his behavior.

From one viewpoint, these data are hardly surprising, for it has long been known that elements are given meaning by the context in which they occur. Gestalt psychology made this point vigorously, and Asch[13] demonstrated that there are "central" personality traits (such as "warm" versus "cold") which are so powerful that they markedly color the meaning of other information in forming an impression of a given personality.[14] "Insane," "schizophrenic," "manic-depressive," and "crazy" are probably among the most powerful of such central traits. Once a person is designated abnormal, all of his other behaviors and characteristics are colored by that label. Indeed, that label is so powerful that many of the pseudopatients' normal behaviors were overlooked entirely or profoundly misinterpreted. Some examples may clarify this issue.

Earlier I indicated that there were no changes in the pseudopatient's personal history and current status beyond those of name, employment, and, where necessary, vocation. Otherwise, a veridical description of personal history and circumstances was offered. Those circumstances were not psychotic. How were they made consonant with the diagnosis of psychosis? Or were those

diagnoses modified in such a way as to bring them into accord with the circumstances of the pseudopatient's life, as described by him?

As far as I can determine, diagnoses were in no way affected by the relative health of the circumstances of a pseudopatient's life. Rather, the reverse occurred: the perception of his circumstances was shaped entirely by the diagnosis. A clear example of such translation is found in the case of a pseudopatient who had had a close relationship with his mother but was rather remote from his father during his early childhood. During adolescence and beyond, however, his father became a close friend, while his relationship with his mother cooled. His present relationship with his wife was characteristically close and warm. Apart from occasional angry exchanges, friction was minimal. The children had rarely been spanked. Surely there is nothing especially pathological about such a history. Indeed, many readers may see a similar pattern in their own experiences, with no markedly deleterious consequences. Observe, however, how such a history was translated in the psychopathological context, this from the case summary prepared after the patient was discharged.

> This white 39-year-old male . . . manifests a long history of considerable ambivalence in close relationships, which begins in early childhood. A warm relationship with his mother cools during his adolescence. A distant relationship to his father is described as becoming very intense. Affective stability is absent. His attempts to control emotionality with his wife and children are punctuated by angry outbursts and, in the case of the children, spanking. And while he says that he has several good friends, one senses considerable ambivalence embedded in those relationships also. . . .

The facts of the case were unintentionally distorted by the staff to achieve consistency with a popular theory of the dynamics of a schizophrenic reaction.[15] Nothing of an ambivalent nature had been described in relations with parents, spouse, or friends. To the extent that ambivalence could be inferred, it was probably not greater than is found in all human relationships. It is true the pseudopatient's relationships with his parents changed over time, but in the ordinary context that would hardly be remarkable—indeed, it might very well be expected. Clearly, the meaning ascribed to his verbalizations (that is, ambivalence, affective instability) was determined by the diagnosis: schizophrenia. An entirely different meaning would have been ascribed if it were known that the man was "normal."

All pseudopatients took extensive notes publicly. Under ordinary circumstances, such behavior would have raised questions in the minds of observers, as, in fact, it did among patients. Indeed, it seemed so certain that the notes would elicit suspicion that elaborate precautions were taken to remove them from the ward each day. But the precautions proved needless. The closest any staff member came to questioning these notes occurred when one pseudopatient asked his physician what kind of medication he was receiving and began to write down the response. "You needn't write it," he was told gently. "If you have trouble remembering, just ask me again."

If no questions were asked of the pseudopatients, how was their writing interpreted? Nursing records for three patients indicate that the writing was seen as an aspect of their pathological behavior. "Patient engages in writing behavior" was the daily nursing comment on one of the pseudopatients who was never questioned about his writing. Given that the patient is in the hospital, he must be psychologically disturbed. And given that he is disturbed, continuous writing must be a behavioral manifestation of that disturbance, perhaps a subset of the compulsive behaviors that are sometimes correlated with schizophrenia.

One tacit characteristic of psychiatric diagnosis is that it locates the sources of aberration within the individual and only rarely within the complex of stimuli that surrounds him. Consequently, behaviors that are stimulated by the environment are commonly misattributed to the patient's disorder. For example, one kindly nurse found a pseudopatient pacing the long hospital corridors. "Nervous, Mr. X?" she asked. "No, bored," he said.

The notes kept by pseudopatients are full of patient behaviors that were misinterpreted by well-intentioned staff. Often enough, a patient would go "berserk" because he had, wittingly or unwittingly, been mistreated by, say, an attendant. A nurse coming upon the scene would rarely inquire even cursorily into the environmental stimuli of the patient's behavior. Rather, she assumed that his upset derived from his pathology, not from his present interactions with other staff members. Occasionally, the staff might assume that the patient's family (especially when they had recently visited) or other patients had stimulated the outburst. But never were the staff found to assume that one of themselves or the structure of the hospital had anything to do with a patient's behavior. One psychiatrist pointed to a group of patients who were sitting outside the cafeteria entrance half an hour before lunchtime. To a group of young residents he indicated that such behavior was characteristic of the oral-acquisitive nature of the syndrome. It seemed not to occur to him that there were very few things to anticipate in a psychiatric hospital besides eating.

A psychiatric label has a life and an influence of its own. Once the impression has been formed that the patient is schizophrenic, the expectation is that he will continue to be schizophrenic. When a sufficient amount of time has passed, during which the patient has done nothing bizarre, he is considered to be in remission and available for discharge. But the label endures beyond discharge, with the unconfirmed expectation that he will behave as a schizophrenic again. Such labels, conferred by mental health professionals, are as influential on the patient as they are on his relatives and friends, and it should not surprise anyone that the diagnosis acts on all of them as a self-fulfilling prophecy. Eventually, the patient himself accepts the diagnosis, with all of its surplus meanings and expectations, and behaves accordingly.

The inferences to be made from these matters are quite simple. Much as Zigler and Phillips have demonstrated that there is enormous overlap in the symptoms presented by patients who have been variously diagnosed,[16] so there is enormous overlap in the behaviors of the sane and the insane. The sane are not "sane" all of the time. We lose our tempers "for no good reason." We are occasionally depressed or anxious, again for no good reason. And we may find it difficult to get along with one or another person — again for no reason that we can specify. Similarly, the insane are not always insane. Indeed, it was the impression of the pseudopatients while living with them that they were sane for long periods of time — that the bizarre behaviors upon which their diagnoses were allegedly predicated constituted only a small fraction of their total behavior. If it makes no sense to label ourselves permanently depressed on the basis of an occasional depression, then it takes better evidence than is presently available to label all patients insane or schizophrenic on the basis of bizarre behaviors or cognitions. It seems more useful, as Mischel[17] has pointed out, to limit our discussion to *behaviors,* the stimuli that provoke them, and their correlates.

It is not known why powerful impressions of personality traits, such as "crazy" or "insane," arise. Conceivably, when the origins of and stimuli that give rise to a behavior are remote or unknown, or when the behavior strikes us as immutable, trait labels regarding the *behavior* arise. When, on the other hand, the origins and stimuli are known and available, discourse is limited to the behavior itself. Thus, I may hallucinate because I am sleeping, or I may hallucinate because I have ingested a peculiar drug. These are termed sleep-induced hallucinations, or dreams, and drug-induced hallucinations, respectively. But when the stimuli to my hallucinations are unknown, that is called craziness, or schizophrenia — as if that inference were somehow as illuminating as the others.

The Experience of Psychiatric Hospitalization

The term "mental illness" is of recent origin. It was coined by people who were humane in their inclinations and who wanted very much to raise the station of (and the public's sympathies toward) the psychologically disturbed from that of witches and "crazies" to one that was akin to the physically ill. And they were at least partially successful, for the treatment of the mentally ill *has* improved considerably over the years. But while treatment has improved, it is doubtful that people really regard the mentally ill in the same way that they view the physically ill. A broken leg is something one recovers from, but mental illness allegedly endures forever.[18] A broken leg does not threaten the observer, but a crazy schizophrenic? There is by now a host of evidence that attitudes toward the mentally ill are characterized by fear, hostility, aloofness, suspicion, and dread.[19] The mentally ill are society's lepers.

That such attitudes infect the general population is perhaps not surprising, only upsetting. But that they affect the professionals — attendants, nurses, physicians, psychologists, and social workers — who treat and deal with the mentally ill is more disconcerting, both because such attitudes are self-evidently pernicious and because they are unwitting. Most mental health professionals would insist that they are sympathetic toward the mentally ill, that they are neither avoidant nor hostile. But it is more likely that an exquisite ambivalence characterizes their relations with psychiatric patients, such that their avowed impulses are only part of their entire attitude. Negative attitudes are there too and can easily be detected. Such attitudes should not surprise us. They are the natural offspring of the labels patients wear and the places in which they are found.

Consider the structure of the typical psychiatric hospital. Staff and patients are strictly segregated. Staff have their own living space, including their dining facilities, bathrooms, and assembly places. The glassed quarters that contain the professional staff, which the pseudopatients came to call "the cage," sit out on every dayroom. The staff emerge primarily for caretaking purposes — to give medication, to conduct a therapy or group meeting, to instruct or reprimand a patient. Otherwise, staff keep to themselves, almost as if the disorder that afflicts their charges is somehow catching.

So much is patient-staff segregation the rule that, for four public hospitals in which an attempt was made to measure the degree to which staff and patients mingle, it was necessary to use "time out of the staff cage" as the operational measure. While it was not the case that all time spent out of the cage was spent mingling with patients (attendants, for example, would occasionally emerge to watch television in the dayroom), it was the only way in which one could gather reliable data on time for measuring.

The average amount of time spent by attendants outside of the cage was 11.3 percent (range, 3 to 52 percent). This figure does not represent only time spent mingling with patients, but also includes time spent on such chores as folding laundry, supervising patients while they shave, directing ward cleanup, and sending patients to off-ward activities. It was the relatively rare attendant who spent time talking with patients or playing games with them. It proved impossible to obtain a "percent mingling time" for nurses, since the amount of time they spent out of the cage was too brief. Rather, we counted instances of emergence from the cage. On the average, daytime nurses emerged from the cage 11.5 times per shift, including instances when they left the ward entirely (range, 4 to 39 times). Late afternoon and night nurses were even less available, emerging on the average 9.4 times per shift (range, 4 to 41 times). Data on early morning nurses, who arrived usually after midnight and departed at 8 A.M., are not available because patients were asleep during most of this period.

Physicians, especially psychiatrists, were even less available. They were rarely seen on the wards. Quite commonly, they would be seen only when they arrived and departed, with the remaining time being spent in their offices or in the cage. On the average, the physicians emerged on the ward 6.7 times per day (range, 1 to 17 times). It proved difficult to make an accurate estimate in this regard, since physicians often maintained hours that allowed them to come and go at different times.

The hierarchical organization of the psychiatric hospital has been commented on before,[20]

but the latent meaning of that kind of organization is worth noting again. Those with the most power have least to do with patients, and those with the least power are most involved with them. Recall, however, that the acquisition of role-appropriate behaviors occurs mainly through the observation of others, with the most powerful having the most influence. Consequently, it is understandable that attendants not only spend more time with patients than do any other members of the staff—that is required by their station in the hierarchy—but also, insofar as they learn from their superiors' behavior, spend as little time with patients as they can. Attendants are seen mainly in the cage, which is where the models, the action, and the power are.

I turn now to a different set of studies, these dealing with staff response to patient-initiated contact. It has long been known that the amount of time a person spends with you can be an index of your significance to him. If he initiates and maintains eye contact, there is reason to believe that he is considering your requests and needs. If he pauses to chat or actually stops and talks, there is added reason to infer that he is individuating you. In four hospitals, the pseudopatient approached the staff member with a request which took the following form: "Pardon me, Mr. [or Dr. or Mrs.] X, could you tell me when I will be eligible for grounds privileges?" (or ". . . when I will be presented at the staff meeting?" or ". . . when I am likely to be discharged?"). While the content of the question varied according to the appropriateness of the target and the pseudopatient's (apparent) current needs the form was always a courteous and relevant request for information. Care was taken never to approach a particular member of the staff more than once a day, lest the staff member become suspicious or irritated. In examining these data, remember that the behavior of the pseudopatients was neither bizarre nor disruptive. One could indeed engage in good conversation with them.

The data for these experiments are shown in Table 1, separately for physicians (column 1) and for nurses and attendants (column 2). Minor differences between these four institutions were

TABLE 1 Self-Initiated Contact by Pseudopatients with Psychiatrists and Nurses and Attendants Compared to Contact with Other Groups

Contact	Psychiatric Hospitals		University Campus (non-medical)	University Medical Center		
	(1) Psychiatrists	(2) Nurses and Attendants	(3) Faculty	(4) "Looking for a Psychiatrist"	(5) "Looking for an Internist"	(6) No Additional Comment
Responses						
Moves on, head averted (%)	71	88	0	0	0	0
Makes eye contact (%)	23	10	0	11	0	0
Pauses and chats (%)	2	2	0	11	0	10
Stops and talks (%)	4	0.5	100	78	100	90
Mean number of questions answered (out of 6)	[a]	[a]	6	3.8	4.8	4.5
Respondents (No.)	13	47	14	18	15	10
Attempts (No.)	185	1283	14	18	15	10

[a]Not applicable.

overwhelmed by the degree to which staff avoided continuing contacts that patients had initiated. By far, their most common response consisted of either a brief response to the question, offered while they were "on the move" and with head averted, or no response at all.

The encounter frequently took the following bizarre form: (pseudopatient) "Pardon me, Dr. X. Could you tell me when I am eligible for grounds privileges?" (physician) "Good morning, Dave. How are you today?" (Moves off without waiting for a response.)

It is instructive to compare these data with data recently obtained at Stanford University. It has been alleged that large and eminent universities are characterized by faculty who are so busy that they have no time for students. For this comparison, a young lady approached individual faculty members who seemed to be walking purposefully to some meeting or teaching engagement and asked them the following six questions.

1. "Pardon me, could you direct me to Encina Hall?" (at the medical school: ". . . to the Clinical Research Center?")

2. "Do you know where Fish Annex is?" (there is no Fish Annex at Stanford).
3. "Do you teach here?"
4. "How does one apply for admission to the college?" (at the medical school: ". . . to the medical school?")
5. "Is it difficult to get in?"
6. "Is there financial aid?"

Without exception, as can be seen in Table 1 (column 3), all of the questions were answered. No matter how rushed they were, all respondents not only maintained eye contact, but stopped to talk. Indeed, many of the respondents went out of their way to direct or take the questioner to the office she was seeking, to try to locate "Fish Annex," or to discuss with her the possibilities of being admitted to the university.

Similar data, also shown in Table 1 (columns 4, 5, and 6), were obtained in the hospital. Here too, the young lady came prepared with six questions. After the first question, however, she remarked to 18 of her respondents (column 4), "I'm looking for a psychiatrist," and to 15 others (column 5), "I'm looking for an internist." Ten other respondents received no inserted comment (column 6). The general degree of cooperative responses is considerably higher for these university groups than it was for pseudopatients in psychiatric hospitals. Even so, differences are apparent within the medical school setting. Once having indicated that she was looking for a psychiatrist, the degree of cooperation elicited was less than when she sought an internist.

Powerlessness and Depersonalization

Eye contact and verbal contact reflect concern and individuation: their absence, avoidance and depersonalization. The data I have presented do not do justice to the rich daily encounters that grow up around matters of depersonalization and avoidance. I have records of patients who were beaten by staff for the sin of having initiated verbal contact. During my own experience, for example, one patient was beaten in the presence of other patients for having approached an attendant and told him, "I like you." Occasionally, punishment meted out to patients for misdemeanors seemed so excessive that it could not be justified by the most radical interpretations of psychiatric canon. Nevertheless, they appeared to go unquestioned. Tempers were often short. A patient who had not heard a call for medication would be roundly excoriated, and the morning attendants would often wake patients with, "Come on, you m——f——s, out of bed!"

Neither anecdotal nor "hard" data can convey the overwhelming sense of powerlessness which invades the individual as he is continually exposed to the depersonalization of the psychiatric hospital. It hardly matters *which* psychiatric hospital— the excellent public ones and the very plush private hospital were better than the rural and shabby ones in this regard, but, again, the features that psychiatric hospitals had in common overwhelmed by far their apparent differences.

Powerlessness was evident everywhere. The patient is deprived of many of his legal rights by dint of his psychiatric commitment.[21] He is shorn of credibility by virtue of his psychiatric label. His freedom of movement is restricted. He can-

not initiate contact with the staff, but may only respond to such overtures as they make. Personal privacy is minimal. Patient quarters and possessions can be entered and examined by any staff member, for whatever reason. His personal history and anguish is available to any staff member (often including the "gray lady" and "candy striper" volunteer) who chooses to read his folder, regardless of their therapeutic relationship to him. His personal hygiene and waste evacuation are often monitored. The water closets may have no doors.

At times, depersonalization reached such proportions that pseudopatients had the sense that they were invisible, or at least unworthy of account. Upon being admitted, I and other pseudopatients took the initial physical examinations in a semipublic room, where staff members went about their own business as if we were not there.

On the ward, attendants delivered verbal and occasionally serious physical abuse to patients in the presence of other observing patients, some of whom (the pseudopatients) were writing it all down. Abusive behavior, on the other hand, terminated quite abruptly when other staff members were known to be coming. Staff are credible witnesses. Patients are not.

A nurse unbuttoned her uniform to adjust her brassiere in the presence of an entire ward of viewing men. One did not have the sense that she was being seductive. Rather, she didn't notice us. A group of staff persons might point to a patient in the dayroom and discuss him animatedly, as if he were not there.

One illuminating instance of depersonalization and invisibility occurred with regard to medications. All told, the pseudopatients were administered nearly 2100 pills, including Elavil, Stelazine, Compazine, and Thorazine, to name but a few. (That such a variety of medications should have been administered to patients presenting identical symptoms is itself worthy of note.) Only two were swallowed. The rest were either pocketed or deposited in the toilet. The pseudopatients were not alone in this. Although I have no precise records on how many patients rejected their medications, the pseudopatients frequently found the medications of other patients in the toilet before they deposited their own. As long as they were cooperative, their behavior and the pseudopatients' own in this matter, as in other important matters, went unnoticed throughout.

Reactions to such depersonalization among pseudopatients were intense. Although they had come to the hospital as participant observers and were fully aware that they did not "belong," they nevertheless found themselves caught up in and fighting the process of depersonalization. Some examples: a graduate student in psychology asked his wife to bring his textbooks to the hospital so he could "catch up on his homework" — this despite the elaborate precautions taken to conceal his professional association. The same student, who had trained for quite some time to get into the hospital, and who had looked forward to the experience, "remembered" some drag races that he had wanted to see on the weekend and insisted that he be discharged by that time. Another pseudopatient attempted a romance with a nurse. Subsequently he informed the staff that he was applying for admission to graduate school in psychology and was very likely to be admitted, since a graduate professor was one of his regular hospital visitors. The same person began to engage in psychotherapy with other patients — all of this as a way of becoming a person in an impersonal environment.

The Sources of Depersonalization

What are the origins of depersonalization? I have already mentioned two. First are attitudes held by all of us toward the mentally ill — including those who treat them — attitudes characterized by fear, distrust, and horrible expectations on the one hand, and benevolent intentions on the other. Our ambivalence leads, in this instance as in others, to avoidance.

Second, and not entirely separate, the hierarchical structure of the psychiatric hospital facilitates depersonalization. Those who are at the top have least to do with patients, and their behavior inspires the rest of the staff. Average daily contact with psychiatrists, psychologists, residents, and physicians combined ranged from 3.9 to 25.1 minutes, with an overall mean of 6.8 (six pseudopatients over a total of 129 days of hospitalization). Included in this average are time spent in the admissions interview, ward meetings in the presence of a senior staff member, group and individual psychotherapy contacts, case presentation conferences, and discharge meetings. Clearly, patients do not spend much time in interpersonal contact with doctoral staff. And doctoral staff serve as models for nurses and attendants.

There are probably other sources. Psychiatric installations are presently in serious financial straits. Staff shortages are pervasive, staff time at a premium. Something has to give, and that something is patient contact. Yet while financial stresses are realities, too much can be made of them. I have the impression that the psychological forces that result in depersonalization are much stronger than the fiscal ones and that the addition of more staff would not correspondingly improve patient care in this regard. The incidence of staff meetings and the enormous amount of record-keeping on patients, for example, have not been as substantially reduced as has patient contact. Priorities exist, even during hard times. Patient contact is not a significant priority in the traditional psychiatric hospital, and fiscal pressures do not account for this. Avoidance and depersonalization may.

Heavy reliance upon psychotropic medication tacitly contributes to depersonalization by convincing staff that treatment is indeed being conducted and that further patient contact may not be necessary. Even here, however, caution needs to be exercised in understanding the role of psychotropic drugs. If patients were powerful rather than powerless, if they were viewed as interesting individuals rather than diagnostic entities, if they were socially significant rather than social lepers, if their anguish truly and wholly compelled our sympathies and concerns, would we not *seek* contact with them, despite the availability of medications? Perhaps for the pleasure of it all?

The Consequences of Labeling and Depersonalization

Whenever the ratio of what is known to what needs to be known approaches zero, we tend to invent "knowledge" and assume that we understand more than we actually do. We seem unable to acknowledge that we simply don't know. The needs for diagnosis and remediation of behavioral and emotional problems are enormous. But rather than acknowledge that we are just embarking on understanding, we continue to label patients "schizophrenic," "manic-depressive," and "insane," as if in those words we had captured the essence of understanding. The facts of the matter are that we have known for a long time that diagnoses are often not useful or reliable, but we have nevertheless continued to use them. We now know that we cannot distinguish insanity from sanity. It is depressing to consider how that information will be used.

Not merely depressing, but frightening. How many people, one wonders, are sane but not recognized as such in our psychiatric institutions? How many have been needlessly stripped of their privileges of citizenship, from the right to vote and drive to that of handling their own accounts? How many have feigned insanity in order to avoid the criminal consequences of their behavior, and, conversely, how many would rather stand trial than live interminably in a psychiatric hospital—but are wrongly thought to be mentally ill? How many have been stigmatized by well-intentioned, but nevertheless erroneous, diagnoses? On the last point, recall again that a "type 2 error" in psychiatric diagnosis does not have the same consequences it does in medical diagnosis. A diagnosis of cancer that has been found to be in error is cause for celebration. But psychiatric diagnoses are rarely found to be in error. The label sticks, a mark of inadequacy forever.

Finally, how many patients might be "sane" outside the psychiatric hospital but seem insane in it—not because craziness resides in them, as it were, but because they are responding to a bizarre setting, one that may be unique to institutions which harbor neither people? Goffman calls the process of socialization to such institutions "mortification"—an apt metaphor that includes the processes of depersonalization that have been described here. And while it is impossible to know whether the pseudopatients' responses to these processes are characteristic of all inmates—they were, after all, not real patients—it is difficult to believe that these processes of socialization to a psychiatric hospital provide useful attitudes or habits of response for living in the "real world."

Summary and Conclusions

It is clear that we cannot distinguish the sane from the insane in psychiatric hospitals. The hospital itself imposes a special environment in which the meanings of behavior can easily be misunderstood. The consequences to patients hospitalized in such an environment—the powerlessness, depersonalization, segregation, mortification, and self-labeling—seem undoubtedly countertherapeutic.

I do not, even now, understand this problem well enough to perceive solutions. But two matters seem to have some promise. The first concerns the proliferation of community mental health facilities, of crisis intervention centers, of human potential movement, and of behavior therapies that, for all of their own problems, tend to avoid psychiatric labels, to focus on specific problems and behaviors, and to retain the individual in a relatively nonpejorative environment. Clearly, to the extent that we refrain from sending the distressed to insane places, our impressions of them are less likely to be distorted. (The risk of distorted perceptions, it seems to me, is always present, since we are much more sensitive to an individual's behaviors and verbalizations than we are to the subtle contextual stimuli that often promote them. At issue here is a matter of magnitude. And, as I have shown, the magnitude of distortion is exceedingly high in the extreme context that is a psychiatric hospital.)

The second matter that might prove promising speaks to the need to increase the sensitivity of mental health workers and researchers to the *Catch 22* position of psychiatric patients. Simply reading materials in this area will be of help to some such workers and researchers. For others, directly experiencing the impact of psychiatric hospitalization will be of enormous use. Clearly, further research into the social psychology of such total institutions will both facilitate treatment and deepen understanding.

I and other pseudopatients in the psychiatric setting had distinctly negative reactions. We do not pretend to describe the subjective experiences of true patients. Theirs may be different from ours, particularly with the passage of time and the necessary process of adaptation to one's environment. But we can and do speak to the relatively more objective indices of treatment within the hospital. It could be a mistake, and a very unfortunate one, to consider that what happened to us derived from malice or stupidity on the part of the staff. Quite the contrary, our overwhelming impression of them was of people who really cared, who were committed, and who were uncommonly intelligent. Where they failed, as they sometimes did painfully, it would be more accurate to attribute those failures to the environment in which they, too, found themselves than to personal callousness. Their perceptions and behavior were controlled by the situation, rather than being motivated by a malicious disposition. In a more benign environment, one that was less attached to global diagnosis, their behaviors and judgments might have been more benign and effective.

NOTES

1. P. Ash, *J. Abnorm. Soc. Psychol.* 44, 272 (1949); A. T. Beck, *Amer. J. Psychiat.* 119, 210 (1962); A. T. Boisen, *Psychiatry* 2, 233 (1938); N. Kreitman, *J. Ment. Sci.* 107, 876 (1961); N. Kreitman, P. Sainsbury, J. Morrisey, J. Towers, J. Scrivener, *ibid.*, p. 887; H. O. Schmitt and C. P. Fonda, *J. Abnorm. Soc. Psychol.* 52, 262 (1956); W. Seeman, *J. Nerv. Ment. Dis.* 118, 541 (1953). For an analysis of these artifacts and summaries of the disputes, see J. Zubin, *Annu. Rev. Psychol.* 18, 373 (1967); L. Phillips and J. G. Draguns, *ibid.* 22, 447 (1971).

2. R. Benedict, *J. Gen. Psychol.* 10, 59 (1934).

3. See in this regard H. Becker, *Outsiders: Studies in the Sociology of Deviance* (Free Press, New York, 1963); B. M. Braginsky, D. D. Braginsky, K. Ring, *Methods of Madness: The Mental Hospital as a Last Resort* (Holt, Rinehart & Winston, New York, 1969); G. M. Crocetti and P. V. Lemkau, *Amer. Sociol. Rev.* 30, 577 (1965); E. Goffman, *Behavior in Public Places* (Free Press, New York, 1964); R. D. Laing, *The Divided Self: A Study of Sanity and Madness* (Quadrangle, Chicago, 1960); D. L. Phillips, *Amer. Sociol. Rev.* 28, 963 (1963); T. R. Sarbin, *Psychol. Today* 6, 18 (1972); E. Schur, *Amer. J. Sociol.* 75, 309 (1969); T. Szasz, *Law, Liberty and Psychiatry* (Macmillan, New York, 1963); *The Myth of Mental Illness: Foundations of a Theory of Mental Illness* (Hoeber Harper, New York, 1963). For a critique of some of these views, see W. R. Gove, *Amer. Sociol. Rev.* 35, 873 (1970).

4. Goffman, *Asylums* (Doubleday, Garden City, N.Y., 1961).

5. T. J. Scheff, *Being Mentally Ill: A Sociological Theory* (Aldine, Chicago, 1966).

6. Data from a ninth pseudopatient are not incorporated in this report because, although his sanity went undetected, he falsified aspects of his personal history, including his marital status and parental relationships. His experimental behaviors therefore were not identical to those of the other pseudopatients.

7. A. Barry, *Bellevue Is a State of Mind* (Harcourt Brace Jovanovich, New York, 1971); I. Belknap, *Human Problems of a State Mental Hospital* (McGraw-Hill, New York, 1956); W. Caudill, F. C. Redlich, H. R. Gilmore, E. B. Brody, *Amer. J. Orthopsychiat.* 22, 314 (1952); A. R. Goldman, R. H. Bohr, T. A. Steinberg, *Prof. Psychol.* 1, 427 (1970); unauthored, *Roche Report 1* (No. 13), 8 (1971).

8. Beyond the personal difficulties that the pseudopatient is likely to experience in the hospital, there are legal and social ones that, combined, require considerable attention before entry. For example, once admitted to a psychiatric institution, it is difficult, if not impossible, to be discharged on short notice, state law to the contrary notwithstanding. I was not sensitive to these difficulties at the outset of the project, nor to the personal and situational emergencies that can arise, but later a writ of habeas corpus was prepared for each of the entering pseudopatients and an attorney was kept "on call" during every hospitalization. I am grateful to John Kaplan and Robert Bartels for legal advice and assistance in these matters.

9. However distasteful such concealment is, it was a necessary first step to examining these questions. Without concealment, there would have been no way to know how valid these experiences were; nor was there any way of knowing whether whatever detections occurred were a tribute to the diagnostic acumen of the staff or to the hospital's rumor network. Obviously, since my concerns are general ones that cut across individual hospitals and staffs, I have respected their anonymity and have eliminated clues that might lead to their identification.

10. Interestingly, of the 12 admissions, 11 were diagnosed as schizophrenic and one, with the identical symptomatology, as manic-depressive psychosis. This diagnosis has a more favorable prognosis, and it was given by the only private hospital in our sample. On the relations between social class and psychiatric diagnosis, see A. deB. Hollingshead and F. C. Redlich, *Social Class and Mental Illness: A Community Study* (Wiley, New York, 1958).

11. It is possible, of course, that patients have quite broad latitudes in diagnosis and therefore are inclined to call many people sane, even those whose behavior is patently aberrant. However, although we have no hard data on this matter, it was our distinct impression that this was not the case. In many instances, patients not only singled us out for attention, but came to imitate our behaviors and styles.

12. J. Cumming and E. Cumming, *Community Ment. Health 1,* 135 (1965); A. Farina and K. Ring, *J. Abnorm. Psychol. 70,* 47 (1965); H. E. Freeman and O. G. Simmons, *The Mental Patient Comes Home* (Wiley, New York, 1963); W. J. Johannsen, *Ment. Hygiene 53,* 218 (1969); A. S. Linsky, *Soc. Psychiat. 5,* 1966 (1970).

13. S. E. Asch, *J. Abnorm. Soc. Psychol. 41,* 258 (1946); *Social Psychology* (Prentice-Hall, New York, 1952).

14. See also I. N. Mensh and J. Wishner, *J. Personality 16,* 188 (1947); J. Wishner, *Psychol. Rev. 67,* 96 (1960); J. S. Bruner and R. Tagiuri, in *Handbook of Social Psychology,* G. Lindzey, Ed. (Addison-Wesley, Cambridge Mass., 1954), vol. 2, pp. 634–654; J. S. Bruner, D. Shapiro, R. Tagiuri, in *Person Perception and Interpersonal Behavior,* R. Tagiuri and L. Petrullo, Eds. (Stanford Univ. Press, Stanford, Calif., 1958), pp. 277–288.

15. For an example of a similar self-fulfilling prophecy, in this instance dealing with the "central" trait of intelligence, see R. Rosenthal and L. Jacobson, *Pygmalion in the Classroom* (Holt, Rinehart & Winston, New York, 1968).

16. E. Zigler and L. Phillips, *J. Abnorm. Soc. Psychol. 63,* 69 (1961). See also R. K. Freudenberg and J. P. Robertson, *A.M.A. Arch. Neurol. Psychiatr. 76,* 14 (1956).

17. W. Mischel, *Personality and Assessment* (Wiley, New York, 1968).

18. The most recent and unfortunate instance of this tenet is that of Senator Thomas Eagleton.

19. T. R. Sarbin and J. C. Mancuso, *J. Clin. Consult. Psychol. 35,* 159 (1970); T. R. Sarbin, *ibid. 31,* 447 (1967); J. C. Nunnally, Jr., *Popular Conceptions of Mental Health* (Holt, Rinehart & Winston, New York, 1961).

20. A. H. Stanton and M. S. Schwartz, *The Mental Hospital: A Study of Institutional Participation in Psychiatric Illness and Treatment* (Basic, New York, 1954).

21. D. B. Wexler and S. E. Scoville, *Ariz. Law Rev. 13,* 1 (1971).

QUESTIONS FOR DISCUSSION

For further discussion of this topic, see the Wadsworth Sociology Resource Center, "Virtual Society," *http://sociology.wadsworth.com,* under *Sociological Footprints,* by Cargan and Ballantine. You can respond to the discussion questions there or enter your own comments in the online chat forum.

SUGGESTED READINGS AND
SOCIOLOGY INTERNET RESOURCES

See the Wadsworth Sociology Resource Center, "Virtual Society," *http://sociology.wadsworth. com,* for additional links, suggestions for further reading, and learning tools related to this chapter.

Either from the "Virtual Society" website or directly from your web browser, you may access InfoTrac College Edition, an online university library that includes over 700 popular and scholarly journals in which you can find articles related to the topics in this chapter.

CHAPTER 13

The Human Environment

LIVING TOGETHER ON SPACESHIP EARTH

POPULATION EXPANSION AND RELATED ECOLOGICAL PROBLEMS are with us today and will remain with us for many years to come—if we survive. This is the verdict of many experts, called *demographers,* who study changes in human populations. Three variables are crucial to population change: fertility (the birth rate), mortality (the death rate), and migration (population movement). In studying populations, demographers focus on how these variables affect three different areas:

1. Population growth and decline: size of the population.
2. Population distribution: where the population is located.
3. Population composition and structure: characteristics of the population, such as age, sex, education, and so forth.

Each of these areas must be examined to understand the population in the world today. The readings in this chapter deal with these factors and with the results of environmental protection.

We must be concerned with the rapid growth rate that many countries are experiencing, both for humanitarian reasons and because of the increasing demand on scarce resources. The Population Reference Bureau estimates that in developing countries population can double every 20 to 35 years. This means that natural resources, food-production capacity, and other essentials must double in the same period in order to maintain present lifestyles. Yet most countries are demanding more food, improved communications, better education, scientific advances, and higher standards of living at the same time that resources and capacities to produce are already severely strained.

The themes running through these readings are (1) factors that change the population of nations, such as birth rates and migration; (2) the aging of populations in developed countries and the resulting increase in the dependency ratio; and (3) the human consequences of population policies. As you read these articles, consider the implications of population conditions and their effects on our lives and the benefits and value of protecting our environment.

A major factor in the depletion of the earth's resources is overconsumption by developed countries. Exploitation of resources by the world's richest fifth of humanity threatens the well-being of all. In the first reading, Alan Durning argues for changes in consumption habits to protect the environment and make the world livable for all humans.

The problems caused by overpopulation and overconsumption are the topic of Lester Brown's "Overview." He outlines the positive and negative trends taking place in the world.

Problems caused by overpopulation are serious now, but Mitchell looks at the "next doubling," which will occur in the twenty-first century, and at some of the preventive measures to slow population growth.

Cities developed centuries ago as agricultural surplus freed people from farming. Those not needed in the vital task of feeding the population were attracted to urban areas both by work opportunities and the excitement of city living. This process of migration to cities continues today in many parts of the world.

However, the rapid expansion of cities has caused strains on communications, transportation, and distribution of needed goods and services, making them less desirable. Rapidly growing urban areas struggle to meet the needs of swelling populations. At the same time, those who can afford to move to the suburbs, cause increased problems for cities. Raymond Flynn discusses why more attention should be focused on urban problems. The pulls and attractions of urban living are choking large cities around the world. Unable to cope with deteriorating infrastructures and the demands of newcomers, cities are decaying in both physical structures and social control.

One result of these problems is housing shortages and dilapidated housing. In the last reading (an excerpt from his book *Down and Out in America*), Peter Rossi discusses the housing problem at the root of homelessness, and other problems that perpetuate this situation.

As you read these selections, consider how population dynamics relate to economic crises and city problems. How might we change our consumption patterns to influence and help solve world problems?

51

Limiting Consumption: Toward a Sustainable Culture

ALAN DURNING

The environment on which humans depend is suffering under the weight of population growth and an increasing worldwide demand for resources. In addition to rapid population growth in developing countries, which is putting increasing strain on the world's resources, developed countries are taking their toll on earth's vital signs. The developed world consumes resources at many times its rate of population, using nonrenewable resources, creating pollution, and leaving its mark on the health of the world. In this reading, Durning points out the causes of this overconsumption and the importance of limiting it.

Consider the following issues:

1. Why can the developed countries with low population growth be as harmful to the earth as rapidly growing nations?

2. How can human consumption be controlled?

GLOSSARY

Overconsumption Disproportionate use of world resources by the richest fifth of the world's population.
Underconsumption Poverty, or lack of necessities.
Sustainability Exists when each generation meets its needs without jeopardizing those of future generations.

OVERCONSUMPTION BY THE WEALTHIEST fifth of humanity is an environmental problem unmatched in severity by anything but perhaps population growth. The surging exploitation of resources threatens to exhaust or unalterably disfigure forests, soils, water, air, and climate.

Of course, *under*consumption—poverty—is no solution. Poverty is infinitely worse for peo-ple and bad for the natural world, too. Dispossessed peasants slash-and-burn their way into the rain forests of Latin America, and hungry nomads turn their herds out onto fragile African rangeland, reducing it to desert.

If environmental decline results when people have either too little or too much, we must ask ourselves: How much is enough? What level of consumption can the earth support? When does consumption cease to add appreciably to human satisfaction?

Answering these questions definitively is impossible, but for each of us in the world's consuming class, seeking answers may be a prerequisite to transforming our civilization into one the biosphere can sustain.

Consumption's Rapid Rise

Skyrocketing consumption is the hallmark of our era. The headlong advance of technology, rising earnings, and consequently cheaper material goods have lifted overall consumption to levels never dreamed of a century ago.

In the United States, per capita energy use has climbed 60 percent since 1950, car travel has more than doubled, plastic use has multiplied 20-fold, and air travel has jumped 25-fold.

Japanese consumption has increased even more rapidly. As in the United States, the 80s were a particularly consumerist decade in Japan, with sales of BMW automobiles rising tenfold over the decade. Ironically, in 1990 a *reja bumu* (leisure boom) combined with concern for na-ture to create two new status symbols: four-

Reprinted from The Futurist, *July–August, 1991, pp. 11–15.*

wheel-drive Range Rovers from England and cabins made of imported American logs.

Still, Japan has come to the high-consumption ethos hesitantly. Many older Japanese still hold to their time-honored belief in frugality. Yorimoto Katsumi of Waseda University in Tokyo writes, "Members of the older generation . . . are careful to save every scrap of paper and bit of string for future use."

The consumption levels of West Europeans have also increased markedly in recent decades. Just in the first half of the 1980s, per capita consumption of frozen prepared meals—with their excessive packaging—rose more than 30 percent in every West European country except Finland; in Switzerland, the jump was 180 percent. As trade barriers come down in the move toward a single European market in 1992, prices for many goods will likely fall and product promotion grow more aggressive, boosting consumption higher.

The collapse of socialist governments in Eastern Europe, meanwhile, has unleashed a tidal wave of consumer demand that had gone unsatisfied in the region's ossified state-controlled economies. A young man in a Budapest bar captured his country's mood when he told a Western reporter: "People in the West think that we in Hungary don't know how they live. Well, we do know how they live, and we want to live like that, too."

Seventy percent of those living in the former East Germany hope to enter the world's automobile class soon; they bought 200,000 used Western cars in the first half of 1990 alone. Western carmakers' plans for Eastern Europe promise to give the region the largest number of new car factories in the world.

The late 1980s saw some poor societies, such as China and India, begin the transition to consuming ways. Few would begrudge anyone the simple advantages of cold food storage or mechanized clothes washing. But long before all the world's people could achieve the American dream, the planet would be laid waste.

The world's one billion meat eaters, car drivers, and throwaway consumers are responsible for the lion's share of the damage that humans have caused to common global resources. Their appetite for wood is a driving force behind destruction of the tropical rain forests and the resulting extinction of countless species. Over the past century, their economies have pumped out two-thirds of the greenhouse gases that threaten the planet's climate, and each year their energy use releases perhaps three-fourths of the sulfur and nitrogen oxides that cause acid rain. Their industries generate most of the world's hazardous chemical wastes, and their air conditioners, aerosol sprays, and factories release almost 90 percent of the chlorofluorocarbons that destroy the earth's protective ozone layer.

Ironically, abundance has not brought personal fulfillment. According to opinion polls, the percentage of Americans who report that they are "very happy" has been relatively stable since 1957, despite phenomenal growth in consumption. Indeed, most psychological data show that the main determinants of happiness in life are not related to consumption at all: Prominent among them are satisfaction with family life, especially marriage, followed by satisfaction with work, leisure, and friendships.

If money can't buy happiness, what prompts us to consume so much?

The Compulsion to Consume

"The avarice of mankind is insatiable," Aristotle declared 23 centuries ago, setting off a debate that has raged ever since among philosophers over how much greed lurks in human hearts. But whatever share of our acquisitiveness is part of our nature, the compulsion for more has never been so actively promoted, nor so easily acted upon, as it is today.

We are encouraged to consume at every turn by the advertising industry (which annually spends nearly $500 per U.S. citizen), by the commercialization of everything from sporting events to public spaces, and, insidiously, by the spread of the mass market into realms once dominated by family members and local enterprises. Cooking from scratch is replaced by heating prepared foods in the microwave; the neighborhood baker and greengrocer are driven out by the 24-hour supermarket at the mall. As our day-to-day interactions with the economy lose the face-to-face character that prevails in surviving communities, buying things becomes a substitute source of self-worth.

Traditional measures of success such as integrity, honesty, skill, and hard work are gradually supplanted by a simple, universally recognizable indicator of achievement—money. One Wall Street banker put it bluntly to *The New York Times*: "Net worth equals self-worth." Under this definition, there is no such thing as enough. Consumption becomes a treadmill with people judging their status by who's ahead of them and who's behind.

Technologies of Consumption

In simplified terms, an economy's total burden on the ecological systems that undergird it is a function of three factors: the size of the human population, people's average consumption level, and the broad set of technologies—everything from mundane clotheslines to the most sophisticated satellite communications systems—that the economy employs to provide for those consumption levels.

Transformations of agricultural patterns, transportation systems, urban design, energy use, and the like could radically reduce the total environmental damage caused by the consuming societies, while allowing those at the bottom of the economic ladder to rise without producing such egregious effects.

Japan, for example, uses one-third as much energy as the Soviet Union to produce a dollar's worth of goods and services, and Norwegians use half as much paper and cardboard apiece as their neighbors in Sweden, though the Norwegians equal the Swedes' literacy and are richer in monetary terms.

Eventually, though, technological change will need to be complemented by curbing our material wants. Robert Williams of Princeton University and a worldwide team of researchers conducted a careful study of the potential to reduce fossil-fuel consumption through greater efficiency and use of renewable energy.

The entire world population, Williams concluded, could live with the quality of energy services enjoyed by West Europeans—things like modest but comfortable homes, refrigeration for food, and ready access to public transit, augmented by limited auto use.

The study had an implicit conclusion, however: The entire world population decidedly could *not* live in the style of Americans, with their larger homes, more numerous electrical gadgets, and auto-centered transportation systems.

The details of such studies will stir debate among specialists for years to come. What matters for the rest of us is the lesson to hope and work for much from technological and political change, while looking to ourselves for the values changes that will also be needed.

Ethics for Sustainability

When Moses came down from Mount Sinai, he could count the rules of ethical behavior on his fingers. In the complex global economy of the late twentieth century, in which the simple act of turning on an air conditioner affects planetary systems, the list of rules for ecologically sustainable living could run into the hundreds.

The basic value of a sustainable society, the ecological equivalent of the Golden Rule, is simple: Each generation should meet its needs without jeopardizing the prospects of future generations. What is lacking is the practical knowledge—at each level of society—of what living by that ethical principle means.

In a fragile biosphere, the ultimate fate of humanity may depend on whether we can cultivate a deeper sense of self-restraint, founded on a widespread ethic of limiting consumption and finding nonmaterial enrichment.

Those who seek to rise to this environmental challenge may find encouragement in the body of human wisdom passed down from antiquity. To seek out sufficiency is to follow the path of voluntary simplicity preached by all the sages from Buddha to Mohammed. Typical of these pronouncements is this passage from the Bible: "What shall it profit a man if he shall gain the whole world and lose his own soul?"

Living by this credo is not easy. As historian David Shi of Davidson College in North Carolina chronicles, the call for a simpler life is perennial through the history of the North American continent: the Puritans of Massachusetts Bay, the Quakers of Philadelphia, the Amish, the Shakers, the experimental utopian communities of the 1830s, the hippies of the 1960s, and the back-to-the-land movement of the 1970s.

None of these movements ever gained more than a slim minority of adherents. Elsewhere in the world, entire nations such as China and Vietnam have dedicated themselves to rebuilding human character—sometimes through brutal techniques—in a less self-centered mold, and nowhere have they succeeded with more than a token few of their citizens.

It would be hopelessly naive to believe that entire populations will suddenly experience a moral awakening, renouncing greed, envy, and avarice. The best that can be hoped for is a gradual widening of the circle of those practicing voluntary simplicity. The goal of creating a sustainable culture, that is, a culture of permanence, is best thought of as a challenge that will last several generations.

Political Measures to Combat Consumption

Voluntary simplicity, or personal restraint, will do little good, however, if it is not wedded to bold political steps that confront the forces advocating consumption. Beyond the oft-repeated agenda of environmental and social reforms necessary to achieve sustainability, such as overhauling energy systems, stabilizing population, and ending poverty, action is needed to restrain the excesses of advertising, to curb the shopping culture, to abolish policies that push consumption, and to revitalize household and community economies as human-scale alternatives to the high-consumption lifestyle.

Such changes promise to help both the environment, by reducing the burden of overconsumption, and our peace of mind, by taming the forces that keep us dissatisfied with our lot.

ADVERTISING

The advertising industry is on the march around the world, but it is already vulnerable where it pushes products demonstrably dangerous to human health. Tobacco ads are or soon will be banished from television throughout the West, and alcohol advertising is under attack as never before.

Advertisers' influence can be further diluted by limiting their access to the most-vulnerable consumers. In late 1990, the U.S. Congress, for example, wisely hemmed in television commercials aimed at children, and the European Communities' standards on television for Europe after 1992 put strict limits on some types of ads.

At the grass-roots level, the Vancouver-based Media Foundation has set out to build a movement boldly aimed at turning television to anti-consuming ends. The premier spot in their "High on the Hog" campaign shows a gigantic animated pig frolicking on a map of North America while a narrator intones: "Five percent of the people in the world consume *one-third* of the planet's resources . . . those people are us."

Commercial television will need fundamental reorientation in a culture of permanence. As religious historian Robert Bellah put it, "That happiness is to be attained through limitless material acquisition is denied by every religion and philosophy known to humankind, but is preached incessantly by every American television set."

THE SHOPPING CULTURE

Some countries have resisted the advancing shopping culture, though only rarely is the motivation opposition to consumerism itself. England and Wales have restricted trading on Sundays for 400 years, and labor groups beat back the most recent proposal to lift those limits. Similarly, the protected "green belts" around British cities have slowed the pace of development of suburban malls there.

As in much of Europe, German stores must close most evenings at six o'clock and have limited weekend hours as well. In Japan, most shopping continues to take place in neighborhood shopping lanes, which are closed to traffic during certain hours to become *hokoosha tengoku,* literally "pedestrian heavens."

All these things help control the consumerist influence of marketing on the shape and spirit of public space. Shopping is less likely to become an end in itself if it takes place in stores thoroughly knit into the fabric of the community rather than in massive, insular agglomerations of retail outlets, each planned in minute detail to stimulate spendthrift ways.

GOVERNMENT POLICIES

Direct incentives for overconsumption are also essential targets for reform. If goods' prices reflected something closer to the environmental cost of their production, through revised subsidies and tax systems, the market itself would guide consumers toward less-damaging forms of consumption. Disposables and packaging would rise in price relative to durable, less-packaged goods; local unprocessed food would fall in price relative to prepared products trucked from far away.

The net effect might also be lower overall consumption as people's effective purchasing power declined. As currently constituted, unfortunately, economies penalize the poor when aggregate consumption contracts: Unemployment skyrockets, and inequalities grow. Thus arises one of the greatest challenges for sustainable economics in rich societies: finding ways to ensure basic employment opportunities for all without constantly having to stoke the fires of economic growth.

WEAK HOUSEHOLD AND COMMUNITY ECONOMIES

Ultimately, efforts to revitalize household and community economies may prove the decisive element in the attempt to create a culture less prone to consumption. At a personal level, commitment to nonmaterial fulfillment is hard to sustain without reinforcement from family, friends, and neighbors. At a political level, vastly strengthened local institutions may be the only counterweight to the colossus of vested interests—ranging from gas stations to multinational marketing conglomerates—that currently benefit from profligate consumption.

Unwanted Consumption

There could be many more people ready to begin saying "enough" than prevailing opinion suggests. After all, much of what we consume is wasted or unwanted in the first place. How much of the packaging that wraps products we consume each year—462 pounds per capita in the United States—would we rather never see? How many of the distant farms turned into suburban housing developments could have been left in crops if we insisted on well-planned land use inside city limits?

How many of the miles we drive—almost 6000 a year apiece in the United States—would we not happily give up if livable neighborhoods were closer to work, a variety of local merchants closer to home, streets safer to walk and ride bicycles on, and public transit easier and faster? How much of the fossil energy we use is wasted because utility companies fail to put money into efficient renewable energy systems before building new coal plants?

In the final analysis, accepting and living by sufficiency rather than excess offers a return to a true materialism that does not just care *about* things but cares *for* them. Maybe Henry David Thoreau had it right when he scribbled in his notebook beside Walden Pond, "A man is rich in proportion to the things he can afford to let alone."

52

Overview: New Records, New Stresses

LESTER R. BROWN

Human consumption and use of resources is causing other environmental problems. As the demand for food—especially animal protein—and energy resources grows, other changes in natural systems—such as climate and land use—occur. Some countries are taking steps to reduce the human impact on the environment.

Answer the following questions as you read:

1. *What are some environmental stresses facing the world?*
2. *What can countries do to reduce these environmental stresses?*

GLOSSARY

Environmental stress Problems causing deterioration of environmental conditions.
Global economy The interdependent world economy.

THE WORLD TODAY IS warmer, more crowded, more urban, economically richer, and environmentally poorer than ever before. This past year was one of near-record global economic growth—and of disturbing new signs of environmental stress.

In 1997, the Earth's average temperature was the highest since recordkeeping began in 1866. With each additional year of record or near-record temperature, the evidence of human-induced climate change becomes more convincing. In December 1997, government representatives gathered in Kyoto, Japan, to negotiate an agreement to reverse the rise in carbon emissions from human activities, with the hope of eventually checking the increase in temperature.

At the end of 1997, we shared the Earth with 80 million more people than a year earlier. Of this total, nearly 50 million people were added in Asia, the region that is already home to more than half of humanity. Each month, the world adds the equivalent of another Sweden. And it becomes more urban with each passing day: In 1800, only London had a million people. Now there are 326 cities that are at least that size. Sometime in the next decade, the number of people living in cities is expected to surpass those in the countryside.

Despite financial turmoil in Southeast Asia, the global economy expanded by 4.1 percent in 1997, marking the third consecutive year with growth of 4 percent or more. Economic output per person jumped by 2.6 percent. If the global economy continues to expand as projected, output per person worldwide will top $5,000 for the first time in 1998.

Signs of environmental stress continue to accumulate. Among the more disturbing in 1997 was the uncontrolled burning of Indonesia's rainforests, a conflagration that filled the air in the region with smoke for several months—smoke so intense at times in Indonesia and Malaysia that it caused acute respiratory stress, leaving millions physically sick. It led to the cancellation of 1,100 flights and a precipitous drop in earnings from tourism. The economic mismanagement in Indonesia that has led to bad debt, failing banks, and a falling currency has also weakened the rainforests to the point where they now burn out of control during droughts like the one induced by El Niño.

In Lester R. Brown, Michael Renner, and Christopher Flavin, Vital Signs, 1998: The Environmental Trends That Are Shaping Our Future. *New York: W. W. Norton & Company, 1998, pp. 15–24.*

The Yellow River, the northernmost of China's two major rivers, was drained dry by withdrawals from upstream provinces for several months, failing to make it to the sea for the thirteenth consecutive year. The river ran dry for longer than ever before, and in 1997 failed to reach the sea for 226 days out of 365. Farmers in the lower reaches of the river, deprived of irrigation water, saw their grain output fall.

Food: Surpluses to Scarcity

In 1997, the world's farmers harvested a record 1,881 million tons of grain, narrowly eclipsing 1996's record harvest of 1,869 million tons. Although the harvest rose, it did not keep up with population growth, so per capita grain output dropped from 324 kilograms to 322 kilograms. The drop in per capita grain production worldwide of more than 6 percent since its all-time peak in 1985 is one indication that the half-century dominated by food surpluses may be coming to an end.

As recently as 1990, the world had two food reserves to call upon—carryover stocks of grain (the amount in the bin when a new harvest begins) and cropland idled under U.S. farm commodity programs that were designed to avoid price-depressing world grain surpluses. In 1995, the farm support programs were dismantled, letting the set-aside land be returned to production in 1996. The 11 million hectares of grainland held out in 1990, assuming a yield of 4 tons per hectare, represented a reserve of 44 million tons—nearly nine days of world consumption.

Even with this land back in production in 1996 and 1997, however, the world was not able to rebuild its depleted grain stocks. With carryover stocks of grain remaining below 60 days of world consumption, the world has little more than pipeline supplies. One poor harvest could lead to chaos in world grain markets.

Along with the scarcity of productive new land to bring under the plow and the diminishing response to the use of additional fertilizer in many countries, water scarcity is emerging as a serious constraint on efforts to expand world food production. For example, in North Africa and the Middle East—from Morocco in the west through Iran in the east—water shortages are making it impossible for farmers to keep up with the growth in demand. As countries in the region push against the limits of their water supplies, the growing demand by cities is typically satisfied by diverting irrigation water from farmers. Countries then are forced to import grain to offset the loss of irrigation water. Importing a ton of wheat is the same as importing a thousand tons of water. In 1997, the water required to produce the grain imported into this region was equal to the annual flow of the Nile River.

Under China's north central plain, which supplies nearly 40 percent of the country's grain harvest, the water table is falling by a reported 1.5 meters per year. At some point in the not distant future, aquifer depletion in this region will lead to sharp cutbacks in irrigation water supplies. The bottom line is that if the world is facing a future of water scarcity, it is also facing a future of food scarcity.

An Appetite For Protein

Perhaps the single most important distinguishing feature of dietary changes over the last half-century has been the growing appetite for animal protein. It is hunger for protein that spurred an increase in the world fish catch of nearly fivefold, boosting it from 19 million tons in 1950 to 93 million tons today. (See Figure 1.) This has pushed the oceans to their limits and in some cases beyond. Marine biologists at the U.N. Food and Agriculture Organization report that almost every oceanic fishery is now being fished at or beyond capacity.

As we reach the limits of the oceans to supply animal protein, many countries are turning to aquaculture, or fish farming. The disadvantage of fish farming is that fish in ponds or cages have to be fed—just like chickens in coops. Fish farmers are now competing with poultry and pork producers for grain and protein meal supplements, such as soybean meal.

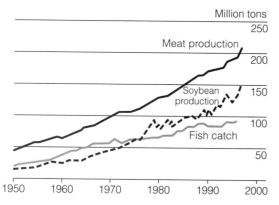

Figure 1 World Protein Trends, 1950–97.
Source: FAO, USDA

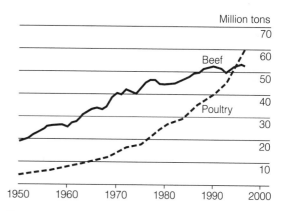

Figure 2 World Poultry and Beef Production, 1950–97.
Source: FAO, USDA

Worldwide, the production of beef and mutton, like that of fish, depends heavily on a natural system — rangelands. And, like oceanic fisheries, rangelands are being pushed to the limits of their carrying capacity and harvest. Once rangelands are fully exploited and substantial growth in beef production can come only from feedlots, then the competition with pork and poultry for grain intensifies. Chickens, which require scarcely 2 kilograms of grain concentrate to produce a kilogram of live weight, have a decided advantage over cattle in the feedlot, which require nearly 7 kilograms of grain per kilogram of weight gain. As a consequence, world poultry production has now overtaken beef for the first time in history. (See Figure 2.)

Beef and pork production, which were running neck and neck from mid-century until 1978, have now separated: in 1997, pork production was easily a third higher than beef. Much of this surge in world pork production came in China, where half the world's pork is now produced and consumed.

One consequence of the growing demand for animal protein has been a dramatic growth in world soybean production over the last 50 years, since pork and poultry producers depend heavily on soybean meal as a supplement to grain in their feed rations. At 152 million tons, the world soybean harvest in 1997 was nine times larger than in 1950.

Although the soybean, the world's leading source of high-quality protein, originated in China, it has found its agronomic and economic niche in the United States, which produces half of the global harvest. Grown largely in rotation with corn, especially in the Corn Belt, the U.S. soybean harvest is now worth far more than the wheat harvest.

Energy Revolution Under Way

Although the changes in the world protein economy are dramatic, to say the least, even more far-reaching changes are in prospect in the world energy economy. Energy historians may remember 1997 as the year in which two of the world's largest oil companies announced they were making major investments in solar and wind energy. With the commitment of $1 billion and $500 million, respectively, by British Petroleum and Royal Dutch Shell to the development of wind, solar, and other renewable energy resources, these leading oil companies have, in effect, become energy companies. And they have indicated that they take the threat of global warming seriously.

From a commercial point of view, it is not too surprising that oil companies are beginning to look at renewable energy resources. Thus far during the 1990s, sales of coal and oil have grown just over 1 percent a year. (See Table 1.) The sale

TABLE 1 Trends in Energy Use, by Source, 1990–97[1]

Energy Source	Annual Rate of Growth (percent)
Wind power	25.7
Solar photovoltaics	16.8
Geothermal power[2]	3.0
Natural gas	2.1
Hydroelectric power[2]	1.6
Oil	1.4
Coal	1.2
Nuclear power	0.6

[1]Energy use measured in varying units: installed generating capacity (megawatts or gigawatts) for wind, geothermal, hydro, and nuclear power; million tons of oil equivalent for oil, natural gas, and coal; megawatts for shipments of solar photovoltaic cells.
[2]1990–96 only.

of natural gas, regarded by many as a transition fuel from the fossil fuel era to the solar/hydrogen age, has been growing at 2 percent a year since 1990. Wind power, meanwhile, has grown at an amazing 26 percent a year. And sales of solar cells, averaging 15 percent annually from 1990 through 1996, jumped by a phenomenal 43 percent in 1997. At the end of the year, an estimated 400,000 homes, most of them in Third World villages, were getting their electricity from solar cell arrays.

Advancing technology is also fueling this growth in solar cell use. The use of a photovoltaic roofing material developed in Japan is now growing by leaps and bounds. The Japanese government plans to have in place 4,600 megawatts of rooftop generating capacity by 2010, an output comparable to the electricity generation of a country the size of Chile.

Corporations in the energy business that are interested in growth are starting to shift investments from oil, coal, and nuclear power, where growth is at a near standstill, to wind and solar, which have rather spectacular growth rates. Once thought of as fringe energy sources, wind and photovoltaic cells are seen increasingly as mainstays of the new energy economy now emerging.

A wind resource survey by the U.S. Department of Energy, for example, concluded that North Dakota, South Dakota, and Texas had enough harnessable wind energy to meet all U.S. electricity needs. Today, the world gets roughly one fifth of its electricity from hydropower, but its potential is dwarfed by that of wind.

The energy revolution is not limited to new sources of energy. It also involves some dramatic gains in the efficiency of energy use. One of these involves the compact fluorescent light bulb, which provides the same amount of light as traditional incandescents, but with less than one fourth as much electricity. Sales of compact fluorescent bulbs have climbed from 45 million in 1988 to 356 million in 1997, an eightfold increase, with China now the leading manufacturer. The estimated 980 million compact fluorescent bulbs in use today lower electricity needs by the output of roughly 100 coal-fired power plants.

The Desire For Mobility

Evidence of the human desire to become more mobile is reflected in sales of vehicles, such as bicycles, motorbikes, and automobiles. Although world production of bikes and cars was roughly the same in 1969, at just over 20 million, the gap between the two has widened dramatically since then. (See Figure 3.) Now more than 100 million bicycles come off the assembly lines each year, compared with fewer than 40 million automobiles. In 1997, car production increased more than 5 percent over 1996. Bicycle production, meanwhile, suffering from too much capacity and excessive inventories, dropped in 1996 (the latest year for which data are available) to 101 million from 109 million the year before.

The enormous differences in the sales volume of bicycles and automobiles reflects more than anything else the number of people reaching the level of affluence that lets them buy bicycles versus the much smaller number who can afford an automobile. In addition, those living in cities, particularly crowded Asian cities, have discovered that they can often be more mobile with a

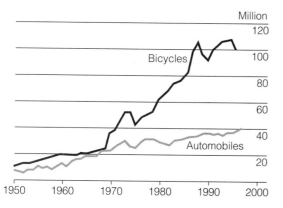

Figure 3 World Bicycle and Automobile Production, 1950–97.
Source: Interbike Directory, UN, AAMA, Standard & Poor's DRI

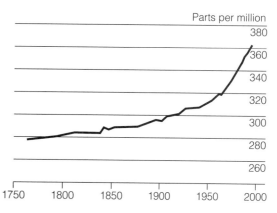

Figure 4 Atmospheric Concentrations of Carbon Dioxide, 1764–1997.
Source: Nature (20 Nov. 1986), Scripps Inst. of Oceanography

modest investment in a bike than with a far larger investment in a car.

Several countries in Europe systematically try to increase bicycle use. In Danish and Dutch cities, an estimated 20 percent and 30 percent respectively of all trips are taken by bicycle. Bikes have also been strongly encouraged in Germany, where use has increased by 50 percent over the last two decades.

In recent years, electric bicycles have begun to attract attention. Relying on a small battery, these provide electrical assistance on hills and in other situations that enable the average speed of the bicycle to increase. The technology is particularly attractive to older riders, to those who have to contend with hilly terrain, or to those who have a particularly long daily commute.

World Getting Warmer

In 1997, carbon emissions, carbon dioxide (CO_2) concentrations in the atmosphere, and the Earth's average temperature all climbed to record highs. Carbon emissions in 1997 totaled 6.3 billion tons, up 1.5 percent from the 6.2 billion tons of 1996. Atmospheric concentrations of CO_2 climbed to 364 parts per million — the highest in 160,000 years. (See Figure 4.) The Intergovernmental Panel on Climate Change, a body of

some 1,500 of the world's leading meteorologists and other scientists, estimates the annual carbon emissions will have to drop below 2 billion tons by 2050 if atmospheric concentrations of CO_2 are to stabilize.

With the record temperature of 1997, the 14 warmest years since recordkeeping began in 1866 have all occurred since 1979. (See Figure 5.) And the 5 warmest have come during the 1990s. Although this strong warming trend over the last two decades does not provide absolute proof of CO_2-induced climate change, it is yet

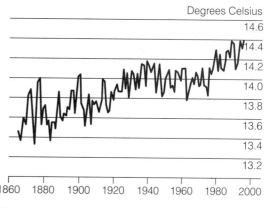

Figure 5 Average Temperature at the Earth's Surface, 1866–97.
Source: Goddard Institue for Space Studies

another piece of evidence that global warming is indeed under way.

Additional evidence can be found in melting icecaps in the Andes, shrinking glaciers in the European Alps, and the shrinkage in the sea ice around Antarctica. The combination of ice melting and the expansion of water from warming has raised average sea level between 10 and 25 centimeters over the last century.

Altering Natural Systems

By far the most visible human alteration of the planet has been the destruction of forests. Almost half the forests that once covered vast expanses of the Earth are already gone. Between 1980 and 1995, the world lost at least 200 million hectares of forest—an area three times as large as Texas. In recent years, the world has experienced an estimated net loss of 16 million hectares a year.

The amount of nitrogen fixed in forms that plants can use through fertilizer manufacturing, the burning of fossil fuels, and the extensive planting of leguminous crops such as soybeans now exceeds the amount fixed by nature. Synthesized nitrogen fertilizer, the use of which has increased ninefold since 1950, is the major form of nitrogen fixation as a result of human activities. Wherever it leads to excessive nutrient runoff, as it does in the Midwest and the lower Mississippi Valley, it often leads to vast algae blooms that then decay, absorbing the free oxygen in the water and depriving fish of oxygen. The hypoxic region, or "dead zone," now formed through this process each year in the Gulf of Mexico is roughly the size of New Jersey.

Closely associated with the burning of fossil fuels is the emission of sulfur dioxide and nitrous oxides, which combine with moisture in the atmosphere to form acid rain. Although emissions of these two pollutants have been sharply reduced in North America and Western Europe, they are still climbing rapidly in Asia. Acid deposition in parts of China is now far higher than the levels reached in Japan in 1975 before that nation established stringent emission limits. Acids can eliminate fish in freshwater lakes, rendering them lifeless.

Another economic activity that is particularly disruptive of the environment is mining. In recent years, mineral exploration has expanded dramatically in developing countries as mines have been depleted in industrial nations. Gold mining is especially disruptive. The 2,400 tons of gold produced in 1997 generated 725 million tons of waste—one ton for every eight people on the planet. In addition to the physical disruption that gold mining brings, the resulting waste includes large quantities of cyanide solution and mercury, which are used to separate gold from the ore.

One of the consequences of the many alterations in the environment just described is an accelerating loss of species. The most recent study of the state of life on Earth by the World Conservation Union-IUCN estimates that 11 percent of all bird species are threatened with extinction. For fish, the figure is far higher—34 percent. In the U.S. Colorado River basin, 29 of 50 native fish species are either endangered or already extinct. Among the 233 species of primates, of which humans are one, half are now threatened with extinction. The surviving populations of some primate species are measured in the hundreds.

Changing Social Conditions

As noted earlier, at the end of 1997 we shared the planet with 80 million more people than at the beginning of the year. Close to 60 percent of these people were added in Asia, in countries that are already densely populated. If recent urbanization trends continue, in a few years—for the first time in human history—more people will live in cities than in the countryside.

Educational levels are rising worldwide. Among the more prominent gains in recent years has been the increase in female education in developing countries. Between 1990 and 1995, female enrollment in some 47 developing countries surveyed by the U.S. Agency for International Development increased from 226 million to 254 million. As a result, nearly 70 percent of girls of primary-school age worldwide were in school in 1995. Notwithstanding this progress, a third of all children in the developing world fail to complete even four years of education.

In industrial countries, the big difference between men and women in educational achievement traditionally has been in graduate degrees in professional schools. But now this, too, is changing. Law and business school enrollments are approaching gender parity. In medical schools in the United States and Canada, more than 40 percent of students are female. For veterinary schools, it is nearly 70 percent. In engineering and architecture schools, however, men still greatly outnumber women.

Of the social trends that affect human health most directly, the spread of HIV is among the most destructive. In 1997, nearly 6 million people were newly infected with the virus that causes AIDS, bringing the total infected worldwide to 42 million. Although a majority of HIV infections are found in Africa, the number of new infections is growing fastest in Asia. Some countries, such as Uganda and Thailand, have made impressive progress in checking the spread of HIV. In sheer numbers, the principal threats today are in India and China: prostitutes in Bombay, India, and intravenous drug users in parts of China have infection rates over 50 percent. If the virus cannot be contained in these early centers of infection, it could spread rapidly in these huge populations, infecting record numbers. With 2.3 million fatalities in 1997, this new disease now claims more than twice as many lives as malaria.

One threat to health that affects far more people than AIDS is cigarette smoking. Roughly half of those who smoke will eventually be killed by the effects of this habit, either through heart disease, stroke, or lung cancer or through one of the many other life-threatening illnesses associated with smoking.

In 1997, the world produced some 5.8 trillion cigarettes, roughly 1,000 for each of its 5.8 billion men, women, and children. The one encouraging sign is that production is not expanding as fast as population. As a result, the number of cigarettes manufactured per person has fallen 4 percent from the all-time high reached in 1990. China is by far the largest manufacturer of cigarettes, followed by the United States.

While cigarette smoking has been declining in industrial countries, it has been expanding rapidly in developing ones. If recent smoking trends continue, tobacco-related deaths, now estimated at 3 million annually, could reach 10 million in 2020, with 70 percent of them being in the developing world. Raising taxes on cigarettes in many countries has helped reduce smoking and the soaring health care costs associated with this deadly habit. In some countries, including Norway, the United Kingdom, and Denmark, the tax per pack of cigarettes now exceeds $4. This compares with a cigarette tax in the United States of 66¢ per pack.

One developing country that is making progress in reducing cigarette smoking is the Philippines, where a combination of an aggressive educational effort and a stiff new tax on cigarettes lowered cigarette production by 16 percent. In some countries, higher cigarette taxes have achieved dramatic results. For example, in New Zealand an increase in the tax per pack of nearly $2 between 1980 and 1991 reduced the cigarettes smoked per person from 4,100 a year to just over 1,500. And in Belgium, a 46-percent increase in cigarette prices between 1985 and 1995 appears to have cut cigarette consumption by one fourth. A number of national governments, as well as California and Massachusetts in the United States, use the cigarette tax proceeds to fund educational programs to discourage smoking.

A Wired World

Although more than a century has passed since Alexander Graham Bell invented the telephone, most people in the world do not yet have ready access to this modern mode of communication. The good news is that the number of new telephone hookups is increasing at 7 percent a year, bringing the total number of hookups to 740 million in 1996. Since 1960, the telephone network has expanded eightfold.

The number of telephones per 100 people varies widely from country to country. The United States, for example, has a telephone density of 60 phones per 100 people, while China has 4. Most of the growth in phone installations is now coming in developing countries, where the number of telephones is increasing by 19 percent a year.

Even more exciting in terms of facilitating communications among people is the worldwide boom in cellular phones, the number of which has increased by more than half each year since 1991. By 1996, there were 135 million cellular phone subscribers worldwide.

Although this technology was first adopted in industrial countries, it is moving even more rapidly in developing ones. In those that have not yet invested in a vast network of telephone lines strung along poles, the cellular phone — linked either by relay towers or, within the next year or two, by a network of satellites — will conserve millions of tons of copper and wood. It enables developing countries to literally leapfrog into the future, avoiding investment in traditional equipment and networks.

Increasing even more rapidly than the number of telephones is the number of computers linked together electronically. In 1997, there were more than 30 million host computers on the Internet. The number of Internet users was far greater, since one host computer could plug several computers into the global network. The number of personal computers now linked together is estimated at more than 100 million.

Of the 100 million or so people who are online, more than half are in the United States. Most of the rest are in Australia, Europe, Canada, and Japan. Thus far, developing countries have only 8 percent of the Internet hookups. But like the telephone network, the Internet is now expanding rapidly in poorer countries. For example, Internet access has grown fivefold in Brazil and Russia in the last two years alone. Hookups are also increasing rapidly in China and India, where the numbers in 2000 are projected at 4 million and 1.5 million, respectively.

One technology facilitating the explosion in electronic communications is satellites. The number of satellites launched annually has exceeded 100 in all but a few years since 1965, typically ranging between 100 and 150. Thirty years ago, just over half the satellites launched were for military purposes — reconnaissance, surveillance, and other military uses. With the end of the cold war, this shifted dramatically. In 1997, only 8 percent of satellites launched were for military uses, while 69 percent were for communications. Over the next decade, some 1,700 additional communications satellites — 10 times the number now in orbit — are scheduled for launch. This new generation of satellites, mostly for low orbit, is expected to revolutionize global communications.

Demilitarization Continues

After peaking in 1984 at $1,140 billion (in 1995 dollars), global military expenditures dropped to $701 billion in 1996, a decline of 39 percent. The United States still accounts for one third of the total. But at $243 billion in 1997, U.S. military outlays were down from some $370 billion in the late 1980s. The most precipitous drop has occurred in Eastern Europe and the former Soviet republics, where expenditures have fallen from $247 billion in 1985 to just $21 billion in 1995.

The number of armed conflicts is also declining. In 1992, the number exceeded 50, but by 1997 it had dropped to 24. In contrast to earlier historical periods, nearly all these armed conflicts were taking place within rather than between nations. They involved government forces, insurgent and guerrilla bands, and drug warlords, among others. Unfortunately, civilians are more often the victims of these conflicts than in earlier eras — rising from 67 percent of the victims in World War II to 90 percent in the 1990s. The heaviest fighting in 1996 was in Afghanistan, Algeria, Sri Lanka, Sudan, and Turkey, typically involving ethnic, tribal, or religious conflicts.

This decline in armed conflicts, particularly in countries such as Croatia and Angola, also reduced the U.N. peacekeeping presence. After peaking at $3.3 billion in 1994, U.N. peacekeeping expenditures dropped to an estimated $1.3 billion in 1997.

Environmental Change: The Fiscal Factor

As analysts have focused on the magnitude of changes needed to convert the existing fossil-fuel-based, automobile-centered, throwaway economy into one that is environmentally sustainable, it is clear than an increasingly popular instrument for doing this is fiscal policy. At present, most governments tax income and saving heavily. But working and saving are constructive activities and should be encouraged. Activities that should be discouraged include carbon emissions, sulfur emissions, the generation of hazardous waste, the use of virgin raw materials (as opposed to recycled ones), and the use of pesticides.

Six European countries have begun this tax shifting process. Sweden, Denmark, Spain, the Netherlands, the United Kingdom, and Finland have all begun reducing taxes on personal income and wages while raising taxes on such things as carbon emissions, vehicle ownership, and landfilling.

The other side of this coin is that governments have long subsidized environmentally destructive activities, with the most important activity being the use of fossil fuels. One reason fossil fuel use and carbon emissions have declined so precipitously in the former Soviet republics and Eastern Europe is that subsidies have been sharply reduced during the 1990s. In 1991, subsidies for fossil fuel use in the former Soviet Union and Eastern Europe exceeded $130 billion. By 1995, this had dropped to $40 billion.

Substantial cuts in fossil fuel subsidies have also occurred in China, from $26 billion to $11 billion thus far during the 1990s. These cuts in China have led to higher coal prices and more-efficient energy use. The United Kingdom was able to cut its carbon emissions during this decade in part because it largely eliminated the subsidies to coal mines, many of which were too inefficient to compete on their own.

Although the world is still in the early stages of restructuring fiscal policies to achieve environmental goals, this approach does promise to accelerate the shift to an environmentally sustainable economy. One advantage of tax policy over regulation is that it enables policymakers to steer the economy in the right direction while exploiting the inherent efficiency of the market.

53
Next Doubling

JENNIFER D. MITCHELL

Adding to the environmental crisis is the growing world population. Nearly 6 billion people now inhabit the planet Earth—almost twice as many as in 1960. At some point over the course of the next century, the world's population could double again, and each of these people will have food and resource needs. But we don't have anything like a century to prevent that next doubling; we probably have less than a decade.

Think about the following questions as you consider this reading:

1. Do we really have a population problem? If so, what can be done about it?

2. Can the world accommodate the growing masses of humanity?

GLOSSARY

Doubling Amount of time it will take for the world's population to double in size.

Demographic projections Estimates of birth and death rates.

Family planning Ability to control the number of pregnancies, usually with available contraception.

Population momentum The next reproductive generation; the group currently entering puberty or younger who will increase population growth.

Infant/child mortality Number of deaths among infants/children per 1,000 population.

IN 1971, WHEN BANGLADESH won independence from Pakistan, the two countries embarked on a kind of unintentional demographic experiment. The separation had produced two very similar populations: both contained some 66 million people and both were growing at about 3 percent a year. Both were overwhelmingly poor,
rural, and Muslim. Both populations had similar views on the "ideal" family size (around four children); in both cases, that ideal was roughly two children smaller than the actual average family. And in keeping with the Islamic tendency to encourage large families, both generally disapproved of family planning.

But there was one critical difference. The Pakistani government, distracted by leadership crises and committed to conventional ideals of economic growth, wavered over the importance of family planning. The Bangladeshi government did not: as early as 1976, population growth had been declared the country's number one problem, and a national network was established to educate people about family planning and supply them with contraceptives. As a result, the proportion of couples using contraceptives rose from around 6 percent in 1976 to about 50 percent today, and fertility rates have dropped from well over six children per woman to just over three. Today, some 120 million people live in Bangladesh, while 140 million live in Pakistan— a difference of 20 million.

Bangladesh still faces enormous population pressures—by 2050, its population will probably have increased by nearly 100 million. But even so, that 20 million person "savings" is a colossal achievement, especially given local conditions. Bangladeshi officials had no hope of producing the classic "demographic transition," in which improvements in education, health care, and general living standards tend to push down the birth rate. Bangladesh was—and is—one of the poorest and most densely populated countries on earth. About the size of England and

World Watch. *January/February 1998, pp. 21–27 (excerpts). Reprinted by permission of the publisher.*

Wales, Bangladesh has twice as many people. Its per capital GDP is barely over $200. It has one doctor for every 12,500 people and nearly three-quarters of its adult population are illiterate. The national diet would be considered inadequate in any industrial country, and even at current levels of population growth, Bangladesh may be forced to rely increasingly on food imports.

All of these burdens would be substantially heavier than they already are, had it not been for the family planning program. To appreciate the Bangladeshi achievement, it's only necessary to look at Pakistan: those "additional" 20 million Pakistanis require at least 2.5 million more houses, about 4 million more tons of grain each year, millions more jobs, and significantly greater investments in health care—or a significantly greater burden of disease. Of the two nations, Pakistan has the more robust economy—its per capita GDP is twice that of Bangladesh. But the Pakistani economy is still primarily agricultural, and the size of the average farm is shrinking, in part because of the expanding population. Already, one fourth of the country's farms are under 1 hectare, the standard minimum size for economic viability, and Pakistan is looking increasingly towards the international grain markets to feed its people. In 1997, despite its third consecutive year of near-record harvests, Pakistan attempted to double its wheat imports but was not able to do so because it had exhausted its line of credit.

And Pakistan's extra burden will be compounded in the next generation. Pakistani women still bear an average of well over five children, so at the current birth rate, the 10 million or so extra couples would produce at least 50 million children. And these in turn could bear nearly 125 million children of their own. At its current fertility rate, Pakistan's population will double in just 24 years—that's more than twice as fast as Bangladesh's population is growing. H. E. Syeda Abida Hussain, Pakistan's Minister of Population Welfare, explains the problem bluntly: "If we achieve success in lowering our population growth substantially, Pakistan has a future. But if, God forbid, we should not—no future."

The Three Dimensions of the Population Explosion

. . . About 5.9 billion people currently inhabit the Earth. By the middle of the next century, according to U.N. projections, the population will probably reach 9.4 billion—and all of the net increase is likely to occur in the developing world. (The total population of the industrial countries is expected to decline slightly over the next 50 years.) Nearly 60 percent of the increase will occur in Asia, which will grow from 3.4 billion people in 1995 to more than 5.4 billion in 2050. China's population will swell from 1.2 billion to 1.5 billion, while India's is projected to soar from 930 million to 1.53 billion. In the Middle East and North Africa, the population will probably more than double, and in sub-Saharan Africa, it will triple. By 2050, Nigeria alone is expected to have 339 million people—more than the entire continent of Africa had 35 years ago.

Despite the different demographic projections, no country will be immune to the effects of population growth. Of course, the countries with the highest growth rates are likely to feel the greatest immediate burdens—on their educational and public health systems, for instance, and on their forests, soils, and water as the struggle to grow more food intensifies. Already some 100 countries must rely on grain imports to some degree, and 1.3 billion of the world's people are living on the equivalent of $1 a day or less.

But the effects will ripple out from these "front-line" countries to encompass the world as a whole. Take the water predicament in the Middle East as an example. According to Tony Allan, a water expert at the University of London, the Middle East "ran out of water" in 1972, when its population stood at 122 million. At that point, Allan argues, the region had begun to draw more water out of its aquifers and rivers than the rains were replenishing. Yet today, the region's population is twice what it was in 1972 and still growing. To some degree, water management now determines political destiny. In Egypt, for example, President Hosni Mubarak has announced a

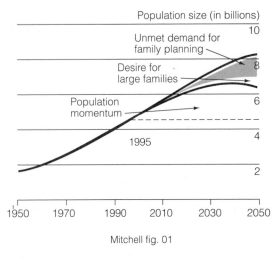

Mitchell fig. 01

Population of Developing Countries, 1950–95, with Projected Growth to 2050.
Source: U.N., *World Population Prospects: the 1996 Revision* (New York: forthcoming); and John Bongaarts, "Popultion Policy Options in the Developing World," *Science,* 11 February 1994.

$2 billion diversion project designed to pump water from the Nile River into an area that is now desert. The project—Mubarak calls it a "necessity imposed by population"—is designed to resettle some 3 million people outside the Nile flood plain, which is home to more than 90 percent of the country's population.

Elsewhere in the region, water demands are exacerbating international tensions; Jordan, Israel, and Syria, for instance, engage in uneasy competition for the waters of the Jordan River basin. Jordan's King Hussein once said that water was the only issue that could lead him to declare war on Israel. Of course, the United States and the western European countries are deeply involved in the region's antagonisms and have invested heavily in its fragile states. The western nations have no realistic hope of escaping involvement in future conflicts.

Yet the future need not be so grim. . . . The first step is to understand the causes of population growth. John Bongaarts, vice president of the Population Council, a non-profit research group in New York City, has identified three basic factors. (See figure above.)

UNMET DEMAND FOR FAMILY PLANNING

In the developing world, at least 120 million married women—and a large but undefined number of unmarried women—want more control over their pregnancies, but cannot get family planning services. This unmet demand will cause about one-third of the projected population growth in developing countries over the next 50 years, or an increase of about 1.2 billion people.

DESIRE FOR LARGE FAMILIES

Another 20 percent of the projected growth over the next 50 years, or an increase of about 660 million people, will be caused by couples who may have access to family planning services, but who choose to have more than two children. (Roughly two children per family is the "replacement rate," at which a population could be expected to stabilize over the long term.)

POPULATION MOMENTUM

By far the largest component of population growth is the least commonly understood. Nearly one-half of the increase projected for the next 50 years will occur simply because the next reproductive generation—the group of people currently entering puberty or younger—is so much larger than the current reproductive generation. Over the next 25 years, some 3 billion people—a number equal to the entire world population in 1960—will enter their reproductive years, but only about 1.8 billion will leave that phase of life. Assuming that the couples in this reproductive bulge begin to have children at a fairly early age, which is the global norm, the global population would still expand by 1.7 billion, even if all those couples had only two children—the long-term replacement rate.

Meeting the Demand

Over the past three decades, the global percentage of couples using some form of family planning has increased dramatically—from less than 10 to more than 50 percent. But due to the growing population, the absolute number of women not using family planning is greater today than it was 30 years ago. Many of these women fall into that first category above—they want the services but for one reason or another, they cannot get them.

Sometimes the obstacle is a matter of policy: many governments ban or restrict valuable methods of contraception. In Japan, for instance, regulations discourage the use of birth control pills in favor of condoms, as a public health measure against sexually transmitted diseases. A study conducted in 1989 found that some 60 countries required a husband's permission before a woman can be sterilized; several required a husband's consent for all forms of birth control.

Elsewhere, the problems may be more logistical than legal. Many developing countries lack clinics and pharmacies in rural areas. In some rural areas of sub-Saharan Africa, it takes an average of two hours to reach the nearest contraceptive provider. And often contraceptives are too expensive for most people. Sometimes the products or services are of such poor quality that they are not simply ineffective, but dangerous. A woman who has been injured by a badly made or poorly inserted IUD may well be put off by contraception entirely.

In many countries, the best methods are simply unavailable. Sterilization is often the only available nontraditional option, or the only one that has gained wide acceptance. Globally, the procedure accounts for about 40 percent of contraceptive use and in some countries the fraction is much higher: in the Dominican Republic and India, for example, it stands at 69 percent. But women don't generally resort to sterilization until well into their childbearing years, and in some countries, the procedure isn't permitted until a woman reaches a certain age or bears a certain number of children. Sterilization is therefore no substitute for effective temporary methods like condoms, the pill, or IUDs.

There are often obstacles in the home as well. Women may be prevented from seeking family planning services by disapproving husbands or in-laws. In Pakistan, for example, 43 percent of husbands object to family planning. Frequently, such objections reflect a general social disapproval inculcated by religious or other deeply-rooted cultural values. And in many places, there is a crippling burden of ignorance: women simply may not know what family planning services are available or how to obtain them.

Yet there are many proven opportunities for progress, even in conditions that would appear to offer little room for it. In Bangladesh, for instance, contraception was never explicitly illegal, but many households follow the Muslim custom of *purdah,* which largely secludes women in the communities. Since it's very difficult for such women to get to family planning clinics, the government brought family planning to them: some 30,000 female field workers go door-to-door to explain contraceptive methods and distribute supplies. Several other countries have adopted Bangladesh's approach. Ghana, for instance, has a similar system, in which field workers fan out from community centers. And even Pakistan now deploys 12,000 village-based workers, in an attempt to reform its family planning program, which still reaches only a quarter of the population.

Reducing the price of contraceptives can also trigger a substantial increase in use. In poor countries, contraceptives can be an extremely price-sensitive commodity even when they are very cheap. Bangladesh found this out the hard way in 1990, when officials increased contraceptive prices an average of 60 percent. (Under the increases, for example, the cheapest condoms cost about 1.25 U.S. cents per dozen.) Despite regular annual sales increases up to that point, the market slumped immediately: in 1991, condom sales fell by 29 percent and sales of the pill by 12 percent. The next year, prices were rolled back; sales rebounded and have grown steadily since then.

Additional research and development can help broaden the range of contraceptive options. Not all methods work for all couples, and the lack of a suitable method may block a substantial amount of demand. Some women, for instance, have side effects to the pill; others may not be able to use IUDs because of reproductive tract infections. The wider the range of available methods, the better the chance that a couple will use one of them.

Planning the Small Family

Simply providing family planning services to people who already want them won't be enough to arrest the population juggernaut. In many countries, large families are still the ideal. In Senegal, Cameroon, and Niger, for example, the average woman still wants six or seven children. A few countries have tried to legislate such desires away. In India, for example, the Ministry of Health and Family Welfare is interested in promoting a policy that would bar people who have more than two children from political careers, or deny them promotion if they work within the civil service bureaucracy. And China's well-known policy allows only one child per family.

But coercion is not only morally questionable—it's likely to be ineffective because of the backlash it invites. A better starting point for policy would be to try to understand why couples want large families in the first place. In many developing countries, having lots of children still seems perfectly rational: children are a source of security in old age and may be a vital part of the family economy. Even when they're very young, children's labor can make them an asset rather than a drain on family income. And in countries with high child mortality rates, many births may be viewed as necessary to compensate for the possible deaths (of course, the cumulative statistical effect of such a reaction is to *over*-compensate).

Religious or other cultural values may contribute to the big family ideal. In Pakistan, for instance where 97 percent of the population is Muslim, a recent survey of married women found that almost 60 percent of them believed that the number of children they have is "up to God." Preference for sons is another widespread factor in the big family psychology: many large families have come about from a perceived need to bear at least one son. In India, for instance, many Hindus believe that they need a son to perform their last rites, or their souls will not be released from the cycle of births and rebirths. Lack of a son can mean abandonment in this life too. Many husbands desert wives who do not bear sons. Or if a husband dies, a son is often the key to a woman's security: 60 percent of Indian women over 60 are widows, and widows tend to rely on their sons for support. In some castes, a widow has no other option since social mores forbid her from returning to her birth village or joining a daughter's family. Understandably, the fear of abandonment prompts many Indian women to continue having children until they have a son. It is estimated that if son preference were eliminated in India, the fertility rate would decline by 8 percent from its current level of 3.5 children per woman.

Yet even deeply rooted beliefs are subject to reinterpretation. In Iran, another Muslim society, fertility rates have dropped from seven children per family to just over four in less than three decades. The trend is due in some measure to a change of heart among the government's religious authorities, who had become increasingly concerned about the likely effects of a population that was growing at more than 3 percent per year. In 1994, at the International Conference on Population and Development (ICPD) held in Cairo, the Iranian delegation released a "National Report on Population" which argued that according to the "quotations from prophet Mohammad . . . and verses of [the] holy Quran, what is standing at the top priority for the Muslims' community is the social welfare of Muslims." Family planning, therefore, "not only is not prohibited but is emphasized by religion."

Promotional campaigns can also change people's assumptions and behavior, if the campaigns fit into the local social context. Perhaps the most successful effort of this kind is in Thailand, where Mechai Viravidaiya, the founder of the Thai Population and Community Development Association, started a program that uses witty songs, demonstrations, and ads to encourage the use of contraceptives. The program has helped foster widespread awareness of family planning throughout Thai society. Teachers use population-related examples in their math classes; cab drivers even pass out condoms. Such efforts have paid off: in less than three decades, contraceptive use among married couples has risen from 8 to 75 percent and population growth has slowed from over 3 percent to about 1 percent—the same rate as in the United States.

Better media coverage may be another option. In Bangladesh, a recent study found that while local journalists recognize the importance of family planning, they do not understand population issues well enough to cover them effectively and objectively. The study, a collaboration between the University Research Corporation of Bangladesh and Johns Hopkins University in the United States, recommended five ways to improve coverage: develop easy-to-use information for journalists (press releases, wall charts, research summaries), offer training and workshops, present awards for population journalism, create a forum for communication between journalists and family planning professionals, and establish a population resource center or data bank.

Often, however, the demand for large families is so tightly linked to social conditions that the conditions themselves must be viewed as part of the problem. Of course, those conditions vary greatly from one society to the next, but there are some common points of leverage:

Reducing child mortality helps give parents more confidence in the future of the children they already have. Among the most effective ways of reducing mortality are child immunization programs, and the promotion of "birth spacing"—lengthening the time between births. (Children born less than a year and a half apart are twice as likely to die as those born two or more years apart.)

Improving the economic situation of women provides them with alternatives to child-bearing. In some countries, officials could reconsider policies or customs that limit women's job opportunities or other economic rights, such as the right to inherit property. Encouraging "microlenders" such as Bangladesh's Grameen Bank can also be an effective tactic. In Bangladesh, the Bank has made loans to well over a million villagers—mostly impoverished women—to help them start or expand small businesses.

Improving education tends to delay the average age of marriage and to further the two goals just mentioned. Compulsory school attendance for children undercuts the economic incentive for larger families by reducing the opportunities for child labor. And in just about every society, higher levels of education correlate strongly with smaller families.

Momentum: The Biggest Threat of All

The most important factor in population growth is the hardest to counter—and to understand. Population momentum can be easy to overlook because it isn't directly captured by the statistics that attract the most attention. The global growth rate, after all, is dropping: in the mid-1960s, it amounted to about a 2.2 percent annual increase; today the figure is 1.4 percent. The fertility rate is dropping too: in 1950, women bore an average of five children each; now they bear roughly three. But despite these continued declines, the absolute number of births won't taper off any time soon. According to U.S. Census Bureau estimates, some 130 million births will still occur annually for the next 25 years, because of the sheer number of women coming into their childbearing years.

The effects of momentum can be seen readily in a country like Bangladesh, where more than 42 percent of the population is under 15 years old—a typical proportion for many poor countries. Some 82 percent of the population growth projected for Bangladesh over the next half century will be caused by momentum. In other words, even if from now on, every Bangladeshi couple were to have only two children, the country's population would still grow by 80 million by 2050 simply because the next reproductive generation is so enormous.

The key to reducing momentum is to delay as many births as possible. To understand why delay works, it's helpful to think of momentum as a kind of human accounting problem, in which a large number of births in the near term won't be balanced by a corresponding number of deaths over the same period of time. One side of the population ledger will contain those 130 million annual births (not all of which are due to momentum, of course), while the other side will contain only about 50 million annual deaths. So to put the matter in a morbid light, the longer a substantial number of those births can be delayed, the longer the death side of the balance sheet will be when the births eventually occur. In developing countries, according to the Population Council's Bongaarts, an average 2.5-year delay in the age when a woman bears her first child would reduce population growth by over 10 percent.

One way to delay childbearing is to postpone the age of marriage. In Bangladesh, for instance, the median age of first marriage among women rose from 14.4 in 1951 to 18 in 1989, and the age at first birth followed suit. Simply raising the legal age of marriage may be a useful tactic in countries that permit marriage among the very young. Educational improvements, as already mentioned, tend to do the same thing. A survey of 23 developing countries found that the median age of marriage for women with secondary education exceeded that of women with no formal education by four years.

Another fundamental strategy for encouraging later childbirth is to help women break out of the "sterilization syndrome" by providing and promoting high-quality, temporary contraceptives. Sterilization might appear to be the ideal form of contraception because it's permanent. But precisely because it's permanent, women considering sterilization tend to have their children early, and then resort to it. A family planning program that relies heavily on sterilization may therefore be working at cross purposes with itself: when offered as a primary form of contraception, sterilization tends to promote early childbirth.

What Happened to the Cairo Pledges?

At the 1994 Cairo Conference, some 180 nations agreed on a 20-year reproductive health package to slow population growth. The agreement called for a progressive rise in annual funding over the life of the package; according to

U.N. estimates, the annual price tag would come to about $17 billion by 2000 and $21.7 billion by 2015. Developing countries agreed to pay for two thirds of the program, while the developed countries were to pay for the rest. On a global scale, the package was fairly modest: the annual funding amounts to less than two weeks' worth of global military expenditures.

Today, developing country spending is largely on track with the Cairo agreement, but the developed countries are not keeping their part of the bargain. According to a recent study by the U.N. Population Fund (UNFPA), all forms of developed country assistance (direct foreign aid, loans from multilateral agencies, foundation grants, and so on) amounted to only $2 billion in 1995. That was a 24 percent increase over the previous year, but preliminary estimates indicate that support declined some 18 percent in 1996 and last year's funding levels were probably even lower than that.

The United States, the largest international donor to population programs, is not only failing to meet its Cairo commitments, but is toying with a policy that would undermine international family planning efforts as a whole. Many members of the U.S. Congress are seeking reimposition of the "Mexico City Policy" first enunciated by President Ronald Reagan at the 1984 U.N. population conference in Mexico City, and repealed by the Clinton administration in 1993. Essentially, a resurrected Mexico City Policy would extend the current U.S. ban on funding abortion services to a ban on funding any organization that:

- funds abortions directly, or
- has a partnership arrangement with an organization that funds abortions, or
- provides legal services that may facilitate abortions, or
- engages in any advocacy for the provision of abortions, or
- participates in any policy discussions about abortion, either in a domestic or international forum.

The ban would be triggered even if the relevant activities were paid for entirely with non-U.S. funds. Because of its draconian limits even on speech, the policy has been dubbed the "Global Gag Rule" by its critics, who fear that it could stifle, not just abortion services, but many family planning operations involved only incidentally with abortion. Although Mexico City proponents have not managed to enlist enough support to reinstate the policy, they have succeeded in reducing U.S. family planning aid from $547 million in 1995 to $385 million in 1997. They have also imposed an unprecedented set of restrictions that meter out the money at the rate of 8 percent of the annual budget per month—a tactic that *Washington Post* reporter Judy Mann calls "administrative strangulation."

If the current underfunding of the Cairo program persists, according to the UNFPA study, 96 million fewer couples will use modern contraceptives in 2000 than if commitments had been met. One-third to one-half of those couples will resort to less effective traditional birth control methods; the rest will not use any contraceptives at all. The result will be an additional 122 million unintended pregnancies. Over half of those pregnancies will end in births, and about 40 percent will end in abortions. (The funding shortfall is expected to produce 16 million more abortions in 2000 alone.) The unwanted pregnancies will kill about 65,000 women by 2000, and injure another 844,000.

Population funding is always vulnerable to the illusion that the falling growth rate means the problem is going away. Worldwide, the annual population increase has dropped from a high of 87 million in 1988 to 80 million today. But dismissing the problem with that statistic is like comforting someone stuck on a railway crossing with the news that an oncoming train has slowed from 87 to 80 kilometers an hour, while its weight has increased. It will now take 12.5 years instead of 11.5 years to add the next billion people to the world. But that billion will surely arrive—and so will at least one more billion. Will still more billions follow? That, in large measure, depends on what policymakers do now. Funding alone will not ensure that population stabilizes, but lack of funding will ensure that it does not.

The Next Doubling

In the wake of the Cairo conference, most population programs are broadening their focus to include improvements in education, women's health, and women's social status among their many goals. These goals are worthy in their own right and they will ultimately be necessary for bringing population under control. But global population growth has gathered so much momentum that it could simply overwhelm a developing agenda. Many countries now have little choice but to tackle their population problem in as direct a fashion as possible—even if that means temporarily ignoring other social problems. Population growth is now a global social emergency. Even as officials in both developed and developing countries open up their program agendas, it is critical that they not neglect their single most effective tool for dealing with that emergency: direct expenditures on family planning.

The funding that is likely to be the most useful will be constant, rather than sporadic. A fluctuating level of commitment, like sporadic condom use, can end up missing its objective entirely. And wherever it's feasible, funding should be designed to develop self-sufficiency—as, for instance, with UNFPA's $1 million grant to Cuba, to build a factory for making birth control pills. The factory, which has the capacity to turn out 500 million tablets annually, might eventually even provide the country with a new export product. Self-sufficiency is likely to grow increasingly important as the fertility rate continues to decline. As Tom Merrick, senior population advisor at the World Bank explains, "while the need for contraceptives will not go away when the total fertility rate reaches two—the donors will."

Even in narrow, conventional economic terms, family planning offers one of the best development investments available. A study in Bangladesh showed that for each birth prevented, the government spends $62 and saves $615 on social services expenditures—nearly a tenfold return. The study estimated that the Bangladesh program prevents 890,000 births a year, for a net annual savings of $547 million. And that figure does not include savings resulting from lessened pressure on natural resources.

Over the past 40 years, the world's population has doubled. At some point in the latter half of the next century, today's population of 5.9 billion could double again. But because of the size of the next reproductive generation, we probably have only a relatively few years to stop that next doubling. To prevent all of the damage—ecological, economic, and social—that the next doubling is likely to cause, we must begin planning the global family with the same kind of urgency that we bring to matters of trade, say, or military security. Whether we realize it or not, our attempts to stabilize population—or our failure to act—will likely have consequences that far outweigh the implications of the military or commercial crisis of the moment. Slowing population growth is one of the greatest gifts we can offer future generations.

54

America's Cities: Centers of Culture, Commerce, and Community—or Collapsing Hope?

RAYMOND L. FLYNN

As U.S. citizens with resources migrate to suburbs and rural areas, inner-cities become concentrations of poverty and problems stemming from poor conditions. Flynn outlines reasons why cities deserve attention, help, and support—and how they came to their present situation. He then argues for what should be done to help cities.

Think about the following questions as you consider this reading:

1. What factors have contributed to the problems faced by cities?

2. Why does Flynn argue that we should turn our attention to the problems of the cities?

GLOSSARY

Utilitarian approach Considering resources available in the economic system of the city.

PERHAPS THE GREATEST OBSTACLE facing cities today is the changing nature of the definition of *city.* The term *city* formerly signified a social center wherein large populations gathered to live, to exchange goods and ideas, and to develop and sustain a system that provided for the needs of its inhabitants. The very word had connotations of hopelessness, a place where "they" live. People demand greater measures against crime, wel-

Urban Affairs Review, *Vol. 30, No. 5, May 1995, 635–640. © 1995 Sage Publications, Inc. Reprinted by permission of Sage Publications, Inc.*

fare fraud, and illegal immigration. Underlying these demands, however, is the sentiment held by many Washington officials that few resources should be dedicated to urban areas—and to those who dwell within them.

In 1968, the Kerner Commission (U.S. National Advisory Commission on Civil Disorders) issued a warning that America was in danger of being divided into two nations: one white, one black. Presently, the United States faces the prospect of becoming a *gated community*—confining the poor within the city limits, separating them from those better off in the suburbs. Instead of seeking solutions to the problems of the cities, the cities themselves, along with the people living in them, have been incorrectly identified as the problem. If this misperception continues, more will be at stake than our cities. Indeed, the very values on which our nation was founded—equality, and life, liberty, and the pursuit of happiness—will be placed in jeopardy.

The question has been asked, Why should we concern ourselves with cities? It has been suggested by some high-ranking officials and sociologists that cities have outlived their usefulness. It is argued that new technology and the world economy have made cities obsolete and that we should discard them like unproductive units in a company that needs downsizing.

This utilitarian approach to the modern city ignores the reality that cities are made up of much more than material and human resources. The people are the heart of the city and cannot be reduced to a pool of disposable "goods" in an economic system. Cities are much more than economic entities; therefore, the human side of urban life cannot be ignored.

There are many compelling motives for turning our attention to the problems of the modern city. Among them are the following:

1. Cities have always been, and will always be, places of refuge, where those in need seek the support and comfort of others. They are centers for opportunities and hopes, where ideas, talents, and native intelligence are translated into a mutually energizing and life-giving environment conducive to the development of both culture and commerce. The historic roots of our nation remind us that nearly all of our families entered the American mainstream through cities. Most of these families arrived by ship, crossing one border or another, legally or illegally (and, many times, in the "gray area" in between). Cities in the United States kept the promise inscribed at the base of the Statue of Liberty—to receive "Your tired, your poor, your huddled masses yearning to breathe free." No matter how far we may have come since then, we cannot forget the values of the cities that were home to them. To do so would be hypocritical, denying to new immigrants the promise offered to our ancestors by American cities.

2. From a purely economic perspective, it would actually be less expensive to spend more rather than less on cities and the people living within them. The cost of urban misery is astronomic. From furnishing prison beds to caring for low-birth-weight babies, from providing for health care for AIDS victims and the elderly to feeding the urban poor, the cost of the *barely living index* is exorbitant. This growing moral deficit pulls not only on our consciences but also on our economy. The expense of preventive programs can reduce the cost of urban neglect.

3. From a socioeconomic perspective, saving urban America might be in everyone's self-interest. It seems that the rumors of the death—and decrease in importance—of cities are greatly exaggerated. Cities are again seen for what they have always been—economic engines that create and distribute wealth. In an upcoming book, Neil Pierce argues that *city-states* are replacing nations as the key units of production in the modern global economy (Spence 1994, 11). Michael Porter, author of *The Competitive Advantage of Nations* (1990), talks about the "untapped economic potential" of cities, especially as hosts for the "clusters" of industry he sees

as the driving force in the new economy (Porter 1994, 11). Yes, capital is mobile, but it has to land somewhere. Invariably, it is in cities. But which ones? A new school of thought, with proponents such as Paul Romer, an economist at the University of California at Berkeley, Lester Thurow of M.I.T., and Michael Porter of Harvard, holds that cities attract investment to the degree that they can bridge the income gap with their surrounding suburbs. Romer states that "maybe even the rich can be worse off from inequality" (Bernstein 1994, 79).

These sentiments are being echoed on the political front by Democrats and Republicans alike. Labor Secretary Robert B. Reich recently warned that "A society divided between the 'haves' and the 'have-nots' or between the well educated and the poorly educated . . . cannot be prosperous or stable" (Bernstein 1994, 79). Republican theorist Kevin Phillips, who traces the growing inequality to a transfer of wealth from the middle class not *down* to the poor but *up* to the rich (Bernstein 1994, 79), agrees with this assessment. He remarks that economic stratification is contrary to the American sense of fairness and equality.

Where did we go wrong? How did we lose the idea of equal opportunity that has been part and parcel of city life? At the moment, it is fashionable to ascribe the plight of our cities to the failure of the urban policies of the 1960s and 1970s. Fashionable, but false. There are at least four factors that have contributed to the present situation.

1. Even as the urban policies of the 1960s and 1970s were being initiated, the "suburbanization" policies that began in the 1950s were continuing. Superhighway subsidies and low-interest mortgages accelerated the process of urban disinvestment. Cities began to spruce up their front yards and put out the welcome mats while the moving trucks were pulling up to the back door, carrying away not only the furniture but, more important, the families that form the fabric of a strong and vibrant community.

2. Those who did stay to "fight the good (urban) fight" found themselves embroiled in an unproductive and unnecessary civil war (well documented by urban expert Nicholas Lemann, 1991) over whether these new policies should be administered from the bottom up (by community-based organizations) or from the top down (by local government). It is not clear who won that war, but it is clear who lost—the cities and the people in them. It is also clear that with few exceptions, mayors began to see themselves more as CEOs than as community champions, while people in the neighborhoods increasingly found themselves having to fight City Hall.

3. The urban policies of the 1960s and 1970s were preempted by the "What's in it for me?" policies of the 1980s. Tax and investment policies were enacted by an antiurban administration in Washington that favored the wealthy corporations at the expense of the community. This political about-face prevented any progress that had begun in urban areas from taking root.

4. Finally, America still has not dealt with the issue of race. Federal government mandates, quota systems, and reckless policies have divided poor whites and blacks, pitting one against the other. Until we deal with this problem, our urban areas will remain fragmented.

So what are we going to do about it? . . . Let's begin by not repeating the mistakes of the past. Let's recognize the importance of U.S. cities and support them, just as we support any valuable institution in American society, such as home ownership and business investment. It is imperative to encourage ownership and investment in our cities—by individuals and corporations—at least as much as we do in the suburbs. We need to promote policies that will halt the flight of the working middle class, the backbone of our society, from our cities.

Too costly? Many say so. However, those who call for cuts in support to the cities might eventually have to consider equal cuts in the suburbs. No enterprise zones downtown? Fine. But let's stop building express roads to the suburban shopping malls, roads that carry away both shoppers and jobs.

Further, let's not force a false choice between community and local government. During my 10 years as mayor, the city of Boston was able to enjoy unprecedented success in building affordable housing by collaborating with community development corporations, in promoting jobs for Boston residents by working together with employers and unions, in caring for the hungry and the homeless by uniting our efforts with a network of charitable organizations, in providing quality community health care by working with neighborhood-based health centers, and in fighting crime by facilitating cooperation between police and residents to form "crime-watch" groups. Citizens and governments have enough to fight against without fighting each other.

Moreover, mayors should be the leaders in working for economic and social justice. They should be out in the communities, fighting for the rights of their people in the neighborhoods and not just in the boardrooms, up at the State House (where much of the political power has shifted), and down in Washington. The present generation of "button-down" mayors needs to return to a more grassroots approach if they want their constituents to recognize that they are working for their benefit and to avoid the divisiveness of a citizen-versus-City-Hall mentality. Urban America needs players, not spectators; fighters, not promoters; activists, not actors.

I believe that city mayors have some powerful and active allies in their effort to serve the well-being of their citizens. One such ally is the religious community. I have some experience in this area and can personally testify that the Catholic Church, for example, is not motivated by what is considered liberal or conservative or by labels such as Democrat or Republican but, rather, by the quest for Truth and Justice. The Catholic Church may be perceived as conservative on moral issues, but it is liberal and progressive regarding economic and social issues such as strong

concern for working families and the needy (once traditional Democratic voters). This, of course, is true for other religious organizations as well.

You have only to read the documents from the Annual U.S. Bishop's Conference to be convinced that on many social and economic issues, the positions of the Catholic Church are very much like those of the Clinton administration, whose agenda supports working families, the needy, and the American cities. Furthermore, their stated positions are in strict opposition to those set forth in the "Contract with America." Although the Catholic Church does not support the Democratic party platform on abortion, it is equally opposed to the Republican party's policies of taking poor mothers off welfare and putting their kids in orphanages or cutting services to the needy to give big tax breaks to the rich.

It is essential to establish urban policies that join "doing well" with "doing good." Boston's linkage program is a perfect example of this. Through linkages, developers of large, downtown projects contribute to a fund for construction of affordable housing, as well as for job training in the neighborhoods. Between 1984 and 1995, the linkage payments to the city totaled more than $50 million. It was obviously no disincentive to business, because nearly $1 billion a year was being invested simultaneously in downtown development. As a result of linkage and Boston's jobs policy, the city was one of the few that cut its poverty rate in the 1980s. This kind of linkage is effective because it employs the economic principles of Romer, Thurow, and others, applying them to the actual marketplace. The expense of such a program is a reasonable price for doing business because it makes the city a better place in which to conduct business.

It should be recognized that the city has been, and must continue to be, a place that brings people together. Racial and ethnic harmony is the most important ingredient for the health of an urban center. As mayor of Boston, I spent half my time working to promote racial harmony and wished that I could have spent even more. Boston made tremendous progress in this area during my term of office, but as in any city, there is still much to be accomplished. It is essential to recognize the importance of forming a community as fundamental to the life of any city and to realize that racial harmony is a journey, not a destination.

While we're doing that, let's remember the most successful urban social program that I have ever seen — good jobs in the private sector. From the beginning of the industrial age until very recently, jobs were what cities were all about. They provided unskilled or semiskilled workers with the kinds of jobs that allowed them to raise families, buy homes, send children to college, and improve their general well-being. When my father was 16 years old, he went to work on the docks of South Boston and was able to support himself and his family as a longshoreman, and would continue to do so for the next 50 years. He couldn't do that today.

Cities are under siege today because society hasn't yet been able to prepare the people who live in them for the kinds of jobs that are being created in the new economy. A recent statement described the people in our cities today as trying to climb a ladder with the bottom rungs sawn off. We've got to put those rungs back onto the ladder — and the jobs back into our cities.

Finally, we must remake the cities of America into centers of community, commerce, and culture. We must bring cities back because they work — and they make this country work. We must bring cities back if we're going to remember who we are, where we came from, and what we hope to be. We must bring cities back if we're going to continue to care.

REFERENCES

Bernstein, A., 1994. Inequality: How the gap between rich and poor hurts the economy. *Business Week,* 15 August, 78–83.

Lemann, N., 1991. *The promised land: The great black migration and how it changed America.* New York: Alfred A. Knopf.

Porter, M., 1990. *The competitive advantage of nations.* New York: Free Press.

_____, 1994. The competitive advantage of the inner city. Report. Harvard Business School, June.

Spence, H., 1994. A matter of survival: The economic and social well-being of the region is at stake. *Boston Globe* special edition, 21 November, 11.

U.S. National Advisory Commission on Civil Disorders. (Kerner Commission). 1968. Report. Washington, DC: Government Printing Office.

55
Why We Have Homelessness

PETER H. ROSSI

Among the problems posed by urban life are the urban poor, some of whom end up on the streets, as discussed by Peter Rossi. The poorest members of society often do not have basic living necessities such as housing. The world over, people live in shacks constructed of any materials available, without water or sanitary conditions. In this excerpt from Down and Out in America, *Rossi discusses factors related to the reasons for homelessness and the lack of adequate and affordable housing.*

Consider these questions in relation to this reading:

1. What are the causes of homelessness?

2. What can be done to remedy the problem of homelessness?

GLOSSARY

Shantytowns Poverty-stricken living areas, usually near large cities, lacking in basic necessities such as water and sanitary disposal, with shacks improvised from whatever materials are available.

SROs Single-room-occupancy housing.

Episodic homelessness The moving of individuals in and out of poverty and homelessness depending on their current situations.

Chronic homelessness Long-term homelessness often caused by impaired earning capacity.

Blaming the victim Faulting the homeless (for example) for their condition rather than looking to societal problems that cause people to fall into poverty.

Housing and Homelessness

IN DISCUSSING THE DISTINGUISHING characteristics of homeless Americans, it is easy to lose sight of the fact that the essential and defining symptom of homelessness is lack of access to conventional housing. Clearly, if conventional housing were both everywhere abundant and without cost, there would be no homelessness except for those who preferred to sleep in the streets and in public places. That there are homeless people in fairly large numbers means that our housing market is not providing appropriate housing abundantly at prices the homeless can afford. Nor is it providing affordable housing for the extremely poor, who must double up with others.

To be sure, there is no way any housing market dominated by private providers can offer housing at an "affordable price" for those who have close to zero income. But market-offered housing is not the only option. Most of the extremely poor are domiciled, and their housing chances are affected by the supply of low-cost housing generally, a market factor that affects the households they live with. There is abundant evidence that homelessness is related both directly and indirectly to the shortage of inexpensive housing for poor families and poor unattached persons that began in the 1970s and has accelerated in the 1980s.

The decline in the inexpensive segment of our housing stock has been precipitous in the largest cities, such as New York and Los Angeles, but it also has characterized cities of all sizes (Wright and Lam 1987). The Annual Housing Survey, conducted by the Census Bureau for the Department of Housing and Urban Development, has recorded in city after city declines in

From Peter H. Rossi, Down and Out in America: The Origins of Homelessness, *University of Chicago Press, 1988. Reprinted by permission.*

the proportion of housing renting for 40 percent or less of poverty-level incomes. These declines ranged from 12 percent in Baltimore between 1978 and 1983 to 40 percent in Washington, D.C., for 1977 to 1981 and 58 percent in Anaheim, California, in the same period. In twelve large cities surveyed between 1978 and 1983, the amount of inexpensive rental housing available to poor families dropped precipitously, averaging 30 percent. At the same time, the number of households living at or below the poverty level in the same cities increased by 36 percent. The consequence of these two trends is that in the early 1980s a severe shortage occurred in housing that poor households could afford without bearing an excessive rent burden. Note that these calculations assume that such affordable housing rents for 40 percent or less of the poverty level, a larger proportion of income than the customary prudent 25 percent for rent.

Most of the housing I have discussed so far consists of multiroom units appropriate to families. If we restrict our attention to that portion of the housing stock that is ordinarily occupied by poor unattached single persons, then the decline is even more precipitous. Chicago's Planning Department estimated that between 1973 and 1984, 18,000 single-person dwelling units in SRO hotels and small apartment buildings—amounting to 19 percent of the stock existing in 1973—were demolished or transformed for other uses (Chicago Department of Planning, 1985). In Los Angeles a recent report (Hamilton, Rabinowitz and Alschuler, Inc. 1987) indicated that between 1970 and 1985 more than half of the SRO units in downtown Los Angeles had been demolished. Of course there is nothing wrong per se with the demolition of SROs; most were certainly not historical landmarks or examples of any notable architectural style. Nor can they be said to have been of high quality. The problem is that units comparable in function or price were not built or converted in sufficient volume to replace them. . . .

In 1958 about 8000 homeless men were accommodated in such units in Chicago; by 1980 all the cubicle hotels had been removed. In New York, by 1987 only one of the cheap hotels that dominated the Bowery in the 1960s remained (Jackson 1987). Similar changes have occurred in other large cities. Of course it is difficult to mourn the passing of the often dirty and always inadequate cubicle hotels. Like the SROs, they had little or no symbolic or aesthetic value. But only the emergency dormitory shelters have replaced the housing stock they represented. There are virtually no rooms in Chicago today that can be rented for $1.80 to $2.70 a night, today's dollar equivalent of the 1958 rents. The emergency dormitory shelters are arguably cleaner than the cubicle hotels, but they are certainly not much closer to decent housing. Indeed, the old Skid Row residents regarded the mission dormitory shelters as considerably inferior to the cubicle hotels, lacking in privacy and personal safety (Bogue 1963).

The decline in inexpensive housing influences homelessness both directly and indirectly. Indirectly, the effect can be felt through the increased financial housing burden placed on poor families, whose generosity toward their dependent adult members becomes more difficult to extend or maintain. Housing prices partially reflect the amount of housing involved, with larger units commanding higher prices. Faced with declining real income, poor families may have had to opt for smaller dwellings, restricting their ability to shelter adult children.

The direct effects are upon the homeless themselves, putting inexpensive housing, such as SRO accommodations, beyond the reach of most of the new homeless. For example, in a study of SROs in Chicago, Hoch (1985) found that the average monthly rental for SRO hotels in Chicago in 1984 was $195 if rented by the month or $240 ($8 a day), if rented day to day. For most of the homeless, with median monthly incomes of $100, renting an SRO room steadily was out of the question.

Because rents were so high relative to income, the tenants of Chicago's SROs were forced to spend a very large proportion of their income on housing. When some out of the ordinary expense occurred, many had to resort to the shelters and the streets. According to Hoch, about one in ten of the SRO tenants had been homeless for some period during the previous year, apparently too short of funds to pay the rent. Hoch does not tell us whether these SRO tenants lived in shelters or on the streets when they become homeless. But in our survey of the Chicago homeless, both the shelter and the street samples claimed they spent about 10 percent of their nights in rented rooms, presumably in SRO hotels.

Some of the homeless people we interviewed on the streets or in the shelters ordinarily spent most nights in SRO hotels and were just temporarily homeless. Others occasionally spent a night or two in an SRO, perhaps when they received a windfall. Apparently there is a considerable interchange between the homeless and the SRO populations, the latter being a cut above the former in income. Similarly, Piliavin and Sosin (1987–88) found that homeless people in Minneapolis typically moved between having homes and being homeless several times a year.

High rents relative to income also forced some of the SRO tenants to overspend on housing and, accordingly, to skimp on other expenditures. Hoch reports that many SRO residents resorted to the food kitchens, to the medical clinics set up for homeless persons, and to the clothing depots. In a study of the homeless in downtown Los Angeles, one out of every three persons in the soup-kitchen lines was renting a room in an SRO (Farr, Koegel, and Burnham 1986). Further confirmation can be found in Sosin's 1986 study of persons using Chicago food kitchens and day centers (Sosin, Colson, and Grossman 1988), which found that about half were living in SROs and apartments.

The impact of the housing market on homelessness in the aggregate was shown dramatically in a recent analysis by Tucker (1987). There are several deficiencies in Tucker's procedures; nevertheless, some of his findings are both useful and relevant. Using the HUD estimates of the number of homeless in each of fifty cities to compute a homelessness rate for each city, Tucker was able to show a fairly strong negative correlation, −.39, between housing vacancy rates in 1980 and homelessness rates in 1984 across cities. In other words, the higher the vacancy rate in a city, the lower its homelessness rate. Tucker also showed that the vacancy rate is highly sensitive to the presence of rent control measures, but that need not concern us here. The point Tucker's analysis drives home is that the tighter the housing market from the buyer's (or renter's) point of view, the greater the housing burden on poor families and the more difficult it becomes for the extremely poor to obtain housing, and consequently the easier it is to become homeless.

In a perfect unrestricted housing market, the range of housing offered by sellers at equilibrium would supply all buyers who can enter bids. But this statement is more a matter of faith than of fact. The American housing market is neither unfettered nor perfect. Nor would we have it any other way. Our building codes are designed to ensure that the housing industry provides accommodations that meet minimum standards of public health and safety. Zoning laws attempt to regulate the externalities surrounding existing structures. Occupancy laws discourage overcrowding of dwelling units. These regulations also accomplish other ends, some undesirable to many citizens: for example, zoning laws designed to ensure that structures occupy no more than some given proportion of urban land plots, a desirable aesthetic amenity, also make neighborhoods socioeconomically homogeneous. In some cities rent control is an additional restriction whose burden falls heavily on households entering the market and provides a bonus in the form of cheaper rents to long-term residents. These regulations are not the only factors restricting the amount of "affordable housing" available to the poor, but they certainly drive up the prices of even minimum standard housing.

However, there can be no market where there is no effective demand. The market cannot provide affordable housing for the homeless because their incomes are so low and variable that their demand is too weak to stimulate housing providers. The housing market was not always unresponsive to the demand of poor people. The Skid Rows of the nation were such responses, but the old cubicle hotels of the 1950s and 1960s were responding to a much stronger demand. Recall that the constant-dollar income of the Skid Row residents in 1958 was at least three times the income of the current homeless. Even so, as Bogue and the other social researchers observed in the 1950s and 1960s, the cubicle hotels were experiencing high vacancy rates.

The records are silent on whether the cubicle hotel owners and operators welcomed or fought the exercise of eminent domain in the urban renewal of Skid Row areas. Perhaps they welcomed the bulldozers as a way to recover some of the equity they had sunk into an increasingly unprofitable business.

In the past, when the housing industry was unable (or unwilling) to provide homes for the extremely poor they sometimes built their own. In the Great Depression of the 1930s, "shantytowns" consisting of shacks cobbled together out of scrap materials were built on New York's riverfronts and even in Central Park. Similar settlements were erected on Chicago's lakefront, in Washington's Anacostia Flats, and on vacant sites in other cities. In the 1980s no comparable settlements have appeared, unless one counts the cardboard and wooden package cases used as living quarters by a few of the homeless. It may be that vacant land is not as available now or that law enforcement officials are quicker to respond. Whatever has caused the difference, the self-help response of the homeless to market failure has not been as strong as in the past.

As the rents the homeless could afford declined with their incomes during the 1970s and 1980s, housing providers found them an increasingly unattractive set of customers, especially in contrast to others. There is no mystery about why no housing is offered on the unsubsidized market that is affordable to the homeless. If there is a question, it is why local, state, and federal government have not intervened in the market to ensure that such housing is supplied. . . .

An Interpretation of Homelessness

. . . I have described the characteristics of the current homeless, highlighting those that mark off this population from that of the old Skid Rows and from the current domiciled poor. Drawing these various threads together, we can now begin to weave an explanation both of why some people are more likely to be found among the homeless and of why homelessness has apparently increased over the past decade.

First of all, it is important to distinguish between the short-term (episodic) homeless, and the long-term (chronic) homeless who appear likely to remain so. Most of what I have to say . . . concerns the latter group; the former consists primarily of people in the lower ranks of the income distribution who meet short-term reversals of fortune. This is not to deemphasize the problems of the short-term homeless but simply to say that their problems are different.

The "dynamics" of episodic homelessness are distressingly straightforward. So long as there is a poverty population whose incomes put them at the economic edge, there will always be people who fall over that edge into homelessness. Small setbacks that those above the poverty line can absorb may become major disruptions to the very poor. Several homely examples illustrate this point. The failure of an old refrigerator or stove and a subsequent repair bill of $50 can make the nonpoor grumble about bad luck, but for someone whose monthly income is under $500 and whose rent is $300, the bill represents one-fourth of the monthly resources used to buy food, clothing, and other necessities. For a poor person who depends on a car to travel to work, a car repair bill of a few hundred dollars may mean months of deprivation. Renting an apartment increasingly means paying one month's rent in advance and perhaps a security deposit as well and is often why poor people remain in substandard housing. In many states welfare programs make provision for such emergency expenses, but the unattached person who is not eligible for welfare may experience wide swings of fortune, with the downsides spent among the homeless.

The solution is to be found in extending the coverage of the social welfare system and incorporating provisions that would cushion against short-term economic difficulties.

What about the long-term or chronic homeless? Their critical characteristic is the high level of disabilities that both impair their earning capacity and reduce their acceptance by their families, kin, and friends. These are the people who are most strongly affected by shortages of unskilled positions in the labor force, lack of inexpensive housing, and declines in the economic fortunes of their families, kin, and friends. Under these unfavorable conditions, unattached persons with disabilities have increasing difficulty in getting along on their own. And as the living conditions of poor households decline, those disabled by chronic mental or physical illnesses or by chronic substance abuse are no longer tolerated as dependents.

Note that I am using the term disabled in this context to mark any condition that appreciably impairs the ability to make minimally successful connections with the labor market and to form mutually satisfactory relationships with family, kin, and friends. This definition goes beyond the usual meaning of disability to include a much wider set of conditions—for example, criminal records that interfere with employment chances or chronic problems with drinking, as well as physical and psychiatric impairments.

Let me emphasize that this interpretation is not "blaming the victim." It is an attempt to explain who become the victims of perverse macrolevel social forces. If there is any blame, it should be placed on the failure of the housing market, labor market, and welfare system, which forces some people—the most vulnerable—to become victims by undermining their ability to get along by themselves and weakening the ability of family, kin, and friends to help them. . . .

The resurgence of extreme poverty and homelessness in the past two decades should remind us that the safety nets we initiated during the Great Depression and augmented in the 1960s are failing to prevent destitution. The Reagan administration did not succeed in dismantling any significant portion of the net, but it made the mesh so coarse and weak that many fall through. Those who are disabled by minority status, chronic mental illness, physical illness, or substance abuse are especially vulnerable. All the very poor suffer, but it is the most vulnerable who fall to the very bottom—homelessness.

The social welfare system has never been very attentive to unattached disaffiliated men, and now it appears to be as unresponsive to unattached females. Likewise, the social welfare system does little to help families support their dependent adult members. Many of the homeless of the 1950s and early 1960s were pushed out or thrown away by their families when they passed the peak of adulthood; many of the new homeless are products of a similar process, but this one commences at age twenty-five or thirty rather than at fifty or sixty.

As a consequence, homelessness now looms large on our political agenda, and there is anxious concern about what can be done. I have suggested a number of measures to reduce homelessness to a more acceptable level. These include compensating for the failure of our housing market by fostering the retention and enlargement of our urban low-income housing stock, especially that appropriate for unattached persons; reversing the policy that has put personal choice above institutionalization for those so severely disabled that they are unable to make decisions that will preserve their physical well-being; enlarging our conception of disability to include conditions not purely physical in character and, in particular, recognizing that chronic mental illness and chronic substance abuse are often profound disabilities; restoring the real value of welfare payments to the purchasing power they had in the late 1960s; and extending the coverage of welfare benefits to include long-term unemployed, unattached persons.

There is considerable public support in the United States for a social welfare system that guarantees a minimally decent standard of living to all. Homelessness on the scale currently being experienced is clear evidence that such a system is not yet in place. That homelessness exists amid national prosperity without parallel in the history of the world is likewise clear evidence that we can do something about the problem if we choose to. I have stressed that public policy decisions have in large measure created the problem of homelessness; they can solve the problem as well.

REFERENCES

Bogue, Donald B., 1963. *Skid Row in American cities.* Chicago: Community and Family Study Center, University of Chicago.

Chicago, Department of Planning, 1985. *Housing needs of Chicago's single, low-income renters.* Manuscript report.

Farr, Rodger K., Paul Koegel, and Audrey Burnham, 1986. *A survey of homelessness and mental illness in the Skid Row area of Los Angeles.* Los Angeles County Department of Mental Health.

Hamilton, Rabinowitz, and Alschuler, Inc., 1987. *The changing face of misery: Los Angeles' Skid Row area in transition, housing, and social services needs of Central City East.* Los Angeles: Community Redevelopment Agency.

Hoch, Charles, and Diane Spicer, 1985. *SROs, an endangered species: Single-room occupancy hotels in Chicago.* Chicago: Community Shelter Organization and Jewish Council on Urban Affairs, 1985.

Jackson, Kenneth M., 1987. The Bowery: From residential street to Skid Row. In *On being homeless: Historical perspectives,* Ed. Rick Beard. New York: Museum of the City of New York.

Piliavin, Irving, and Michael Sosin, 1987–88. Tracking the homeless. *Focus* 10, 4:20–24.

Sosin, Michael, Paul Colson, and Susan Grossman, 1988. *Homelessness in Chicago: Poverty and pathology, social institutions and social change.* Chicago: Chicago Community Trust.

Tucker, William. 1987, Where do the homeless come from? *National Review,* 25 September, 32–43.

Wright, James D., and Julie Lam, 1987. Homelessness and low income housing supply. *Social Forces* 17, 4:48–53.

QUESTIONS FOR DISCUSSION

For further discussion of this topic, see the Wadsworth Sociology Resource Center, "Virtual Society," ***http://sociology.wadsworth.com,*** under *Sociological Footprints,* by Cargan and Ballantine. You can respond to the discussion questions there or enter your own comments in the online chat forum.

SUGGESTED READINGS AND SOCIOLOGY INTERNET RESOURCES

See the Wadsworth Sociology Resource Center, "Virtual Society," ***http://sociology.wadsworth.com,*** for additional links, suggestions for further reading, and learning tools related to this chapter.

Either from the "Virtual Society" website or directly from your web browser, you may access InfoTrac College Edition, an online university library that includes over 700 popular and scholarly journals in which you can find articles related to the topics in this chapter.

CHAPTER 14

Social Movements and Change
SOCIETY IN FLUX

THERE ARE TIMES WHEN THE NORMS that guide societal behavior break down. At these times, *collective behavior*—behavior that is both spontaneous and unorganized—may predominate. Fads, crazes, riots, and panics are all examples of collective behavior; natural disasters, social movements, public opinion, and crowds can create collective behavior situations.

Most of us have been involved in several forms of collective behavior. When we watch a demonstration at a fair, we are part of a *casual crowd;* when we cheer at a football game, we are part of an *expressive crowd;* when we join a campus confrontation, we are part of an *acting crowd*. In each of these situations, however, we maintain a degree of anonymity, because the activities are both temporary and impersonal.

Many of us have been involved in a form of collective behavior called a social movement, a group of people attempting to resist or bring about change. There are many recent examples: the civil rights movement; various educational reform movements such as "back to basics"; fundamentalist Christian movements; and the women's movement. A movement may have a short life span, or it may become an institutionalized, stable part of the ongoing society. The reading by Barry Adam discusses a worldwide movement that has been influenced by globalization—the mobilization of gay and lesbian communities.

Social change is the alteration, over time, of a basic pattern of social organization. It involves two related types of changes: changes in folkways, mores, and other cultural elements of the society; and changes in the social structure and social relations of the society. These cultural and structural changes affect all aspects of society, from international relations to our individual lifestyles. An example of a change in cultural elements is women increasingly entering the labor force; an example of structural change is the rise of secondary-group relationships with a corresponding increase in formal organization.

Social change happens through the triple impact of *diffusion* (the speed of ideas across culture), *discovery* (unpremeditated findings), and *invention* (purposeful new arrangements). Each of these actions occurs at geometrically expanding rates—that is to say, the period of time over which an initial invention evolves may require centuries, but once it is established, further changes related to that invention are rapid. For example, it took less than fifty years to advance from the Wright brothers' 12-second flight to supersonic travel to trips to the moon. Similarly, the rates of diffusion and discovery are now greater because of advancements in communication and travel.

Change occurs in all societies, but at different rates. While change is rapid and radical in some societies, such as Eastern European countries and the former Soviet Union, and can actually tear societies apart, as in the case of Yugoslavia, change generally occurs more slowly in stable, traditional societies.

Pressures for change affect every corner of the globe because of the interdependence of nations. Though it was once assumed that countries would go through developmental stages, including industrialization, as they modernized, Mayur and Daviss show that some Third World countries are adapting new technologies to improve life in rural agricultural communities.

The final two readings lead us into the twenty-first century. The first reading by Salzman and Matathia provides predictions for life in the twenty-first century.

In the final reading, Allen Tough reviews his findings from students around the world concerning their desires for the future world in which they must live. As you read these readings on collective behavior and social change, consider some collective behavior situations in which you have been involved and whether they fit the patterns described.

1. What are some current social movements and how are they affecting society?
2. Were these predicted changes to take place, what would the strains be in existing structure and roles?
3. What kind of society and world do we want, and how can we attain it?

56

Globalization and the Mobilization of Gay and Lesbian Communities

BARRY D. ADAM

Change occurs in many ways. Stresses from outside the country or organization can bring about change, just as strains from within the country or organization can. Change can be rapid or slow, planned or unplanned. Social movements—groups of individuals organized to push for or resist change in some part of society—are one means of putting pressure on systems to change. The first reading provides an example of change brought about by international globalization processes that can shape movements.

As you read it, consider the following:

1. What is the effect of globalization on social movements within countries? In the world?

2. Do you belong to any social movements (women's rights, animal rights, "greens," etc.), and how do you see your movement bringing about change?

GLOSSARY

Social movements Organized effort to encourage or oppose some dimension of change.

Globalization Economic and political changes spanning many nations of the world.

THERE HAS BEEN A strong tendency to treat gay and lesbian communities and their movements as cultural artifacts unrelated to structure, political economy, or the modern world-system (Seidman 1996). This tendency gives gay and lesbian people over to the "globalization-as-culture" school virtually by default. A good deal of contemporary social movement theory has assimilated lesbian and gay mobilization into the "new social

movement" camp, and into "identity politics" in particular. New social movements now supposedly "seem to exist independently from . . . structural and cultural conditions" (Eder 1993:44) and gay movements are, somewhat inexplicably, all about affirming and expressing identities (Dudink and Verhaar 1994). What is missing from this view are the ways the changes wrought by the world-system articulate with the historical rise of gay and lesbian forms of same-sex connection and how lesbian, gay, bisexual, and transgender identities have become "necessary fictions" (Weeks 1993) in contemporary political environments.

. . . The "gay" and "lesbian" categories of the modern world, when compared to cross-cultural and historical forms of same-sex relations, show several specific characteristics:

1. In societies where kinship has declined as a primary organizing principle determining the survival and well-being of their members, homosexual relations have developed autonomous forms apart from dominant heterosexual family structures.

2. Exclusive homosexuality has become increasingly possible for both partners and a ground for household formation.

3. Same-sex bonds have developed relatively egalitarian forms characterized by age and gender "endogamy" rather than involving people in differentiated age and gender classes.

4. People have come to discover each other and form large-scale social networks because of their homosexual interests and not only in

Forthcoming in Pierre Hamel, Henri Lustiger-Thaler, and Sasha Roseneil, eds. Global Flows. *New York: Macmillan (excerpts). Reprinted by permission of the author.*

the context of pre-existing social relationships (such as households, neighborhoods, schools, militaries, churches, etc.).

5. Homosexuality has come to be a social formation unto itself characterized by self-awareness and group identity (Adam 1995:7).

These characteristics presume a sociological infrastructure characteristic of the modern world-system:

1. The financial "independence" of wage labor. In societies where access to the means of production depends on kin ties and inheritance, homosexuality is either integrated into the dominant kinship order (through gender redesignation or limitation of homosexual ties to a life stage), or subordinated to it as a hidden and "unofficial" activity.

2. Urbanization and personal mobility. The growth of cities, the invention of public spaces, migration away from traditional settings and from the supervisory gaze of families have all created new opportunities, especially at first for men, to form new, nontraditional relationships.

3. Disruption of traditional gender rules. As minorities and women have been entering wage labor in increasing numbers, often occupying newly created locations in the division of labor, more of the choices taken for granted by men have become available.

4. Development of the welfare state. The creation of social services in health, education, welfare, employment insurance, pensions, and so on have supplemented traditional family functions, providing alternatives to reliance on kin.

5. Liberal democratic states. Legal guarantees of basic civil liberties also facilitate the ability of people to love and live with those of their choice, though most liberal democratic states have lengthy histories of violating basic constitutional freedoms of conscience, assembly, and free speech in order to suppress their gay and lesbian citizens; and gay and lesbian communities have also carved out small spaces for themselves in the hostile environments of authoritarian states.

Having enumerated some of these factors, it is important to note that these are structural underpinnings of contemporary gay and lesbian worlds, not of homosexuality itself, which is best conceived as a universal human potential with a wide range of expression across cultures. Even though gay and lesbian worlds (as defined above) have now been documented for at least three centuries in Europe, these social conditions do not in themselves determine that same-sex relationships will necessarily take gay/lesbian forms. Medieval constructions of sodomy (Jordan 1997), Judeo-Christian obsession with narrowing legitimate sexuality to its procreative form, and the western tendency to force sexuality to "confess truth" about the nature of persons (Foucault 1978) all enter into western constrictions of the "gay" and "lesbian." Japan offers an example of a society with the political economy of an advanced industrial society but none of the history of religious or state persecution of homosexuality. Though Japan has developed gay (and to a very limited degree, lesbian) public spaces like its western counterparts, the formation of a popular sense of identity and movement has been less evident (Lunsing, forthcoming). Finally, even among the citizens of advanced industrial societies, many more people have homosexual experiences than identify as "gay," "lesbian," or "bisexual," and these identities may, as well, be less widespread among nonwhite and working class people who nevertheless have homosexual interests and practices (Laumann et al. 1994).

Still, where this sociological infrastructure is lacking, gay and lesbian identities, and the movements built out of a sense of commonality signified by these identities, are also unusual. As Badruddin Khan (1997) remarks about urban Pakistan, widespread practices of sexuality among men tend not to lead to gay identities: where family networks remain major determinants of one's well-being, families make sure their progeny marry regardless of sexual orientation, and very few young people have the financial ability or freedom to form households of their own choosing. Similar conditions apply to much of Latin America with the partial exception of the urban middle classes. The Communist states of eastern Europe, on the other hand, showed a different array of social conditions that both facilitated and inhibited the growth of gay and lesbian cultures. State socialism typically pursued a development model founded on industrialization, resulting in urban migration, conversion of much of the population to wage labor, and improved opportunities for women. As in capitalist societies, state socialism displaced kinship as the primary determinant of people's life chances, disestablished churches, and devolved decisions about family and reproduction to the individual level (Adam 1995:166). It is not surprising, then, that gay and lesbian bars and coffeehouses became part of the urban scene in eastern European capitals well before the fall of Communism and movement organizations came into existence in several countries despite the power of central bureaucracies to control the mass media and administer labor migration, commercial meeting places, and housing (Hauer et al. 1984). When the Communist states collapsed in the early 1990s, gay and lesbian movement organizations rapidly emerged across the region in a pattern similar to western Europe because the sociological infrastructure was already in place. . . .

The Global Movement

Today the International Lesbian and Gay Association federates movement organizations and attempts to monitor civil rights around the world, twin established organizations with new, and link international efforts around asylum, the military, churches, youth, ableism, health, trade unions, AIDS, and prisoners.[1] A primary objective of national movements has been to win legal recourse against discrimination. Beginning with Norway in 1981, sexual orientation had been included in the human rights legislation of Canada, Denmark, France, Iceland, Netherlands, New Zealand, Slovenia, South Africa, Spain, and Sweden, as well as eleven of the fifty United States, two states in Australia, and seventy-three cities in Brazil (as of 1997).[2] Ireland and Israel also legislated workplace protection in the 1990s. Even more recent

is legislation recognizing the spousal status of same-sex relationships. Norway, Sweden, Netherlands, and Iceland have followed the precedent set by Denmark in 1989, and a supreme court decision in Hungary has included same-sex couples in common-law spousal status. Initial breakthroughs in relationship recognition have also occurred in 1997 in the Canadian province of British Columbia and the US state of Hawaii. All of these laws, however, fall short of full equality, often barring gay and lesbian couples from full-fledged marriage, adoption, or access to artificial insemination.

Gay and lesbian movements have emerged along with (and sometimes in explicit coalition with) democratic movements to oppose authoritarian rule in Spain (after the death of Franco), South Africa (in overcoming apartheid), in eastern Europe (with the fall of Communism), and Brazil, Argentina, Uruguay, and Chile (in ending military dictatorships). Once these precedents are in place, gay and lesbian communities may take the opportunity to mobilize in neighboring countries. With the African National Congress in power in South Africa and sexual orientation entrenched in the new constitution, groups in Zimbabwe (1990), Botswana (1996), Namibia (1996), and Swaziland (1997) have declared themselves often in the face of virulent homophobia enunciated by local rulers. The struggle against AIDS has created the first opening for gay mobilization in even less hospitable places such as Kenya, Malaysia, and Ecuador.

This record of civil emancipation must be viewed in the context of ongoing homophobic practices of states and their agents, of reactionary civil and religious movements, and of those in control of cultural reproduction from the schools to television programming. Amnesty International (1994) documents the incarceration or murder of homosexual people in several countries, most notably, Iran, Turkey, and Romania. Movement groups continue to face repeated police raids on gay gathering places and death squad activity in many cities of Latin America, especially in Colombia, Peru, Mexico, and Brazil. Gay and transgender Argentines were among the targets of the "dirty war" perpetuated by the police and death squads and sanctioned by the military and Roman Catholic hierarchies (Jauregui 1987). Sweeping police powers continued after the fall of the military dictatorship into the 1990s resulting in numerous bar raids and arbitrary arrests. The work of the Comunidad Homosexual Argentina, in collaboration with other democratic organizations, resulted in 1996 in the revocation of police powers and inclusion of sexual orientation as legally protected categories in Buenos Aires and Rosario.

Globalization processes, then, have affected the mobilization of gay and lesbian communities in a wide variety of ways. Insofar as globalization refers to processes of incorporation of local communities into the political and economic networks of the capitalist world-system, globalization has accelerated processes of proletarianization, urbanization, family change, and democratization. These processes have created social conditions where indigenous forms of same-sex adhesiveness (to use Walt Whitman's term) have tended to evolve toward gay and lesbian social forms. Contemporary gay and lesbian communities and movements have also, in turn, provided fertile ground for the defense, growth, and reworking of older, more traditional bisexual and trans-gendered forms of sexual life. Where globalization refers to the faster and easier circulation of cultural practices and ideologies, gay and lesbian movements have been participants as well. Though gay and lesbian cultural forms usually have an uneasy or oppositional relation with the established, institutional ideological circuits of religion, nationalism, and "family values," an international gay and lesbian culture has emerged throughout the metropoles and, increasingly around the world, through personal contacts, the gay press, and now more formal, if fragile, organizations like the International Lesbian and Gay Association. Pride celebrations, which originated in political demonstrations against repression, have become so massive, in such cities as Sydney, Toronto, and San Francisco, that they have changed the cultural landscape and attracted commercial interests intent on making inroads into a new market of consumers.[3]

Globalization, as a political and economic process underlying the transition from the welfare state to neoliberalism, has consequences for the well-being and potential mobilization of numerous social movement constituencies. Neoliberal governments, particularly in the United States and the United Kingdom, have drawn on conservative ideologies of work discipline, delayed gratification, racial and national supremacy, and patriarchal definitions of gender and family to "sell" wage restraints and the withdrawal of social services, as well as to re-attribute blame for the declining quality of life. As Stuart Hall (1988b:48) remarks, Thatcherism combined "organic Toryism—nation, family, duty, authority, standards, traditionalism—with the aggressive themes of a revived neoliberalism—self-interest, competitive individualism, anti-statism." Neoconservative governments thereby encouraged and exploited the resentment of social classes damaged by the world economy, channeling their anger toward traditional lightning rods of popular prejudice, including lesbians and gay men (Adam 1995:ch 6). The global economy has struck hard at some social groups, typically rural people, small business people, and workers located in heavy industry. Declining or beleaguered ethnic and class groups have proven to be especially susceptible to right-wing mobilization, from the antisemites of the early twentieth century to the homophobes and xenophobes of the late twentieth century.

At the same time, gay and lesbian movements, with the other "new social movements," have flourished in the neoliberal age. The welfare state both changed and reinforced family and gender requirements in the 1950s not only by (1) relieving families of some of the functions they could not always fulfill but also by (2) restoring the home as women's sphere and by overtly suppressing gay and lesbian life. The social policies of neoliberal governments of the 1980s tended to press for the "restoration" of traditional gender and family scripts (restoring to some degree a family that never was [Coontz 1992]) while reducing the social services available to them. Gay and lesbian mobilization, like the women's movement, is part of the larger wave of "antisystemic" movements since 1968 (Wallerstein

1990), building on and expressing some aspects of structural change—such as women's increasing participation in wage labor—while resisting others—such as explicitly conservative social policy. And as capital has gone global, social movements as well have sought unprecedented worldwide networks.

NOTES

1. The International Lesbian and Gay Association can be contacted at: Antenne Rose & FWH, 81, rue Marché-au-charbon, B-1000 Bruxelles 1, Belgium; website: http://www.ilga.org. The International Gay and Lesbian Human Rights Commission is at: 1360 Mission Street #200, San Francisco, California, U.S.A. 94103; website: http://www.iglhrc.org.

2. Much of this section draws on Adam (1995) and recent issues of the Bulletin of the International Lesbian and Gay Association.

3. For a critique of the commercial cultivation of gay consumerism, see the early work of Dennis Altman (1982) and the more recent work of Mark Simpson (1995).

REFERENCES

Adam, Barry. 1995. *The Rise of a Gay and Lesbian Movement.* Revised Edition. New York: Twayne/ Simon & Schuster Macmillan.

Coontz, Stephanie. 1992. *The Way We Never Were.* New York: Basic Books.

Dudink, Stefan, and Odile Verhaar. 1994. "Paradoxes of identity politics." *Homologie* 4–94:29–36.

Eder, Klaus. 1993. *The New Politics of Class.* London: Sage.

Hall, Stuart. 1988b. *The Hard Road to Renewal.* London: Verso

Hauer, Gudrun, et al. 1984. *Rosa Liebe unterm roten Stern.* Hamburg: Frühlings Erwachen.

Jauregui, Carlos. 1987. *La homosexualidad en la Argentina.* Buenos Aires: Tarso.

Jordan, Mark. 1997. *The Invention of Sodomy in Christian Theology.* Chicago: University of Chicago Press.

Khan, Badruddin. 1997. "Not-so-gay life in Pakistan in the 1980s and 1990s." In *Islamic Homosexualities,* edited by Stephen Murray and Will Roscoe. New York: New York University Press.

Laumann, Edward, John Gagnon, Robert Michael, and Stuart Michaels. 1994. *The Social Organization of Sexuality.* Chicago: University of Chicago Press.

Seidman, Steven. 1996. *Queer Theory/Sociology.* Cambridge, MA: Blackwell.

Wallerstein, Immanuel. 1989. *The Modern World-System.* Volumes 1–3. San Diego: Academic Press.

Weeks, Jeffrey. 1993. "Necessary fictions." In *Constructing Sexualities,* edited by Jacqueline Murray. Windsor, Ontario: University of Windsor Humanities Research Group.

57

The Technology of Hope: Tools to Empower the World's Poorest Peoples

RASHMI MAYUR AND BENNETT DAVISS

Must all countries experience large-scale industrialization as they change and modernize? Mayur and Daviss argue that developing nations can turn to technologies that will bring about a postindustrial form of prosperity without industrialization.

As you read, keep in mind the following questions:

1. *What are alternatives to industrialization for developing countries?*
2. *How can technology help improve the lives of the rural poor in developing countries?*

GLOSSARY

Technology of hope Techniques used by the world's poor, rural peoples to improve their status.
Industrialization Period of transition from agricultural economies to manufacturing economies.
Technologies Application of knowledge to practical tasks of living.
Benign Benefits outweigh drawbacks over time.
Malignant Drawbacks outweigh benefits over time.

SINCE THE INDUSTRIAL REVOLUTION, Western economies have relied on vast supplies of raw material from lesser-developed countries in order to prosper: timber from South America, minerals from Africa, oil from the Middle East. Today, with 20% of the world's population, the Western economies consume more than half of the planet's energy and raw materials. Although industrial production is being globalized swiftly, consumption is not; the same minority of the human species remains the only group able to buy most of the planet's manufactured goods.

The developing countries cannot build their future, as the West did, by consuming more than their proportionate share of the world's resources. The typical resident of Ghana can't outbid the average American for a truck or computer; Nicaragua can't expropriate coal from Germany to fuel its factories. In the new century, the rising economic expectations of the world's poor will have to be met in ways that use resources conservatively and efficiently, not profligately. That necessity forecloses a global economic future rooted in world industrialization.

. . . The idea that today's emerging economies must pass through a phase of mass industrialization as Western nations did is an antiquated premise, rooted in and suited to a period of social and economic history that has passed. An economy in which strong hands and a time card produced an abundant material life for an individual or for an entire society has yielded to an emerging global economy that drives down industrial wages and places increasing premiums on knowledge, environmental stewardship, decentralization, and the personal touch.

Instead of accepting industrialization as the only way to satisfy the economic aspirations of their people, developing nations can turn to several decentralized, relatively inexpensive, and environmentally compatible tools and techniques—all available now—that can help bring a postindustrial form of prosperity to emerging nations. Together, these technologies can begin to redefine the notion of development, showing that it is possible to balance the demands of people with the needs of the biosphere.

From The Futurist. *October 1998, pp. 46–51. Originally published by the World Future Society. Used by permission of the publisher.*

Thanks to these new technologies, it's now possible for countries to develop sound, broad-based economies without industrializing, without draining people from rural areas and concentrating low-paid workers in company towns or urban ghettos, and without degrading and exhausting their land, air, and water.

These technologies will not replace industrial installations around the globe; indeed, many of these benign technologies are produced in manufacturing plants. But they will enable emerging nations to define a future that separates development from industrialization, creating new routes to prosperity.

Benign vs. Malignant Technologies

A "technology" is simply a way to accomplish an objective. A computer is a technology that manipulates data; banking is a technology that allows people to readily store, exchange, and multiply value. But a technology, like any human act, is never free of consequences, often unforeseen, that can render a technology either benign or malignant.

A benign technology is one whose systemic or cyclic benefits outweigh its drawbacks over the life of the technology. The vaccines and inoculation campaigns that eradicated smallpox are an ideal example of a benign technology; so has been the effort, led by Peru's International Potato Center, to genetically engineer a potato containing the basic daily amounts of the amino acids, vitamins, and minerals essential to human nutrition.

A malignant technology is the opposite: one whose drawbacks outweigh its benefits over the life of the technology. There can be no better example than the gasoline engine. Undeniably, its economic benefits have been greater than almost any other invention in human history. But as smog and greenhouse gas accumulate in worrisome amounts and eat away the atmosphere's protective ozone layer, it's clear to most honest observers that the gasoline engine's malignancy is entering a lethal phase. Even automakers themselves are beginning to scramble for alternatives.

A sustainable future can be built only from benign technologies. Fortunately, there is an array of them—simple, relatively inexpensive, and available now—that, taken together, offer a "soft path" to economic development.

Technologies of Energy

Reliable energy supplies are usually a key requirement for emerging nations with poor populations to become self-sufficient. Benign ways to supply electric power underlie all aspects of nonindustrial development and serve as the heart of a prosperous nonindustrial economy.

The industrial-age approach to electrifying an outlying area is to spend hundreds of millions of dollars to build a central generating plant that burns fossil fuels and spews harmful gases—or that dams rivers and disrupts ecosystems—and to run thousands of miles of cable from the plant to all corners of a region. Creating such a system demands that either the national government or a private licensee finance and build the necessary infrastructure. Because emerging nations lack sufficient paying customers to support such an installation, countries typically must finance such projects through foreign debt and repay those loans by exporting hardwoods, oil, and other portions of their natural heritage; usually, these are poor countries' only available assets that fetch the hard currency they must have to repay foreign loans. "Development" then becomes a path to environmental exploitation and depletion, all too often accompanied by resentment and political backlash.

The benign technologies of renewable energy can provide power at far less expense.

Solar technologies are simple enough and inexpensive enough to meet typical household uses in most of the world. Solar ovens use one-way glass with a mirrored inner lid to focus the sun's rays into a black-lined cooking chamber. The devices are the size of a briefcase, require no maintenance, and typically cost no more than $70. A $1,450 solar-powered water pump uses just 200 watts of electricity, as much as three average light bulbs, and can deliver 30 gallons of water—the minimum for health specified by the World Health Organization—for 135 people every day. Cheap solar evaporators can separate clean water from a number of pollutants, making water safe to drink that otherwise would not be potable.

The Washington, D.C.–based Solar Electric Light Fund is equipping village homes in 11 countries with photovoltaic systems to power lights and a television at an installed cost of less than $400 per household. Buyers pay for their systems through a revolving loan fund that finances other customers and expands solar power's availability.

Energy Conversion Devices in Troy, Michigan, makes paper-thin solar collector sheets that roll up like window shades, can be carried anywhere, and deliver electricity even when shot through with bullets. There is no longer any portion of the world too rugged or remote for effective solar technologies.

Another technology, used by 10 million Chinese farms, is the methane digester power plant. The plants capture natural gas rising from garbage and human or animal dung—"biogas"—and concentrate it in sealed chambers so it can be burned to run pumps, boilers, or small generators. The plants are set into pits as shallow as six feet, can be as small as eight feet in diameter, and can deliver power using the dung from as few as two cows. Farmers themselves can be trained to conduct the needed routine maintenance and to make simple repairs, and the "spent" fuel extracted from the digester can still be used as a rich fertilizer. The installed cost for the smallest biogas plant: less than $150. India's Renewable Energy Development Agency, a project of the national government, is subsidizing installation of the plants in farms and villages. The subsidies' costs are more than offset by the money the government saves—everything from regulatory paperwork to construction—by not bringing energy to rural areas through the traditional centralized, industrial approach.

There is no reason to deny millions of people in the developing world a steady supply of power simply because their nation has not yet built massive, fossil-fuel-burning generating plants and run thousands of miles of transmission lines.

Those technologies are no longer synonymous with electric power. Decentralized approaches can not only spark social and economic progress, but also can enable resource-poor people to stop harvesting dung and vegetables to burn. When they can do so, they can begin to use those resources to renew their soils and ecosystems in ways that will enable the land to support them again.

Technologies of Education and Communication

Until now, outlying areas of emerging nations have been cut off from the world not only physically, but also psychically and socially. Neither governments nor broadcasters could economically justify building ground-based transmitters that would beam programs to poor areas; people could neither pay for broadcasting services nor buy goods from advertisers. But now in regions lacking cable, poles, and the elaborate infrastructure of conventional communications, wireless communications and decentralized sources or renewable energy are an ideal match that can be installed quickly to fill the need.

For example, the technology now exists to place in every isolated village an "information kiosk"—a booth containing a cellular telephone, radio, television, videocassette recorder, and even a computer linked to the Internet—all powered by solar energy. Satellite broadcasts could bring villagers information ranging from weather alerts to arithmetic lessons to tips on caring for newborns. With a telephone, farmers could monitor market conditions and avoid selling crops when prices are weak, and parents could phone for medical advice when a child is sick. With the strides being made in telemedicine, distance learning, and similar services, the information kiosk could serve as classroom, agricultural extension office, doctors' examination room, and bulletin board—not just as an entertainment center.

The seeds of these rural information networks are already being planted. The Bombay-based International Institute for Sustainable Future is designing prototypes of the kiosks for its two demonstration "ecovillages" in west central India. Meanwhile, in Bangladesh, Mohammad Yunus, founder of the pioneering microlender Grameen Bank, has launched a venture that will place a cellular phone in 65,000 of the nation's villages. An entrepreneur in each village buys the phone, charges customers by the minute, repays the loan from the fees, and makes a small profit on each call. The venture is investing $25 million to cover a third of the country with 50 to 60 relay towers, sharing the cost with companies in New York, Japan, and Norway.

In Africa, Tanzania's Kibadula Farm Institute uses a small ground station to communicate with the world via satellite. The Institute, operated by the Seventh Day Adventist Church to bring education, health care, and improved farming techniques to the remote African villages surrounding it, set up the equipment to communicate with the church's U.S. headquarters. Soon it was using e-mail over the Internet to find expert advice on an array of practical problems, including the assembly of ultralight aircraft. In 1995, the Solar Electric Light Fund provided Africa's Masai people with solar-powered radio-telephones that they use to communicate not only with each other, but also over the country's public telephone system.

These projects provide a small glimpse of a large future. Teledesic, a company formed in 1990 by Microsoft founder Bill Gates and cellular-phone magnate Craig McCaw, plans to launch as many as 800 low-orbit satellites to bring cellular communications and Internet access to every point on the globe. The company itself will not provide communications services, but rather maintain an open network through which others can offer services. In a key provision, Teledesic has pledged to reserve a number of communications channels specifically for services in poor nations.

This practical, affordable combination of decentralized communications and decentralized energy sources means that no village need remain isolated from knowledge, advice, or opportunity or from their larger societies.

Population Control: A New Opportunity

Unchecked birth rates are a root cause of the spiraling environmental destruction and social desolation rampant in developing nations. Popula-

tions that outgrow their water supplies pollute and deplete what sources remain. Burgeoning families strip the countryside for food and fuel, exhausting soils, razing forests, extinguishing wildlife species, and transforming once-productive land into desert. From the sands of Ethiopia to the burning rain forests of Brazil, the global map is dotted with the consequences of overpopulation.

Clearly, most efforts to check population growth in emerging nations have met with mixed results at best. It's not possible to deliver regular supplies of condoms or birth-control pills to remote regions. In some areas, women whose arms bear the marks of having been fitted with sub-skin time-release fertility controls have been stigmatized. And in too many poor countries, adults still insist that they need flocks of children to help support the family.

The installation of electric lights has been shown to slow birth rates by giving adults new choices of activities after dark, but decentralized energy technologies also enable a new way to achieve the same goal: by changing minds.

The nonprofit, New York-based Population Communications International (PCI) has discovered that one of the most effective routes around the barriers to population control is through an old American invention—the "soap opera," or continuing drama on radio and television.

In Mexico, PCI partnered with sociologists and television producers to create and televise a dramatic series following the struggles of a lower-class Mexican family with two children. The wife has come from a large family that was held in poverty by its sheer size. She aspires to a better life, but her husband's interest in sex is equaled only by his disdain for family planning. After a dramatic crisis, she persuades her husband to go with her to a family-planning clinic. He realizes that they'll have more sex because his wife won't fear the consequences. In the six months following the show, registration at Mexico's family-planning clinics jumped by 33%.

"To change behavior, you have to combine information with culturally tailored emotional content that lets the audience discover the benefits of seeing the world in new ways," says PCI's former executive vice-president, William Ryerson.

The result has been replicated in other countries where PCI works. In Kenya, for example—once considered a hopeless case among family planners—the group helped create a radio serial that reached 85% of the nation's families and was shown to be more popular than broadcasts of soccer matches, the national game. The serial probed issues not only of family planning, but of land inheritance—specifically, the problem of dividing the family farm among more and more children, who then would be unable to support their parents in their old age. In the wake of the serial, use of contraceptives in Kenya rose 58%, the desired family size fell from an average of 6.3 children to 4.4, and the birth rate began to fall for the first time in two decades.

But to show poor people these new ways of seeing, one must be able to reach them. Satellite communications coupled to decentralized sources of renewable energy can provide a means by which population growth—the root cause of so many social and environmental ills—can be effectively addressed.

Technologies of Finance

Benign technologies are not only less toxic than their industrial counterparts, but also less expensive per person served. But bringing these technologies to those who need them still will require enormous investment. Under the industrial-age approach to development financing, transglobal institutions such as the World Bank lent money to build power plants, mining ventures, and other large-scale government projects in unindustrialized nations. Now, new technological solutions to poverty require new funding structures to support them.

Although well-intentioned, many of those traditionally funded projects damaged ecosystems and destroyed cultures. For example, the World Bank loaned Brazil money to build a paved highway through the rain forest state of Rondonia. The road was intended to promote integrated economic development by facilitating the transport of products from ranchers, miners, and farmers. But the road also opened huge areas of

the forest to hordes of land-hungry colonists able to support themselves only through slash-and-burn farming. In effect, the World Bank helped to fund global climate change.

In another Brazilian fiasco, the World Bank underwrote the building of two huge dams that drowned vast tracts of the rain forest—homelands of indigenous peoples—to provide electricity for mining operations that, in turn, obliterated even more of the rain forest.

In 1987, the World Bank announced reforms designed to take social and environmental impacts into account in selecting projects for funding, but those reforms have been slow in appearing. As one observer commented, "The World Bank is an elephant that's now facing in the right direction. How fast it's actually moving in that direction is another question."

While large development projects can be justified in a few cases, a decentralized economy that relies on benign technologies requires decentralized funding. There are two models that are showing new directions.

One is being pioneered by the Solar Electric Light Fund (SELF). To bring solar-energy systems to villagers in emerging nations, SELF doesn't rely solely on the charity of international philanthropies. Instead, it often forges partnerships among nonprofit and for-profit groups. Using public and private grants as seed capital, SELF finances local small businesses that sell and service solar-energy equipment in developing countries. Families buy the power systems using low-interest loans from revolving funds that recycle the payments to finance purchases by other families.

In one such joint venture in India, SELF formed a for-profit subsidiary called SELCO India with local partners. Through India's Renewable Energy Development Agency, SELCO tapped World Bank funds set aside specifically for photovoltaic installations. In part, the company used the money to finance rural co-ops' bulk purchase of solar-energy systems for its members, install the systems, train local technicians to continue the work, then repaid the World Bank's loan from funds SELCO collected from the co-ops. In

1997, SELF began working with several non-profit foundations and the World Bank to evolve an ongoing public-private partnership to fund photovoltaic installations in emerging nations.

SELF's innovative collaborations sketch a new way to fund wide-scale development: pooling funds from several sources and targeting them not only to bring benign, localized technologies to specific areas, but also to develop the support structures and services those technologies need to continue on their own.

The second financial technology that can support nonindustrial economic development is microlending: financial institutions making small loans, usually no more than the equivalent of a few hundred U.S. dollars, to entrepreneurs in emerging nations to launch the kinds of cottage industries that traditional financial institutions would ignore.

The concept is no longer novel; indeed, after having been developed by small institutions such as Bangladesh's Grameen Bank and the Boston-based Accion International, it's become popular. Recent studies have found that more than 98% of microborrowers repay their loans—a higher average rate than commercial borrowers at U.S. banks—and that microlending funds, originally set up to funnel charitable grants to emerging nations, can support themselves just as commercial lenders do.

Technologies of Hope

Together, these benign, inexpensive technologies sketch a vision of economic equity and social stability for hundreds of millions of people throughout the nonurban areas of emerging nations—and even developed ones. It is a vision in which technologies of renewable energy enable people to replenish their soils and ecosystems, allow them to spend less time collecting fuel and water, and give them more abundant and reliable harvests. Renewable energy also would power lights, radios, and televisions in homes; cellular phones and Internet-linked computers as village utilities—perhaps in the local school; and local health clinics with adequate lighting and refrigerators to store vaccines. Villagers could supplement farm income by using microloans to launch small businesses, also using renewable energy, and market their wares over the Internet.

"These technologies can make nonurban areas of emerging nations a part of twenty-first-century global society that is not homogeneous, that celebrates diversity," says Robert Freling, SELF's executive director. "Indigenous people can maintain their culture and connection to the earth, but also feel part of the global society instead of being cut off from it."

We do not suggest that benign technologies will or should purge industrialization or its products from the earth. However, these technologies can lay the foundation of a new kind of economy—one that enables the people of emerging nations to remain in their villages instead of swelling urban slums, to protect their farmland and resources instead of exhausting them, and to prosper and live with dignity in a human, not a mechanical, context.

There still will be villagers whose personal ambitions lead them to cities. But for those hundreds of millions who wish to live within the heritage of their traditional cultures, these technologies represent a "soft path" to economic development and prosperity. It offers emerging nations a new choice—and that is the point.

58

Lifestyles of the Next Millennium: 65 Forecasts

MARIAN SALZMAN AND IRA MATATHIA

As we enter the twenty-first century, change is occurring at a rapid pace because of new technologies. Salzman and Matathia, specialists in predicting the future, look to some new ideas that will bring about change in our way of life.

Consider the following questions as you read:

1. What changes do top trend watchers predict for the twenty-first century?

2. How is your life likely to change as a result of the ideas presented in this reading?

GLOSSARY

New millennium The twenty-first century.

PEOPLE IN ADVERTISING RELENTLESSLY monitor lifestyle changes in order to market products successfully. In 1997, two top trend watchers at advertising agencies, Marian Salzman and Ira Matathia, published their book *Trends for the Future* in a Dutch edition, which became a bestseller in The Netherlands. The book is now being revised under a new title, *Next: The Flow of the Future,* for forthcoming international editions. Here is a sampling of their anticipations for the coming years:

Home

- **Company towns** will reemerge as high-tech companies lure workers to subsidized apartments, houses, and condos that are wired to the workplace. Larger corporations may build campuses, complete with on-site child and elder care, health facilities (dentist, chiropractor, certified nurse practitioner), and personal-service concierges.

- **Community personal assistants** will be hired by community groups and neighbors who will pool their resources. These neighborhood concierges will do things that homeowners no longer have time to do themselves, such as dropping off and picking up dry cleaning, going food shopping, doing yard work and pool maintenance, and performing routine home repairs or overseeing the work of plumbers and other service providers. Services will be provided by homeowners' associations, with fees charged on a menu basis.

- **Living in the global village.** The ability of information technology to bring the rest of the world into our living rooms has made us more aware not only of each other's fashions and preferences, but also of each other's passions and plights. Accompanying this awareness will be a deepening sense that one can both "think globally" and "act globally."

- **"Special purpose" rooms** will be in demand. Home builders will see an increase in requests for sewing rooms, hobby shops, wine cellars, and other rooms to individualize the home.

- **Fully automated bedrooms** are coming, allowing us to control lights, phones, drapes, alarms, media unit, climate, etc. with the touch of a button.

- **Sleeping machines** will be used either to produce restful sleep or to provoke intense dreams. Already in development: Nova-Dreamers, a technology that combines eye masks and circuitry to promote dreaming; alarms awaken the user moments after the dream has ended so he or she can make a record of the dream.

From The Futurist. *August/September 1998. Originally published by the World Future Society. Used by permission of the publisher.*

- **Intelligent wallpapers** using high-tech fabrics will turn every flat surface into an art gallery one moment, a TV/computer screen the next.
- **Intelligent refrigerators** will keep track of butter, orange juice, and other essentials; the homeowner can print out the list prior to a grocery shopping trip or transmit it electronically to a home-delivery service.
- **Virtual aquariums** or scenic vistas on flat screens will add interest to the family room wall.
- **Robotic lawn mowers** will be an increasingly common sight in suburban backyards. These robo-mowers will cut grass within a specified boundary, avoiding obstacles such as bushes and children's playthings.
- **Granny minders.** In-home infrared detectors that monitor an individual's daily routines will allow elderly persons to live independently longer. An absence of activity, such as not opening the refrigerator during the day, will trigger an automatic distress call to a caregiver.
- **Soundproofed rooms will be in demand.** The increased use of home offices and multimedia rooms will make soundproofing an increasingly common option in new homes.

Family and Education

- **Shopping for progeny:** With the increase in infertility and older couples wanting to have children, we'll see the rise of mail-order catalogs with details about the egg and sperm donors, so prospective parents can shop for genetics.
- **Divorce insurance** may become popular. Surety bond-like products will be introduced to cover the cost of divorce.
- **Co-parenting initiatives** will become more common as working moms and dads turn for assistance to childless relatives and friends. Gay couples unable to adopt children will be among those who co-parent, as will retirees, who will serve as surrogate grandparents and field-trip organizers for kids whose own relatives live too far away.

- **Cyber-Spock:** The Internet will offer growing amounts of parenting and child-care advice and resources; many of such sites on the Web will be created by companies hoping to lure customers to their products. Look for the emergence of cyber-based child development experts to challenge the current batch of best-selling authors and advisers.
- **"Nannycams"** hidden inside teddy bears and other security measures will give parents visual access to their children at day care at all times. Some day-care centers already allow concerned parents to check in on their kids via a series of cameras deployed throughout the facility. Images are accessible via a password-protected intranet. Parents of latchkey kids who are concerned about their children being unsupervised in the afternoons may extend camera-surveillance systems to their own homes.
- **Member-only playgrounds may be in vogue.** Parents concerned about their children's safety will buy memberships in local safe havens, such as restricted-access beaches, parks, zoo groups, and museum clubs. Playgrounds may increasingly require electronic-passcard entrance.
- **Ankle or wrist monitors** for kids could be a desperate but popular measure for parents who despair of keeping their kids at home while they work in the afternoon. Security stores will be selling this type of "house arrest" hardware to parents who have given up on controlling their children with low-tech methods.
- **Enforced parental responsibility:** Parents increasingly will pay the price for their children's illegal actions and truancy; penalties will range from fines to mandatory school attendance with the child to prison time. Look for insurance policies that protect parents from financial responsibilities caused by their children's misbehavior.
- **In-home videoconferencing** will catch on as a way for families scattered around the world to stay in touch.
- **Networked schooling will emerge.** As schools around the world get wired, we'll see

increased cooperation—and competition—among schools around the globe. Language classes will be taught via videophone by native speakers in other countries; schools in developed countries will "adopt" schools in impoverished areas, sending school supplies, etc.; students around the world will compete in such activities as online scavenger hunts, art competitions, and general-knowledge quizzes.

- **Cybersex and sexual media literacy** will become part of schools' sex-education lesson plans.
- **Patriot schools will emerge.** Parents unhappy with the trend toward multiculturalism in American schools will form private schools based not on religion but on the supremacy of Western thought and patriotic ideals.
- **Bilingual day-care centers and preschools** will rise in reaction to new findings regarding language development in the early childhood years. Spanish/English schools will be particularly popular in the United States due to the growing Hispanic population.
- **Entrepreneur camps** will give intensive instruction to teenagers on the skills they need to succeed in the decades ahead.
- **Spirituality-based camps and clubs** for kids will be on the rise, with an emphasis on such Eastern religions as Buddhism.
- **Libraries will become virtual.** Libraries will gradually condense their book sections, creating room for larger computer labs providing free Internet access, high-tech classes, etc. Electronic media will not make paper books obsolete anytime soon, but online-only editions will replace some types of books, such as mysteries, romances, and other "read and throw away" paperbacks. Digital books will be downloaded to a small, flat tablet, with a touchpad controlling page turning and an "erase" feature available to make room for new downloads.

Leisure and Socializing

- **Simple pleasures from times past,** such as sewing and quilting bees, will emerge as an antidote to today's chaotic lifestyles. The American Home Sewing and Craft Association reports that nearly a third of American adult women have adopted sewing as a hobby. A 1995 study by the association found that sewing reduces three stress indicators: heart rate, blood pressure, and perspiration rate.

- **Comfort foods** will push out sophisticated gourmet meals on most consumers' menus. Instead, expect more convivial entertainment with "childhood memory" meals, served on everyday china.
- **"Dinner clubs"** will pool the resources of neighbors. Each household in the club will become responsible for providing one dinner a week to all the participating families.
- **Meals-on-wheels for the masses:** Minivans and other family vehicles will come equipped with minifreezers and microwave ovens for meals-on-the-run.
- **Holographic storage** will vastly expand your home video library. Archiving and reviewing the special moments of your life will become more important as our increasingly fast-paced world refuses to slow down long enough for us truly to absorb what happens to us.
- **The loss of privacy** in the age of the Internet and World Wide Web may turn many people into exhibitionists. There is even a family in Sweden who have mounted a camera inside their refrigerator so that visitors to their Web site can monitor the family eating habits.
- **Futuristic fabrics** will allow the ultimate personalization of the wardrobe. Examples include massaging fabrics for relieving stress and fabrics that emit favorite scents and aromatherapies. Nanotechnology will also allow fabrics to be embedded with tiny computers, sensors, and micromachines; possible applications include cooling and heating systems, periodic self-cleaning, and self-repair. Multimedia T-shirts will include earpieces attached to the shirt that play prerecorded clips of favorite songs. An alternative use of this innovation will be navigational jackets that enable us to locate anyone anyplace.

- **Computer-generated friendship circles** will identify people around the world with whom you may share common interests. New tribes of kindred spirits will unite parents who homeschool, gardeners who practice organic farming, mothers who perform community service, upwardly mobile professionals who bowl, etc. Cataloging these hyper-local associations will allow marketers to target these groups with products and pitches designed especially for them. Also certain to increase: online book clubs, political-discussion groups, salons, therapy sessions, and support groups.
- **Matchmakers for friendships** will consist of paid agents in and around cities, who will bring together people interested in forming friendships with other like-minded individuals.

Health and Wellness

- **Food-phobia**—fear of processed foods and pesticides—will lead to greater consumption of organic foods. A desire for maximum nutrition will also lead more people to adopt "raw" diets, in which they eat mainly uncooked foods.
- **Nutrition-on-wheels.** A growing number of companies will deliver an assortment of nutritious frozen meals to busy households once a week. Modeled on old-fashioned ice-cream trucks, meal trucks will circle through neighborhoods at dinnertime each day to offer a selection of entrees and extras.
- **Kid-friendly foods** are on the way. In a bid to get children to eat what's good for them, food researchers will develop tastier vegetables, such as chocolate carrots and pizza-flavored corn.
- **Allergy stores** will blossom, peddling vitamins, homeopathic medicines, foods made without certain ingredients (wheat, nuts, dairy products, etc.) and nutriceuticals for specific ailments. A growing number of mail-order catalogs and cyber-merchants also will be offering a full line of these products.
- **Alternative health providers** will supplement traditional doctor visits. Acupuncturists,

naturopaths, dietitians, and dispensers of traditional Chinese medicines will see a boom in business.
- **Beauty-in-a-pill** supplements will become popular, as the health and beauty industries promote nutriceuticals and vitamins with purported beauty benefits. Vitamin-enriched moisturizers and makeup will also see increased sales.
- **Telemedicine will link** doctors and patients by videophone, with vital signs transmitted electronically.
- **"Smart" pacemakers** will keep heart patients energized longer by adjusting to transmissions from the heart; miniaturization is expected to allow pacemakers to be tucked into the wrist rather than embedded in the chest muscle.
- **Birth defects** will be all but wiped out as breakthroughs in prenatal testing and treatment give doctors the tools they need to diagnose and treat nearly all birth defects *in utero.*
- **The dentist's drill** will be replaced by a combination of laser treatments, new decay-fighting toothpastes, and chewing gum that cleans teeth.
- **Diseases will spread** more rapidly around the globe. Surging international travel means more people will unwittingly transport disease to other countries—on their bodies, luggage, clothing, or even the soles of their shoes. A rise in xenophobia may follow.
- **"Sick building" syndrome** will generate more complaints and lawsuits from affected workers. Some companies may offer telecommuting as an option for those worried about getting sick at work.

Economy and Technology

- **Cybersecretaries** will function as intelligent agents to collect news and information on your favorite subjects and handle your business and social correspondence. Already, a variety of online forums send birthday and anniversary reminders on request. Now, a growing number of companies such as florists, wine merchants, and specialty food stores will send gifts automatically.

- **Cyberdetectives,** or Internet-based private eyes, will be in demand. They'll check up on the criminal, medical, and financial records of potential mates, employees, and so on.
- **House calls by computer repairers:** The increase in home offices will escalate demand for on-site, emergency computer diagnostics and repair.
- **Electronic migrant workers:** Telecommuters will cross national boundaries, as high-tech and other companies recruit worldwide via the Internet.
- **Portable hard drives** will allow computer users to carry the "brains" of their computers with them from work to home to temporary office site to airport lounge, etc., then simply pop them into a hardware receptacle that lets them get right to work.
- **Rent-a-mutt.** In our increasingly security-conscious world, we'll see the growth of a rental market for trained dogs: bomb-sniffing attack dogs for patrolling corporations, guard dogs for homeowners on vacation, "security" dogs for women jogging alone or working late at night, etc.
- **"Packaged" holidays** and other celebrations will help reduce stress in busy families. At Christmas, for instance, consumers will hire entrepreneurs to come in and decorate the home, using the customer's decor or rental decor from the company. Expect to see an increase also in packaged holiday meals, either delivered to the door or picked up at high-end hotels, restaurants, or supermarkets.
- **Personal shoppers** will grow in importance, taking the stress out of shopping. At holiday time, they will buy and wrap gifts for everyone on your list.
- **Nonstop strip malls:** In North America, the success of 24-hour supermarkets, restaurants, copy shops, and the like will lead to a number of other types of retail operations being open around the clock.
- **Drive now, pay later.** Toll roads will become more convenient, as new technologies will allow travelers to run tabs or have their accounts automatically debited as they cross each cost barrier.
- **Robots will dispense the gas** into your tank, based on preprogrammed information regarding tank location and preferred fuel type.
- **Electric cars** will become far more common as environmentally active celebrities join advertising campaigns and as the necessary infrastructure (recharging stations) is put into place.
- **Airports as destinations:** With Amsterdam's Schipol as a model, airports around the world will become entertainment destinations, drawing not just travelers but also residents of nearby areas to dine, shop, gamble, and play.
- **"Underdog" companies** will flourish. These newly formed companies can leapfrog the competition by embracing newly introduced technologies before entrenched organizations have time to make the transition.
- **Values will increasingly influence business decisions.** Businesses will seek partners with compatible political, social, and religious points of view. Corporations will increasingly be scrutinized by potential investors for their underlying social values, and business relationships in general will be based on far more than the bottom line.
- **Sabbaticals will become standard.** Workers will increasingly take "sabbaticals" between career assignments, taking a break to raise a family, go back to school, start their own businesses, or simply recharge themselves spiritually. Many companies will see the benefit of offering partially paid sabbaticals as a benefit in high-stress fields. Taking several such sabbaticals throughout one's working life will allow workers to extend their working years, or even postpone retirement indefinitely.

Questions Ahead

We have only scratched the surface of the vast changes afoot for life in the next century. Because the specific technologies that will develop are uncertain in almost every respect but their unpredictability, we can guarantee that new questions will continue to provoke our thought and concern about the future and our place in it.

For starters, consider these questions:

- Will cloned children of today's geniuses dominate matriculating classes at top universities in the years ahead?
- Will telephone numbers be assigned at birth, replacing Social Security numbers as our main identification numbers?

- Will the funeral industry disappear as cryonics and other forms of preservation keep us all alive until cures can be found for what ails us—even our aging bodies?
- Will vacations become pharmacological retreats for rest, relaxation, and mood enhancement?
- Will printed newspapers become obsolete?
- Will videoconferencing replace school buildings? Will libraries shut their doors and be available only in digitized form online?

59
A Message from Future Generations

ALLEN TOUGH

Some depressing facts and trends have emerged from the articles included in this anthology. There are also many signs of hope. This final article outlines seven priorities to make the world a better place, taken from suggestions of students around the world. They fall into the categories of peace and security, environment, dealing with catastrophes, governance, knowledge, children, and learning. Let us hope generations to come will work on these priorities to improve life for all humankind.

As you consider this reading, think about the following:

1. What are the main points made by students around the world in Tough's survey?

2. What are your priorities to make the world a better place?

GLOSSARY

Futurists Those who study, predict, and plan for the future.

LET'S IMAGINE THAT WE can see humanity many decades from now: We can see real people of all ages playing, working, talking, building, learning, laughing, crying, loving. Let's also imagine that these future generations can speak to us in order to influence our actions today. What would they say? What do they need from us? What would be their "message from the future"?

Trying to understand the needs of future generations broadens the perspective of futurists, planners, and politicians who set agendas and develop strategies for dealing with today's complex issues. And it transforms each of us into an advocate for future generations—and for the planet. We need to speak on behalf of future generations in the policy-making process. As "agents" for tomorrow's generations, we obviously need a clear and empathic understanding of what future generations need from us.

Unfortunately, we cannot actually receive messages from future generations, so we must

From The Futurist, *pp. 30–32, March–April 1995. Reprinted by permission of the publisher.*

use other procedures for determining their needs and priorities. I developed the following "message from future generations" partly through traditional futures-analysis techniques, such as literature scanning, networking, thinking, and getting feedback. But I also used role-playing exercises: I had 13 professors in nine countries arrange for some of their students to play the roles of people who will be alive several decades from now.

The students were asked to contemplate what message or key recommendations future people would want to send back to the people and governments of the 1990s. Some responses were written by individual students, and others were composed by small groups or entire classes, often during a mock council meeting held several decades into the future.

The role-playing exercises produced a rich array of ideas, priorities, and messages. I then selected the key points and blended them into the single message shown here.

The message may not seem very ambitious. The list of seven priorities could be achieved if enough people around the world become strongly motivated to do so. Dozens of other worthy priorities have intentionally been omitted from the list to make the program more feasible.

From another perspective, however, the message is *overly* ambitious. Even if humanity limited its efforts to this list of seven items and suspended its efforts to deal with all unrelated problems, the obstacles and costs would be daunting. Perhaps no one today can see any realistic way (short of an incredibly effective and benevolent world dictatorship) to produce that much individual change and societal improvement in just a few decades. But by focusing our present efforts on the seven core recommendations from future generations rather than scattering our efforts too widely, we will dramatically improve our chances of creating a positive future for tomorrow's generations.

Cordial greetings from the people of the future! We represent your grandchildren's generation, as well as the world of their grandchildren. We are delighted that you, the people of the 1990s, are willing to listen to us.

Please take our needs as seriously as you take your own. Please care about our well-being as much as your own. This is our central plea to you.

In your major decisions and actions, please consider our perspective and welfare along with your own. Our needs and rights are not inferior to yours: Please regard your generation and ours as equals. You might call this principle "intergenerational equity"—equal opportunity across the generations. Even though we live in an era that is very different from yours, we, too, are people, vigorously engaged in a wide variety of activities and projects, just as you are.

We sometimes feel frustrated and angry with you for taking such enormous risks with our security. We want to have opportunities and resources that are equal to yours. As you make choices in the face of competing voices and demands, please shift toward a fair and appropriate balance between your well-being and ours. We appreciate the efforts that you have already made, but you must work faster and make far more dramatic changes.

Humanity's Role in the Cosmos

Be awed by the significance of your role in cosmic evolution. You *and* we are the local manifestations of life in the cosmos. Our role at this stage is to advance our understanding, culture, and harmonious functioning to its maximum potential. The grand perspective of cosmic evolution and our own long-term future can provide people with meaning, purpose, awe, wonder, and the ultimate adventure of contributing to cultural evolution. If humanity's continued well-being becomes your supreme value, your highest priority, the center of your religions and ethics, then there is a good chance that you will make the required changes.

We hope that those of you who are religious will think for a moment about the creator's purposes. Surely the point of creating life in the universe was for it to develop to its maximum positive potential. Surely the deity or supreme being (whether called God, Yahweh, Allah, Brahma, or some other name) wants human civilization and our planet to survive and flourish for many more centuries rather than deteriorate miserably.

You must realize just how deep and pervasive your changes must be in order to give us equal opportunities. Individual behavior, social and economic structures, even paradigms and world views must all change. If you do not change vigorously and successfully, we—your descendants—will be much worse off than you are now. If you continue your shortsighted selfishness, the consequences will be catastrophic, perhaps even beyond the stretch of your imagination.

The Great Transformation

We urge you to celebrate your unique place in human history. The people in your decade face the historic challenge of making the shift from a narrow, self-centered, short-term focus to a long-term global focus that takes into account our needs as well as your own. If you do take vigorous action to achieve deep, long-lasting changes, your actions will earn a permanent name for the decade of the 1990s: The Great Transformation Decade. You will be remembered as the people who saved human civilization and its planet from catastrophe and disintegration, simultaneously building a foundation for a more positive world.

On a more personal level, we hope you will feel caring, love, and even a spiritual connection toward those of us who are members of future generations. We hope each of you—our grandmothers and grandfathers—will get in touch with your inner dream—your altruistic and ethical desire to make a positive difference to the world—and let yourself take the risk of striving for that dream.

Never give up your search for personal contributions that you can make, your search for a path toward a positive future. Feel empowered and vigorous rather than discouraged, disheartened, despairing, dispirited. Look for like-minded people and form a bond with them.

Yes, you live in a difficult and dangerous decade, but all successful social transitions feel difficult and dangerous at times. You know what you have to do, and you have the courage and the ability to do it if you choose to.

Seven Priorities and Recommendations

We offer you these specific priorities and recommendations—seven spheres in which your improvements could vastly improve our lives as well as your own.

1. PEACE AND SECURITY

Eliminate as many of your nuclear and biological weapons as you possibly can, along with any other weapons capable of destroying human civilization. Please reduce warfare and civil unrest by fostering nonmilitary methods and institutions for dealing with disputes. Learn to peacefully resolve your differences arising from greed, ambition, anger, and revenge. Foster widespread human rights and political participation. Promote a spirit of tolerance and cooperation among various religious and ethnic groups. And cut your military forces and budgets by 50–90 percent, thus freeing up resources for other recommendations.

2. ENVIRONMENT

We the people of the future obviously require a planet capable of supporting life. Please develop a sustainable relationship with the planet in your agriculture, forestry, fish, wildlife, and water and energy use, and do so very soon.

In order to accomplish this, you will have to work quickly and globally to:

- Halt or even reverse population growth.
- Reduce frivolous consumption and undue concentrations of wealth throughout the world.
- Live in balance and harmony with the interconnected web of life on Earth.
- Minimize the most burdensome types of environmental deterioration—global warming, ozone depletion, pollution of drinking water, toxic and radioactive waste, and loss of wilderness species.

3. CATASTROPHES

Please don't dismiss potentially devastating catastrophes or trends that might harm or end civilization. Study the possibilities and impacts of an asteroid or comet collision, a runaway epidemic, a biological experiment gone awry, a reversal of the earth's magnetic poles, or another ice age. Then vigorously take all necessary precautions to avoid catastrophes.

4. GOVERNANCE

Your present governance is incapable of coping with global problems. Please build an effective foundation for public priority setting and decision making at the local, regional, and global levels. In particular, develop governance traditions that emphasize the long-term global perspective, that address interlocking problems, that promote participation, that are free from corruption, and that accept the best ideas and individuals regardless of gender, race, socioeconomic status, or cultural background.

Balance your budgets instead of leaving us with burdensome debts. Try to get ahead of your major problems instead of letting them outstrip your efforts. Please do not leave us with horrendous problems that could have been alleviated far more easily in your era than in ours.

5. KNOWLEDGE

Conserve, enlarge, and widely disseminate your most significant knowledge, insights, and ideas. Double your efforts to understand humanity's broad contexts, such as our sources of meaning and purpose, our place and significance in the universe, and our potential relationship, someday, with other civilizations in this galaxy.

We value knowledge and understanding as one of humanity's most treasured possessions. Please develop steps to ensure that the core of human culture, knowledge, literature, music, and art will survive.

6. CHILDREN

Please reduce child poverty, hunger, neglect, and abuse. This is imperative to stop the physical, intellectual, and emotional stunting of children's growth and development. The competence and creative problem solving of tomorrow's adult population depend on your success at raising children with good mental and physical health, adequate self-esteem, and excellent skills in learning and thinking.

7. LEARNING

From early childhood to late adulthood, learning opportunities should be widespread. They can include libraries, discussion groups, informal education, support groups, printed materials, mass media, and self-planned learning projects, as well as various educational programs. This range of learning opportunities should help people of all ages to feel concern for others, understand global issues, and grasp the significance of future generations.

We want to end this message with our gratitude for your awareness of our needs and for your efforts that are already well under way. Many of you already grasp our perspective, and we hope that more and more of you during the 1990s will understand our needs and will take them into account. If this widespread change in perspective occurs, there is an excellent chance that the necessary actions we've offered you will flow easily from the new perspective. That is our urgent hope!

QUESTIONS FOR DISCUSSION

For further discussion of this topic, see the Wadsworth Sociology Resource Center, "Virtual Society," *http://sociology.wadsworth.com,* under *Sociological Footprints,* by Cargan and Ballantine. You can respond to the discussion questions there or enter your own comments in the online chat forum.

SUGGESTED READINGS AND SOCIOLOGY INTERNET RESOURCES

See the Wadsworth Sociology Resource Center, "Virtual Society," *http://sociology.wadsworth. com,* for additional links, suggestions for further reading, and learning tools related to this chapter.

Either from the "Virtual Society" website or directly from your web browser, you may access InfoTrac College Edition, an online university library that includes over 700 popular and scholarly journals in which you can find articles related to the topics in this chapter.